AF183426

Designing Farms in Space

Bryce L. Meyer

Designing Farms in Space

Balances for Life Support, Food and Settlement

Springer

Bryce L. Meyer
National Space Society
O'Fallon, MO, USA

ISBN 978-3-032-04394-8 ISBN 978-3-032-04395-5 (eBook)
https://doi.org/10.1007/978-3-032-04395-5

© The Editor(s) (if applicable) and The Author(s), under exclusive license to Springer Nature Switzerland AG 2026

This work is subject to copyright. All rights are solely and exclusively licensed by the Publisher, whether the whole or part of the material is concerned, specifically the rights of translation, reprinting, reuse of illustrations, recitation, broadcasting, reproduction on microfilms or in any other physical way, and transmission or information storage and retrieval, electronic adaptation, computer software, or by similar or dissimilar methodology now known or hereafter developed.
The use of general descriptive names, registered names, trademarks, service marks, etc. in this publication does not imply, even in the absence of a specific statement, that such names are exempt from the relevant protective laws and regulations and therefore free for general use.
The publisher, the authors and the editors are safe to assume that the advice and information in this book are believed to be true and accurate at the date of publication. Neither the publisher nor the authors or the editors give a warranty, expressed or implied, with respect to the material contained herein or for any errors or omissions that may have been made. The publisher remains neutral with regard to jurisdictional claims in published maps and institutional affiliations.

This Springer imprint is published by the registered company Springer Nature Switzerland AG
The registered company address is: Gewerbestrasse 11, 6330 Cham, Switzerland

If disposing of this product, please recycle the paper.

This book is a love letter for generations yet born. It is axiomatic in biology that a species that does not continuously expand its range becomes extinct. As humans, we have essentially saturated our planetary range. Even though the Green Revolution, and continuous increases in technology, combined with a dropping reproductive rate, allows humanity to feed itself for the long term, it does not mean that humanity will have a long-term happy future if we stop our populations at the edge of the atmosphere. We are happiest with plenty of elbow room. The more room and possibilities, the less stress each person experiences, and the happier we are on average. While some will decry any expense that is not local (the classic "why spend money to explore when we have problems here and now" argument), there will always be others willing to roll the dice on expanded possibilities. We are all descended from the ones who went forth. Humans began as a small population, in a small area, likely in Africa. The ones who went out were the ones who peopled the rest of Africa and the whole planet. No doubt the

genes from these explorers overwhelmed the stay put faction. As a result, we all have the potential to go forth. We need a distant horizon, and to put boots on new terrain.
I love people and want people to survive and thrive!

Foreword

Have you ever wondered about living on the moon or Mars or other planets? What would you eat? Where would the food come from? Could I farm on the moon?

Farming to provide food for humans to survive has been around for eons. From sowing seeds with hand tools (stones, sticks, etc.) to very large machines, we till the soil, plant, and harvest grains, grow vegetables, and grow animals to feed ourselves and our families. When I look back at the farm I grew up on, I see many ways we used to grow food for our family. We grew grains from the ground such as corn, wheat, oats, and soybeans. We had a garden to grow many vegetables such as beans, sweet corn, melons, and greens. We fed the grain to swine, milk cows, and chickens which provided meat and milk, and eggs. To survive the winter months, we had to process the food and store it.

To raise a family, it took many acres of land, many hours of manual labor and machinery to produce the food required to live a healthy life. What will the resource requirement be to sustain a food supply on the space station, moon, or other planets? There are limited resources for food in space.

In this book, we will look at all aspects and methods of producing the food required to live over long periods of time. It is interesting to learn about all the ideas and methods that can be developed for our future in space!

O'Fallon, MO, USA — Leonard C. Meyer

Preface

This book is written to be a step on the path to human settlement of space, i.e., the universe beyond Earth's atmosphere. This need for expansion leverages very real and accountable benefits for all humanity. Space provides an infinite source of resources, and new places to live, without destroying populations of native complex species, and without shoving aside other people. Imagination is the only limitation of how people will live, and how they can choose to be, providing avenues for freedom yet unimagined, and resources to feed and sustain beyond billions. Further technologies to enable settlement can be used to enhance life on Earth. Spinoffs from previous space exploration are common and pervasive already, in our pockets, on our walls, in our transportation, and in our hospitals. Everyone can be happier due to the expansion of people into space to stay, and space exploration will lead settlement, enhancing knowledge. In the long term, trillions of future happy, smiling, babies await.

Some would argue that we should never leave anything on any planetary body or in free-floating space and we have no right to expand the species. If you believed that hokum, you would not be reading this. There is a long set of arguments against Mars and the moon. Personally, my argument for space is that I am human, and I want humans to have every possible place to live. You make up your own mind. That said, the farms described herein can be built on any option, from rotating space stations to Mars underground, to moon domes, and anything in between. This is an engineering book, not a philosophical one. I just ask that future generations do not sacrifice human future expansion at the altar of someone's science project, or someone else's negative view of our species. I choose to be a hopeful optimist! My dreams are dominated by trillions of happy, smiling babies, of every race and creed, all over the stars.

O'Fallon, MO, USA Bryce L. Meyer

Acknowledgements

No book like this comes into existence in a vacuum, or without help from many people. As noted above, my parents and grandparents taught me the basics of how we get food, and how to apply effort over time to achieve a goal. My wife Andrea encouraged me endlessly and tolerated my numerous presentations and conferences. The same can be said for my 'space family': my space godmother Anita Gale, my space godfather Ron Kohl, in addition to many, many space brothers and sisters, mostly in the National Space Society, AIAA Space Settlement Technical Committee, and St. Louis Space Frontier. I tried to reference as much of their effort in the bibliography and references below as possible. If I forgot you, it is accidental. I owe many beers and dinners. I can blame a few astronauts I have met over the years, too, a few are moonwalkers.

Inspiration for any space work like this stems from a few sources such as Genesis in the Bible, and similar books from other faiths. It can be noted that multiple popes from the 1960's on encouraged space exploration and settlement. Science fiction, especially Robert Heinlein (another Missourian) in *The Moon is a Harsh Mistress* and *Farms in the Sky* talks about farming in space. Other works by Robinson, Clarke, and Asimov and many other authors fit here too. Too many to name.

My Reviewers are:
First Copy Edit: Judy Tippet. Thank you for being the first!

Various reviewers include many members of the National Space Society (NSS) and the American Institute of Aeronautics and Astronautics (AIAA) Space Settlement Technical Committee. The list is a long one and includes: Anita Gale, Ron Kohl, Al Globus, Doug Plata, Jack Gregg, Freda Taub, Madhu Thangavelu, and Richard Soilleux.

Their comments improved the book and lead to a better experience for everyone.

About This Book

The book is arranged to build up an understanding. See Figs. 1 and 2. First, we delve into the basics of why you want a space farm. Next, we cover the simplified chemistry and how mass flows in organisms. After that, we take a walk through some of the possible crops that might be in a farm in space. Continuing, we cover human-made machines you will need to make the farm work. With the basics of biology and technology in hand, the book covers combining species and machines to make a farm element called a stage. The book then shows how you can put the stages together to make farms. On this journey the book covers many tradeoffs, challenges, and potential in growing crops for future space settlements to make interesting meals for those who choose to explore and live in the final frontier.

The topics are built to allow an understanding of the factors and design tradeoffs to design a farm for various needs and scenarios. There is an accumulation of subjects, that lead to the ability to decide crops, and menus for a human diet, that then lead to space farms that have energy needs, initial mass requirements, area and volumes, and mass flows between the various parts of the farm.

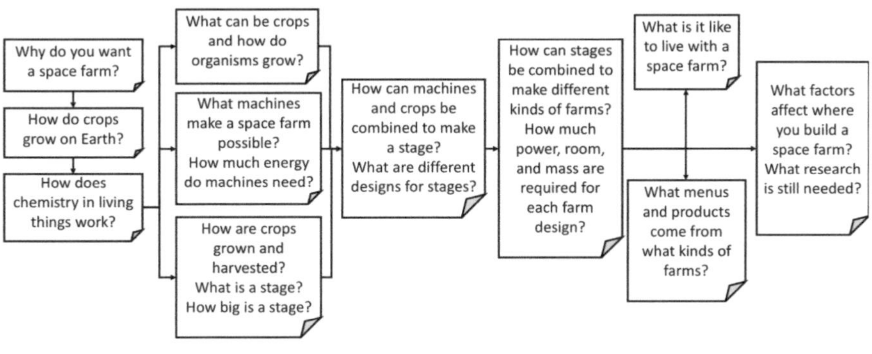

Fig. 1 Question map for this book

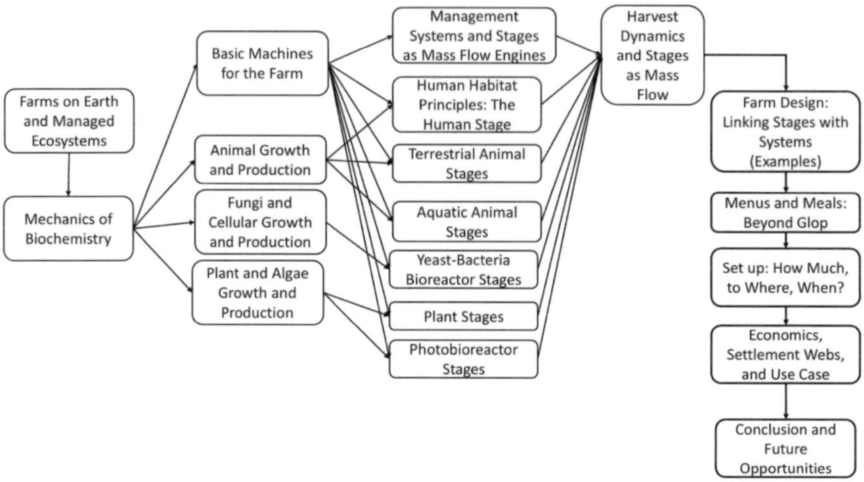

Fig. 2 Topic map for this book. Follow the arrows to follow your interests

Intended Audience

I set out to write this book to consolidate a wide variety of thoughts on technologies and techniques to feed and recycle for future space settlements. It is meant however for a wide group, not just scientists and engineers. This book should be readable by any technically interested adult with some college or late high school student, who may have some very basic knowledge of current events in space. You might have some exposure to near term space fiction and maybe someone who has grown a garden or worked on a farm. While engineers, agricultural scientists, and other space technologists will likely derive the most value from this book, it is broad enough to bring together the multidisciplinary nature of future agriculture in space from other fields (pun intended). If you are into cooking, there are numerous notes about menu options, including a chapter on food prep. Psychologists would inform the reader that working with crops and a diverse diet are very important to those living far from Earth. The book is rich in pictures, diagrams, and tables to make understanding easier for the non-technical reader, such as managers and educators. If you are reading this, you will have to tolerate my Midwestern language, coarse farmer analogies, and dry humor. I scatter some silliness in this book to add flavor. You can read just the words and learn quite a bit. You could even just scan the figures and tables and find use. I limit the math to just a few spots, but the tables would allow initial design and scaling, especially for funding level or project/contest level designs. If you want, skip those parts that do not interest you. Otherwise, enjoy it all to get the full picture. The science portions are digested so anyone can figure out the basics of space farming without a deep science background. If you like the math, or deep engineering, my technical papers (see Bibliography) are likely something you want to explore in addition to this work. I also limit the details that force one particular design over another, because I want YOU to be free to design as you see fit! Even if you never consider going into space, the technologies herein also apply to feeding people on Earth with minimal environmental impact and efficiency. Earth farmers will gain by referring to this book.

Conventions

Units herein are metric, i.e., SI. Kilograms (kg) and metric tons for mass, meters for distance, square meters or hectares for area, volume, meters for Newtons for force, etc. Occasionally, the Missourian in me will sneak out and I will drop into English units then translate back out, for effect. Farms around the USA use acres, in the rest of the planet Earth, hectares (which are around 2.4 acres).

I had to put in some biochemistry, but I do try to limit it to the important end results. Organisms are viewed largely in aggregate and as biochemical factories. There are oodles of reaction pathways that happen in the beautiful complexity that is life, but to make pessimistic calculations, I just assume the worst case, a beginning set of chemicals, and an ending set. Occasionally, I skip subscript hell, and list things like carbon dioxide as CO2 in lieu of CO_2. I also do not delve into the isomer, enantiomer, and organic chemistry nebula, and instead focus on the masses of elements, not necessarily how they look in three-dimensional space. If two sugars for example have 6 carbons, 12 hydrogens, and 6 oxygens, I am concerned with the mass of the elements, not if they code for fructose versus glucose. You might care about those, but mass, volume, and energy are the big worries in this book. There is a lot of guesswork to arrive at mass flows for crops, and these need to be explored in detail by experiment before we build a farm on Mars.

Critically, all numbers, figures, and calculations are notional until confirmed by experiment and not to be taken as authoritative.

Contents

List of Figures

List of Tables

Chapter 1
Introduction to Space and Farming

Abstract This chapter explores the differences between farming on Earth and in space. The subject starts with the space environment, then examines closed and open mass flow cycles, then the evolution of settlements from outposts to fully functioning settlements with complex space farm needs. Finally, the chapter begins to examine how mass and energy are used to produce Earth crops.

Keywords Space farms · Mass recycling · Crop production · Closed systems

1.1 Chapter Introduction

This chapter explores why you need a farm in space, how food is grown on Earth in farms, then the differences between farming on Earth and farming in space. To understand space farms, we begin by exploring the space environment and need for efficiency and the cost of enclosing areas against vacuum or harsh atmospheres. Next, we will cover the evolution of settlements from outposts, and how the food needs shift as the settlement becomes more self-sustaining and more complex. Further, the chapter covers the concepts of closed and open mass cycles and how closed cycles lead to less resupply. Finally, the chapter covers how mass and energy are used to produce Earth crops.

1.2 The Space Environment and Factors for Settlement Efficiency and Design

The rest of the universe beyond the thin atmosphere of Earth is a harsh place. Aside from planetary bodies like Mars or large moons, there is vacuum. Outside the atmosphere, temperatures fluctuate from near absolute zero to hundreds of degrees Celsius. Radiation increases above the atmosphere, and you go past the Van Allen belts, the

© The Author(s), under exclusive license to Springer Nature Switzerland AG 2026

B. L. Meyer, *Designing Farms in Space*, https://doi.org/10.1007/978-3-032-04395-5_1

environment includes an even harsher radiation regime, punctuated by cosmic rays, solar events, and radiation from other stars.

As a result of this harsh place, space habitation comes at a cost. The costs of survival are getting material, and enclosing volume and area. Getting mass to Low Earth Orbit (LEO) costs on the order of thousands of US dollars per kilogram as of the second decade of the twenty-first century. If you lived in LEO, a large, fully outfitted pizza delivered to you would run the same as getting a nice house on Earth. Every kilo matters. Getting mass is part of the problem, creating a safe environment to live is another. In the harshness of vacuum and radiation, volume must be enclosed, at exponential costs in materials. A hemisphere (i.e., a dome) has the volume of $(2/3) * \pi * r^3$, and a surface area of $2*\pi*r^2$. If that hemisphere is the classic dome over a flat area, that flat area is circle of only $\pi * r^2$. That means you must have twice as much area in the dome to cover a circle of ground. If you had to use aluminum at ten centimeters thickness, covering a hectare (10,000 m^2, a bit over 2 US acres) in classic sci-fi mode, you will need 5400 metric tons of aluminum. In short, volume and area have a high cost in space, so every cubic meter counts.

Therefore, if we wish to live in space, we must be efficient. Efficiency does not have to mean a dinner of green glop, washed down with chemically filtered pee, while breathing chemically processed stale air. Poop to plate indeed! There is hope, trust me. A wide variety of technologies have emerged on Earth from vertical farming, indoor gardening, fermentation, aquaculture, and other areas to enable a much happier space existence. These technologies are melded by borrowing ideas from a much more complex optimization: nature. Ecosystems on Earth are massively efficient at recycling matter, and we will need to follow some of the principles from nature, using organisms that are good at giving us what we need, when we need it. Farms in space, or space farms, can provide a variety of dietary options, while fully recycling air and water, and provide a greener living space. We can achieve these goals by creating a space farm that is a managed ecosystem that includes humans in the habitat. These highly efficient biological and technological farm constructions can be arranged to maximize production, while minimizing volume and area and using minimal energy. Different space habitats will have different goals, and these farms will be created to meet these goals.

As this book progresses, you will find your author's best ideas using current and maybe slightly advanced methods to feed future space settlers, defining settlement, technologies and species involved, timeframes, designs, and calculations. This book gives a reader building blocks, that they can assemble as they want. Get ready for a wild ride through space farms!

1.3 Where Would You Want a Farm, and Why?

So, what is a space farm? Anything that by design uses living organisms to recycle mass in space from the von Karman line (100 km) and out is a space farm. Simply put, a space farm is a combination of organisms and machines to grow food and recycle mass, in space. A farm is beyond simply experimental plant grows, such in the ISS science modules. Farms in space are made to perform a practical function, usually to provide food, though simply recycling air and water would be a good start as in Fig. 1.1.

For example, for the ISS or any space station or long duration mission, a farm could add resiliency, and lower costs. To start, adding a bed of plants, preferably plants that also provide food, would recycle air and water. The plants would lower costs by reducing the need for zeolite or similar chemicals in the air and waste processors. Plants convert carbon dioxide to oxygen while capturing the carbon in the plant itself. Plant can also take water fed to roots, then clean it by both releasing it to the atmosphere, and through filtering water as it flows past the roots. In a low gravity environment, vine crops could be used to line walls and use both normal lighting and supplemental lighting while minimizing volume. Waste liquids could be pumped through the roots after a grind and coarse filter process. Since the human mind works better when we see green and we smell air processed by plants, the astronauts would feel better too. Even if the vines were inedible we obtain benefits, though there is no reason a few tasty melons, beans, or similar couldn't be produced from such vines, as in the cartoon in Fig. 1.2.

Once this little first plant step, maybe a farm gets a whole volume not to just recycle air and water but add a few fresh tidbits to the diet of a space tourist, explorer, or worker. There are many stages that lead to settlements. The first steps and further stages are covered below and in Gerard O'Neill's works, including *Space Settlements: A Design Study* (Johnson et al., 1977), and *The High Frontier: Human Colonies in Spac*e (O'Neill, 1976), and Heppenheimer's *Colonies in Space* (Heppenheimer, 1977), and more recently Strickland's *Developing Space* (Strickland et al, 2021a) and *Settling Space* (Strickland et al, 2021b) among many others.

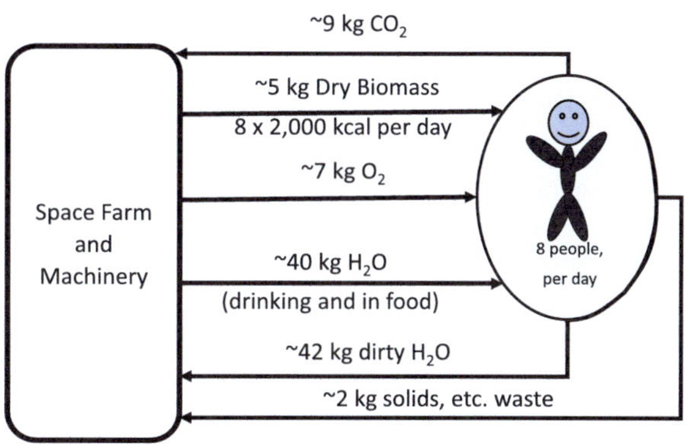

Fig. 1.1 Humans to space farm flow. A small crew could use a space farm and machinery to recycle all their carbon dioxide and wastes, to provide food, oxygen, and clean water. Very loosely calculated from data in (NASASTD-3001, 2015), (NASA-TD9702A, 1988), and (NASA-SP-2010-340, 2014) then extrapolated

Fig. 1.2 Cartoonish subsurface lunar base with in-habitat plants

1.4 Space Settlement Definitions and Levels

To understand how space farms will enable the settlement of space, we need to explore what we mean by space settlement. A space settlement as used in this book is a place where people go to stay. A mature settlement includes a diversity of age groups, and reproducing family groups, ideally growing in population. Initial space habitats have not quite reached mature states and are more outposts than settlements, as Table 1.1. Menus that may emerge from these levels are covered later in this book.

Our first toes in the water are initial outposts like the International Space Station (ISS) or the Earth versions: South Polar Station, Everest Base Camp, or remote North Sea Oil Rigs. These small footholds are where supplies are shipped in, and only hardy individuals, largely scientists, explorers, engineers, and technicians, rotate stays. Some foods may be produced, but they are limited and not the bulk of the diet of the residents. Life support is largely via chemical processing, resulting in limited recycling, and need for constant resupply to cover inefficiencies. In these places,

Table 1.1 Comparison of evolving settlement levels and Earth equivalents

Settlement level	Space equivalent	Earth analog	Food source	Recycling and farming
1	Improved orbital station, basic factory	Antarctic bases	All from Earth	Minimal chemical recycling of gases and water. Plants (largely salad greens and vegetables) as compliment to diet. Requires extensive Resupply
2	Small residential base, with attached factory or resort	Factory town, retirement village (no families)	Most from Earth, some local	Organic waste and water largely recycled. Hydroponic growth for salads and some staples. Minimal bioreactors supplement diet. Requires import for some elements of menus, and some staples. Features gravity
3	Significant residential development, business ecosystem	Farming transient or developing town (few families, but many adults)	Some items Earth, though staples and spice items local	Complete recycling of solids, liquids, gases. Hydroponics, some soil growth, and bioreactors, with initial aquaponics or small animals. Complex menus combining local sources with luxury items from off-site
4	Small settlement as part of a web of economic and trade relationships. Age diversity	Permanent Growing Town	Almost all foods are locally grown. Self-sufficient for all but guests and added children	Complete recycling of solids, liquids, gases, very efficiently. Mix of hydroponic and soil growth, and bioreactors, many aquatic species (fish, shrimp), possibly other animal species. Complex menus from local sources, though some resupply for luxury items and inefficiencies
5	Full space community	Growing city or village with surrounding farms, business	Self sufficient for all but export. Access to resources for excess production. Part of a trade web	Complex farms with diverse species, including crop species for export. Parks with extensive in-soil planting, including tree and bush crops and ornamental plants

food is just calories for survival, just beyond what old people like me think of as TV dinners or MREs! Maybe you get a special meal for high holidays just to keep you from the utter despair of binging romantic comedies.

A teeny step beyond these initial outposts will be stations where space farms of bioreactors and very limited hydroponic gardens increase efficiency. A major portion of the diet consists of locally grown crops of algae, yeast extracts, and plant crops. That said, imports are needed to balance the diets for variety or luxury. You know it is luxury when it isn't glop and salad and has more than five ingredients. The crews on these next level stations will have extended tour lengths, improving the labor efficiency of the operation, and lowering the costs to tourists. The pay may be good, the view great, but the diet will be less than wonderful except for those prized imports. The composition of residents is still adults who are explorers and experts and likely focus on their jobs just to avoid thinking about dinner.

The next step is nearly complete self-sustainment, with a substantial space farm that has not only bioreactors, but a very substantial diverse hydroponic stage, and possibly a very small aquatic stage to improve farm efficiency and add a rare treat of animal protein. Fish or shrimp on the holidays? Maybe yes! Finally, beyond the glop! For reasons illustrated later in this text, even if the aquatic part contains just pets, it will improve the diversity of effectiveness of the hydroponic assets, though here it is assumed that some fish or shrimp will end up on the menu. It is also in this settlement that the human population starts diversifying in age and background. Older families might find this place home. It is also possible a retirement community may form, with a few older children in pre-teen to teen age groups. Get off my lawn might have a different meaning here. If the station is somewhere where humans can gestate and grow given technology available, the age dynamic may include families with younger children, but that is more likely in further evolved settlements. Tourists, explorers, scientists, or visiting workers are likely still a key source of income, but settlers here are long term, with many of those permanent residents.

Leveling up, the settlement has a complete space farm, with a diversity of crops, where bioreactors produce feeds for animals, and act as aerobic producers of carbon dioxide for extensive, diverse, hydroponic, and aquatic facilities. There may be enough enclosed volume for parks and the beginnings of soil crops and trees. The menu is very diverse, with animal protein an option for every meal. Further development will result in soil growth and possibly in the long-term terrestrial animal crops such as poultry or other livestock.

This is not to say that Fido would be unwelcome at any stage. Space puppies and kitties might be along for the ride very early in the settlement's evolution for psychological benefits. Like the plants that will be grown on grey water in the human habitat, including many spices, pets provide clear emotional stability to people in tight confines and close contact, making the human habitat a home in lieu of just a workplace. Petting pets soothes the soul. Space can be lonely enough, so why not have a furry companion to ease the isolation? Unlike the spices, pets constitute a 'partial person' based on mass and metabolism and will need to be counted in the mass balance but are well worth the effort.

It is implied that settlements are as close to self-sustaining as possible, and part of an economic ecosystem, a web, of trade and monetary links. This is so people can deal with the long delays or high expenses in getting materials from Earth, and at the same time have access to a wide variety of foods, especially those foods that make life more enjoyable. Cinnamon, olives, black pepper, dairy, red meats, and especially chocolate will be very lucrative food trade items for reasons you will find later in this book. In turn, in addition to tourism and science travel, settlements in later stages can use their positions in the gravity well to ship fresh foods and specialty items and value added foodstuffs (like distilled spirits, coffees, teas, or luxury pre-made meals) to less evolved settlements and outposts. It is the latter stages where settlements have a city or town population dynamic, where families are the bulk of the population, government exists, and excess production is produced using inputs via in situ or import, and possibly excess population to further explore and settle. It is likely that these settlements will supply resorts and factories and act as waypoints on the edge of further space expansion.

1.5 Earth Farm as Recycling Ecosystem

On Earth, we are truly blessed with an environment that makes growing food efficient. Plants are one of the core dietary items, especially grains like maize as in Fig. 1.3.

Modern grain farmers can expect roughly three tons of wheat per hectare per year in the Great Plains or Midwest region of North America. Figure 1.4 shows a typical multispecies Midwestern farm, for grains, beans, hogs, and cattle.

While this bounty is due to a plethora of technologies from improved genetics, data, and fertilizers, it is also because the Earth provides many inputs for free.

Fig. 1.3 Field of maize. My father with a nearly mature crop of corn (maize) at the Meyer farm in Camp Point, IL, USA. Photo by author

Fig. 1.4 The Meyer Farm, Camp Point, Illinois, USA. Photo by author

Plants grow by cycling mass using solar energy. Essentially a plant in a farm is a mass converter that takes sunlight, air, water, and dirt to make dinner as in Fig. 1.5. They take in carbon dioxide, water, soil nutrients, and a very small amount of soil oxygen, then use sunlight to produce more plants that end up either as harvest or plant wastes, and send oxygen and transpired water to the air.

Energy in the form of sunlight illuminates plants with very roughly 10 MW per hectare for on average 12 h a day. In Missouri, USA, for example, the sun provides 51,000 kilowatt * hours of per bushel of wheat (27 kg) per season (loosely calculated via (Massey & Brown, 2021) and (PVWatts®, 2022)), grown on around 75 m^2 of very good land. Not all that sun gets used, but that is what hits the ground during the growing season. Rain provides water (either during season, or in past seasons as ground and river water), to make those 27 kg (kg) of edible wheat grain. If you are wondering how 27 kg relates to dinner, at 3400 kilocalories (kcal) per kg grain (loosely via USDAARS, 2022), each bushel provides roughly enough calories for 45 days for one adult at 2000 kcal/day. Assuming 36 kg of stem, root, leaves for the 27 kg grain (a very rough assumption), about 63 kg total mass for the plant grows that season. All 63 kg comes from: rain or ground water for the 41 kg of water is retained by the plant (though about 200 times that gets pulled through and evaporated in transpiration), 73 kg carbon dioxide from the air minus 53 kg of oxygen released back to the air (some oxygen produced is reabsorbed by the roots), and traces of

Fig. 1.5 Plants are mass conversion machines

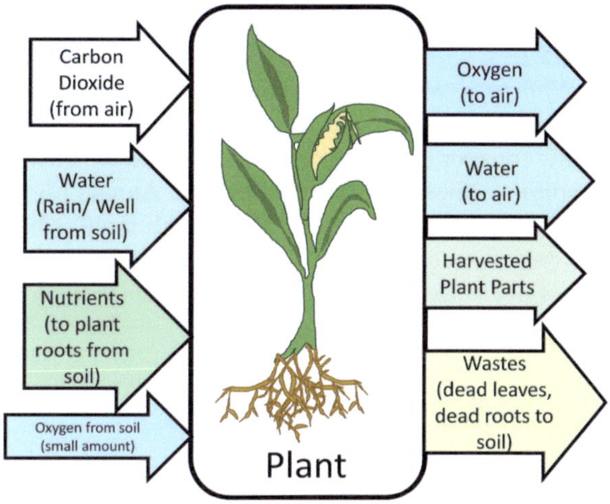

Fig. 1.6 Mass inputs and outputs for wheat. Rough mass inputs (in kg) to grow the whole live plant on the left, to get a bushel of harvested wheat, as shown on right, calculated loosely from data in (Massey & Brown, 2021)

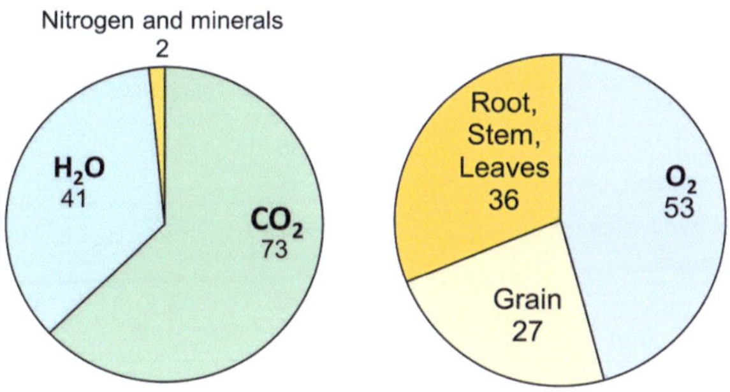

minerals and nitrogen from the soil and air (around 2 kg) (again starting loosely with (Massey & Brown, 2021)). See Fig. 1.6.

The atmosphere provides carbon dioxide for photosynthesis, oxygen for roots and the soil ecosystem, and nitrogen that can be fixed by bacteria to become proteins in plants. To feed the soil, farmers can plow under dead material from plants, where bacteria and fungi release the carbon dioxide in the cellulose and other chemical parts of plant cells over time. Soil also provides support and minerals, and the open air provides room. The ozone layer and magnetic field around the Earth shield plants and animals from radiation. Earth farmers must have it easy? If you believe that, I got a big bridge for sale to you in New York. In space, it is harder yet, since each of these gifts must be paid for in resources and energy. Table 1.2 has the rough basics for many plant crops in Earth soil, loosely calculated and extrapolated from these references: (Keene et al., 2022); (Keene et al., 2023); (TxA&M, 2022); (Farag et al., 2012); (Ransom & Eisinger, 2017); (Holmstrom et. al., 2008); (OrSUCAS, 2010); (Tourte et. al., 2009); (Molinar et al., 2006); (FP-Peanut, 2022); (Endres et al., 2018); (Caravati, 2021); (USDAERS-RiceYB, 2024); (Kandel et al., 2018); (Habib et al., 2008); (AMRC, 2024); (Coolong & Boyhan, 2020); (Harper & Orzolek, 2006) (Table 1.2).

While it looks like vegetables produce more food (that is, edible wet mass) per square meter, they produce fewer calories than grains in the same space. A good farmer tries for one kilogram per square meter. Also note that plants change in ratio of edible to inedible parts as they age. Annual plants like grains dump everything into making seeds, though in the growth months before harvest, the ratio of shoot and root to seed favors the shoot and root. Plants must have leaf area to make the sugars in the seeds and roots to bring in the water to make the same sugars.

Table 1.2 Soil-based growth of selected plant crops and food energy

Species	Earth						Theoretical ideal in space	
	kg food/ m^2 soil/ crop	kcal/ kg food wet mass	kcal food/ m^2 soil/ crop	People * year/ hectare/ crop	People * year/ acre/ crop	Days to harvest	Crops/ year ideal	People * year/ ideal
Barley	0.74	3780	2813	38.5	15.6	90	4.1	63.2
Bell peppers	1.68	290	486	6.7	2.7	60	6.1	16.4
Corn (dried grain)	0.86	3650	3127	42.8	17.3	100	3.7	63.3
Lettuce (Arugula)	0.89	250	224	3.1	1.2	45	8.1	10.1
Lettuce (leaf)	2.10	140	294	4.0	1.6	45	8.1	13.2
Oat	0.55	3790	2085	28.5	11.6	88	4.2	47.9
Onion (raw)	3.58	400	1431	19.6	7.9	100	3.7	29.0
Peanuts (unroasted but dried)	0.26	5670	1466	20.1	8.1	175	2.1	17.0
Pinto beans (dried)	0.34	3399	1162	15.9	6.4	90	4.1	26.1
Potatoes	6.82	700	4774	65.4	26.4	110	3.3	87.8
Quinoa	0.20	3680	718	9.8	4.0	90	4.1	16.1
Rice	0.70	3570	2486	34.0	13.8	105	3.5	47.9
Rye	0.26	3380	875	12.0	4.8	150	2.4	11.8
Soybeans	0.40	4460	1825	25.0	10.1	100.0	3.7	37.0
Strawberry	7.60	320	2433	33.3	13.5	45	8.1	109.4
Teff	0.29	3670	1064	14.6	5.9	100	3.7	21.5
Tomato	4.19	180	755	10.3	4.2	75	4.9	20.4
Wheat	0.36	3390	1220	16.7	6.8	100	3.7	24.7

For anyone reading into the table, note that potatoes are a great crop given limited space, which is why the Incas domesticated the plants, and grew them in terraces in the Andes. Potatoes also made the population boom possible among small farm holds in Europe after they were brought across the Atlantic. However, also since potatoes were not adapted to a certain set of diseases, the Potato Famine resulted in a mass migration of poor European farmers all over. Soil can give and soil can take away.

In the last two columns, you will find my mapping to the plan of O'Neill's thoughts as in (Johnson et al., 1977), i.e., given the ability to make a great big rotating station, and control of conditions, I could make the crops grow in many cycles per year. Normally a farmer in the temperate zone of Earth gets one, maybe two, crop growth cycles per year, before winter freezes the ground or frosts the plants.

Plants, here vascular plants, aren't the only farmed items on Earth or (eventually) in space. It is a common mistake to assume that a magic hydroponic garden will provide a horn of plenty of food and recycle air and water. Humans do not exhale enough carbon dioxide to provide enough calories to feed themselves, though the explanation for this is in later chapters. In short Midwestern coarse language, all the carbon gets locked up in the fiber in your poo, and you die because you are not a perfect cow. As I said above, more in later chapters. Bacteria, fungi, and animals can make up the carbon deficit. Even then best farmers on Earth need to raise crops not just to feed themselves, but to sell. While subsistence farming might fill your belly with staples in good years, it won't buy a tractor or pay for parts for it. In a truly diverse farm, some of these crops include terrestrial or aquatic livestock, which provide dietary variety, and a source of cash to pay the dentist to get the kids teeth fixed.

A farm on Earth can really be thought of as a cycle of mass and energy (Fig. 1.7).

Terrestrial livestock such as poultry, rodents, and cattle get free air from the atmosphere and plenty of room if free range. Free range poultry, such as chicken, turkeys, or geese, supplements their protein by consuming insects, worms, and plants from the surrounding ground. This mass flow at a top level for animals is shown in Fig. 1.8. Remember that soil arrow for later, it is the next topic after animals.

Many chicken farmers can describe how chickens keep yards tick free. Cattle get fed from grass that grows readily in pastures and are supplemented with grains from other fields, or cut grasses in the form of hay and straw (a.k.a. forage), again from other fields. Water from wells or ponds is consumed by terrestrial livestock, at the cost of energy for the well water or irrigation, or from rain in ponds.

Put the human wastes straight into the soil, and this cycle has been used by farmers, including First Nations from North America, for millennia. My grandparents' farm was only the latest farm on their land in Illinois, the Illini, and others, used the same land thousands of years before them to grow crops. This is true in farms all over Earth, from South America, to Africa, to Europe, and to Asia. The soil is a living mix that needs to be fed, as depicted in Fig. 1.9. It is quite alive ideally and consumes oxygen to produce carbon dioxide, water, and plant ready nutrients.

In space, the soil's role as cellulose digester might be played by a bioreactor. On Earth the soil is alive and is like an air-breathing 'animal' that breathes oxygen, eats

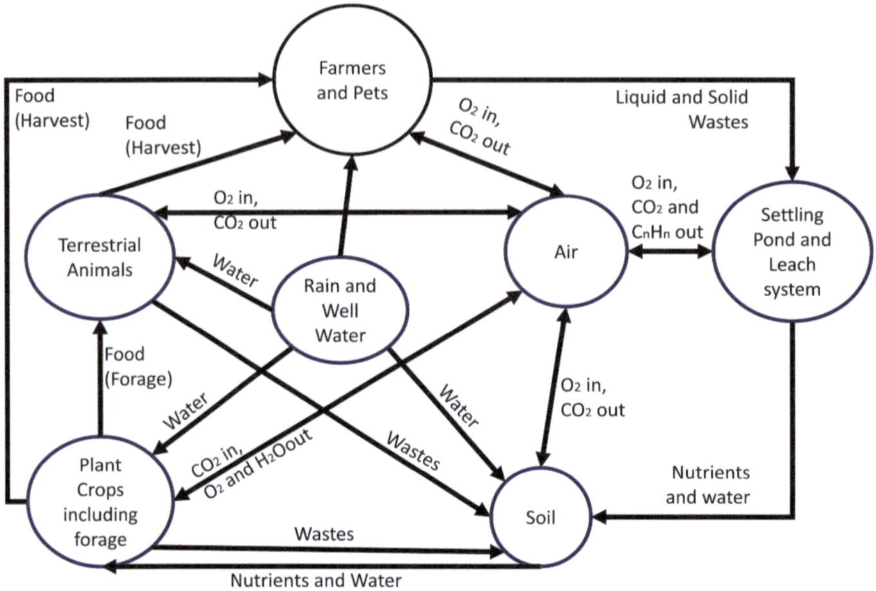

Fig. 1.7 Mass cycles on an Earth soil farm

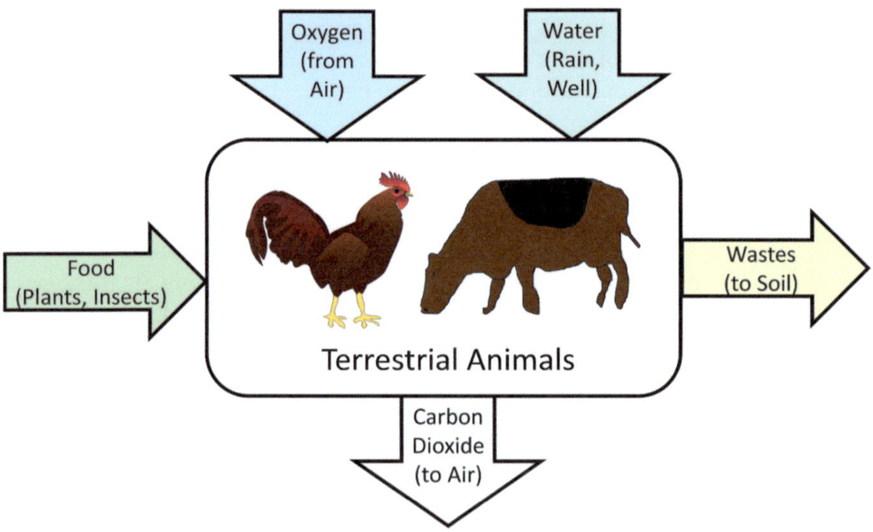

Fig. 1.8 Terrestrial animal ins and outs as on an Earth farm

Fig. 1.9 Close up of wet soil. Soil is a mix of root debris, dead leaves and plants, insects, bacteria, fungi, worms, and other organisms. Photo by author

cellulose and wastes, and provides carbon dioxide to the air and nutrients and water to plant roots. A schematic of these mass flows is shown in Fig. 1.10.

Another factor often ignored when talking about herbivore animals is how the pasture, and its soil, duplicates a natural prairie. As previously noted, the soil contains an ecosystem of life forms. Grazing cattle stimulate the growth of grasses. One way is by trimming excess growth, exposing greenery to sunlight, but other benefits are less visible. Large herbivores (i.e., animals that mostly eat plants) excrete lots of waste. As anyone who has walked after cattle, horses, elephants, or similar, can say,

Fig. 1.10 Soil as mass flow. Soil is a living ecosystem that acts as an aerobic animal that digests waste and cellulose

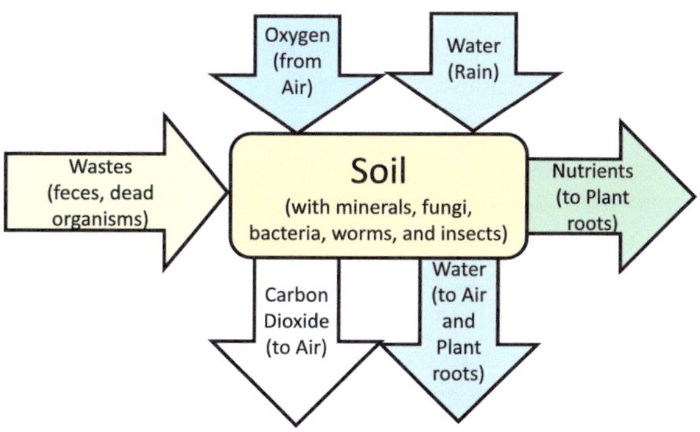

these wastes contain a mix of digested and undigested grasses. The digested part is like a stinky brown glue that stains anything and smells bad. The undigested rest is more solid. Dried buffalo droppings ('chips') provided fuel to the First Nations and Sodbusters alike, and droppings from cattle still provide fuel in many parts of the Earth, because the wastes still contain mostly fiber. That fiber is cellulose and maybe some lignin. Ruminants are only partially efficient in digesting cellulose. The digested part is immediately available as a buffet to bacteria and insects, while the cellulose needs further processing by fungi or bacteria, for a few months to a year. This processing on a savanna or prairie happens when the big herbivore steps on dried manure and drives it into the soil, where fungi and bacteria await to eat it. Crunching steps of hooves also stir the soil and aerate it to accelerate fiber digestion in the soil. These hoof steps provide benefit to the grass consumed, in driving undigested seeds into the moist soil to sprout where digested fiber is now a nutrient. Fiber digested aerobically by bacteria and fungi releases heat and carbon dioxide to the atmosphere, and water and nutrients to the soil, to feed plants in the cycle again. The plains also featured a different recycling mechanism: wildfire. Wildfire works faster than bacteria and cattle but produces less-desirable byproducts.

Another type of livestock on the farm could be aquatic. Fish are the most common outdoor aquatic crop, usually these species: catfish, tilapia, pacu, salmonids, perch, or carp. In addition, shrimp and crayfish can be raised in the same ponds or raceways, rice paddies, or shallow ponds, as shown in Fig. 1.11. Raceways are artificial streams where water enters one side of a trough and exits the other side. The water can be recirculated using pump energy. In other systems, the raceway is supplied from a stream or river, where the water is diverted upstream of the raceway, then reenters the stream at the down current end of the raceway, pushed through by gravity, as in many older trout and salmon hatcheries in the USA. Algae and bacteria on the bottom help balance the chemistry of the waters, when not in excess. In any case, Earth again is bountiful, providing gas exchange between the ponds or raceways and atmosphere, where oxygen is added, and carbon dioxide and ammonia removed.

Fish consume food and oxygen, release carbon dioxide, and output nitrogen rich wastes in liquid form, such as ammonia and urea, and output solids such as feces. The liquid and dissolved gas animal outputs are consumed by plants and algae, while the solid wastes and dead plants are digested by the myriad of organisms in soil and pond bottom. That stinky muck is yummy chow for algae and plants. The aquatic animal mass flow schematic in Fig. 1.12 shows this mass flow.

Ponds and raceways are also typically exposed to sunlight, which grows algae along the walls of the raceway or ponds, and phytoplankton in the open water. Raceways are essentially artificial streams, fed by water that flows through a gutter-like shape from an input pipe through the fish or shrimp holding volume, to an output. Raceways in most large-scale farms or hatcheries are made from soil or building materials like concrete, as in Fig. 1.13. Note the screens to prevent fish loss at the outlet. This hatchery receives water at pressure from Table Rock Dam far above, which supplies cold oxygenated water at a nearly constant flow.

Raceways, streams, and ponds, however, don't just hold the crop animals. Algae, phytoplankton, and water plants consume carbon dioxide and nitrates from the water,

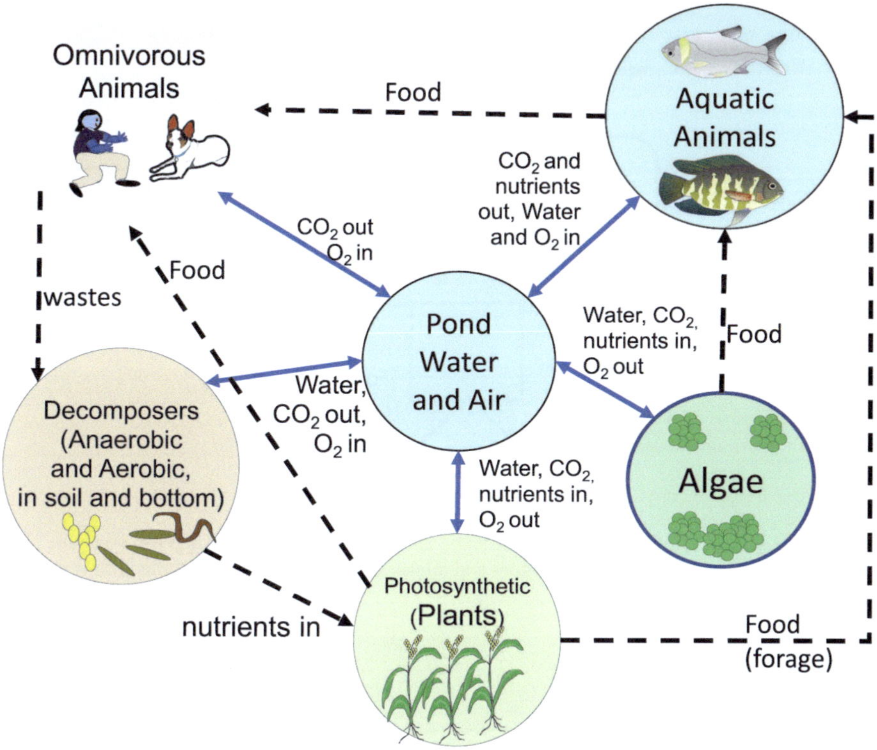

Fig. 1.11 Rice paddy and fish farm mass cycle

such as urea and ammonia from the fish. This alga is consumed by the fish or consumed by zooplankton, crustaceans, or mollusks, then by the fish. The 'free food' supplements any feed provided by the farmer, though it is key to the growth and health of the aquatic crop. Often small crustaceans break up solid wastes making the wastes part of the slime. In any case the slime layer is a habitat for aerobic and anaerobic bacteria that help balance the chemistry of the water by digesting waste. Like on Earth, rocks such as limestone or other soluble rocks also make the water mineral rich and provide a buffer solution for pH level. Such limestone buffering in karst topography makes for great trout and bass fishing in the spring fed waters of the Ozarks. Rain leaches calcium carbonate (limestone) and nutrients, which flow into groundwater then to springs. The mineral and nutrient rich springs in these streams in turn enable a rich bed of plants and algae, fed upon by crustaceans, mollusks, and insects, as in Fig. 1.14 which make chunky, happy, fish, and in turn happy anglers, herons, vultures, and otters.

The fish poop and some die, the nutrients thereof feed the plants, and on it goes. The closer our farm can get to a complete cycle, the more efficient it will be.

So why do we need to know how Earth farms and ecosystems work to discuss space farms? In space nothing is free, but to make an efficient space farm, which

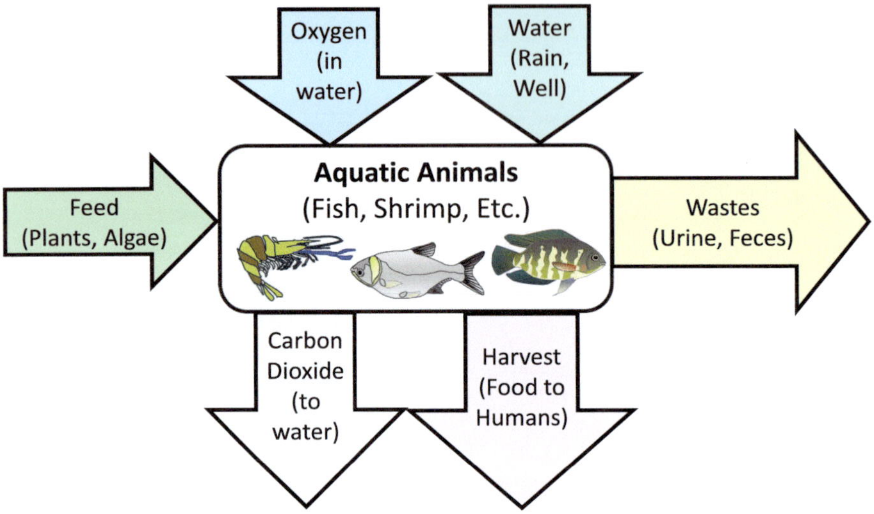

Fig. 1.12 Aquatic animals as mass flow. Aquatic animals including shrimp and fish consume plant forage and algae, take in water and oxygen, and release carbon dioxide, urine which has water, and feces

Fig. 1.13 A raceway for fish growth, in this case for rainbow trout. This picture was taken at Shepherd of the Hills fish hatchery near Branson MO, from the outlet side

Fig. 1.14 Stream bottom. Rocks buffer the water, and the slime layer of bacteria and algae balances the chemistry of water for fish. Mollusks such as these snails manage the slime layer by eating it. Photo by author

with the human habitat (i.e., where the people live) we need to make a synthetic ecosystem. We will need to duplicate many of the cyclic processes noted on Earth. In this book, each stage of the space farm resembles the processes for the Earth farms or ecosystems above.

1.6 Vertical and Urban Farming

Cities and other urban areas have had farms in vacant lots or on rooftops for centuries, but modern technology has taken urban farming to new heights. If you live near cities in the USA, Canada, Europe, Japan, and many other places, you might have had a fresh tilapia dinner, and a fresh lettuce and tomato salad, even mid-winter. If so, you have already had a preview of the technologies in space farms. Vertical farms take advantage of limited floor space by stacking shelves of soil or hydroponic plants, with loops in many cases through a fish tank. Dr. Dickson Despommier, of the Environmental Health Science of Columbia University, largely developed the vertical farm idea, though many others came up with similar concepts and many sci-fi shows have shown various space-born concepts. Indoor high efficiency farms have become a growing industry in urban areas (see (Birkby & Soto-Velez, 2016)). In cities, land area is expensive and limited, and vertical farms allow fresh crops in the small areas available, often using vacant buildings, atriums, and old factories. Such a farm is depicted in Fig. 1.15.

Many of the fresh tomato and lettuce dishes in northern Europe are grown under artificial lights in highly controlled conditions, as implemented in a big indoor farm

Fig. 1.15 Two examples of urban (vertical) farms from the US Department of Agriculture (USDA), Agriculture Research Service (ARS). Lettuces on left, cucumbers on right. Image credits: Photos by James Altland, USDA ARS (see refs: USDAARS-D4686-1, USDA-ARS-D4687-1)

in Denmark (see Peters, 2020) using either sourced nutrients, processed wastewater, or recycling wastewater from an attached fish farm (e.g., tanks with tilapia, trout, shrimp, catfish, or other fish). Combining an aquatic animal crop with a hydroponic crop is called aquaponics. Sushi grade shrimp are grown indoors in at shrimp farms in round tanks even in the northern Midwest, and in many cities vertical farms have taken root in old factories or mills. These farms exist because of the demand for fresh food nearby, and the efficiency of the farm, even in frigid northern hemisphere winters. A nice side effect for the farms is that they use a fish-plant cycle that requires less input mass than a plant-only farm, for reasons discussed later in this book. In the far north, natural gas or propane heaters are used to warm the farm and double as sources of carbon dioxide to speed plant growth as in (Blom et al., 2002). That is right, releasing CO_2 into the greenhouse from fossil fuel combustion not only grows food faster, but recycles the carbon too. It is worth noting that we need carbon dioxide in an indoor farm and therefore in a space farm too. For reasons you will see later in this book, carbon dioxide, not oxygen, is the core gas to survival.

1.7 Open Systems Versus Closed Systems

The open-air Earth farm examples noted above are examples of mostly open systems, as in the top diagram in Fig. 1.16.

Open systems are where a continuous influx of mass and energy is used to produce an output external to the habitat and farm together. Closed systems in contrast balance mass inputs and outputs, creating as close to a zero sum as possible, in a cycle where inputs required by the system of habitat and farm are nearly zero. As of publication time for this book, space outposts are open systems, requiring constant resupply to operate. The International Space Station (ISS) and other spacecraft and space stations are open systems. In Fig. 1.17, is a semi-notional mass flow for current and near-future space stations with values roughly computed from data in (NASASTD-3001, 2015), (NASA-TD9702A, 1988), and (NASA-SP-2010-340, 2014) and some extrapolation.

Space farms in this text are designed to minimize external inputs to the settlement, creating a nearly closed system, or self-sustaining settlement, and replacing the flows in Fig. 1.17 with a ecosystem cycle as in Fig. 1.7. Once in full operation, apart from settlements that produce food for export (food factory settlements), the closed system would cycle mass. In the real world, inefficiencies would require some flow of supply from in situ resources, or shipments. Seals are not perfect, and things just happen to upset balances. It might even be those darn tourists who upset the balance, ask any resident of Orlando, FL, USA. Despite the best engineering, nature and chaos may intervene. In engineering, it is best to look for a worst case. Here I error on the side of caution and assume the worst and will seek to oversize farms beyond the farms that may actually be built. This oversize is very critical to account for significant educated guesses. Better to be a little too big than to suffocate or starve.

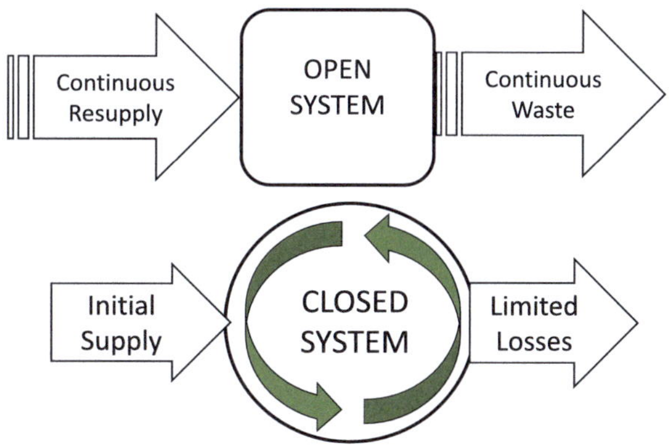

Fig. 1.16 Schematic for an open system (top) and a closed system (bottom)

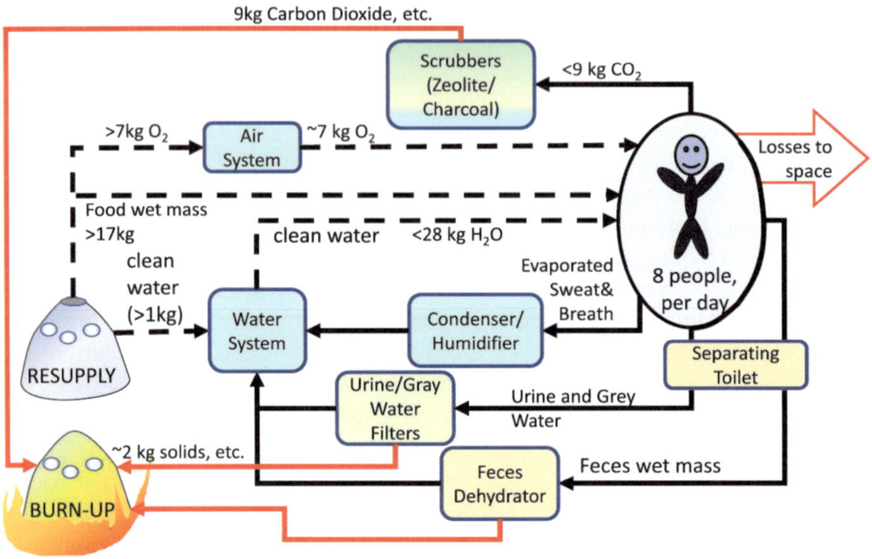

Fig. 1.17 Rough mass flow for current and near-future orbital space stations

1.8 Seasons and Timing

Seasons are critical to growth for plants and animals. In temperate regions, freezing temperatures dictate growing season for plants, while monsoons or dry season dictate growth in tropical regions. In the Missouri wheat example above, winter wheat is planted in the fall, and the seeds mostly sit dormant until emergence from the soil in March, for harvest in late July. In the last few weeks, the grain plant dries out, and mass is pushed into the grains at the cost of the rest of the plant. Light levels are key for triggering life events in crops, and in mating seasons for animals.

Figure 1.18 shows the life cycle for winter wheat. Starting with seeds planted in autumn, the seeds sprout and develop shoots and roots under the snow in winter and then emerge from the soil in early spring. These plants develop to mature plants in the early summer and dry out before harvest in late summer or early fall. In harvest, the above ground part is cut. Grain is separated from the shoot, leaves, and chaff and stored for use or sale. The shoot and leaves become fodder for animals as straw. The remainder of roots and lower shoot are left for a while to hold the soil against runoff and then eventually disked under to feed the soil for the next crop.

As a result of this cycle, many crop species evolved to grow during specific seasons and times. This is a different ‘gearing’ than human needs for food, water, and air, which is every day, regardless of season. Cultures have dealt with seasonality by developing calendars to track optimal times for planting and harvest and methods to store food during off-season periods, i.e., when food is not ready to harvest. The Woodhenge in Cahokia (see Fig. 1.19) is such a calendar. The posts of this structure align with various key times during the year. The ancient Cahokians were

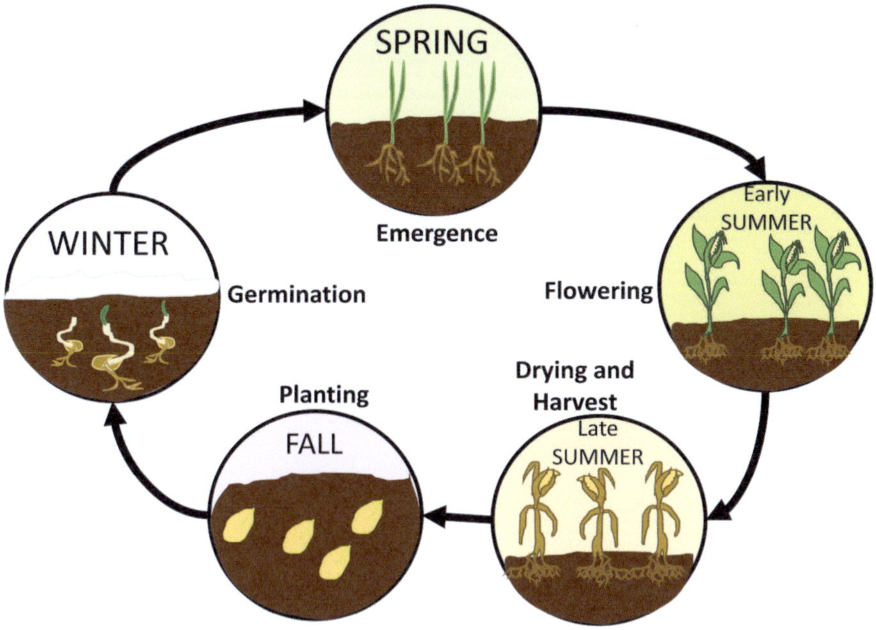

Fig. 1.18 Seasons for winter wheat

accomplished farmers and had a complex organized culture that rerouted waterways for irrigation and transportation, built mounds (see the hills in the background of figure), and even had sporting arenas. It is likely that the Woodhenge was used to time planting for crops, among other events. Knowing when to plant maize, beans, and squash was critical to the health of the people in this society. Plant too early, and the ground might be frozen, or a late frost will kill the plant shoots. Plant too late and miss growing time and some of the food that would have been grown. Not harvesting at the right time can lead to crop loss.

Not all crops are seasonal, however. Some vegetables can grow year-round given the right conditions, as can some animals. Further, it is likely using a combination of light and hormones can trick crops into producing year-round by cohort, to allow a continuous harvest. Since these farms are assumed to have a completely controlled environment and are part of a complex artificial ecosystem, many controls can be used to get fresh chow every day. Do not worry, continuous harvest is further discussed later in this book. Regardless of freshness, storage technology will be critical for space farms, if for no other reason than to allow variety and surges due to human events like those crazy tour groups, or wild parties on Yuri's Day. It can be assumed, as on Earth now, food can be stored for at least a year without significant loss of nutrition.

Fig. 1.19 Woodhenge at Cahokia, IL, USA. Photo by author

1.9 Summary

This chapter covered many core concepts that will be used in later chapters of this book, in designing space farms. First we discussed why space is a difficult environment. Space settlements, which will grow from earlier space base, have to be enclosed, at a cost, to protect living things from the radiation, temperature swings, and vacuum of space. Things like food and water have to be either shipped into the settlement (in an open cycle) or recycled (in a closed cycle). Space farms can become the recycler of air, water, and food, creating an ecosystem with the humans in the settlement. We explored how Earth farms are part of natural ecosystems (ideally) and how in space the full closed mass cycle replicates many natural cycles. This chapter also covered how crop growth is timed, often at a different rate than when people need to eat. All these factors and concepts will be expanded in later chapters.

1.10 Questions

1. What conditions make the space environment harsh for Earth life, including humans and crops?
2. What is a space farm? What is a space settlement?
3. Why is efficiency critical in space settlement?
4. What are the differences in food sources between an outpost, and a fully developed space community?
5. What are the key differences between growing crops on Earth versus growing crops in a space base?
6. What is a closed system, and how is it different than an open system? Is the International Space Station an open system or closed system, and why?

7. What is the seasonal cycle for most grass-based (grain based) crops on Earth? How is this different than human needs? What crops are non-seasonal?

References

AMRC. (2024). *Strawberries.* Revised April 2024© 2024 Ag Marketing Resource Center, Iowa State University and USDA, found at: https://www.agmrc.org/commodities-products/fruits/strawberries

Birkby, J., & Soto-Velez G. (2016). Vertical farming. *Newsletter*, ATTRA—Sustainable Agriculture Program, SKU: IP516, Copyright © NCAT 1997–2022, found 2022 at: https://attra.ncat.org/product/vertical-farming/

Blom, T. J., Straver, W. A., Ingratta, F. J., Khosla, S., & Brown, W. (2002). *Carbon dioxide in greenhouses.* Factsheet, Ontario Ministry of Agriculture, Food and Rural Affairs (OMAFRA), Index: 290/27, Dec 2002, last Reviewed 2009, Reorder#: 00-077.

Caravati, C. (2021). *Potato production in USA*, Kenosha Potato Project, seedsaver@curzio.com, found online at: http://kenoshapotato.com/Potato_Production_USA.htm

Coolong, T., & Boyhan, G. E. (2020). Commercial tomato production handbook. *Bulletin* 1312. University of Georgia © 2020, found at: https://extension.uga.edu/publications/detail.html?number=B1312&title=Commercial%20Tomato%20Production%20Handbook

Endres, G., et al. (2018). *Pinto Bean response to phosphorus starter fertilizer in East-central North Dakota* (A1883, April 2018). North Dakota State University Extension, Fargo ND.

Farag, I., Chaput, G., & Charmanski, K. (2012). Sustainable production of microalgae oil feedstock using municipal wastewater and CO_2 fertilization. *Journal of Engineering Science and Technology (IJEST), 4*, 3489–3499. ISSN: 2141-2820.

FP-2022-Peanut. (2022). *Peanut production 101: Rotation key to bigger yields.* FarmProgress, Copyright © 2022. Informa Markets, a trading division of Informa PLC, found at: https://www.farmprogress.com/peanuts/peanut-production-101-rotation-key-bigger-yields

Habib, M. A. B, Huntington, T. C., & Hasan, M. R. (2008). *FAO fisheries and aquaculture circular No. 1034, A review on culture, production and use of spirulina as food for humans and feeds for domestic animals and fish.* ISSN 2070-6065, Food And Agriculture Organization of the United Nations, Rome.

Harper, J. K., & Orzolek, M. (2006). *Agricultural alternatives: Tomato production, 2006* (p. 16802). Agricultural Research and Cooperative Extension, Pennsylvania State University, University Park, PA.

Heppenheimer, T. A. (1977). *Colonies in space.* Stackpole Books. ISBN-100811703975

Holmstrom, K. E. (2008). *Crop profile for arugula in New Jersey.* National Integrated Pest Management Database, Rutgers New Jersey Agricultural Experiment Station.

Johnson, R., Holbrow, C., & O'Neill, G. K. (1977). *Space settlements: A design study.* NASA-SP-413, LC-76-600068. NASA.

Kandel, H., et al. (2018). *North Dakota soybean variety trial results for 2018 and selection guide (A843-18, Dec. 2018).* North Dakota State University, Fargo ND.

Keene, C., Horsley, R. McMullen, M., Friskop, A., Baldwin, T., Zwinger, S., Ostlie, M., Martin, G., Rickertsen, J., Eriksmoen, E., Hanson, B., & Pradhan, G. (2022). *A1049-22, North Dakota barley, oat and rye variety trial results for 2022 and selection guide.* North Dakota State University, Fargo, North Dakota November 2022, © 2022, found at: https://www.ndsu.edu/agriculture/sites/default/files/2022-12/a1049-22.pdf

Keene, C., Horsley, R., McMullen, M., Friskop, A., Baldwin, T., Simons, K., Martin, G., Rickertsen, J., Bortolon, L., Kraklau, A., Hansen, J., Justin Jacobs, J., & Hanson, B. (2023). *A1049-23 North Dakota barley, oat and rye variety trial results for 2023 and selection guide*, Dec. 2023, North

Dakota State University Extension, Fargo ND 58102 found at: https://www.ndsu.edu/agriculture/sites/default/files/2023-11/a1049-23.pdf

Massey, R., & Brown, B., (2021) *Winter wheat planning budget, extension.* University of Missouri, g656. Revised 10/2021.

Molinar, R. H. (2006) *Sample costs to produce onions, red onion*, San Joaquin valley—South, ON-VS-06. University of California Cooperative Extension.

NASA-SP-201-03407 *Human integration design handbook (HIDH)*, Revision 1—June 5, 2014, NASA/SP-2010-3407/REV1, found at:https://www.nasa.gov/sites/default/files/atoms/files/human_integration_design_handbook_revision_1.pdf

NASA-STD-3001 *Volume 2, revision a NASA space flight human-system standard, Volume 2: Human factors, habitability, and, environmental health*, Approved: 02–10–2015, National Aeronautics and Space Administration, Washington, DC 20546-0001.

NASA-TD9702A. (1998). *International space station familiarization*, Manual TD9702A, NASA, July 31, 1998, found at: https://er.jsc.nasa.gov/seh/td9702.pdf

O'Neill, G. K. (1976). *The high frontier: Human colonies in space.* William Morrow & Co. Inc., ISBN 0-688-03133-1.

OrSUCAS. (2010). *Oregon vegetables: Arugula*, Feb 2010, College of Agricultural Sciences, Oregon State University, Corvallis, Oregon 97331, Copyright ©2022 Oregon State found at: https://horticulture.oregonstate.edu/oregon-vegetables/arugula-0

Peters, A. (2020). This vertical farm in Denmark will grow 1,000 tons of local greens a year. *Fast Company*, 12-08-2020. https://www.fastcompany.com/90582905/this-vertical-farm-in-denmark-will-grow-1000-tons-of-local-greens-a-year

PVWatts ®. (2022). *Calculator, National Renewable Energy Laboratory.* U.S. Department of Energy, Office of Energy Efficiency and Renewable Energy, operated by the Alliance for Sustainable Energy LLC., checked 2022, found at https://pvwatts.nrel.gov/pvwatts.php

Ransom, J., & Eisinger, D. (2017). *North Dakota corn hybrid trial results for 2017*, A793-17. North Dakota State University Extension, Fargo, North Dakota.

Strickland, J., Spencer, S., & Nesterova, A. (2021a). *Developing space.* Apogee Books, paperback, 354 pp., illus., ISBN 978-1-989044-14-8.

Strickland, J., Spencer, S., & Nesterova, A., (2021b). *Settling space.* Apogee Books, paperback, 398 pp., illus. ISBN 978-1-989044-16-2.

Tourte, L., et al. (2009). *Sample costs to produce organic leaf lettuce, double-cropped central coast region Santa Cruz & Monterey counties*, LT-CC-09-O, University of California Cooperative Extension.

TxA&M. (2022). *Texas A&M aglife extension, bell peppers crop guide.* Aggie Horticulture® College Station, TX 77843, 2022, found at: https://aggie-horticulture.tamu.edu/smallacreage/crops-guides/vegtables/bell-peppers/

USDAARS. (2022). *FoodData Central.* U.S. Department of Agriculture (USDA), Agricultural Research Service (ARS), found at: https://fdc.nal.usda.gov/

USDAERS-RiceYB. (2024). *Rice yearbook.* U.S. Acreage, Production, Yield, and Farm Price, Economic Research Service, U.S. Department of Agriculture, updated 3/28/2024, found at:https://www.ers.usda.gov/data-products/rice-yearbook/

Chapter 2
Mechanics of Biochemistry

Abstract This chapter focuses on key biochemistry and processes from a mass perspective. The mass percentages of key elements such as carbon, oxygen, hydrogen, and nitrogen are examined in core nutrition components, such a carbohydrates, proteins, fats, water, carbon dioxide, and others, and how these become wastes and inputs for other processes. Key reactions are examined at a high engineering level, to estimate how organisms move masses of elements from input to output, and especially how aerobic metabolism feeds photosynthesis and the reverse.

Keywords Mass balance · Elements · Chemical reactions · Ash

2.1 Introduction

This chapter shows the cookbook of masses required for life and how tracking elements through biochemistry shows what must go into, and out of, each organism in the farm, and humans in the habitat. The mass percentages of key elements such as carbon, oxygen, hydrogen, and nitrogen are examined in core nutrition components, such a carbohydrates, proteins, fats, water, carbon dioxide, and others, and how these become wastes and inputs for other processes. Key reactions are examined at a high engineering level, to estimate how organisms move elements from input to output, and especially how aerobic metabolism feeds photosynthesis and the reverse, at an engineering estimate level.

2.2 Nothing Basic About Biochemistry!

Biochemistry is the science of how living things use various compounds and elements. Looking too deep at biochemistry will make the average space farm engineer's eyes bug out. It has lots of three-dimensional wildness that is not fully understood. If you had to take biochemistry in college, you will have noted the wild hand

© The Author(s), under exclusive license to Springer Nature Switzerland AG 2026

B. L. Meyer, *Designing Farms in Space*, https://doi.org/10.1007/978-3-032-04395-5_2

signs used by students (looks like an odd sign language) to imagine the molecules. That said, there are four very key elements a farmer must track, and several others that are important to a lesser extent: carbon, hydrogen, oxygen, and nitrogen or 'CHON' for short.

Carbon is the building block of (almost) all complex Earth life. It is the multi-tool of biochemistry due to its ability to make lots of stable bonds. Calories in food for people essentially come from breaking carbon bonds of various types, whether carbon to carbon, or carbon to hydrogen, or carbon to oxygen, and every so often nitrogen to carbon or to oxygen or to hydrogen. Sugar tastes so particularly good because it has lots of carbon bonds that we are really good at converting quickly to energy (i.e., calories). Ditto for fats. Doughnuts taste good and are high in calories, because they are loaded with carbohydrates and fats. This is the same reason peanut butter is irresistible. For your information, human dietary 'calories' are really kilo-calories, as measured by the energy to raise 1 kg of water 1 °C. If you go outside for hours in midwinter like your author, you notice when breakfast burns off because you get colder. Carbon is also used in plants to make cellulose, which is a fiber that makes our bowels work and trees stand tall. Especially in wood, trees make a wickedly harder to break down cellulose cousin called lignin. Plants also assemble complex carbohydrates from sugars to make starch to store energy. Fats also feature lots of carbons (see doughnut note above), as do many other biological molecules. Unfortunately, as known to date, carbon is hard to find on the Moon, but plentiful on Mars and many asteroids and comets. Hard to find in space, means it takes energy and fuel to get it.

Oxygen is what we breathe from the air. Oxygen gas that we breathe in is two oxygen atoms. That oxygen is used to break those carbon bonds. Carbon bonds to two oxygen atoms, making carbon dioxide, and moving energy for our cells to use. The reaction is complex (see Fig. 2.2), but that is part of it. Water is one oxygen and two hydrogens. Water is formed when hydrogen is oxidized, fast as for some chemical rockets, or slowly in animal bodies when digesting carbohydrates. In either case, making water releases a bit of heat, and it takes energy to break up water to get oxygen. Oxygen gas is produced when plants use light to break down carbon dioxide and water to make those carbohydrates (a.k.a. photosynthesis), moving the oxygen out as oxygen gas.

Hydrogen is a super-bonder. Carbons are capped with hydrogen at the edges of organic molecules. One important exception to the tavern patron or sommelier is when oxygen sneaks between that carbon and hydrogen, as in alcohol. That sneaky oxygen helps make the buzz from said alcohol too. Thank you, Mr. Sneaky Oxygen! Otherwise, carbon-hydrogen bonds shape biomolecules. Water is the major source of hydrogen for many biological reactions, including photosynthesis. Hydrogen bonds feature heavily in fats (pun intended). Hydrogen is also used in cell walls and to facilitate many important biological reactions.

Nitrogen is like carbon, in that it bonds well, and is the key in proteins. Proteins are the workhorses of organisms and are made from amino acids, which have nitrogen and hydrogen ends. Muscles are rich in proteins, which is why body builders focus on them. Animals can synthesize and break down proteins, but get nitrogen from dietary amino acids, likely from proteins. Deoxyribonucleic Acid (DNA) and Ribonucleic Acid (RNA) are made from nitrogen too.

While the 'CHON' elements are the most critical, others are important too. See Table 2.1. Sulfur and phosphorous are in some amino acids and nucleic acids. Calcium, potassium, and sodium are in nerves and bones, and in plant cell functions. Iron is in muscles and blood. Many other elements make up what is in us, and in our crops.

This is not an exhaustive list of important elements, just the biggest hitters in your body to keep you alive. These are also important for looking at key compounds, i.e., combinations of elements.

Note in the Table 2.2, I skip the subscript hell and just show the numbers after the elements in the chemical formulas.

Next are some key compounds (in Table 2.2) that you likely know about already. If you notice, this table begins to track how the biggest biological elements show up. Tracking element masses is one way to make sure your farm can balance out. More in later chapters.

Some of these items like the gases and simple compounds can be found in comets and planetary atmospheres. Urea is how complex animals contain ammonia in the body. ATP is how cells power reactions and why plants and you need phosphorous.

Next (in Table 2.3) are the sweet and intoxicating compounds, the carbohydrates, and alcohols.

Many of the readers of this book and your author know the mixed blessings of carbohydrates. They are the sweet things we like to eat, then must work off

Table 2.1 Key biological elements

Name	Chemical formula	Molecular weight (rounded)
Carbon	C	12.01
Hydrogen	H	1.01
Oxygen	O	16.00
Nitrogen	N	14.01
Sulfur	S	32.06
Potassium	K	39.10
Calcium	Ca	40.10
Iron	Fe	55.85
Magnesium	Mg	24.31
Phosphorus	P	30.97
Sodium	Na	22.99
Zinc	Zn	65.38
Copper	Cu	63.55
Selenium	Se	78.96
Chlorine	Cl	35.45

Loosely from CIAAW (2022)

Table 2.2 Key simple compounds and ATP

Name	Chemical formula	Molecular weight (rounded)	% mass by element					
			C (%)	H (%)	O (%)	N (%)	S (%)	P (%)
Pure gases								
Hydrogen gas	H_2	2.02	**0**	**100**	**0**	**0**		
Oxygen gas	O_2	32.00	**0**	**0**	**100**	**0**		
Nitrogen gas	N_2	28.02	**0**	**0**	**0**	**100**		
Key compounds								
Water	H_2O	18.02	**0**	**11**	**89**	**0**		
Carbon dioxide	CO_2	44.01	**27**	**0**	**73**	**0**		
Ammonia	NH_3	17.04	**0**	**18**	**0**	**82**		
Urea	CH_4N_2O	60.07	**20**	**7**	**27**	**47**		
Methane	CH_4	16.05	**75**	**25**	**0**	**0**		
Table salt	NaCl	58.44	**0**	**0**	**0**	**0**		
Acetic acid	$C_2H_4O_2$	60.05	**40**	**7**	**53**	**0**		
Adenosine triphosphate (ATP)	$C_{10}H_{16}N_5O_{13}P_3$	507.18	**24**	**3**	**41**	**14**	**0**	**18**

Loosely calculated from PubChem (2022) and CIAAW (2022) data

Table 2.3 Key carbohydrates and percentages by mass by element

Name	Chemical formula	Molecular weight (rounded, Std)	% mass by element				
			C (%)	H (%)	O (%)	N (%)	S (%)
Key sugars and starches							
General form	*CnH2nOn*		*40*	*7*	*53*	**0**	**0**
Fructose/glucose/galactose	$C_6H_{12}O_6$	180.16	**40**	**7**	**53**	**0**	**0**
Cellulose	$C_6H_{10}O_5$	162.14	**44**	**6**	**49**	**0**	**0**
Sucrose/lactose/maltose	$C_{12}H_{22}O_{11}$	342.30	**42**	**6**	**51**	**0**	**0**
Starches	$(C_6H_{10}O_5)n - (H_2O)$	(Est. large chain)	***44***	***6***	***49***	***0***	***0***
Alcohols							
General form	CnH(n + 2)OH		*52*	*13*	*35*		
Ethanol	C_2H_5OH		***52***	***13***	***35***	***0***	***0***
Methanol	CH_3OH		***37***	***13***	***50***	***0***	***0***
Others							
Lignin (roughly)	$C_{10}H_{12}O_3$ et al	180.22 (rough, partial)	***67***	*7*	***27***		

Loosely calculated from PubChem (2022) and CIAAW (2022) data later. Chemical energy in the form of carbon–carbon and carbon-hydrogen bonds makes cells work, which is why they taste good. Sugars are monosaccharides like fructose in fruits or disaccharides like sucrose the fun sugar stuff in ice cream and candy. That Disaccharides are made from two monosaccharides. Starches, as in bread and potatoes, are how organisms store sugars by linking chains of sugars together in compact form. In our mouths, salivary amylase breaks up the starch into sugars that taste good. Lignin is the stuff of wood and needs fungi digest it. Related to those sugars are alcohols, and ethanol is the type that is in wine, beer, and other bar drinks. Methanol, on the other hand, is toxic to people and will make you blind and dead if you drink it. Alcohols have the extra oxygen and hydrogen group (i.e., OH) attached. Lignin is a pesky complex compound already noted above, that is found mostly in wood and woody stems, and that can only be broken down in nature by wildfire or fungi.

Now we get to the oily, greasy, stuff in Table 2.4.

Fats are critical to many biological functions, and unfortunately accumulate exactly where you do not want them. Fats are how organisms store food energy longer term by linking chains of carbons and hydrogens together. If you want to add lots of carbon to your diet, add fats. Fats are also used in messenger compounds in our bodies. For calculations, it is easiest to use the percentages in triglycerides to substitute for food fats unless you know exactly which fats are involved, and in what amount.

Proteins (in Table 2.5) are combinations of amino acids, which are like carbohydrates and fats, except that they add in nitrogen and sometimes sulfur or other elements. Like fats add carbon to a diet, proteins add nitrogen. Proteins are critical for growth and can be used for food fuel as well, as the keto crowd can relate. If you only know there is a certain percentage of proteins, the average percentages are based on the weighted average of the frequencies that the amino acids are found in living things in nature.

A reason phosphorous is important in your diet, or the intake for any organism is in Deoxyribonucleic Acid (DNA). Table 2.6 is the breakout for the DNA bases and the backbone. The deoxyribose phosphate backbone in living things is assembled

Table 2.4 Key fats by element mass percentage

Name	Chemical formula	Molecular weight (rounded, Std)	% mass by element				
			C (%)	H (%)	O (%)	N (%)	S (%)
Fats/fatty acids/lipids							
Triglyceride (or for fats in general)	$C_{55}H_{98}O_6$	*855.53*	***77***	***12***	***11***	***0***	***0***
Glycerol	$C_3H_8O_3$	92.1	***39***	***9***	***52***	***0***	***0***
Cholesterol	$C_{27}H_{46}O$	386.7	***84***	***12***	***4***	***0***	***0***

Loosely calculated from PubChem (2022) and CIAAW (2022) data

Table 2.5 Proteins and amino acids by mass percentage

Name	Chemical formula	Molecular weight (rounded, Std)	% mass by element				
			C (%)	H (%)	O (%)	N (%)	S (%)
Proteins/amino acids							
Avg for amino acids	n/a	n/a	***45***	*8*	*32*	*15*	***1***
Alanine	$C_3H_7NO_2$	89.1	***40***	*8*	***36***	***16***	***0***
Arginine	$C_6H_{14}N_4O_2$	174.2	***41***	*8*	***18***	***32***	***0***
Asparagine	$C_4H_8N_2O_3$	132.1	***36***	*6*	***36***	***21***	***0***
Aspartic acid	$C_4H_7NO_4$	133.1	***36***	*5*	***48***	***11***	***0***
Cysteine	$C_3H_7NO_2S$	121.2	***30***	*6*	***26***	***12***	***26***
Glutamic acid	$C_5H_9NO_4$	147.1	***41***	*6*	***43***	***10***	***0***
Glutamine	$C_5H_{10}N_2O_3$	146.1	***41***	*7*	***33***	***19***	***0***
Glycine	$C_2H_5NO_2$	75.1	***32***	*7*	***43***	***19***	***0***
Histidine	$C_6H_9N_3O_2$	155.2	***46***	*6*	***21***	***27***	***0***
Isoleucine	$C_6H_{13}NO_2$	131.2	***55***	***10***	***24***	***11***	***0***
Leucine	$C_6H_{13}NO_2$	131.2	***55***	***10***	***24***	***11***	***0***
Lysine	$C_6H_{14}N_2O_2$	146.2	***49***	***10***	***22***	***19***	***0***
Methionine	$C_5H_{11}NO_2S$	149.2	***40***	*7*	***21***	*9*	***21***
Phenylalanine	$C_9H_{11}NO_2$	165.2	***65***	*7*	***19***	*8*	***0***
Proline	$C_5H_9NO_2$	115.1	***52***	*8*	***28***	***12***	***0***
Serine	$C_3H_7NO_3$	105.1	***34***	*7*	***46***	***13***	***0***
Threonine	$C_4H_9NO_3$	119.1	***40***	*8*	***40***	***12***	***0***
Tryptophan	$C_{11}H_{12}N_2O_2$	204.2	***65***	*6*	***16***	***14***	***0***
Tyrosine	$C_9H_{11}NO_3$	181.2	***60***	*6*	***26***	*8*	***0***
Valine	$C_5H_{11}NO_2$	117.1	***51***	*9*	***27***	***12***	***0***

Protein average is weighted by incidence in nature. Loosely calculated from PubChem (2022) and CIAAW (2022) data

into a long repeating scaffold that holds the DNA bases, for humans and dinosaurs. Queue cool sounding movie music, right before the dinosaurs eat the scientists.

All that said, animals, algae, fungi, bacteria, and plants rely on a few important chemical reactions. To save your minds from a slow death, I have added cartoons.

- **Photosynthesis**: Carbon dioxide (CO_2), water (H_2O), and light energy used make carbohydrates (CnHnOn, which are further reacted to make sugars, cellulose, starches) and release oxygen gas (O_2). Since sugars can be combined to make other carbohydrates, we can skip the in between steps to see the high-level result.
  - Pictorially (in Fig. 2.1), the reaction is an exchange of elements between input and outputs using light energy.

Table 2.6 DNA bases and backbone mass percentages by element

Name	Chemical formula	Molecular weight (rounded, Std)	% mass by element				
			C (%)	H (%)	O (%)	N (%)	P (%)
Adenine	$C_5H_5N_5$	135.1	***44***	***4***	***0***	***52***	***0***
Guanine	$C_5H_5N_5O$	151.1	***40***	***3***	***11***	***46***	***0***
Cytosine	$C_4H_5N_3O$	111.1	***43***	***5***	***14***	***38***	***0***
Thymine	$C_5H_6N_2O_2$	126.1	***48***	***5***	***25***	***22***	***0***
Deoxyribose + phosphate	$C_5H_{10}O_4 + P$	165.1	***36***	***6***	***39***	***0***	***19***

Loosely calculated from PubChem (2022) and CIAAW (2022) data

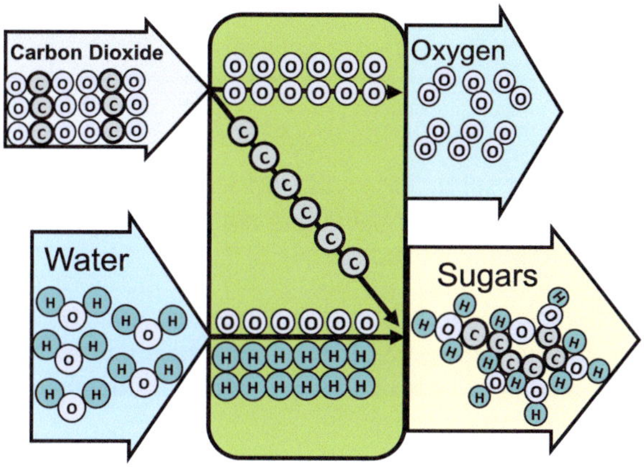

Fig. 2.1 Photosynthesis (simplified) takes light and juggles carbon, hydrogen, and oxygen atoms to turn carbon dioxide and water into sugars and oxygen. Oxygen atoms might move differently than shown

- Numerically (sorry) and roughly, the top-level reaction to make fructose (a sugar) looks like this:

  $6CO_2 + 6H_2O \rightarrow C_6H_{12}O_6 + 6O_2$.
  [Math follows:] Or if the molecular weight of carbon is about 12, oxygen about 16, and hydrogen about 1, then carbon dioxide (CO_2) is 12 + 2*16 = 44, water (H2O) is 2*1 + 16 = 18, oxygen gas (O_2) = 2*16 = 32, and fructose (or glucose, etc.) ($C_6H_{12}O_6$) is roughly 6*12 + 12*1 + 6*16 = 180. Then the equation above looks like: 6*44 + 6*18 = 180 + 6*32 (or cutting to the chase, 372 = 372). Reduce the equations like a cookbook, take about 1.5 measure (or 6*44/180) of carbon dioxide, add 0.6 (6*18/180) of water, put in the light for a while using chlorophyll, and you get 1 measure of fructose and about 1.1 measure (or 6*32/180) of oxygen gas.
  Keep this thought pattern in your head for a bit…we will get back to it.

- **Aerobic Metabolism**: Organisms use oxygen and foods (fats, carbohydrates, proteins) together to release energy (such as heat), waste (anything not sent in CO_2 or H_2O), water (H_2O), and carbon dioxide (CO_2). In animals, waste can be

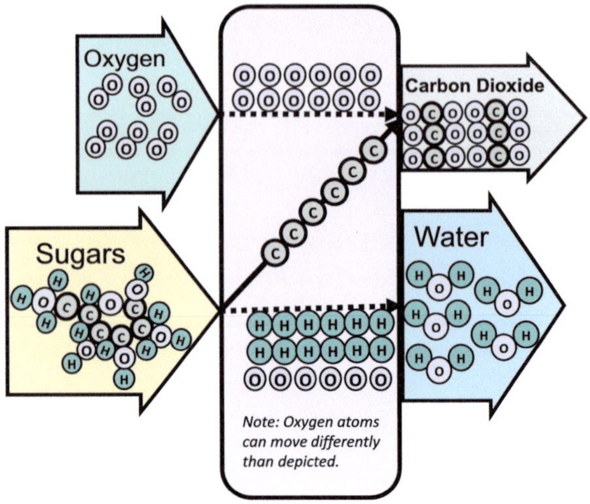

Fig. 2.2 Aerobic metabolism (simplified) takes oxygen and sugars and converts them to carbon dioxide and water to produce cellular energy

liquid or solid. These all start with the oxygen and sugar conversion to carbon dioxide and water. Anything not a carbohydrate gets broken into something like one to fit the reaction in Fig. 2.2.

- Note: While bacteria and fungi may digest cellulose using aerobic metabolism as in the figure, or using enzymes and water, most animals cannot, so the cellulose ends up in solid wastes. Remember that cellulose is a water short of a digestible sugar.
- Liquid wastes for animals have urea or ammonia in their wastes because of nitrogen containing foods like amino acids and proteins. The process is called deamination. It is in Table 2.7 and explains why wastes smell and why the room clears after a chili eating contest. Another product of deamination is methane. Deamination is the magic that fits amino acids and proteins into aerobic metabolism.

- **Anaerobic Metabolism**: Foods (carbohydrates primarily) and enzymes combine to make carbon dioxide, water, and alcohols (see Fig. 2.3) or simple organic molecules like methane, and release energy to cells (with heat). This is where we get some of nature's best gifts for dinner: beer, saki, bread, and wine. Usually, the alcohol is then further broken down to water and carbon dioxide by the cell if oxygen can be found (as in your body).
- A couple of other important reactions:
  - Amino acids are assembled into complex proteins. 'Complex' as in wild three-dimensional structures. DNA tells cells which amino acids to put where, then enzymes assemble them from there.
  - Carbohydrates get strung into fats (as I look down at my own tummy) and reverse. Proteins get broken into amino acids, thence to carbohydrates, and then again into fats. No escape from too much dinner for you!

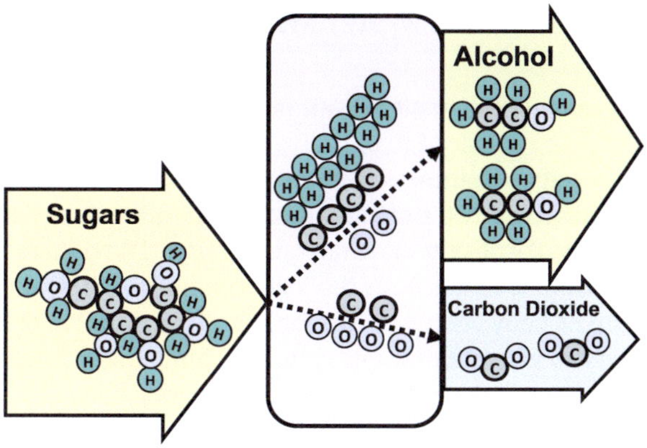

Fig. 2.3 Alcohol anaerobic metabolism (simplified). Simple sugars are broken down to produce alcohol and carbon dioxide

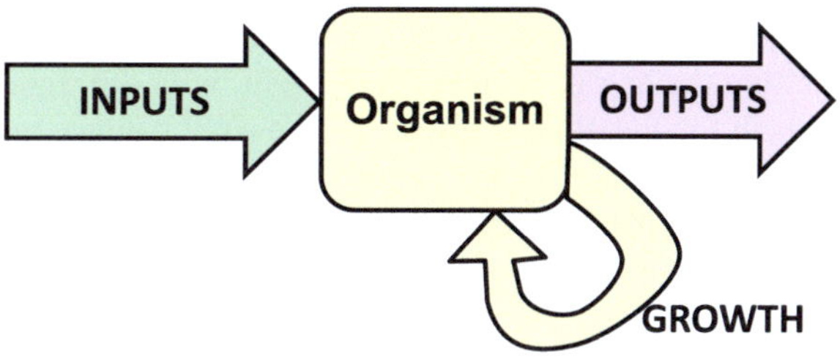

Fig. 2.4 Basic ins and outs. When a living thing takes in mass, it either releases that mass after use, or uses it to grow

- **Cellulose** is just a water away from fructose or glucose, i.e., fructose minus a water and some chemical trickery = a cellulose (unit). Cellulose units are combined to make plant cell walls, and add the crunch in your salad and veggies.
- **Lignin** is what plants use to make a stem tough and is very resistant to bacteria. Trees, bushes, and to a smaller degree, grass like grains, have it. It needs to be burned, fed to mushrooms (like oyster mushrooms), fed to worms, ground into soil, or broken down using tailored fungi or genetically modified bacteria in a bioreactor. Lignin adds a complication to closing the mass flow cycle because it is resistant to bacteria.
- **Solid wastes** include cellulose, lignin, undigested proteins, carbohydrates, and fats since animals are note even close 100% efficient. Cow patties have enough leftovers including cellulose and lignin that they can be dried and burned as fuel. Some food just plain passes through.

If you were paying attention, you noticed that the top two reactions above, photosynthesis and aerobic metabolism, feed each other. That is how a space farm and the human habitat close the cycle. It is why I can have a 5-gallon closed tank with fish, bacteria, snails, and plants.

For a few common reactions, the 'cookbook' looks like the values in Table 2.7.

2.3 Salts and Water Hardness

For aquatic life, water is not just water. Organisms live in a chemical solution that should be balanced to optimize how they grow. Usually, this extra stuff is expressed as salinity or hardness in parts per thousand, and acidity in pH. Freshwater for most fish has roughly 8 g of various salts and minerals for each liter of water. Saltwater has at least 20 g of salts and minerals per liter. The right mineral balance is key to keeping the osmotic balance for the fish, to minimize how much energy the fish expends to keep its body chemistry working. Too much salt in the water for a fish forces the fish to try to get water into its cells and dump out extra salts in its body. The wrong pH is similar. Too low a pH (too acidic) and fish and crustaceans can't harden bones or shells. Too high (too alkaline) and many reactions in cells do not work.

Plants also grow best in the right pH and hardness. Algae and cyanobacteria, while tough, will just die if the pH, salinity, or hardness is wrong. The reverse happens for water that is too fresh, when a fish must dump excess water and struggle to get salts. Getting the hardness and salinity right, with the right minerals and salts, optimizes growth for any aquatic organism; the right balance means growth in lieu of burning energy to balance its biochemistry.

Acidity, pH, is tied to the hardness. Acidity, or pH, is a measure of the positive hydrogen ions dissolved in the water. Water can disassociate to form hydrogen cations and hydroxide anions. Fish have ideal levels they prefer in their surrounding water for growth, though most are in the 6.5–7.5 levels. The right balance of minerals buffers the acidity of the water, though too much hardness can force the pH to trend alkaline, as can too much ammonia or urea. Like hardness and salinity, the wrong pH stresses the fish, causing each fish to expend energy to balance its internal biochemistry.

Chasing the right biochemistry in a tank is a constant battle, that is helped by bacteria that line a tank's walls, and any plants in the tank. Fish release urine and feces into the water, which affect both pH and hardness. Plants and algae with bacteria can take in some of the urine and extras to balance the tank.

Hardness and acidity are also key in soil for plants. Just like in the fish and water plant examples, each plant needs a unique set of root media pH and hardness. The root bed, root media, root fluids, or soil will have to be measured and controlled to optimize harvest. As in the slime layer, hydroponic or aeroponic fluids need to be balanced to keep the plant roots working well and keep the surrounding root bacteria and fungi happy too.

2.4 Key Mass Flow Concept: Goesintos and Goesoutofs

Sir Isaac Newton made western civilization officially aware of something that is common sense to the world: what goes in, must come back out, somehow. In more complex form: matter is neither created nor destroyed under normal conditions. Aside from nuclear reactions, this means that mass that enters a system either leaves it or is retained by it. In USA Midwestern farmer terms: the mass of the 'Goesintos' is the same as the mass of the 'Goesoutofs', minus whatever gets stuck inside. In more relatable terms for a single organism, if you (an animal) eat a kilo of dinner during a holiday, and guzzle down a kilo of liquids with that, the combined two kilos comes out of you in various ways. Some mass comes out in your breath (no onions I hope) in carbon dioxide and water vapor. Some mass comes out in your sweat as water vapor, salts, and urea/ammonia/etc. (a stinky part of body odor) Some mass comes out in your urine and in your feces (some of that kilo was fiber I hope) Lastly, and this is likely, given a big enough meal followed by an afternoon of video entertainment on the couch, some mass is retained around the middle and rump.

The conservation of mass principle applies not only to reactions in an organism, and to organisms, but also: to parts of the farm, the whole farm, and the combined system made by the human habitat and farm. In the introduction of this book in Fig. 2.8, the inputs for wheat were at least 116 kg to get 27 kg of grain, leaving 89 kg of other stuff as output. For a balanced settlement that is a mostly closed system, the inputs and outputs balance. If all the ins and outs balance, that is called being mass neutral. Therefore, the outputs from the habitat (carbon dioxide, wastewater and liquids, waste solids) are the inputs to the mature farm, and the mass products from the farm are inputs to the habitat (oxygen gas, clean water, food). Knowing that elements are not made or destroyed by the farm, tracking elements will help you balance the ins and outs.

Mass flow is the movement of mass around the parts of farm, and from and to the habitat, usually measured in kilograms per day. The reason for using a day is to match the way food calories, water, and oxygen are needed for the humans in the habitat. People need not only a mass of food, but food that contains around 2000 kilocalories (kcal) per person per day, with a healthy nutrient load of fats, proteins, carbohydrates, various minerals, vitamins, and salts. Everything into and out of the farm is averaged over a day, especially given some tricks for harvest and growth, but that comes later.

Every organism in the farm can be looked at as a mass-energy machine, with a mass flow into and out of, and energy required or released by that mass flow. Organisms can be added up into stages, stages into the farm.

A note on infection by diseases and pests should be mentioned. Since the space farm consists of isolated environments and crowded volumes, disease will be a major concern. Mass flows should be sterilized using ultraviolet light or other means in both recirculation and circulation between stages. Given that, it is also wise to mitigate infection using these other tools: a sensor-surveillance apparatus to catch diseases early, a reservoir of separated organisms and seed bank to restock lost stages, and separation inside stages to make separate populations when footprint (volume) allows. Farms are hard work, and any experienced farmer or gardener can relate to extra work and lost revenue from lost crops due to infections. Since space farms are by nature isolated, it will be critical to decontaminate incoming materials and living organisms bound for the farm. It should also be noted that having many species together in a stage, as well as highly controlled mass flows, may increase resistance to infections.

2.5 Organism Mass Flows

This section discusses how mass, and elements in that mass, enters and either leaves or is incorporated into organisms.

2.5.1 How Things Grow

As mentioned before, mass input either goes back out or is accumulated in each living thing as in Fig. 2.4. What happens is we, and all other living things, are balloons with holes, filled with dirty water. Most of the mass that enters is released, but a little makes the balloon a bit larger, and then more water is needed to inflate the new bit.

For most animals, proteins and fats are made from food, and then water inflates the growth. In plants, cellulose and other carbohydrates make up most of the added part, with some fats and amino acids, that get inflated with water. In single-celled or simple organisms, the balloon gets bigger and then divides into two cells or two simple organisms, each a balloon filled with mostly water. Living mass (biomass) is also wet biomass, meaning that it is mostly water. Dry mass is the mass of everything that isn't water. In algae for example, 90% of wet living mass is water, 10% everything else. In aquatic animals it is 80% to 60% water to 20–40% dry mass (loosely calculated form a sampling of animals). In people and most terrestrial animals, it is closer to 60% water/40% dry mass (adjusted from values in (Mitchell et al., 1945)).

When thinking about the ins and outs, we need to think of energy and growth. Analysis for the farm needs to include not just metabolism but also growth. Metabolism is a fancy word for how living things convert mass from one form into another to get energy to live. Most living things grow, taking in water for the growth in wet mass, in addition to just living by converting chemicals which is mass neutral. For crops, puppies, or people, we can find out the constituents of the dry mass, how much water is in the wet mass, and then back out what is needed to grow that crop, in addition to any wastes that come out. Anyone who has raised a baby or puppy is VERY familiar with the inputs, energy (chew toys, lost sleep, sore feet), growth (felt in shoulders and back after a day carrying either), or of course, the messy, smelly, outputs. Note also, babies and puppies make fools of even the wisest people, so completely worth any effort.

2.5.2 How Animals Live and Grow

As mentioned in the sections above, animals primarily aerobically metabolize food using oxygen for energy and consume food and water to grow to add mass as in Fig. 2.4. Each animal has nutritional requirements, based on its age, stage in life, and habitat. The nice thing about crop animals is that they can eat a variety of foods to result in a similar output, i.e., similar nutrient content food leads to similar time to harvest, and nutritional content as human food. However, this assumes the right conditions. In most cases, animals will be fed whatever the habitat releases or farm produces, and animals can be a flex point in waste disposal, taking leftover human foods, inedible portions of plants, and algae, and converting these foods into high protein outputs for people. The wastes from animals (almost all the mass they consume) include ammonia or urea which become nitrogen sources for plants, algae, and cyanobacteria. Animals can make a space farm smaller, by making the plants grow better.

The combination of metabolism and growth makes the simplified equations for the animal as for a mass balanced living machine, using many of the reactions in Table 2.7:

1. Mass in food + mass in added water + oxygen gas → mass added in growth + solid wastes released + liquid wastes released + carbon dioxide.
2. Mass in added water = water used to add mass in growth—water from aerobic metabolism.
3. Mass of animal = mass of animal before growth + mass added in growth.
4. Mass of animal = dry mass (all the stuff not water) + water.
5. Mass added in growth = Dry mass added + water to hydrate dry mass.
6. Mass in added water = water to hydrate dry mass (growth) + excess water to facilitate biochemistry (released in liquid wastes). Since 'excess water to facilitate biochemistry' is a zero sum between in and out, I usually ignore it and the water in liquid wastes for aquatic animals. However, for terrestrial animals (e.g., chickens, insects, you, etc.) both reactions imply a load on waste processing.
7. Food required (dry mass) between time A and B = Added mass of animal dry mass over A to B + dry mass required for animal to survive (a function of living mass of animal).

    - From our farmer perspective I really care about how many living kilograms add how many more living kilograms per day, and how much mass I must add, and how much mass I have to extract, to get those living kilograms.

8. Since elements are not created or destroyed, we can track the core carbon, nitrogen, hydrogen, and oxygen paths too:

    - Mass of carbon in food consumed to date = mass of carbon in organism (largely fats, proteins, and the like)—mass of carbon in wastes (liquid or solid, and a teeny amount in farts and burps) to date—mass of carbon released in carbon dioxide to date.
    - Mass of nitrogen in food consumed to date = mass of nitrogen in organism (proteins mostly)—nitrogen released in wastes (urine, ammonia, and any proteins in solid wastes, or via skin).
    - Mass of hydrogen in water and food consumed = mass of hydrogen in organism (i.e., hydrogen in dry mass + hydrogen in water for wet mass)—hydrogen released in wastes (i.e., water + other compounds, including loss to evaporation and vapor).

- Mass of (salts, minerals, etc.) consumed to date (from food, drinks, and for aquatic organisms, via gills and skin) - mass of (salts, minerals, etc.) released in wastes.
- Mass of oxygen inhaled or absorbed + mass of oxygen in food consumed + mass of oxygen in water consumed = mass of oxygen in organism (i.e., oxygen in dry mass + oxygen in water in wet mass)—mass of oxygen in wastes and water released—oxygen in carbon dioxide exhaled.

2.5.3 Humans and a Human Mass Model

People are metabolically animals, though not a crop outside of some very dark sci-fi movies. What humans release is the input to the farm, and human needs are the outputs of the farm as shown roughly in Fig. 2.5. While everyone eats differently, exercises, or works differently, and these variables change in the day, it is easier to assume a standard set of needs and outputs. Assuming a population that is average adult in size, and not growing or shrinking, the US RDA figure is 2000 kilocalories for most middle-aged adults, scattered across a set of minimum needs for minerals, vitamins, carbohydrates, fiber, and protein. That said, there is lots of flexibility when it comes to human diets, and I decided to error on the teenaged side of humanity. We all know teenagers are just short of a black hole for calorie-rich foods.

Food is eaten in terms of wet mass, not powders. Fresh food needs to be mostly water to be tasty and edible. Ever try to eat dry powder? Yuk! This wet mass therefore includes some of the water we take in every day. One menu possible in a space farm, albeit a diet fit for a highly active 20-something, is like Table 2.8. This menu implies hydroponic or soil plant stages, and an aquatic animal stage. I picked a diet like I would want to eat. This menu also has a very fatty oil content! Nutrition data again is loosely calculated from USDAARS (2022).

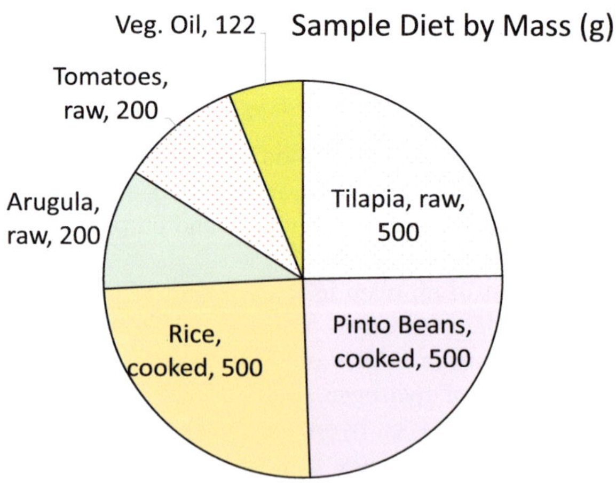

Fig. 2.5 Sample balanced menu with breakout and budget. About three-quarters of the wet mass is rice, beans, and fish

In more real terms, the total for this person's day includes nine tablespoons of oil, likely in salad and in food preparation for the beans and fish. The salads are together two medium tomatoes and a healthy plate or 4 handfuls of arugula greens. The fish is full-grown tilapia, which could be cooked with the beans or fried. As you will find out later, there are reasons for picking tilapia. Four cups of cooked beans in your day's food round out the total and provide amino acids and fiber. The percentage of mass is shown in Fig. 2.5.

All these food items are composed of elements. If you remove all the water, the remaining dry mass, is roughly 53% carbon, 35% oxygen, 8% hydrogen, 4% nitrogen, and less than a percent of all the rest as in Table 2.9. To get that information, one can dig into the details for each food item, or just the nutrient level and use average percentages from the tables: Tables 2.2, 2.3, 2.4, 2.5, and 2.7. Table 2.2 covers the oxygen, water, and carbon dioxide. Table 2.3 covers the carbohydrates and fiber. Use the percentages for fructose for the simple carbohydrates, and the percentages for starches for complex carbohydrates, and cellulose for the fiber percentages. Use Table 2.4 and the triglyceride percentages for the fats. Use Table 2.5 for the proteins and the "Avg. for Proteins" percentages unless you have the amino acid breakdowns. Adding up all the masses for protein, carbohydrates (except fiber, use general form for the simple stuff like fructose, or sucrose, and starches for the starches), fiber (use cellulose), and fat for each food item, then multiplying by the percentages in the tables, gets Table 2.9. Note the richness of fats for calories and carbons. This menu is just shy of 2,900 kcal, with over a third of the input carbons coming from the oil (fat). Other menus could also hit a similar nutrient and element total. You, the space farm engineer, can make your own tradeoffs. As eaten, add ~ 1.45 kg in water mass, which is hydrogens and oxygens, to this food.

NASA (see NASA-STD-3001 (2015)) has also figured out mass inputs and outputs for people in terms of oxygen (0.84 kg/day/per person), water (at least 5 kg drinking and food water input per person per day), and carbon dioxide (1.01 kg/per person/ per day output). You do not have to use their estimates, but whatever you use, must balance inputs and outputs per person, per day, and balance with the whole space farm inputs and outputs. There are also a lot of details on habitat design to make sure partial pressures, humidity, and other things work out, but that is beyond this book. I took these values and made a model for people using the menu above. On Earth, we take for granted the 0.84 kilograms of oxygen gas from the atmosphere and the release of 1.01 kg carbon dioxide to the atmosphere. The carbon is from our foods, around 0.6 kg of dry mass (i.e., not including the water in the food). Since you eat wet mass, the wet mass is roughly 72% water by mass, with the dry mass 28% of that total mass. Since the term 'foods' includes water that makes food edible, that mass is added to the liquids we drink, summing to around 5 kg water total. Assuming no growth, the mass of wastes we release includes water taken in, and the mass of food, minus a fraction of carbon in the 1.01 kg of carbon dioxide we exhale, or ~ 1.4 kg in dry wastes dissolved, exhaled, and mixed with 4 kg water, assuming a high fiber diet. Aerobic metabolism also produces a bit of water, extracted by breaking carbohydrates or other molecules, which is why the water in is ~ 0.1 kg less than the water out. Totaled up, the human mass balance per person per day looks like Fig. 2.6

Table 2.7 Common (and averaged) biological reactions and element percentages

Name	Chemical formula (simple, result, minus enzymes and ATP)	Molecular weight (rounded, one side)	% mass by element (either side)			
			C (%)	H (%)	O (%)	N (%)
Photosynthesis	$6CO_2 + 6H_2O \rightarrow C_6H_{12}O_6 + 6O_2$	372.2	*19*	*3*	*77*	*0*
Aerobic metabolism	$C_6H_{12}O_6 + 6O_2 \rightarrow 6CO_2 + 6H_2O$	372.2	*19*	*3*	*77*	*0*
Hydrolysis of cellulose (and reverse)	$C_6H_{10}O_5 + H_2O \leftrightarrow C_6H_{12}O_6$	180.2	*40*	*7*	*53*	*0*
Hydrolysis of cellulose with aerobic digestion	$C_6H_{10}O_5 + H_2O + 6O_2 \rightarrow 6CO_2 + 6H_2O$	372.2	*19*	*3*	*77*	*0*
Anaerobic metabolism (alcohol) and reverse	$C_6H_{12}O_6 \leftrightarrow 2CO_2 + 2C_2H_5OH$	180.2	*40*	*7*	*53*	*0*
Methane production (example)	$C_6H_{12}O_6 \rightarrow 2CH_4 + C_2H_4O_2$	180.2	*40*	*7*	*53*	*0*
Ammonia from deamination (example)	$C_6H_{14}N_2O_2 + 4H_2O \rightarrow 2NH_3 + 4H + C_6H_{12}O_6$	218.3	*33*	*10*	*44*	*13*
Methane combustion or digestion	$CH_4 + 2\,O_2 \rightarrow CO_2 + 2\,H_2O$	80.1	*15*	*5*	*80*	*0*
Cellulose combustion or digestion	$C_6H_{10}O_5 + 6O_2 \rightarrow 6CO_2 + 5H_2O$	354.2	*20*	*3*	*77*	*0*
Ammonia from gaseous nitrogen (aka nitrogen fixation)	$N_2 + 8H + \rightarrow 2NH_3 + H_2$	36.1	*0*	*22*	*0*	*78*
Ammonia to urea	$2NH_3 + CO_2 \rightarrow CH_4N_2O + H_2O$	78.1	*15*	*8*	*41*	*36*

Loosely calculated from PubChem (2022) and CIAAW (2022) data and standard biology reactions

Table 2.8 Sample 2900 kcal menu that meets US minimums and balances the necessary carbons to meet the NASA carbon dioxide output

Food	Grams wet mass	kcal
Tilapia, raw	500	480
Pinto beans, cooked	500	715
Rice, cooked	500	505
Arugula, raw	200	50
Tomatoes, raw	200	32
Veg. oil	122	1080
Total	2022	2862

Table 2.9 Sample menu's mass and element breakout

Element by nutrient	Mass by nutrient	Mass of elements in grams				
		C	H	O	N	S
Protein	173	77.7	13.3	55.2	25.4	1.5
Carbohydrate (starch)	185	82.2	11.5	91.3	0.0	0.0
Carbohydrates (simple)	10	4.0	0.7	5.3	0.0	0.0
Dietary fiber	61	27.1	3.8	30.1	0.0	0.0
Total fat	144	111.2	16.7	16.2	0.0	0.0
Total dry mass	573	302.2	46.0	198.0	25.4	1.5

and Table 2.10. Needless to say, couch potatoes, tourists, athletes, and kids all vary from this model, but it works to find numbers for the farm.

All the extra water not in these charts for showers, hand washing, and general hygiene can be considered grey water, and a slick space engineer would pipe and sterilize it to grow vine crops and spices as the house plants of the habitat. I leave it out of the farm human model (mostly) for that reason. Plants like these make the air smell better and people feel better, in addition to making food taste better. More on that later. In Table 2.10 is one guess at the human model. The model in that table is a bit different then in Fig. 2.6 but uses the menu above. This is very rough and there are lots of things that can change it, but we can use it for now. I also revise this model further down in this book.

All those extra elements are lost in the decimal places here, but it would pay a good engineer to track them anyhow. The big trick, as stated before, is to balance the elements and make sure that all the inputs equal all the outputs, regardless of

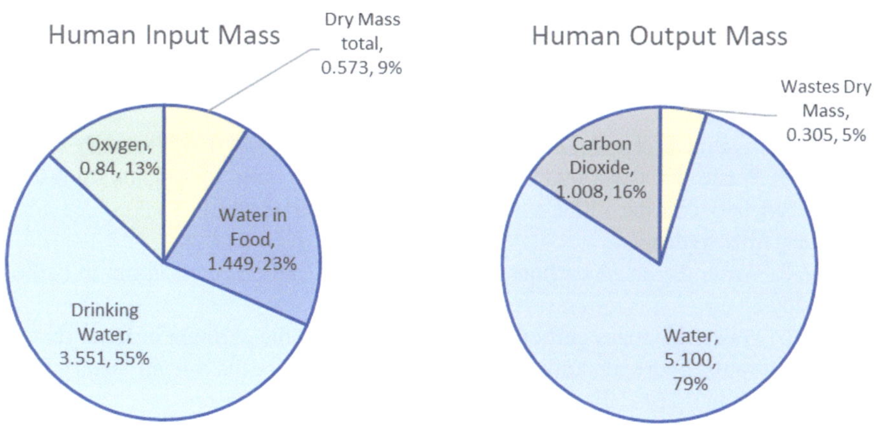

Fig. 2.6 Human model mass budget, per day, per person. Very loosely derived and extrapolated from NASA-STD-3001 (2015), NASA-TD9702A (1998), and NASA-SP-2010-3407 (2014), with some guesses

Table 2.10 Altered human model for the menu above, with all inputs and outputs that go to and from the farm for one person, per day

Component	Mass (kg)	Mass of elements by kilogram				
		C	H	O	N	S
Inputs						
Dry mass	0.57	0.302	0.046	0.198	0.025	0.001
Water (all sources)	5.00	0.000	0.560	4.440	0.000	0.000
Oxygen	0.84	0.000	0.000	0.840	0.000	0.000
Total in	***6.41***	***0.302***	***0.606***	***5.478***	***0.025***	***0.001***
Outputs						
Waste dry mass	0.31	0.027	0.035	0.216	0.025	0.001
Water (all types)	5.10	0.000	0.572	4.528	0.000	0.000
Carbon dioxide	1.01	0.275	0.000	0.733	0.000	0.000
Total out	***6.41***	***0.302***	***0.606***	***5.478***	***0.025***	***0.001***

menu and model. Next, we talk about the fluffier or slimmer world of animal mass balances.

The way I did those graphs and charts really followed basic accounting, and you could use these rules even if you don't know exactly what specific items are on the menu, as long as you know a few basics. These are very rough rules of thumb, just for engineering the space farm, not for biochemistry work, and should eventually be replaced by experimental calculations.

Master Rule: Mass in Must Equal Mass Out

- Rule 1: Nitrogen and sulfur goes from food dry mass into waste dry mass.
- Rule 2: Unless the human is growing, assume water out = water in + a small fraction (I use 0.1 kg). So, if the input water is 5 kg between food water and drinking water, output is 5.1 kg between urine, feces, sweat, and breath, tracked as water output.
- Rule 3: Assume dietary fiber (or any other fiber) is all added to waste dry mass.
- Rule 4: Waste dry mass hydrogen mass = food dry mass hydrogens minus the extra bit between the input water and output water (usually 0.1 kg * 11%) + dietary fiber hydrogen.
- Rule 5: Waste dry mass carbons = food dry mass carbons-carbon out in carbon dioxide.
- Rule 6: Waste dry mass carbon must be larger than the carbons in input dietary fiber. Ditto for hydrogen and oxygen in waste output dry mass from dietary fiber, since you can assume dietary fiber is passed through into feces.

To make the rules work, since oxygen input, carbon dioxide output, and water in and out are assumed, the flex point is in food, i.e., food dry mass.

If you wanted another example that fits the dietary guidelines, just starting at protein, fat, fiber, and carbohydrates, Table 2.11 fits the bill. You will note the amounts here are larger than the dietary minimums (56 g protein, 130 g carbohydrate, 31 g fiber, 35 g fat) (loosely calculated from USDADGA (2020)).

Of course, food must have at least the minimum calories for the population average, usually 2000 kcal.

You can spend hours chasing your tail to balance the inputs and outputs by element. A simple route is to start with food dry mass then make the menu fit the element composition of the food dry mass. Figure 2.7 is one map of nutritional inputs to outputs. Fiber goes as elements directly to waste dry mass in many cases for this model. Anything that is not carbon, hydrogen, or oxygen also goes into waste dry mass. You will still need to track all the CHON elements to make sure they balance in the farm.

The menu (really the averaged inputs to the average menu for the day or year) dictates what kinds of stages you need in your farm, which is the mix of plants, algae, animals, fungi, etc. In practice it is easier to make the whole farm and habitat balance for an assumed human population and large possible set of crops, then optimize for size, then make a variety of menus from those chosen crop masses.

Table 2.11 Another model assuming protein, carbohydrate, fiber, and fat all meet US minimums for 31-year-old humans

	Mass (kg)	Mass of elements by kilogram				
		C	H	O	N	S
Inputs						
Food (dry mass)	*0.588*	*0.294*	*0.044*	*0.233*	*0.016*	*0.001*
Protein	0.110	0.049	0.008	0.035	0.016	0.001
Carbohydrate starch	0.330	0.147	0.021	0.163	0.000	0.000
Carbohydrate simple	0.013	0.005	0.001	0.007	0.000	0.000
Dietary fiber	0.035	0.016	0.002	0.017	0.000	0.000
Total fat	0.100	0.077	0.012	0.011	0.000	0.000
Water	5.000	0.000	0.560	4.440	0.000	0.000
Oxygen	0.84	0	0	0.84	0	0
Total inputs	***6.428***	***0.294***	***0.604***	***5.513***	***0.016***	***0.001***
Outputs						
Wastes dry mass	0.318	0.018	0.032	0.250	0.016	0.001
Water	5.100	0.000	0.572	4.528	0.000	0.000
Carbon dioxide	1.010	0.276	0.000	0.734	0.000	0.000
Total outputs	***6.428***	***0.294***	***0.604***	***5.513***	***0.016***	***0.001***
Net	***0***	***0***	***0***	***0***	***0***	***0***

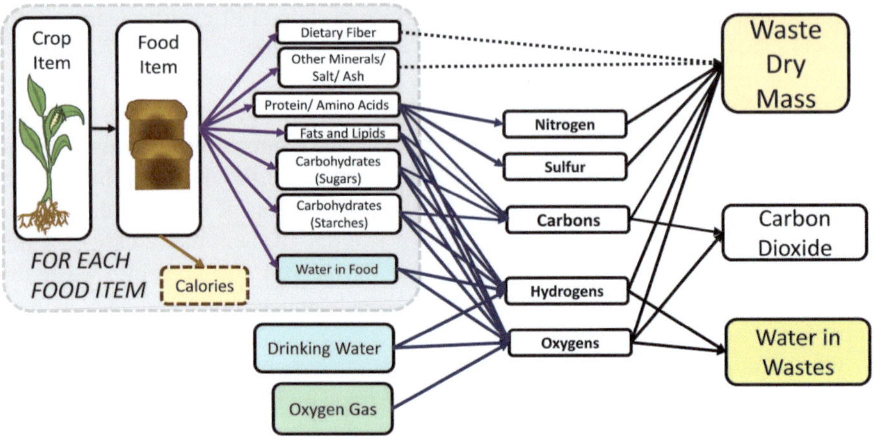

Fig. 2.7 Flow of mass from crop to wastes through people

Fig. 2.8 Human input mass, calculated from the various tables and graphs in this chapter

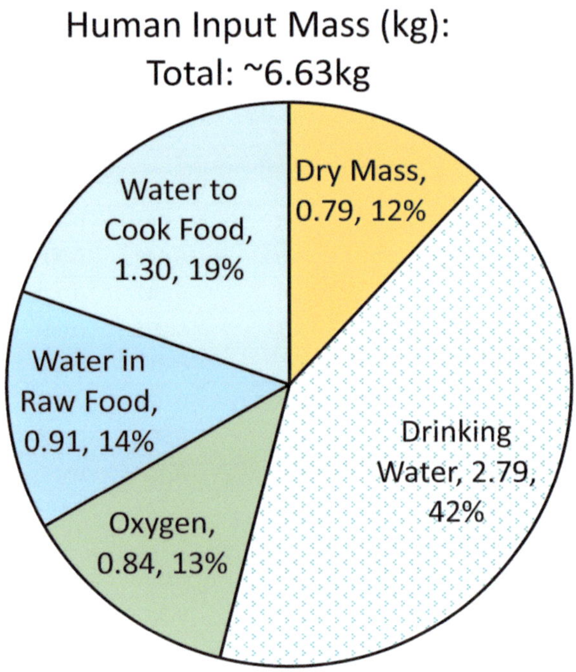

2.6 More on Wastes

You could of course use a balance through a human to get a more detailed look at wastes. To do that we will have to guess a menu for one day. Let's use the menu in Table 2.12, a bit different then the menu in Table 2.8.

Table 2.12 Sample menu and energy

Food	Grams wet mass	kcal
Tilapia, raw (whole)	500	480
Pinto beans, cooked	1000	1430
Rice, cooked	1000	1010
Arugula, raw	200	50
Tomatoes, raw	300	48
Salt	3	0
Total	3003	3018

Very loosely calculated from data in USDAARS (2022)

This menu works somewhat well for a very active 18–35-year-old. Food is not the only thing you take in of course, so we need to look at oxygen and water. Everyone ingests water two ways: by drinking and in the food we eat. We eat wet mass, which means it contains water, either by its nature since it is raw, or from food preparation. In the rice examples, rice absorbs water when it is steamed or boiled. The overall budget in this example, roughly using (NASA-STD-3001), and standard recipes as below, looks like Fig. 2.8.

More generally the food preparation looks like Fig. 2.9. Assuming we use the whole tilapia, by eating the fillets and using the rest to make meal and sauces, and that the tomatoes and arugula are eaten raw like a salad, there is prep water for the rest. Rice as you eat it, absorbs 72% of the final mass from cooking water, starting with 28% of the final mass in dry grains. If you overcooked it to paste while watching a show, well, that took a bit more water (ugh). The same holds for pinto beans, which start with 42% of the mass in dried beans, and 58% of the final mass in cooking water. For our menu, the cooking water adds up to around 1.3 kg. We might want some salt, maybe 3 g, and an ignored mass of spices to make all this food taste good. Beyond the meal, we also need to add 2.79 kg drinking water to wash the food down. Also beyond the meal is 0.84 kg of oxygen from the habitat's air for our human to breathe.

Given all the inputs, the elements look like graphs in Figs. 2.10, 2.11, 2.12, and 2.13. Note that oxygen dominates the input elements, and therefore will also dominate the outputs. Assume here every element that goes in, eventually goes out, and assume the humans are not growing or gaining weight. Hydrogen and carbon are next, with oxygen constituting the 'CHO' acronym. Oxygen is in most of the biological compounds we eat, from water, to sugar, to protein, to fats.

If we dehydrate the input mass, called 'dry mass', the water comes out, along with some of the hydrogen and oxygen, as in Fig. 2.11.

If we really want to see all the 'ash' stuff and nitrogen, we need to pull out the 'CHO' parts to expose the rest of all those elements as in Fig. 2.12. Nitrogen is next in criticality in our inputs, largely in proteins including amino acids. With nitrogen 'CHO' becomes 'CHON' the biggest elements tracked in a space farms inputs and outputs.

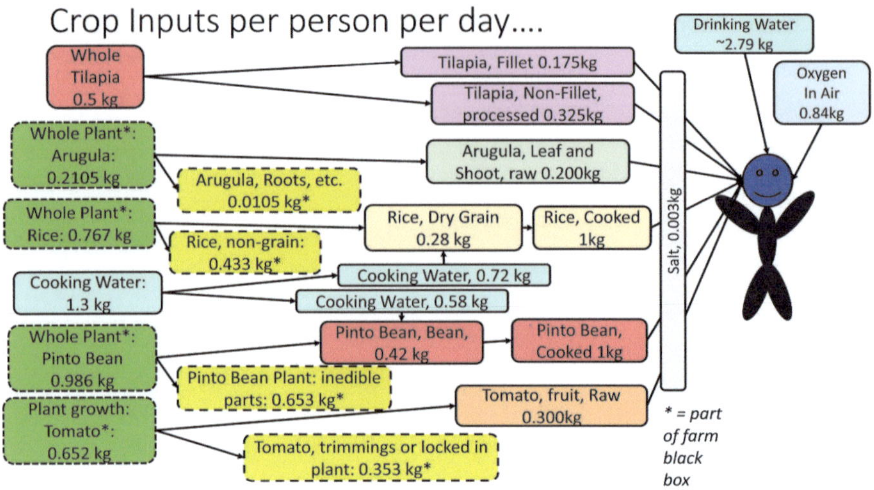

Fig. 2.9 Meal inputs from farm to dinner

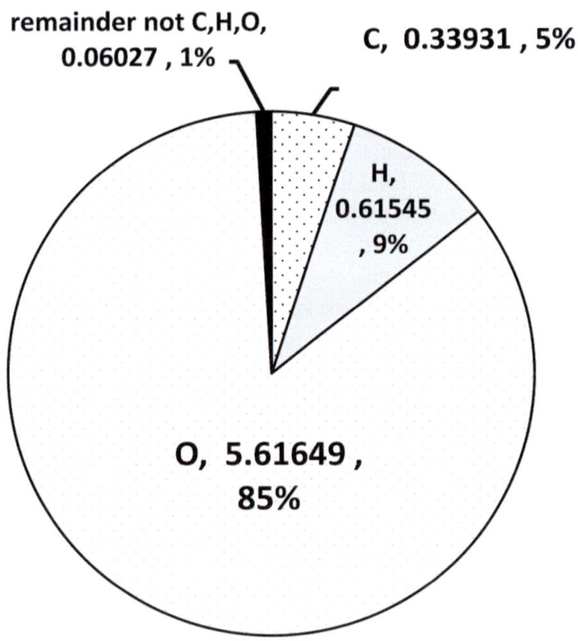

Fig. 2.10 Elements in human inputs for the menu above (total wet mass)

All those inputs result in outputs. Over the course of the day, each of us exhaled carbon dioxide and water, and we sweat. The water in our breath should be recaptured by condensation, and the carbon dioxide sent for photosynthesis, either inside the habitat or in our farm. Our sweat is both a little dry mass and lots of water. The water

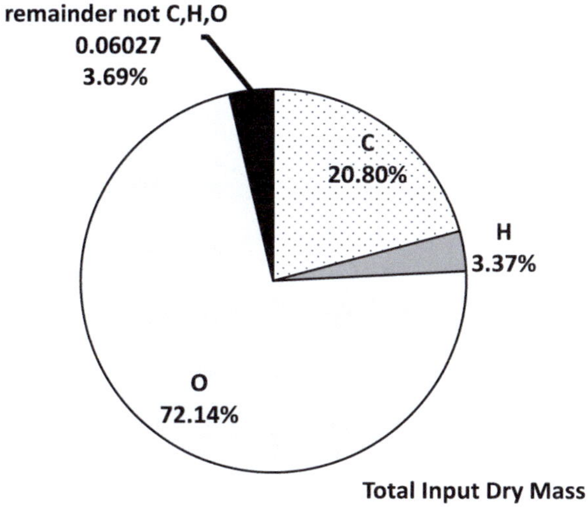

Fig. 2.11 Input elements, minus water (i.e., dry mass)

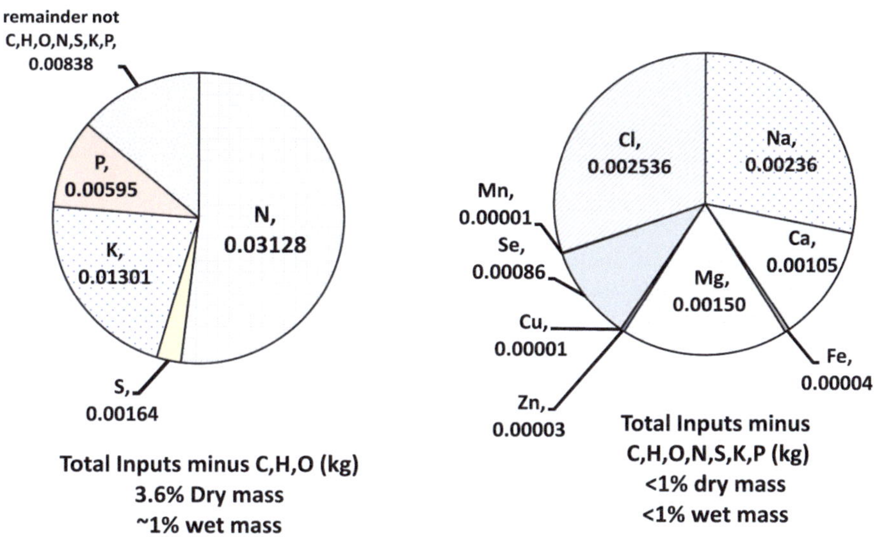

Fig. 2.12 Input elements minus C, H, O and elements minus C, H, O, N, S, K, P

evaporates into the air to be recaptured by condensation. The little bit of dry mass gets washed away in the showers or wash cloth, and in water used to clean clothes. That dry mass includes ammonia which is why it smells. It is a surprise to many people to discover that for an active person, more water comes out in breath and sweat, than in urine and feces. For example, the outputs look like Fig. 2.13.

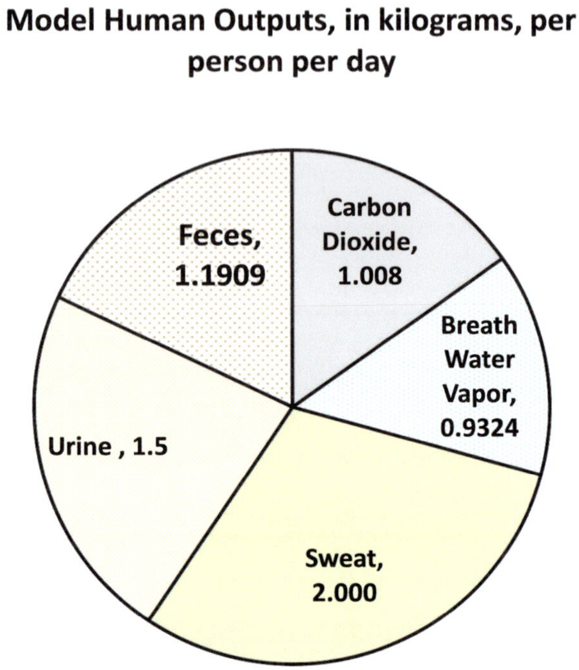

Fig. 2.13 Human output masses. Loosely derived from (Putnam, 1971) via (Meyer, 2020)

2.7 Ash 101

In nutritional data for foods as from the United States Department of Agriculture (USDA), the term 'ash' includes all the minerals, vitamins, and other compounds that are not either water, carbohydrates, proteins, or fats. It is called 'ash' because it is common to heat foods to find out what is in them. The stuff that is left is ash. Ash can contain elements like carbon, nitrogen, oxygen, and hydrogen, so it takes careful consideration to get a mass balance by element, if nutritional data is the source. A trick of the trade for space farm engineers for quick calculations is to pass the ash mass through into the 'nutrient' or 'feed' dry mass inputs for the organism to make the mass flow balance (Fig. 2.14).

The same trick works for plants and algae too. Part of the nutrient inputs usually includes the part that becomes ash, though there is a qualifier here: an exceedingly small fraction of the water and carbon dioxide might also end up as 'ash'. I usually ignore the qualifier when roughing out masses and sizes for a space farm, since I am an engineer, not a biochemist. Given the assumption that all the ash is in the nutrient input dry mass, we get Fig. 2.15.

If you had to know, the ash for shrimp for example looks something like Fig. 2.16. Note that 'Other' also includes some amounts of carbon, oxygen, hydrogen, and nitrogen combined into compounds, which is why it is easier from a mass flow perspective to just pass the ash completely into either nutrient dry mass for photosynthetic organisms, or as mass into feed for aerobic organisms.

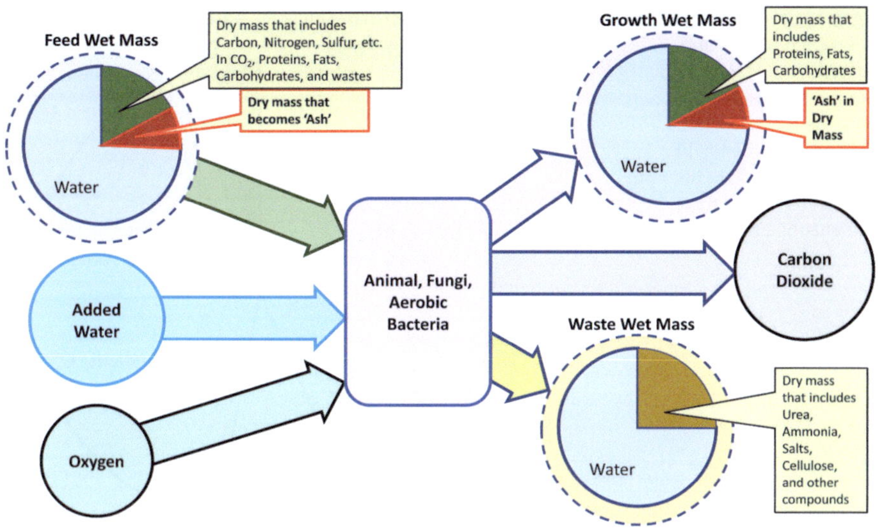

Fig. 2.14 Assumed movement of ash dry mass in aerobic organisms. Ash ingested is assumed to flow to either the dry mass component of growth wet mass or the dry mass component of the waste wet mass, or both

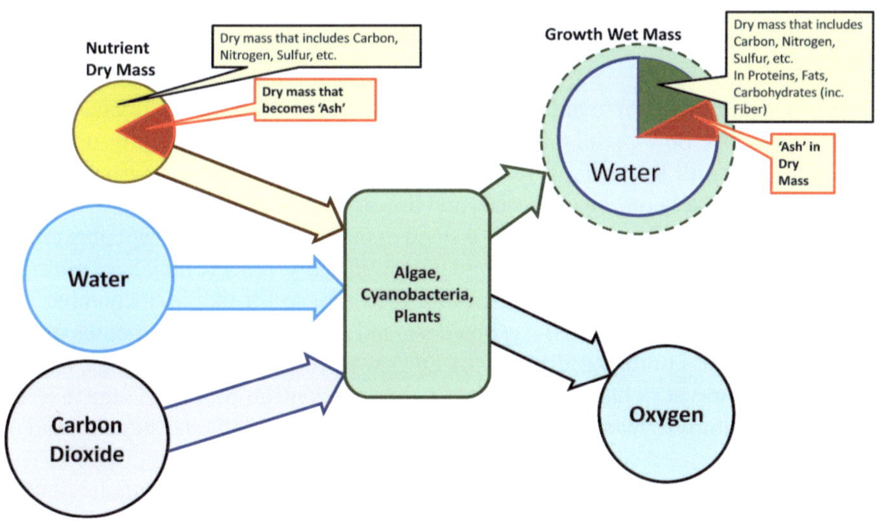

Fig. 2.15 Assumed movement of ash dry mass in photosynthetic organisms

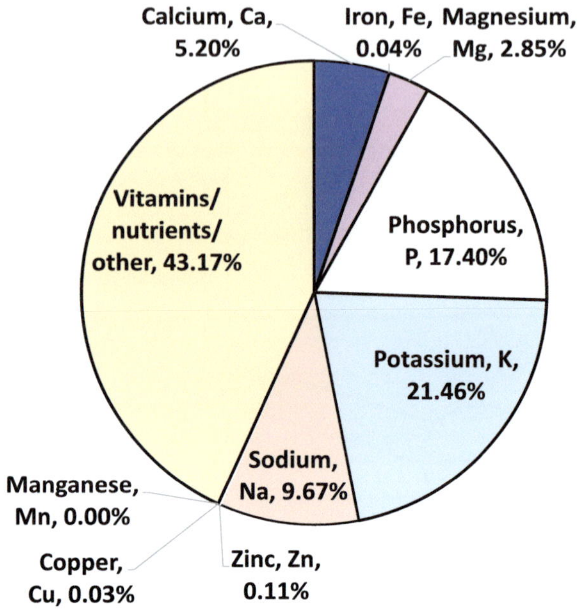

Fig. 2.16 Rough ash breakdown for Shrimp derived from USDAARS (2022). The 'Vitamins/ nutrients/other' section includes compounds like Vitamin C, choline, thiamin, some CHON, and trace elements like selenium

2.8 Introducing the Stage Concept

To perform the math for many organisms, and to add up how the organisms are part of the mass flow picture for the space farm, requires a hierarchy of views. Adding together all the shelves, fans, lights, pumps, and structure to grow a group of organisms, usually one crop species, and the crop species itself, makes a stage for the terms of this book. In other words, a stage in the space farm is a crop, plus all the support equipment that allows the crop to grow optimally. A stage may have multiple growth chambers for each growth period of the crop, so for tilapia, a chamber is a tank, and the stage has tanks for fry, breeders, and tanks for each few months of fish growth to harvest. For maize, this may mean growth chambers for seedlings, and for each 30 day interval to harvest with lights, thermal control, root beds, and a series of tanks and pumps to manage mass flow between chambers and the stage overall as in Fig. 2.17.

Combining multiple species in one growth chamber is possible, essentially putting two stages in the same spot, but for calculation, each stage is one species. The idea behind the stage concept for all crops in the farm, plant or not, is to allow calculating mass and energy flows for calories and products, and to determine how much stuff and room I need to build the farm. Each stage contains all the material and volume to grow organisms to harvestable levels, including breeding and grow-out, nurseries, and spare cultures. There will be a physical representation of each stage as in a greenhouse with tanks and pumps and lights, or set of tanks and pumps, or a battery of bioreactors, cages and fans, pens, or a field of soil. Again, the concept is to relate

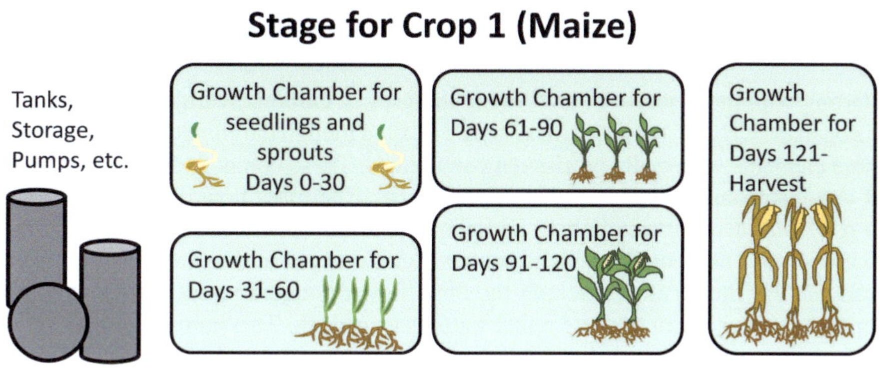

Fig. 2.17 Stage in the space farm is all the growth chambers and machinery to grow a crop, plus the growing organisms too

calculations and chemical changes to each crop item. Connecting all the stages and the human habitat are systems that will manage mass flow for each type of biochemical, wastes, and harvested items, as in Fig. 2.18.

Mass distribution systems such as those used in factories or chemical plants (or might be people moving around buckets, or pumps, etc.) move mass between stages and to and from the human habitat to keep balance. The math of the biochemical changes inside an organism can then be added together to calculate the mass flow and balance for the whole farm, as will be expanded in later chapters.

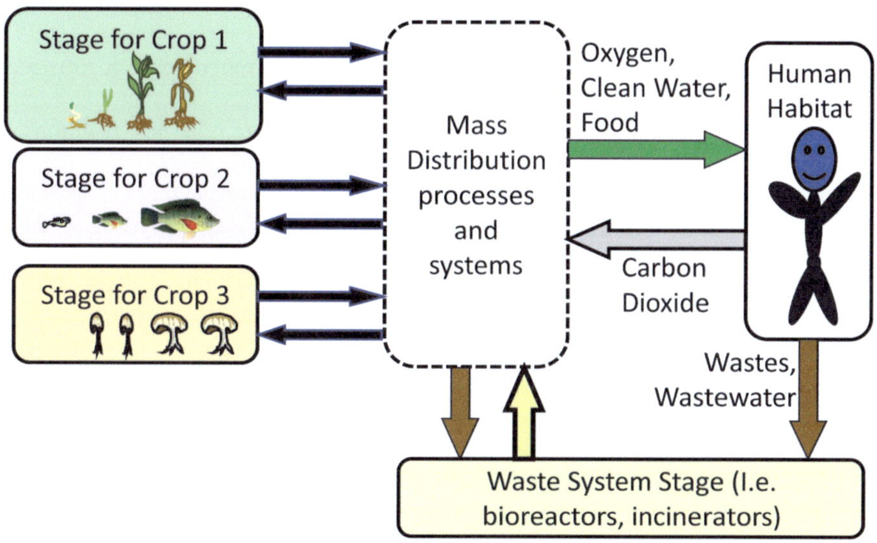

Fig. 2.18 Space farm has a network of mass flows, to and from each stage

2.9 Summary

Starting with the key elements of living things, i.e., carbon, hydrogen, oxygen, and nitrogen or 'CHON' for short, mass can be tracked though the compounds made from these elements, especially nutritional compounds. These compounds in turn are part of organisms and reactions inside of organisms. The mass flow in these reactions, in particular aerobic metabolism and photosynthesis, is key to understanding how to link crops and people together in an artificial ecosystem. For crop species, these reactions for the basis of tracking the mass flow through a set of crop organisms. Inputs to an organism become either outputs or growth. This human linkage to the farm crops can begin with dietary choices, that can result in a human model, which can be used to approximate the mass flow through a population of humans in a habitat. The crop species are viewed in the farm per each stage, which is the machinery and organisms for a single crop species.

2.10 Questions

1. How would you describe in a sentence or two the role of each element: carbon, oxygen, nitrogen, hydrogen, and nitrogen.
2. How would you describe a key difference in how animals process sucrose versus cellulose?
3. What 3 elements compose fats?
4. What key element makes proteins different from fats or carbohydrates? I.e., what element does protein add to our diet?
5. Deoxyribonucleic Acid (DNA) has at least carbon, hydrogen, oxygen, nitrogen, and what other element?
6. The input of aerobic metabolism are the outputs of what other key reaction?
7. What are concerns with pH and salts for living things like fish?
8. What is the conservation of mass, and being mass neutral?
9. Assuming and organism does not grow, and takes in 10 kg of hydrogen, what is the mass of hydrogen in wastes by the eight rules in this chapter?
10. Using rule 8 for an animal or person, what is the source of the carbon exhaled in carbon dioxide?
11. If I feed an animal 10 kg of feed, and 2 kg of water, and the animal breathes in 2 kg of oxygen, then grows 1 kg, what is the mass of the remaining outputs? Of these outputs, if the animal exhales 1.3 kg carbon dioxide, what is the sum of solid and liquid wastes released?
12. What is the role of human diet in predicting the mass of wastes produced?
13. Given a diet of tilapia, beans, rice, arugula, tomatoes, and vegetable oil, where does the nitrogen in urine come from?
14. In the human model of outputs in this chapter, which is larger: feces + urine or sweat + breath.

15. In the human model of outputs in this chapter, excluding carbon dioxide, which device would be more critical to recovering water in outputs: a dehydrator for urine and feces or a dehumidifier to recapture evaporated sweat and water vapor from breath?
16. If the elements of ash become either growth dry mass or waste dry mass, how does an animal or fungus get them?
17. If the elements of ash become growth dry mass in algae (or a plant), where does the organism get these elements?
18. How would you define a stage for the purposes of this book? What might a stage look like for tomatoes? What about shrimp? What about wheat?

References

CIAAW. (2022). Standard atomic weights. In *Commission on isotopic abundances and atomic weights (CIAAW), international union of pure and applied chemistry (IUPAC)*, 2022. https://www.ciaaw.org/

Meyer, B. L. (2020). The messy details of going from excrement to Entrée. AIAA ASCEND 2020, November 16–18, 2020 (Virtual), TECHPRES-23, Control ID 3418246.

Mitchell, H. H., Hamilton, T. S., Steggerda, F.R., & Bean, H. W. (1945). The chemical composition of the adult human body and its bearing on the biochemistry of growth: *Journal of Biological Chemistry*, *158*(3), 625–637. https://www.jbc.org/article/S0021-9258(19)51339-4/pdf

NASA-SP-2010-3407. (2014). Human Integration Design Handbook (HIDH), Revision 1–June 5, 2014, NASA/SP-2010-3407/REV1. https://www.nasa.gov/sites/default/files/atoms/files/human_integration_design_handbook_revision_1.pdf

NASA-STD-3001. (2015). NASA-STD-3001, Volume 2, Revision A NASA Space Flight Human-System Standard, Volume 2: Human Factors, Habitability, and, Environmental Health, Approved: 02-10-2015, National Aeronautics and Space Administration, Washington, DC 20546-0001.

NASA-TD9702A. (1998). NASA, International Space Station Familiarization, Manual TD9702A, NASA, July 31, 1998. https://er.jsc.nasa.gov/seh/td9702.pdf

PubChem. (2022). PubChem, National Center for Biotechnology Information, Bethesda, MD, 20894 USA, 2022. https://pubchem.ncbi.nlm.nih.gov/

Putnam D. (1971). Composition and concentrative properties of human urine (1971), NASA. https://ntrs.nasa.gov/api/citations/19710023044/downloads/19710023044.pdf (see also https://ntrs.nasa.gov/citations/19710023044)

USDAARS. (2022). FoodData Central, U.S. Department of Agriculture (USDA), Agricultural Research Service (ARS), 2022. https://fdc.nal.usda.gov/

USDADGA. (2020). U.S. Department of Agriculture and U.S. Department of Health and Human Services. *Dietary Guidelines for Americans, 2020–2025* (9th ed.) December 2020. https://www.dietaryguidelines.gov/sites/default/files/2020-12/Dietary_Guidelines_for_Americans_2020-2025.pdf

Chapter 3
Animal and Fungi Growth and Production

Abstract This chapter covers how non-photosynthetic organisms grow, including how they process and accumulate mass. These organisms largely take in carbon rich substances, oxygen, and water, and release carbon dioxide and wastes. Crop mass flow is estimated and modeled notionally for a range of species, from fish and shrimp, through aquatic mollusks and snails, through traditional crops such as chickens and other birds, through hogs, goats, sheep, and cattle, through less traditional crops such as insects and rodents. The chapter also explores fungal growth, from macroscopic species such as mushrooms, to microscopic species such as yeasts. This chapter also introduces bioreactors.

Keywords Animal growth · Fungi growth · Bacteria · Bioreactor · Birds · Fish · Mammals · Mushrooms · Yeast

3.1 Introduction

Crops in farms fall into two giant categories, those that you feed and pen, and those that you plant. This chapter is about the kinds of crops that are fed and penned, i.e. those crops that are not photosynthetic. The key reaction that governs mass flow for these crops is mostly aerobic metabolism, where inputs are oxygen, wet biomass, and water, and the outputs are more living biomass of the crop, carbon dioxide and water in exhalation, and waste wet biomass. For example animals take in feed or forage (wet biomass), water, and oxygen. Animals then release waste wet biomass in the form of urine, sweat, or feces, carbon dioxide and water vapor in breath, and add more animal living mass. Animals in this chapter include fish, shrimp, mollusks, insects, birds, and mammals. Each has a lifecycle, environment, and mass flow, and there are tables to show days to maturity and other mass affecting factors. Then the chapter covers fungi, including various terrestrial types that are lumped together as mushrooms such as oyster mushrooms, and then microscopic yeasts that can be grown for food, or to simply process wastes from other parts of the habitat-farm ecosystem.

© The Author(s), under exclusive license to Springer Nature Switzerland AG 2026

B. L. Meyer, *Designing Farms in Space*, https://doi.org/10.1007/978-3-032-04395-5_3

3.2 Modeling Animal Growth and Production

Animals grow from a fertilized egg (essentially zero mass) to a harvestable size over time. The growth is exponential. The larger an animal is, the quicker it adds mass, at the cost of more food and water to add the mass, and more food just to keep the animal alive. In Fig. 3.1 is the rough flow. Food and water are budgeted to either growth or energy, for each aerobic organism, from fungi to bacteria to animals. Reproduction is a combination of both growth and energy, though tilted to energy for the parents, with reproductive growth mass sent to offspring.

A rough rule for animal crops is that the bigger an individual animal, the more food is needed just for basic metabolism (i.e., just to stay alive). When an animal grows, see Fig. 3.2, it has a portion of its lifecycle where growth is mostly exponential, then as the animal matures growth slows down. Typically, an animal crop is harvested after the growth rate peaks but before the growth slows significantly. A few animals are let to grow to full maturity so they can produce offspring for future crops.

A way to look at the growth of an animal is that mass today plus the percentage of current mass added per day, equals mass tomorrow. The mass to just add the dry mass is then mass tomorrow minus mass now, or percentage added per day times mass now. I like to use the letter 'i' for the percentage added per day on average over a farm animal's life from egg to harvest. (I use a similar thought for plants too.) This equation over time looks very suspiciously familiar to accountants and bankers or anyone with a loan to pay off or a retirement account as the compound growth equation. Since this book avoids equations, we will stick with graphs and tables in this book. This book goes into some more interesting analysis on growth curves in later chapters. As a farmer I really want to know when I must feed what, so how much I need to feed per day. Inputs for an animal crop looks like Fig. 3.3. Feed wet mass has dry mass and water. Animals take the inputs and make more animal living wet mass, and some of this is harvested as whole animals. The remaining output mass is in waste wet mass and carbon dioxide. The total input feed wet mass includes a part for growth and a part for metabolism. The percentage feed for growth shadows the exponential graph of mass accumulation (growth) in the animals, minus one day. The feed part for metabolism is a multiple of the total mass of animals in the stage.

If we know how much oxygen is taken in and carbon dioxide and wastes are released, we can calculate how much mass is processed for energy versus how much mass went into growth. The reverse is true also. If a fish goes from 0 g to 1 kg in a

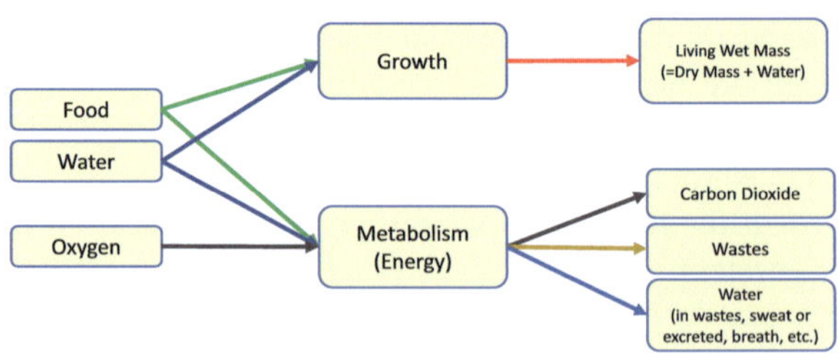

Fig. 3.1 Animal metabolism versus growth

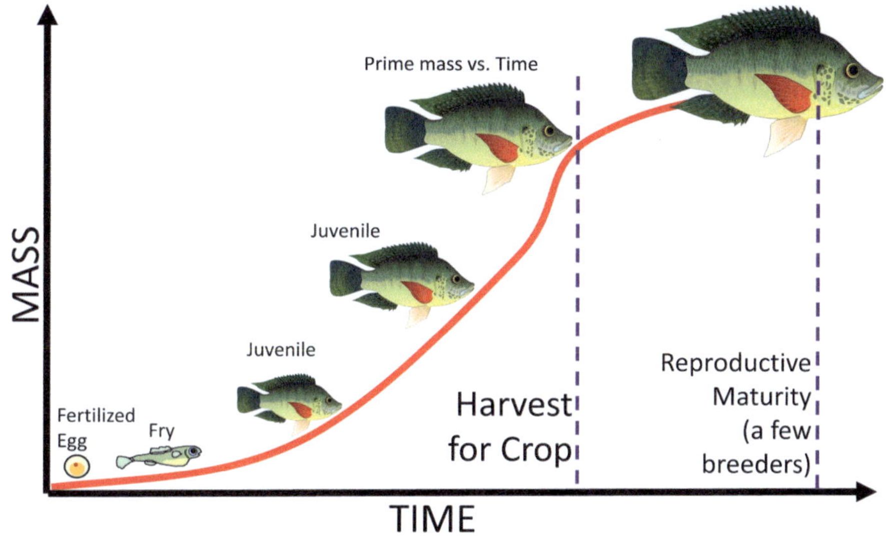

Fig. 3.2 Animal growth is exponential for most of its life, then slopes off before full reproductive maturity

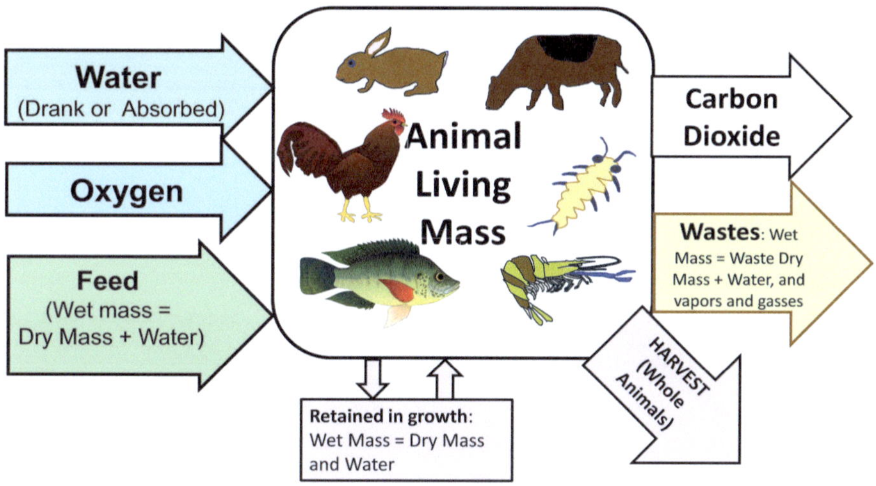

Fig. 3.3 Animals take in water (from the environment or drank), oxygen, and feed which is wet mass

year, and the fish is 75% water, then in a year the fish incorporates 250 g of dry mass in that year, and another 750 g of water to 'inflate' that dry mass, plus some of that water to convert the food into fish. Animals add mass quicker if they are bigger (you included), though the percentage of mass added is greater in younger animals, so the mass intake would also be larger with mass of animal. Plotted against living biomass, the inputs and outputs and growth (added living mass per animal kilogram) looks like a series of lines. The plots of inputs and outputs to tilapia mass is in Fig. 3.4. As mass accumulates, so do inputs and outputs. Note the graph is food against animal mass, not time. This graph is roughly calculated using data in (FAOAFFRIS-Tilapia, 2022).

Note that volume, footprint, and energy are also variable in animal growth. Energy is required to keep the environment optimal for the animals at each life stage. For example, many juvenile warm-blooded animals need heaters to keep them warm as they grow, then can handle cooler temperatures once big. This is one reason for most animals (and plants) that a nursery area is a must have, and is more footprint

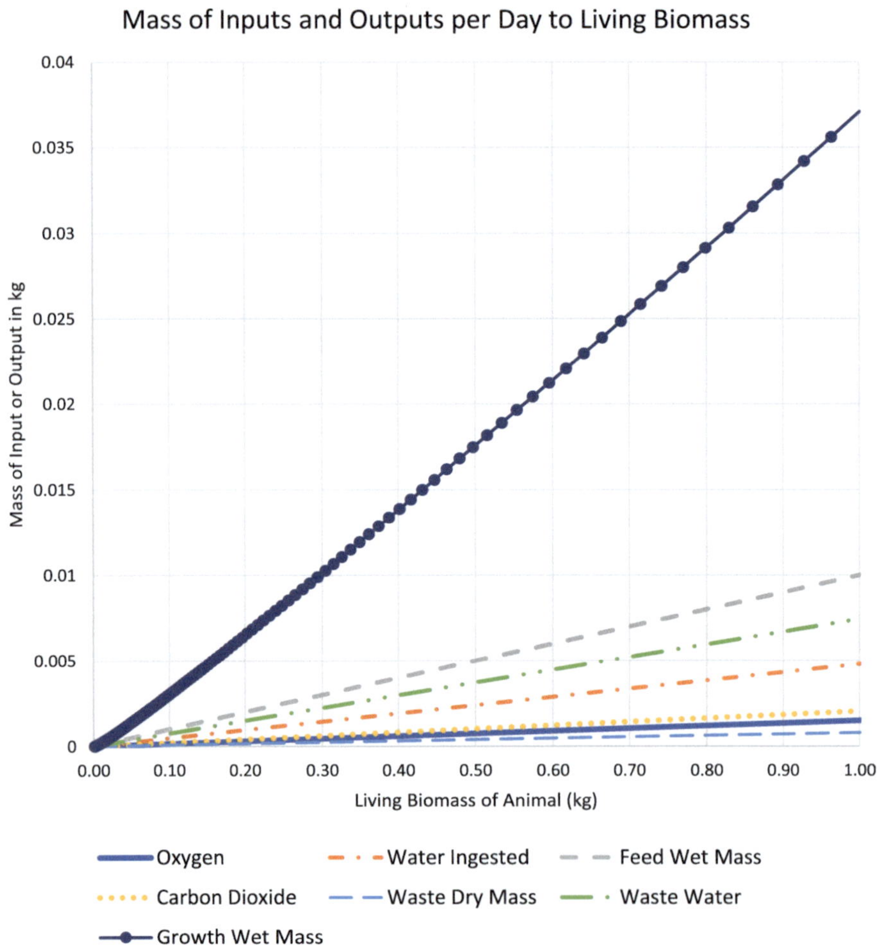

Fig. 3.4 Plot of input and output masses against living biomass of an animal (in this case a tilapia fish)

and volume efficient, than trying to put juvenile animals (or seedlings) in the full-sized growth areas. Lighting and light timing can trigger feeding and life events. Air circulation, fluid movement, auto feeders, and condensers all take energy. Room to move allows healthier animals. The bottom line is that the less energy animals need to adapt to a non-optimal environment, the more mass they accumulate in growth.

That said, a farmer can't just accelerate growth by feeding beyond the most an animal can take in and use. The amount an animal burns for energy, and can use for growth, is tied to the chemistry and temperature of the water (for aquatic animals) or air (for terrestrial animals). Excess food accumulates and rots in the animal's tank (for aquatic animals) or makes a mess on the floor for terrestrials. It therefore is a careful task to maximize growth while not wasting food. Collecting data for how animals metabolize in various gravities and conditions can be difficult to gather. That said in this book are some examples calculated from various sources and some rule of thumb estimations from personal experimentation in Earth gravity.

As noted, before, the mass amount for feed is exponential, but what is in the mass inputs is a concern too. The feed for growth can be reversed out by using the nutritional data for the food made of harvested edible animals. The nutrition data from edible animal parts tell me what mass was retained over the life of the animal, and what elements were eaten and retained in the animal's body (or in the cases of eggs and milk, in the edible outputs). Animals do convert foods of many types into protein and fats, but the food must contain the base elements to get there. All elements beyond hydrogen and oxygen come from food, so the inputs must contain all the carbon exhaled, the salts and masses released in wastes, and the part accumulated in animal mass. This is where the tables of biochemicals come in from the last chapter. Nutrient assays of food contain amino acid profiles, sugar profiles, and fat profiles of dry mass. Assuming the nature of fats and carbohydrates, and knowing the rest of the chemical profiles, we can reverse how much of what elements need to be in feed if there is no other experimental or agricultural source. Now to get a bit more detailed for animal types, we can use the model in Fig. 3.5, though it would be far better to really measure the inputs and outputs by experiment. Figure 3.5 uses pathways of masses and is but one way to calculate all the mass models in this chapter, given nutrition data for animals. For analysis, nutritional parts are broken into percentages of elements C, H, O, N,(together called 'CHON') and the rest are ash. If the animal can at least partially digest cellulose, like cattle, some of the fiber in food is broken down by enzymes using water to be used like other carbohydrates in aerobic metabolism, otherwise that fiber ends up as wastes. If the product is either eggs or milk, the output wet mass includes carbohydrates, otherwise meat has no carbohydrates. Ruminants and other aerobic organisms can use a little anaerobic digestion, but from a mass flow perspective the rough output masses are similar to a guess for aerobic output. Not a perfect model, and this model would make a vet yell, but the model works for getting just close enough numbers for mass flow for a first guess at a space farm. Engineers call this kind of guess a 'back of the envelope' analysis. Again, experimental data would be a far better route to numbers you should use to start construction.

Armed with a guess at how mass flows for various animals and fungi, we can then examine candidate species for the farm.

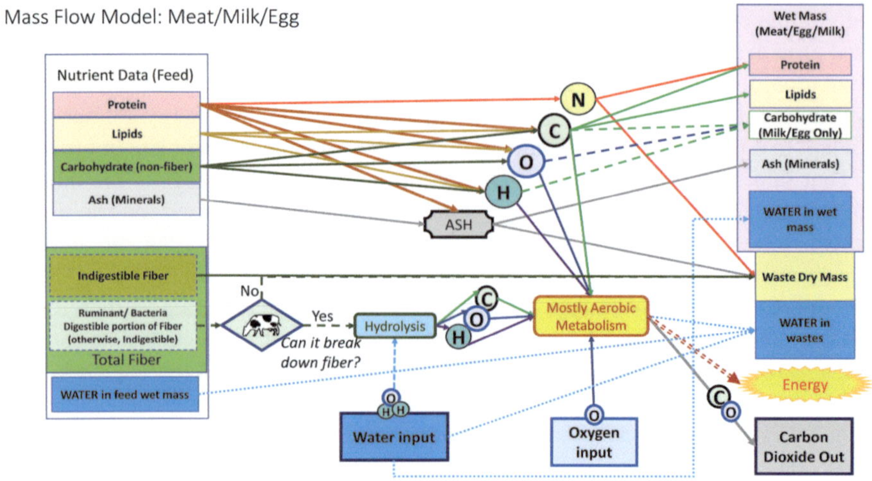

Fig. 3.5 Rough mass flow process pathways (to create mass flow models) for meat, milk, and egg production by animals

3.3 Fish Growth and Production

Fish are narrow ellipsoids for the most part, and data on feeding, spacing, and growing most common aquaculture species can be found courtesy of Food and Agriculture Organization of the United Nations (FAO.org), and chemical composition as food from the United States Department of Agriculture, Agricultural Research Service.

Most fish are roughly 1.3 times as dense as water. Fish grow from egg to fry, which are odd shaped, to small adult looking fish, and finally to adult, see Fig. 3.2, and the photos in Fig. 3.6.

Feeding rates vary based on fish age, where very young fish are fed about 10% of their mass and old fish at 1% of their mass as growth levels off, though ideally, fish are harvested right as their best growth rate levels off. It is easier to calculate inputs to a living kilogram of fish as percentages, however. For a typical fish, the ins and outs look like Fig. 3.7. Around 6% of the input mass is the dry mass component of the feed, though 85% of the mass a fish takes in is water. Thirty-seven percent of the output mass is in growth of the fish, while 51% of the output mass is in waste wet mass (46% water, 5% dry mass). Very roughly modeled from data in (USDAARS, 2022).

You need to pick some fish to produce the next generation of fish, and the best adapted fish given your farm's unique characteristics are likely the biggest and fastest growing. Therefore, the best growers are preserved as a breeder population, adapting to the settlement's conditions. Over time, as an example, a moon settlement might have a moon tilapia optimized for lunar space farms. It is important to note that breeders need to eat well and be extra happy to make more fish, therefore will require more food, oxygen, and importantly, tank volume with special conditions,

Fig. 3.6 Silver carp grow roughly 2 cm per month under ideal conditions from an egg to odd looking larvae, to post larval juveniles (as in top image), to full grown adults (bottom image). Photos by author

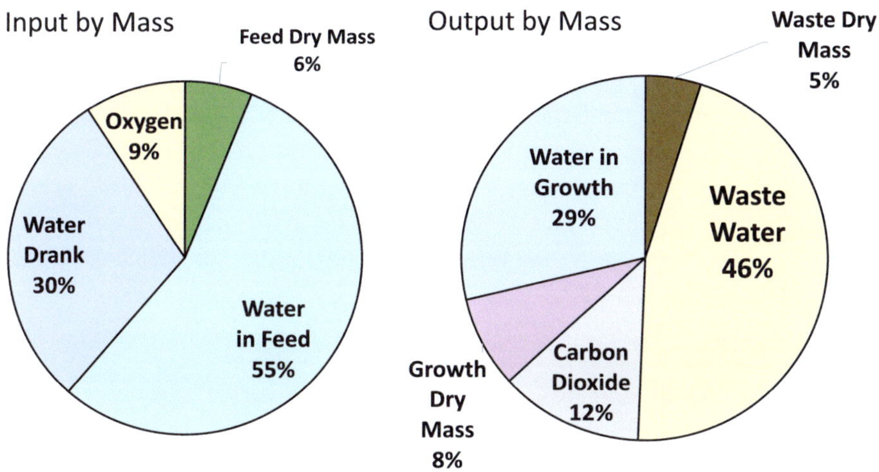

Fig. 3.7 Input and output ratio percentages for a typical fish

and will also need more time to grow to optimal breeding size. Tilapia for example like sandy bottoms (no pun intended) to dig craters to drop eggs and to raise fry. You may one day have craters dug by tilapia in a space farm that is situated in a crater dug by an ancient asteroid impact on Mars. Breeder tilapia might also be two or more years old, so they are twice as old as crop fish, and more than twice as big. For any animal, the reproductive rate determines how many breeders you may need to keep up production. Fortunately, egg laying fish are very effective breeders, producing dozens to hundreds of fry every cycle. The ratio of breeders to crop population affects mass flow and energy to a degree but affect footprint and initial farm mass heavily. The more fry per breeder in breeding tanks, the smaller breeding population, so less tank size and water that have to be added to the crop tank sets. Fish species selection for continuous production also favors fish that can be tricked into continuously breeding without seasonality. For example, many fish species breed on Earth on moon cycles (e.g. full moon), or annually based on seasons. Steelhead trout breed in the spring but can be tricked into breeding an extra time in the fall. Tilapia on the other hand breed continuously if the conditions allow, as in indoor farms on Earth, or in farms in space. Combinations of species selection, environmental conditions, and overtime, selective breeding or direct genetic manipulation, will be used to get fish and other animals in the mood all the time.

Due to the nature of the space farms in this book, it works out to be very efficient to favor herbivorous fish that grow well eating algae, vegetable wastes (i.e. leaves, stems, etc.), excess human food, and cyanobacteria products. A pioneer in using tilapia in space is Dr. Frieda B. Taub, whose research on tilapia as astronaut food, fed on freeze dried astronaut feces, was done in 1963 (Taub, 1963). Her work was inspirational to my own work. However, the pathway from algae to fish to plant is also highly efficient for any space farm and covered later in this book. Greatly beneficial is that many herbivore/omnivore fish can be put together tightly in tanks and raceways (since that is how they swim in the wild, see Fig. 3.8).

The smaller the tank, the less water is required to make the farm, and the smaller the volume and area of the farm. Likewise, as cyanobacteria and algae are volume

Fig. 3.8 Dark crowd of silver carp in the Kaskaskia River, Illinois, USA. Note tight swimming even in wild open water. Picture taken by author

and time efficient at producing high nutrient feed for aquatic animals, the overall farm efficiency is magnified by growing herbivores.

Mass accumulation in a crop is the goal. As stated before, mass accumulation is exponential, cubic with relation to length, and many fish grow at near-linear rates for length. A rule of thumb for our select herbivores is roughly 1 cm/month in wild conditions for round average growth fish (like tilapia), and 2 cm for longer high-speed growth fish in farms (like silver carp).

Given this data, we can reverse feed dry mass and waste dry mass, then reverse oxygen intake assuming food that resembles dried chlorella for the plant and algae eating fish. The result is a picture of the chemical machine that is a kilogram of living biomass of fish, in this case a tilapia, as in Fig. 3.9. This model assumes algae are the primary feed. Very loosely reversed using the methods in Fig. 3.5, and data in (USDAARS, 2022) for tilapia and algae, and (FAOAFFRIS-Tilapia, 2022).

Note that salinity, pH, oxygen level, and temperate are also key to achieving perfect growth. As mentioned before, fish must spend energy to balance water chemistry or temperature, so the more optimum the conditions, the more inputs are budgeted to growth instead of survival. Volume for fish is another condition that affects growth. Stress in crowding costs growth, so tank sizes must include the right balance of room for fish, at the cost of more water, and more enclosed volume. Like I said, if the kind of fish like to be crowded then less tank volume is needed to avoid stress.

A note for carnivorous fish follows. If a fish cannot be fed pellets made from waste products, or algae or plant products, it will be a long while before those fish appear in a space farm. Ditto if that species has a complex life cycle. You may have catfish for example since they can also eat plants and algae, but it will be a long while before sharks, tuna, or pacific salmon hit the menu. To close the mass cycle efficiently, the best fish are those that don't need an extra animal stage to make feed and can take

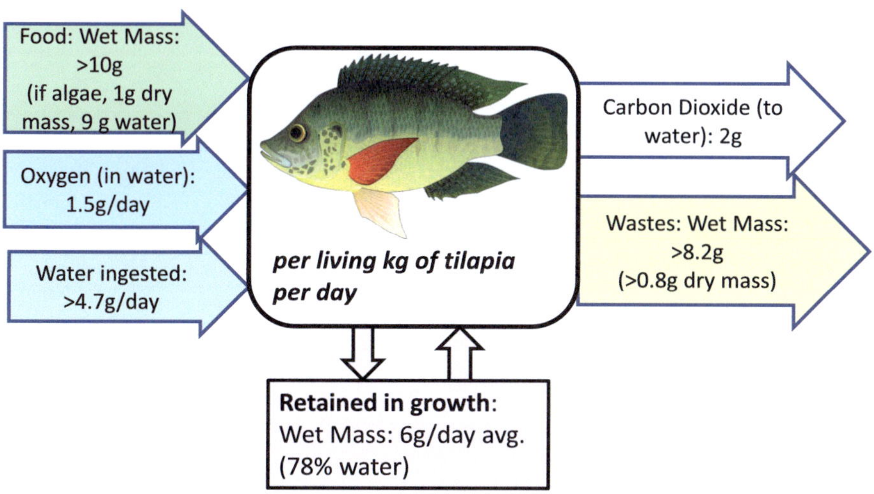

Fig. 3.9 Tilapia inputs and outputs per day per living kilogram of fish

crowding and tanks well. Carnivore fish need lots of room and would require a bigger farm to grow dinner for the fish. There are some experiments in pellet fed yellowtail jacks, yellowtail snappers, and ahi tuna worth following that might make these sushi fish possible in a very mature farm using meal pellets from scraps, but herbivore fish make the numbers work best. In Table 3.1 are some numbers for omnivorous but mostly herbivore fish. These species are already farmed on Earth, and in some cases have been farmed for millennia. Fish like these are essentially protein making machines that take algae or plant matter and convert it into human food. These are my rough guesses via data from my own previous work and via (FAOCAS, 2022). A great reference for pacu is also (Kapetsky & Nath, 1997).

Note that it is very important to consider the whole fish as human food, not just the fillets. To Western palettes the fillet is the only edible portion, but in reality, the entire fish is edible once processed. Fats and minerals are lost if only the fillets are consumed. Note also that fish sauce and other tasty edibles are made from fermented ground fish. As a practical matter, it may be that the non-fillet portions end up as fodder for other animals in the farm, but there is really no reason not to make soup broths, fish sauce, and fish meal from the heads, skin, fins, etc. In space farms it is important to use the whole fish, possibly grinding the non-fillet parts into a fish meal

Table 3.1 Candidate herbivore (or omnivore) farm fish

Species (common name)	Days to harvest	Mass at harvest (kg/fish) ideal	Days to breeding	Feed
Tilapia	150–180	0.3	365	Algae, plant wastes, animal wastes
Silver Carp	365	1	365	Algae, plant wastes
Pacu and Tambaqui	365	1.2	365	Plant wastes, animal wastes
Milkfish	150	0.6	1825	Algae, sea weeds, plant wastes or pellets
Channel Catfish	730	0.6	730	Algae, plant wastes, animal products, soy
Common Carp	1095	3	1460	Algae, plant wastes, animal products, soy
African Catfish	300	1	365	Algae, plant wastes, animal products, soy
Grass Carp	550	1.5	1460	Algae, plant wastes, animal products, soy

for soups and fish balls, to retain the fats and minerals. See Figs. 3.10 and 3.11 to show why. The figure data is loosely calculated via data in (USDAARS, 2022).

It is critical to put all the elements in the fish, head and all, to calculate the mass flow in the stage and then farm. Looking at the whole fish elementally, dry mass is mostly carbon and oxygen as in Fig. 3.12. Note 'Ash + Other' can also have C, N, H, O, S, and minerals in it. Calculated from nutritional data, ash estimates in whole fish, from methods in Chap. 2.

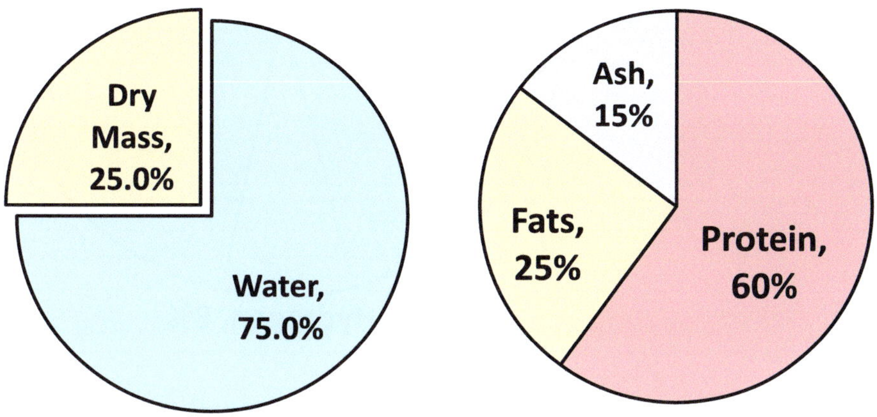

Fig. 3.10 Tilapia whole fish mass components, with dry mass versus water on the left, and rough nutritional breakout on the right

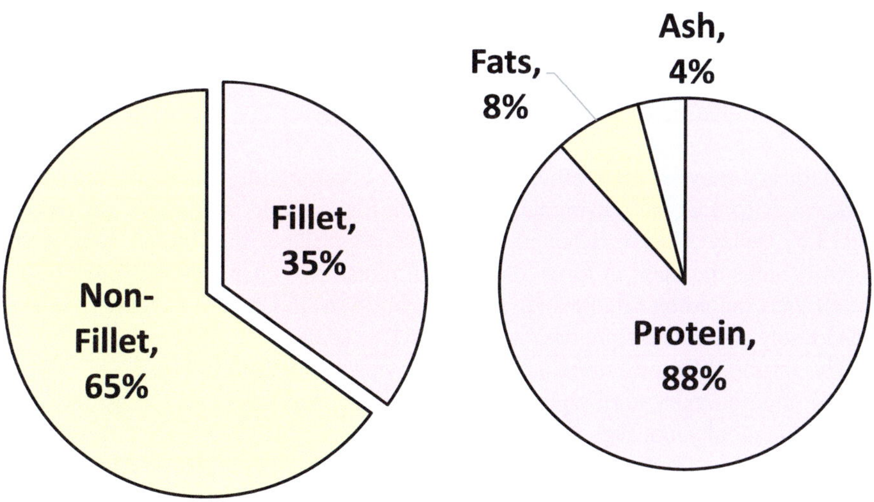

Fig. 3.11 Tilapia mass components by fillet: left is proportion of fish mass that is fillet, right is the rough nutritional component breakdown of the fillet alone

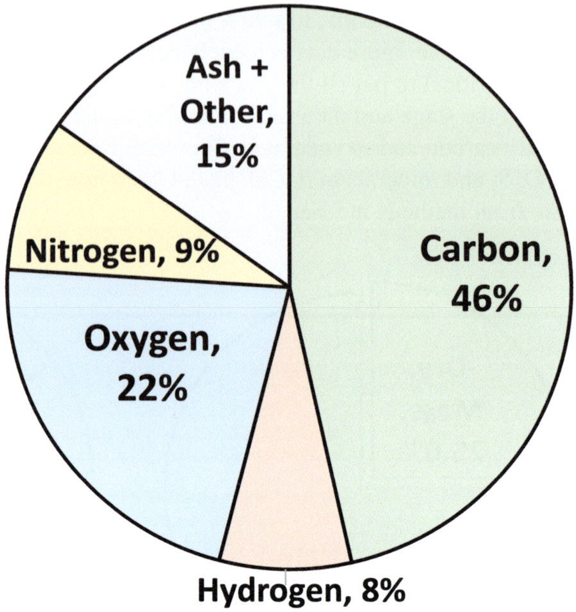

Fig. 3.12 Element breakdown of whole fish dry mass

3.4 Shrimp and Crustacean Growth and Production

Crustaceans, including shrimp, have shells on the outside that act as exoskeletons. They are very sensitive to pH and hardness as a result, as over acidity dissolves shells, and over alkalinity pushes their biochemistry to lethal levels. Crustaceans release ammonia in waste, and dying crustaceans release vast levels, which can cause cascade die offs. Your author knows the cascade die off well from experiments with prawns, that made a smell that could gag a hog, and a mushy mess that was very un-fun to clean up. That said, despite the more delicate nature of crustaceans to biochemistry, they are more efficient than fish in accumulating proteins in the choice portions. A look at the nutritional comparison for example for prawns and carp in Fig. 3.13. Many fish are richer in fats, crustaceans richer in proteins. Ash in this book includes the mass of minerals, salts, nitrates, and other organic elements and compounds including vitamins (See section on 'Ash' in Chap. 2). Figure data here was loosely calculated from data in (USDAARS, 2022).

The shrimp, like most animals, are mostly water, as shown in Fig. 3.14.

The remaining dry mass has components as for the left diagram in Fig. 3.13, and by elements looks like Fig. 3.15.

There are many candidate species useful for the space farmer that add high protein fast growing food sources. Like fish, shrimp and other crustaceans have been farmed or at least encouraged on Earth for centuries. Table 3.2 has a sampling of commonly farmed species. Ideally, the best species are those that quickly accumulate mass, and can grow fast on algae, plant wastes, or animal and food wastes. Krill are not farmed

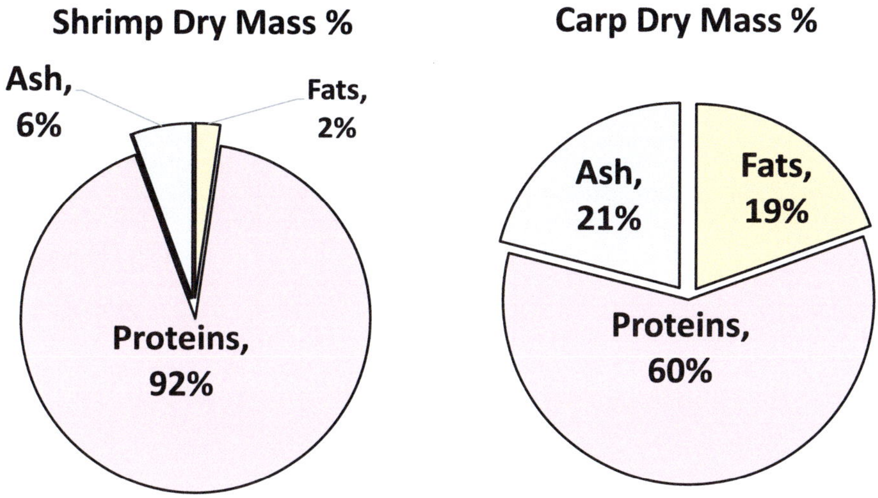

Fig. 3.13 Nutritional comparison between shrimp (left) and carp (right) by dry mass

Fig. 3.14 Wet-dry mass for shrimp. Crustaceans are mostly water, as for shrimp in this chart, via nutritional data

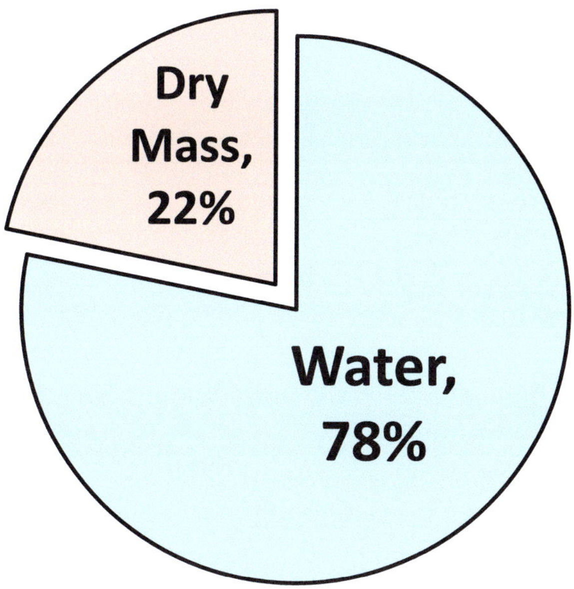

yet per se but might be in future. Brine shrimp are small but accumulate protein and fat amazingly fast, though grow in very saline water. Table 3.2 data is loosely calculated from data in: (FAOCAS, 2022; Meyer, 2019; Hicks & Pierce, 2022; FAO-Artemia, 2022; BSD, 2022; Quetin & Ross, 2001).

Crustaceans have a series of larval stages that transform a hatchling into their mature adult shape, and in many cases crustacean larvae are planktonic, and key

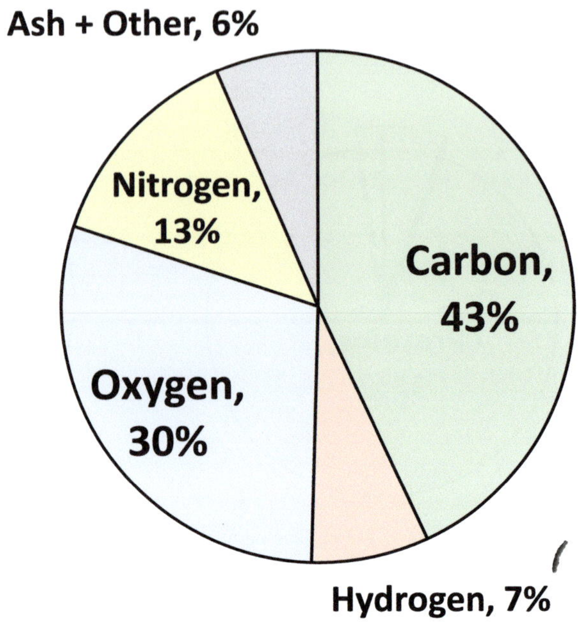

Fig. 3.15 Major element breakdown for shrimp dry mass

Table 3.2 Candidate space farm crustaceans (arthropods). Krill data here is very speculative

Species	Days to harvest	Mass at harvest (kg/animal) ideal	Days to breeding	Feed
Red Claw Crayfish	240	0.02	365	Algae, plant wastes, animal wastes
Freshwater Prawns	180	0.124	365	Algae, plant wastes, animal wastes
White Shrimp	120	0.02	365	Algae, plant wastes, animal wastes
Brine Shrimp	21	0.001264	21	Algae, manure, powdered plant wastes
Northern Krill	365	0.00632	365	Algae

components of Earth aquatic ecosystems. Shrimp (and most crustaceans) start as an egg, then hatch to a series of larvae that vary in shape and size. The shrimp larva eventually changes shape into a recognizable shrimp (or crab or lobster). Once they are mature, the adults can then breed and lay more fertilized eggs. This cycle is shown

Fig. 3.16 Crustacean lifecycle

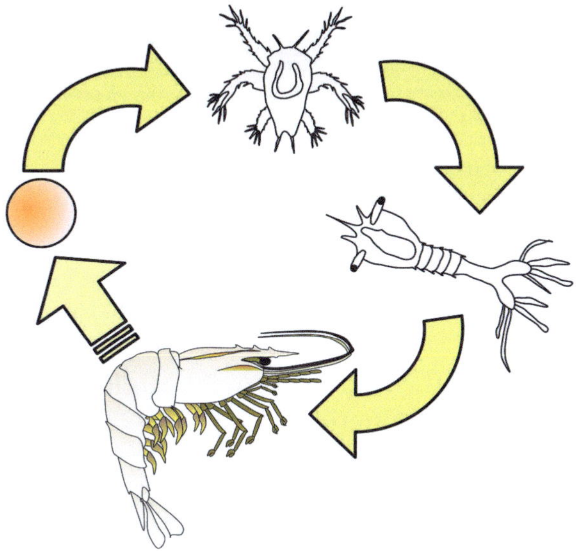

in Fig. 3.16. Growing them requires nursery tanks for each stage, eventually putting the larger animals in size separated tanks.

These animals take on a small adult shape swiftly, called post-larval stage, and from then on, a series of cage sizes and tanks structures will be needed to optimize growth, and minimize cannibalism. Yes, crustaceans are viciously cannibalistic, and in a heathy tank, loss is expected. Often, the first notice you will have of such loss and cannibalism is a simple decrease in population, the carcasses disappearing into the maws of other shrimp (though sometimes you might find a piece of shell). In most cases, decapod farmers have at least three types of tanks: a nursery, a breeder tank, and growth tanks. More on growth tank engineering later in this book. Note that crustaceans also grow by molting. In molting, the crustacean sheds their shell, and for a while has a soft new shell that takes time to harden. These shells can create filtering and clogging problem in tanks. It is also during molting that crustaceans are very vulnerable to water hardness and each other, which is why volume and water chemistry are doubly important. For some crustaceans like prawns, lobsters, and crayfish, it may make sense to have a caging or rock structure in the tank where molting animals can shelter, at the cost of more complex water flow (i.e. extra energy and the risk of unbalanced pockets, see note of messy cascade die-off experiment above). Figure 3.17 shows a guess at mass flow for prawns. A living kilogram of prawns are fed about 58 g of live algae cells per day, and they breathe in around 9.6 g of oxygen dissolved in the tank water per day. These prawns then grow by 5.6 g per day, release 13.2 g of carbon dioxide per day, and release about 0.97 g waste dry mass with 47.8 g water. This growth is depicted as a diagram in Fig. 3.17, values loosely calculated from the data in Table 3.2, nutritional data, and (FAOCAS, 2022; Meyer, 2019; Hicks & Pierce, 2022).

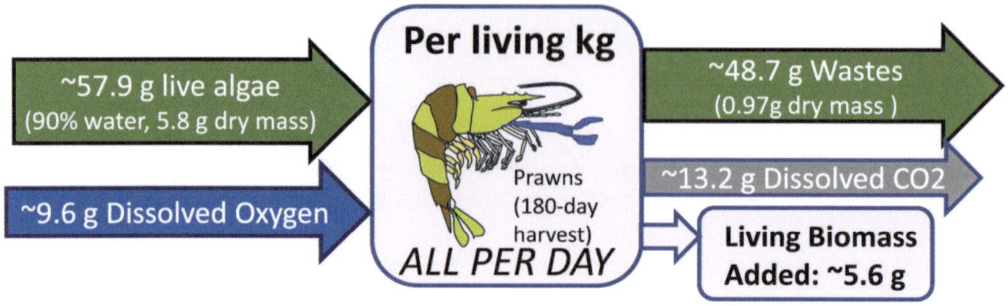

Fig. 3.17 Prawns as modeled living machine

3.5 Aquatic and Terrestrial Mollusk Growth and Production

Mollusks come in a wide variety of fast-growing shapes and sizes and can be a crop, or an add-on to a tank for another crop. Snails for example are great at cleaning surfaces of algae and bacterial mats. Clams and mussels are also particularly good at filtering water, and can provide an extra crop, either on the bottom of a tank, or in raceway flow diverted from the tail of a larger tank. Asian clams are extremely fast growing and an input to a variety of dishes, and can survive a wide variety of water conditions, and can act as a last stage biofilter before sending tank water to a photosynthetic stage. They are an invasive species in many USA waterways, testifying to their resilience. Like the Asian carp above, nature's invasive species make well tested space crops. African Giant Land Snails are also an invasive species that grows amazingly fast, can eat plant wastes, and provide a high protein food. Note that these species are often invasive species on Earth, which is shows they are robust and therefore useful in a space farm. This rule applies to most plants and animals in a space farm, so a species that is a lemon on Earth can be lemonade in space, and in space ecological problems become useful traits in the ecosystem of the farm and habitat. A few candidate space farm species are in Table 3.3. Data in Table 3.3 is very loosely modeled and calculated using the methods for animals above, and data derived from: (Kumbhar, 2020; FAO-Snails, 1986; Munywoki, 2022; Ortmann & Grieshaber, 2003; Lach & Schwartz, 2008; Hamwi & Haskin, 2008; RIRDC, 2001; Benson et al., 2022; UFIFAS-fa152, 2021).

Mollusk life cycles are varied, depending on where the adults live. Most aquatic mollusks have free-swimming larval stages, even if they eventually settle to the bottom like clams. Land mollusks like escargot snails have eggs that hatch into small snails. As for other organisms, the farm would need areas for breeding, juveniles, and adults. Yes, mollusks are cannibal too. Land snails can take two to three years to mature, while many aquatic snails live just one year. The mollusk life cycle is in Fig. 3.18. In short, mollusks like bivalves, cephalopods, and water snails emerge from the egg as free swimmers. Some clams and mussels are parasitic and attach to fish or other organisms. The larvae of snails and bivalves eventually settle and become a shelled adult.

Table 3.3 Sample of potential space farm mollusks, with rough growth data

Species	Days to harvest	Mass at harvest (kg/animal) ideal	Days to breeding	Feed	Habitat
Roman (helix) Snail	365	0.012	750	Plant wastes	Terrestrial
African Giant Land Snail (Lissachatina)	240	0.19	365	Plant wastes	Terrestrial
Apple Snail (Pomacea)	365	0.15	60	Algae, plant wastes	Freshwater
Asian Clam (Corbicula)	175	0.025	90	Algae, powdered (Sterilized) wastes	Freshwater
Freshwater Mussel (Unionidae)	365	0.0101	180	Algae, powdered (Sterilized) wastes	Freshwater
Atlantic Northern Oyster (Crassostrea)	540	0.18	120	Algae, powdered (Sterilized) wastes	Saltwater
Clam (Venerupis sp.)	1095	0.015	1095	Algae, dissolved wastes	Saltwater
Octopus	365	2	730	Crustaceans, other mollusks	Saltwater
Quagga and Zebra Mussels (Dreissena)	365	0.00544	730	Plankton, algae, wastes	Fresh and Brackish water

Many mollusks are more difficult to manage than those in Table 3.3, though these species may be worth studying at some point. Most mollusks are even more sensitive to water chemistry than crustaceans, though they grow extremely fast, and therefore convert food to protein swiftly. Many squids and octopus live one year, though require animal foods like crustaceans or fish. The shells of bivalve and snail mollusks however do provide a challenge. It may be good to grind the shells to balance the calcium levels in both aquatic and hydroponic growth, or to add to developing soil, but either way there is work required to handle the shell pile. As previously mentioned, at a minimum snails or clams can be used to clean tanks for crustaceans and fish, possibly with clams or mussels as a filter stage for tanks (Fig. 3.19).

Zebra and quagga mussels are a class of attaching mollusks, and are well known for filtering, but also for clogging. Using mussels as a filter will require care to keep mussel larvae from attaching to inlets and outlets of pipes and blocking them. Larvae are naturally pulled and attracted to areas of high flow, so intakes become a clog risk for highly productive attaching mussels. Using coatings on intakes and using ultraviolet (UV) sterilizers exiting the mussel beds can eliminate most clogging concerns. Many species of mollusks are also good food for fish and crustaceans, providing a high protein source once ground.

Land snails could be a very useful crop in low gravity, since they only require a confined space, maybe with pipes and struts, and can use all three dimensions. In

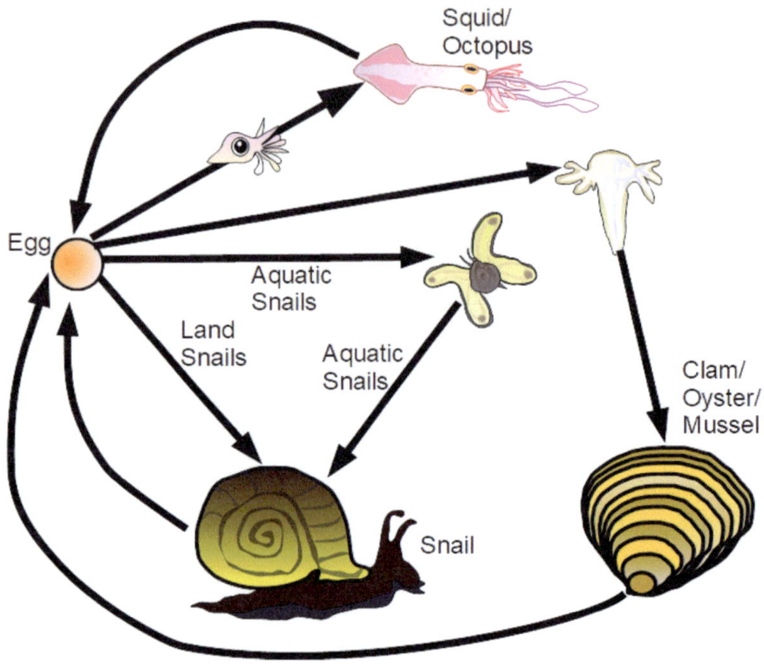

Fig. 3.18 Mollusk life cycles

California, people from Europe liked escargot so much that they released helix land snails. These escargot snails resulted in many curses by gardeners over the years.

As a protein and fat source, they are like fish by wet mass, once the shells are removed. Mollusks however shine as protein and carbohydrate source by dry mass, though not as eaten due to high water content. We eat wet mass, not dry mass. That said, ground filtering mollusks could be a food for crustaceans or fish. Adding filtering zebra or quagga mussels, then grinding them up as food pellets, can add rainbow trout or perch fish as crops. Shells also make a great base for soil starter once ground fine and could be a calcium supplement for people.

In Fig. 3.20, the left graph is the edible mass versus the shell (inedible) mass. The shell is the major mass component in many bivalves, including oysters and clams. The right graph in Fig. 3.20 shows the edible dry mass versus water ratio, 81% of which is water, which is higher than most animals.

In Fig. 3.21, the left graph is the nutrient break out of the dry edible mass, which shows the protein and carbohydrate content of bivalve mollusks. In the right graph in Fig. 3.21 the element breakout of the dry (edible) mass is similar for many types of animals.

The diagram in Fig. 3.22 shows how mollusks become a mass flow machine. Assuming the clam takes in 51 g live algae or similar detritus, and roughly 3.8 g oxygen dissolved in water, the clam will filter out the dry mass and release 44 g of water with some waste mass, roughly 5.3 g carbon dioxide dissolved in water, and

Fig. 3.19 Mussels and shells. On the top left, a live mussel feeds by filtering large quantities of water through its siphon. The top right picture is a shell from an Asian clam. On the bottom is a cluster of quagga mussels. Photos by author

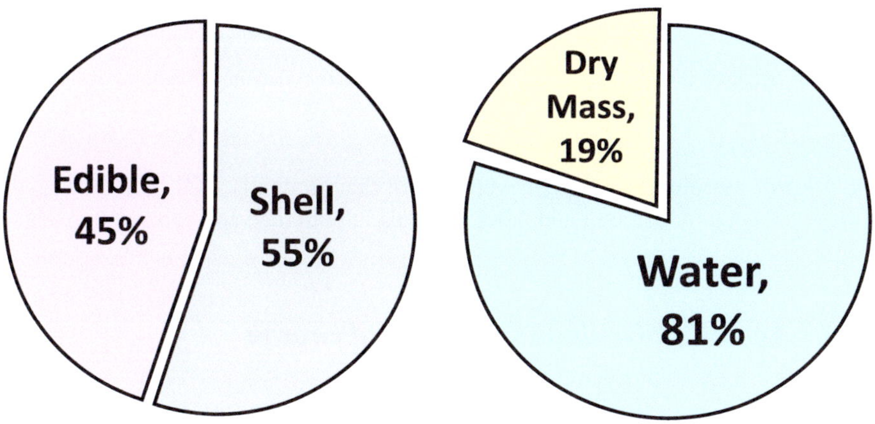

Fig. 3.20 Mollusk (Typical Bivalve) mass breakout. On the left is the shell to meat proportion for live animal. On the right, is the portion of water in the wet mass of the edible portion of the animal

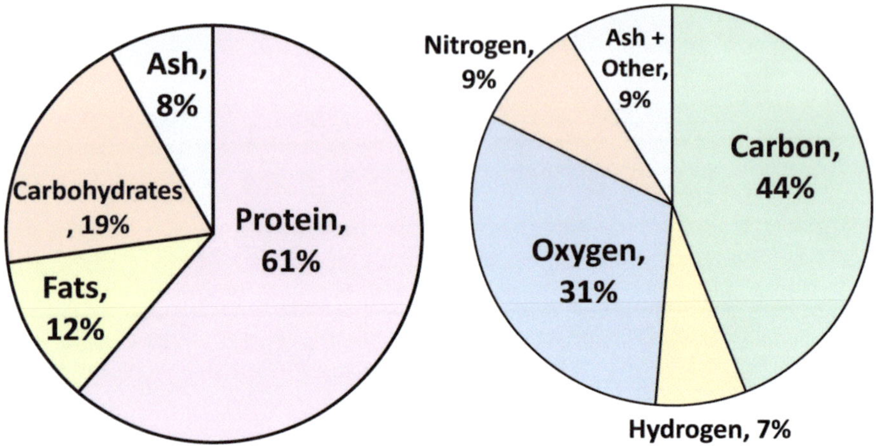

Fig. 3.21 Mollusk edible dry mass breakout. On the left is the nutritional breakout for the edible parts. On the right is the elemental breakout of the dry mass of the edible parts

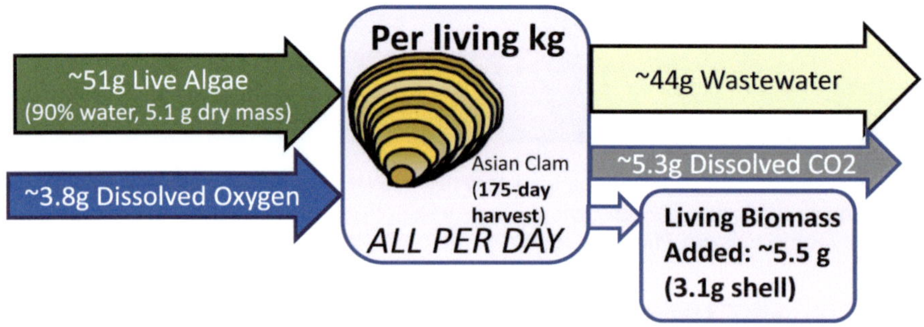

Fig. 3.22 Asian clam rough mass flow model. Calculated from data in tables and graphs in this and last chapter

add 5.5 g of wet biomass, though over 3.1 g of that growth is shell. This flow again hints at the need for calcium and other minerals in both feed and surrounding water.

3.5.1 Aquatic Polyculture in Combined Growth

Growth tank architectures need not be for just a single species. Fish tanks should include rough areas for bacterial colonies to balance pH, and possibly water plants and algae for the same purpose. Given this design, small crustaceans such as amphipods or isopods can reduce cleaning work by trimming algae and eating excess food and settled waste, but water snails can help too. Multispecies grows are called polycultures, and in China a four carp system with rice paddies is common where common carp are fed wastes, grass carp eat excess algae from carp wastes, silver carp clean and filter wastes from other fish from the water, and black carp keep snails in check. All the carps become human dinner.

3.6 Growth and Production of Terrestrial Arthropods: Insects

Insects are essentially crustaceans in air. Ick factor aside for western pallets, insects are a high protein food source, and some are even kosher. I remember sitting in church as a kid and hearing about Saint John the Baptist eating locusts and honey, then gasping. Later, I got to try hoppers myself via the Air Foce, sans honey, but that is another story. Would describe the taste as somewhere between popcorn and shrimp. The shells are only a minor percentage of wet mass (as opposed to clams and mussels) and are efficient in protein conversion similar to shrimp. A list of candidate insects is in Table 3.4. Most grow in similar conditions to humans and are traditional pest species. Silkworms and Mealworms are well understood insects and can be genetically manipulated to produce materials or mediate waste, in addition to being dinner. All the mealworm and insect data in the graphs and figures in this section were calculated roughly using data from: (Jones et al., 1968; Renault et al., 2003; Grau et al., 2017; Xu et al., 2019; Lyn et al., 2011; Valles, 2021; Antani & Burgeson, 2011; Busolutely, 2021; Kayton, 2022; USDAARS, 2022).

Humidity and gas levels are key to growing insects, though they are very resilient, and can tolerate a wide variety of conditions, in most cases their environment is not unlike that for humans (e.g. cockroaches). Insects can also be crowded together using a variety of three-dimensional shapes, such as rows of meshes with just enough room to allow free movement. Crickets are a high nutrient-based food, that grow and reproduce well on simple cellulose egg-crates and fed table scraps. Your author tormented his parents providing cricket homes in a basement screened cage. Unfortunately, in addition to being noisy, they produce strong ammonia and other smells, like most insects. The gases can be injected into water bound for photosynthetic stages, or aerially sent to the same plant stages. Like crustaceans, insects have an egg-larvae-adult lifecycle, except with a pupae stage, with molting, as in Fig. 3.23. Most of the time, like crustaceans, molted shells or dead animals are eaten by other living insects, but if not can be treated as any other solid waste. There is typically a high mortality rate for insects as well, but most insects are also cannibal and will consume dead insects likely before you notice.

As mentioned earlier, insects are high in protein as a human food. Mealworms have loosely a quarter more protein than shrimp or ground beef, twice as much as fish or yeast, and four times the protein of algae by wet mass. Shells can be ground into a digestive aid for people, like fiber, or just crunched like popcorn, so edible mass is nearly 100%. Like other foods, the protein content can vary by diet but given the varied possibilities of modified beetles or silkworms, bugs are a protein powerhouse

Table 3.4 Candidate crop insects

Species	Days to harvest	Mass at harvest (g/ animal) ideal	Days to breeding	Feed	Notes
Silkworm moth (larvae) (Bombyx)	42	2	56	Fresh leaves from mulberry similar bushes, possibly engineered to eat other ag. wastes	Larvae consumed from 5th instar on, pre-pupae. Cocoons can be engineered to be a variety of useful materials
Mealworm Beetle (Tenebrio)	70	0.14	91	Variety of plant or dried wastes, including paper, some plastics, and Styrofoam	Eaten as Larva. Can be hormonally manipulated to grow large as larvae and prevent maturity. Good food source for other animals
Cricket (Gryllidae, et al. likely Acheta)	60	0.2	60	Wide variety of wastes	Eaten as adult
Cockroach (Blattella et al.)	103	0.1	103	Wide variety of wastes	Eaten as adult. Good for waste mediation

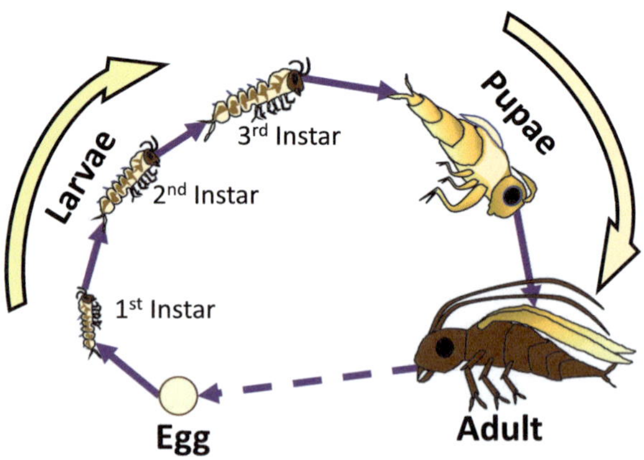

Fig. 3.23 Insect life cycle. Insects hatch from an egg as a larva, then go through several molting events (instars) to reach the pupa stage. The pupa emerges as the adult stage. Adults mate and lay eggs continuing the cycle

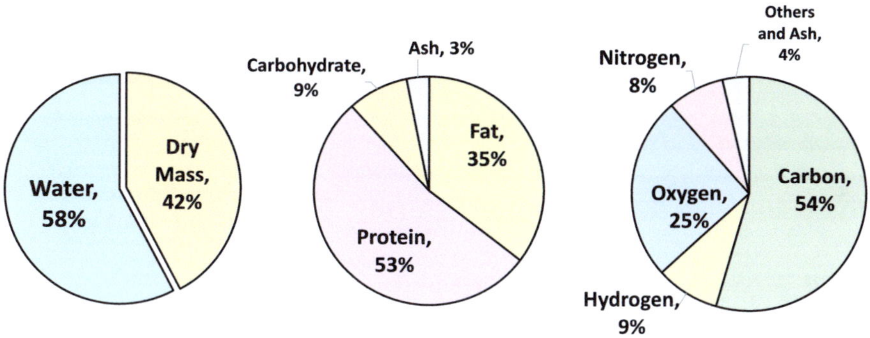

Fig. 3.24 Nutrition and element composition of mealworm larvae. Loosely calculated

like most animals. Mealworms are roughly 58% water, and over half the dry mass is protein (or roughly a quarter of wet mass). See Fig. 3.24. As a result most dry mass is carbon, with around 8% nitrogen.

Some insects are critical too, not so much as a crop per se, but for what they do. Parasitic wasps are very useful to control problematic insects. Bees, moths, and others are key pollinators for many plant crops. A space farm with a large vegetable growth area might consider a beehive which also provides honey. Your author loves to turn honey into a wonderful drink to make any event interesting: mead. Absent pollinators, humans or robots would have to be used. Many Earth workers can vouch for the pollination work using paint brushes. For the soil farmer, insects are exceptionally useful in soil-based farms, as part of the soil ecosystem, where the insects aerate the soil, break down wastes, and control bacteria and fungi.

A further class of insects and arachnids are useful as bio-machinery in lieu of food. As mentioned, insects can be used to process wastes, but also to produce materials like silk. Bioengineered insects or spiders could be used to clean farm areas, to trim and tend plant crops, to protect crops against infection or infestation, to produce pharmaceuticals, or a variety of custom materials. Mealworm beetles can eat plastics and convert them into non-toxic compounds as per (Brandon, 2019). Instrumented insects could be a tool for troubleshooting. Indeed, a space farm with the right insects could become an exporter of insect-made products. For each living kilogram of mealworms, they will need at least 0.2 kgwet mass of food, 0.01 kg oxygen, and 0.1 kg water (vapor or liquid), and will output at least 0.28 kg wastes, and 0.0143 kg carbon dioxide, producing 0.0143 kg more mealworms. This is a very rough estimate however! (Fig. 3.25).

We could also take a guess at how the elements move through the insect mass flow model for mealworms as in the rough mass accounting in Table 3.5.

Mass flow accounting tables will become more important in the later chapters of this book, but here it shows how the critical CHON elements move from input, through the organism, to living mass and outputs.

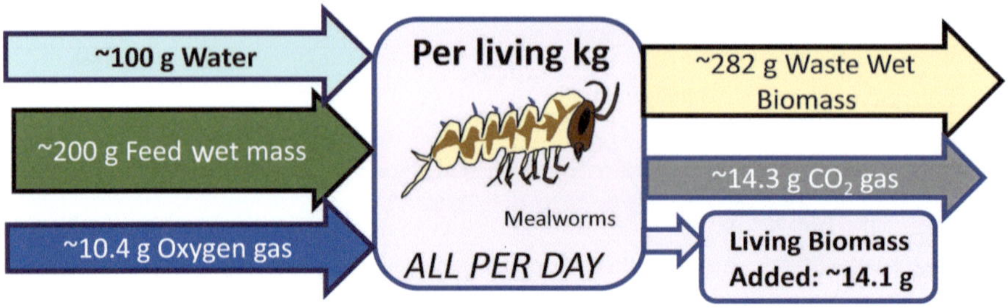

Fig. 3.25 Mealworm rough daily metabolism estimate

Table 3.5 Very rough estimated element mass flow accounting table for mealworms

Component	Grams per day per living kg					Total
	C	H	O	N	Ash et al	
Feed wet mass	56	16	116	4	8	**200**
Water		11	89			**100**
Oxygen			10.4			**10.4**
Total input	***56***	***27***	***215***	***4***	***8***	***310***
Waste wet mass	45	25	201	3	7	**282**
Carbon dioxide	3.9		10.4			**14.3**
Living biomass	7.6	1.3	3.5	1.1	0.6	**14.1**
Total output	***56***	***27***	***215***	***4***	***8***	***310***

3.7 Terrestrials: Traditional Farm Animals

Birds and mammals are in many cases a less efficient option for volume or footprint when traded against aquatic crops or plant crops. That said, traditional animals do provide a series of dietary variations, including secondary products like eggs and dairy. Initial space farms for economic reasons will be providers for luxury visitors or residents, and therefore those farms in gravity may have at least chickens, guinea pigs, rabbits, and possibly goats, and with very mature large farms, larger ruminants like cattle. Like other animals, terrestrial mammals even if retained as pets provide waste processing, converting plant waste or waste food into nutrients for photosynthesis. Some animals double as plant pollinators, especially hummingbirds and bats, and could provide welcome visual variety to a larger mature settlement even if not eaten per se.

3.7.1 Bird Growth and Production

In addition to the hummingbirds above, poultry could be used as primary or secondary crops. The main reason to raise birds might be just for eggs. See Table 3.6. Many species may be more valuable as egg layers than as meat birds. Many breeds are optimized for laying versus meat, so the space farmer would have to pick a ratio of meat to egg layers, especially for chickens. All calculations in this section are very loosely calculated using data from these sources: (Biggs, 2022a; Harlow, 2016; Kokoszyński, 2017; Randall, 2006; Jacob, 2022; Bolla, 2007; Cummins, 2018; Clauer & Skinner, 2007; Buckland & Guy, 2002; Kathiresan, 2013; Gujral, 2009; Rao, 2004; Meltzer, 1983; USDAARS, 1949–1573; Jayathilakan et al., 2012; Ellestad, 2020; USDAARS, 2022), and for eggs: (Biggs, 2022b; CUDRL, 2022; Liedel, 2014; Macklin, 2022; Robertson, 1988).

Broiler chickens convert 2 kg of food into 1.5 kg of bird, in 63 days (loosely (Biggs, 2022a)). However, egg laying chicken hens (about double the mass of a broiler) each can produce roughly 13 kg of eggs every year for up to seven years (Ellestad, 2020; Robertson, 1988). While poultry can be crowded into crates in very confined spaces, it may be more advantageous to let the chickens run in mature soil based planted stages, where their waste can fertilize the soil, and the chickens can control insects and worms in the soil biome or hydroponic farms and parks. Hen coops are designed to allow wastes, and eggs, to be removed without bothering the hen. See Fig. 3.26. After a fertilized egg is laid, it takes 21 days to hatch. Chicks grow into broiler sized chickens in about 42 more days, at which point they can be harvested. If left to continue to grow, roosters and hens are ready to breed in roughly another 84 days. Hens (female chickens) lay approximately 250–360 eggs per year; most can be unfertilized.

Crowded coops are not a good smelling place, but runs of chickens are more tolerable, given handling to recapture ammonia in the air. Your author as a child plucked many eggs out the century old coop on my great-grandparents farm. The hens tolerated the activity in the smelly coop, and I had to brush off floating feather fragments and straw from my clothes before being allowed inside for breakfast. On

Table 3.6 Candidate bird species

Species	Days to harvest	Mass at harvest (kg/animal)	Days to breeding	Egg mass/hen/year (kg)	Days to hatch	Minimum mass for hen (estimated, kg)
Chicken	63	1.6	147	12.5	21	3.7
Ducks	70	2.9	210	17.5	28	8.6
Turkey	126	9.0	224	8.0	28	16.1
Quail (Coturnix)	67	0.4	67	2.5	18	0.4
Pigeon	46	0.5	198	0.14	18	2.2
Goose	133	5.0	210	7.5	28	7.9

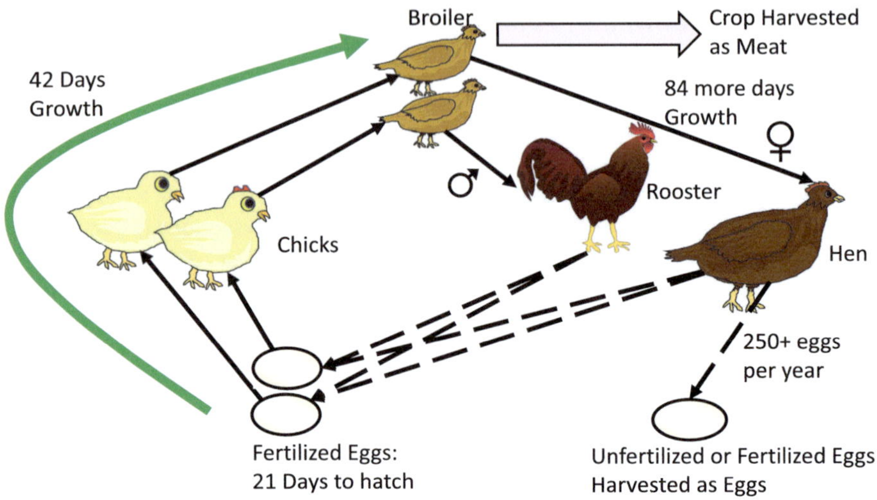

Fig. 3.26 Bird (chicken) life cycle

that note, feathers are a problem, in that they must be collected, and tend to float and fraction into fine aerial fibers. In lower gravity environments, fibers from feathers could be a serious hazard to heath and for fire. These fibers may clog air filters and provide a flame source, in addition to making humans sneeze. Feathers are keratin that must be processed for the farm's mass flow but can be ground and made into plant growth media or soil starter. If there are insect stages, these insects can be fed to birds. Another bird plus, bird guano is an extremely high nitrogen nutrient product that can be liquified and used in soilless or soil-based plant beds or used for algae or cyanobacteria reactors. Little work has been done for bird reproduction and growth in various gravity levels, though fertilized chicken eggs have been sent to freefall space and hatched (Valentine, 2017). It is likely breeding, or genetic manipulation will be required to ensure production in various gravities such as lunar or Martian gravity, or in microgravity. It is uncertain as to date if fertilized eggs can be frozen for travel, then activated to produce the initial crop of birds, so live eggs or birds may have to travel. Unfertilized eggs can keep for months under the right conditions, then fertilized later, providing another transport option. It is also possible that gametes can be frozen separately then thawed and recombined in a yet to be developed decanter. Given an artificial yolk and white, the raw gametes may be the seed, decanted into the artificial egg medium on fertilization for the first generation of birds.

Now back to the engineering part. Mass flow for birds has two routes. There is an initial investment to make the egg itself through the hen, then feed as the chick grows to harvest size, or additional time and feed to reproductive (egg-laying) size. Typically egg laying breeds are genetically separate then broiler harvest breeds in chickens, though in other birds, they can be the same breed, just fed and purposed differently. As a rough estimate, growth mass in a hen once laying begins goes almost all to eggs, versus a meat bird where growth mass goes to increased body mass. The tradeoff in protein production is almost a wash in an ideally managed and timed flock, but as far as footprint and resources, egg laying is more beneficial to smaller settlements due to the smaller space needed for a egg laying hen versus cycles of meat birds, and continuous food production from eggs versus need to have lots of growing chicks for meat. In Fig. 3.27 is the breakout of a kilogram of live chicken from a food perspective. Note that choice pieces are the parts traditionally eaten in Western diets, but everything except the feathers is edible with some processing, so we should think 92% edible, not just 69% that is fast food restaurant edible.

We can see the nutrition and element breakout for choice piece in Fig. 3.28 (since that is what is in the USDA Food Database). Around a fifth of the wet mass, or 56% of the dry mass of the edible choice parts, is protein.

Compared to meat, the eggs (see Fig. 3.29) are more watery but similar in protein. Here we also see an interesting note that helps rough farm calculations: the element percentage in dry mass of most organisms is similar for most parts.

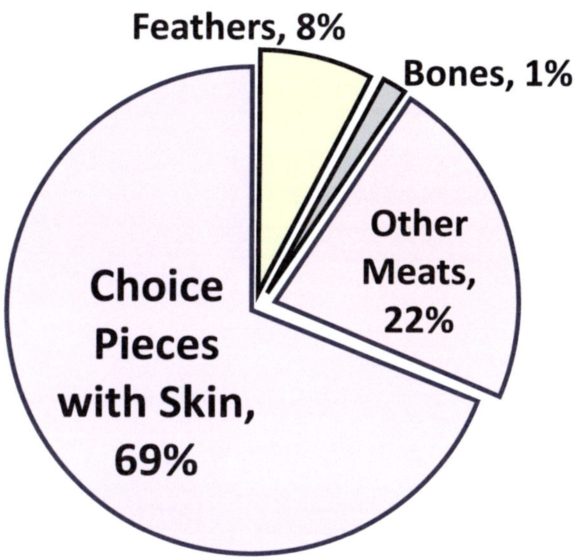

Fig. 3.27 Whole chicken wet mass distribution

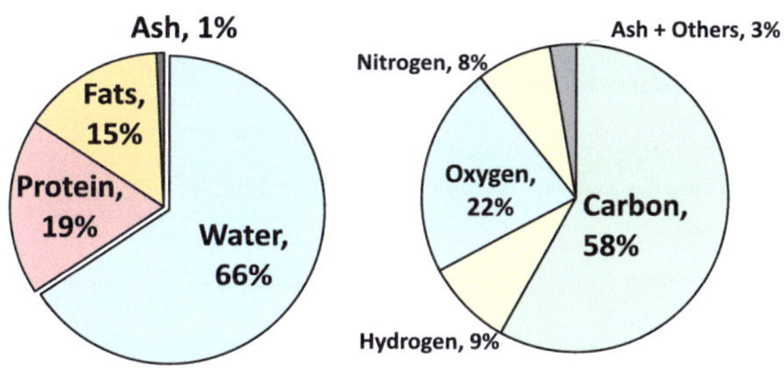

Fig. 3.28 Mass breakdown of data for chicken meat choice pieces by nutrition and element. The graph on left is nutritional breakout for wet mass of choice pieces. The graph on the right is the element breakout for the dry mass of choice pieces

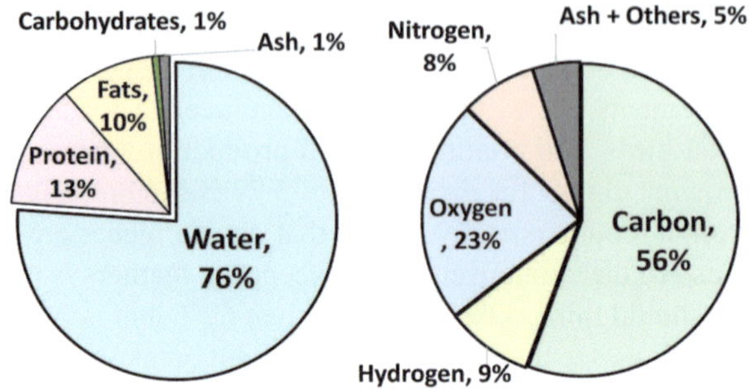

Fig. 3.29 Egg wet mass and dry mass components. The graph on the left is wet mass nutritional components minus shell, while the graph on the right is the element breakout for the egg dry mass minus the shell

The tradeoff between meat producing birds and egg laying hens in terms of productivity is an interesting problem set, which is covered in Figs. 3.30 and 3.31, and with the shocking conclusion in Fig. 3.32.

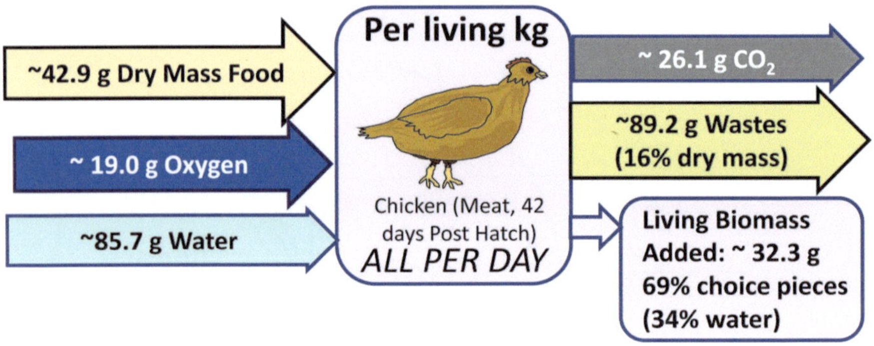

Fig. 3.30 Meat chicken (broiler) very rough mass flow model. This model assumes 42 days of growth post hatching

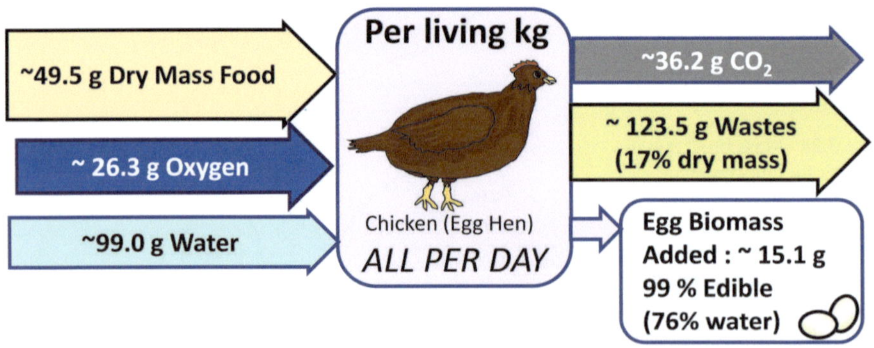

Fig. 3.31 Egg laying hen rough mass flow model. Assumes a White Leghorn or similar breed at near peak production

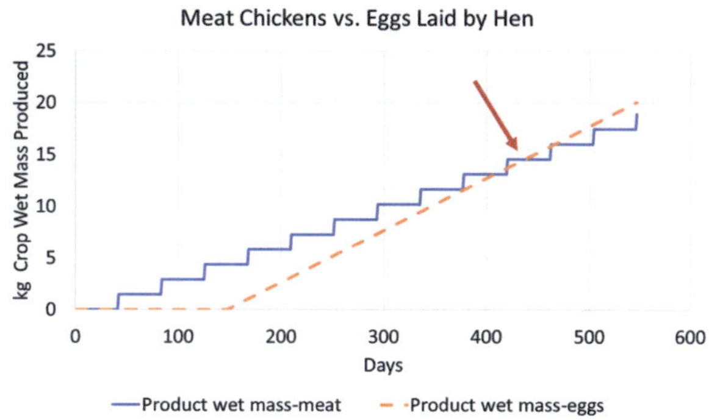

Fig. 3.32 Tradeoff: Food in eggs from hens versus food from meat chickens

Even assuming the same feed ratio by mass, the egg laying option produces more food wet mass versus meat after the first year. This result holds for chickens, fixing the ratio of food input and bird footprint, even assuming egg incubation for meat birds in is parallel with production.

When looking at the trade-off of bird meat to eggs, time from farm start is critical. Egg production outpaces meat production for the same footprint by 13 months and stays ahead after that time. So, it is likely one may have eggs for breakfast in a smaller settlement, while a larger older settlement may then also have roasted chicken for dinner, likely old roosters or hens that can no longer lay eggs.

3.7.2 Mammal Crop Growth and Production

As humans, we know the basics of the mammalian life cycle, though things are not quite the same for many crop species, in the wild or otherwise. For many wild species, parental investment varies widely, but as per usual, most of the work lands on the mother's lap. As mass flow, the assumption of zero mass to full grown shifts through the mother for gestation and lactation. In gestation, the growth component for that animal flows though the mother to the offspring, and then after birth, mass flow again passes partially through the mother to the offspring until weaning. The aggregate result for analysis is still the same, since as a living kilogram for the whole population, the maternal investment is averaged into the crop need to produce mass efficiently based on inputs. Efficiency therefore needs to include the maternal mass flow. Extra mass passes through female into litter during both gestation and nursing phases. Feeds are gradually introduced during later periods of nursing to weening, when the litter converts to feed fully. Most of the litter is harvested as meat for meat animals, while a few specimens are retained to grow as breeders. The ratio of male breeders to female breeders is roughly 1:10 for many rodents. For cows or camels for example, milk can be the primary product in lieu of meat, requiring fewer animals (Fig. 3.33).

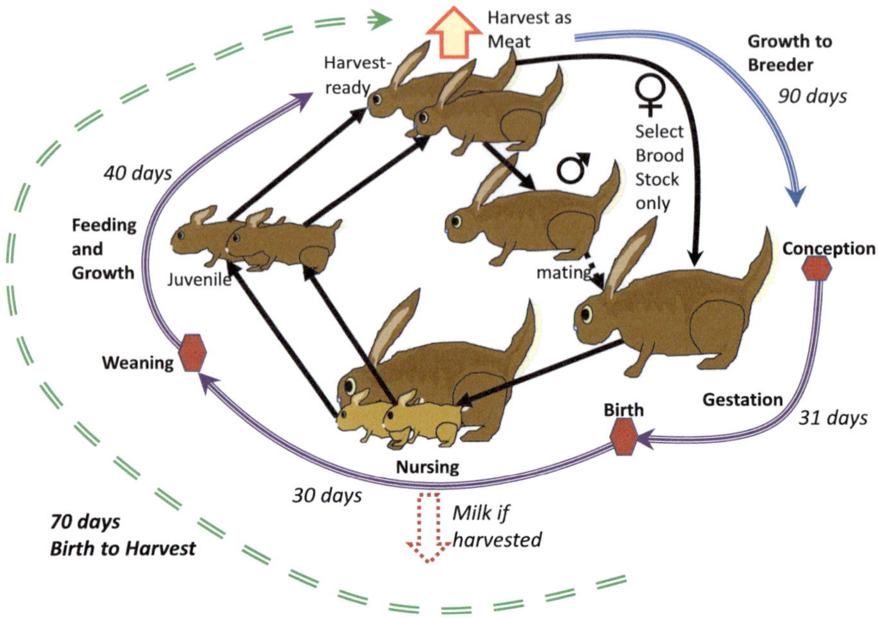

Fig. 3.33 Mammalian life cycle (e.g. rabbit)

Given this life cycle, as a food farmer for larger mammals, the tradeoffs for mammals are around the ratio of meat animals to animals raised for milk and fur or wool. Some animals only purpose might be to eat plant wastes to make wastes better suited to soil and plants. In many scenarios with small areas, there will be a higher ratio of mass to milk and wool, with only rare instances of slaughter for meat, if at all. With a very large settlement, with human diets for residents, export, or tourists that favor meat products, the ratio can become more meat based, with less milk basis. These are among the many considerations in designing and evolving a space farm.

Water management for mammals is critical, especially considering humidity in growth areas. Mammals pant and perspire, and like transpiration in plants, there is a strong water cycle that occurs inside the farm enclosures. Most mammals consume multiples of water incorporated for thermal regulation or other key biological processes, which then come out as water in breath or sweat. This breath and sweat water are most efficiently captured from the air and recycled near the animals by a condenser with sterilizer, then returned as drinking water or used for food processing.

To examine mammalian crop animals, we should start with rodents. Many rodents, contrary to the diets of many Westerners, are good crops for space to add dietary variety. Data in Table 3.7 is very loosely estimated and calculated from: (BSTIDNRC, 2009; Kime & Harper, 2008; Anderson, 2014; Waldo, 2022; OAH, 2019; Pinchao-Pinchao et al., 2024; Pace & Smith, 1984; Lanier, 2018; USDAAPHIS, 2020; Baird, 2015; MDC-Nutria, 2022; Tůmová, 2017; Głogowski & Panas, 2009; Brown-Uddenberg et al., 2004; Garcia, 2015).

Table 3.7 Potential space farm rodents

Species	Days to harvest (birth to harvest)	Mass at harvest (kg/ animal)	Days to breeding	Gestation (days/ Litter)	Days to weaning (after birth)	Offspring per year
Rabbit	70	2.0	160	31	30	49
Guinea Pig	56	1.0	35	63	28	15
Coypu/ Nutria	240	4.5	120	130	56	21
Agouti	180	2.0	360	104	84	4

Rodents can be raised in small cages, eat plant trimmings or silage and other wastes, multiply readily, and are robust

Rodent crops feature a lifecycle that includes mass flow from female to pup in both gestation and nursing (i.e. time from birth to weaning). Some rodents are not so good mothers and can eat their own offspring for the first couple of litters. Likewise, offspring have a mortality rate that lowers overall production. Once weaned, rodents grow extremely fast, and typically are ready to breed in under a year, and in the wild live only a year or two.

Hairless varieties save the problem of dealing with constant floating fibers, and in general rodents reproduce quickly and prolifically, often without seasonality. Like many animals, rodents can consume plant or animal wastes as feed. The cellulose and lignin from plants will mostly still be in the solid wastes, but that used waste will be fully ground up and ready for fermentation or soil starter. Rabbit and Guinea pigs are already in the diet of many cultures and can provide luxury visitors a variety of interesting dishes with a small volume footprint. Rabbits and Guinea pigs grow readily in boxes and can be provided tunnels or other shapes for breeding. Cages often feature a mesh bottom that drops waste to a tray for easy removal.

Rodents have a downside, however, as do most mammalian crops. They can carry some human diseases. Further, rodents will require extensive cage cleaning to prevent human or rodent health issues. Worse yet, overcrowding of animals, including rodents, results in lower growth and reduced reproduction. Rodents also require a more substantial breeding population then birds or fish. Like other mammals, the gametes may be frozen separately for travel, then either inserted into a rodent that made the trip alive or placed into an artificial womb for the first crop set. It is notable that both rabbits and Guinea pigs hibernate in nature, and if kept in this state, may be candidates for long trips to far settlements. Some work has been done by NASA in rodent reproduction and growth in microgravity including a study in 1984 (see (Pace, 1984)) and continuing to current on ISS and beyond.

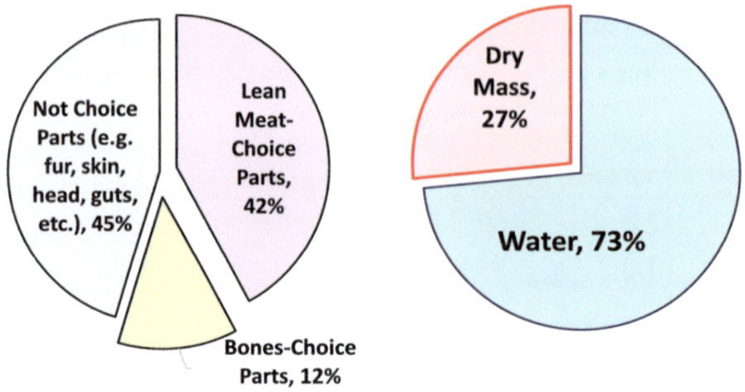

Fig. 3.34 Rabbit meat as harvested. On the left is the choice breakout of the harvested meat. On the right is the wet mass to dry mass ratio for the harvested choice lean meat

Mammals in general have a high percentage of non-meat mass, i.e. guts, bones, etc. As mentioned for other species, non-meat portions may be processed to become edible or used to feed other crops or as compost mass to make soil. Rodent meat is high in a variety of human nutrients however, and a delicacy that might be useful in attracting tourists. Most of the mass harvested from a rodent, or mammal in general, are parts not usually sold at market for food, but could be processed into food at need. See Fig. 3.34. The choice parts (meat with bone) are around 42% of rabbit living mass, but portions of not choice parts are also edible tissue. The bones can be stewed to extract their nutrition. Of the tissue, (also in Fig. 3.34) 73% is water, 27% dry mass.

Looking at the nutritional content of the rabbit meat, over three-quarters of the dry mass is protein, making rodent a lean protein source, and the meat is very carbon rich. See Fig. 3.35.

The mass flow model for rabbits (or rodents in general) shows around a 44% conversion from feed + water to living wet mass, producing a good amount of waste as well. See Fig. 3.36.

Next in considering traditional crop mammals, are goats and sheep. Goats have been proposed by Soilleux (2017) and Heppenheimer (1977) and many others, as at least milk providers and waste processors. Sheep are similar except they are pickier about dinner, better behaved, and make more wool. Goats can consume a variety of wastes, especially plant wastes, as well witnessed by your author, though they do not always have the best attitudes. Watch those pant legs and sleeves! Overall, feed should have a good amount of protein and minerals balanced for either goat or sheep as appropriate, and on Earth is supplemented with hay or other crop leftovers. In space the overall diet will have to be controlled but could look like the guess used for mass flow modeling in Fig. 3.37. Most commercial feeds on Earth have around 16% to 18% protein, but that feed is supplemented with a high protein forage of grass or other crop residue, often partially fermented to increase proteins. Calculations for the tables and diagrams in this section loosely use data from these sources: (Van Saun, 2007; Luginbuhl, 2015; Van Saun et al., 2022; NTAE, 2019; Barkley et al., 2005; Craddock & Steen, 2012; McCune, 2022; AGF, 2022; USDAARS, 2022).

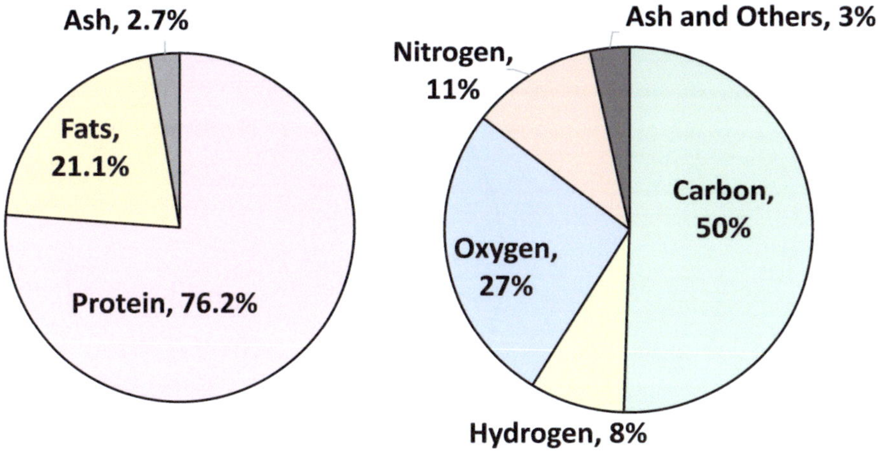

Fig. 3.35 Rabbit meat nutritional and elemental components. On the left are nutritional components of dry mass of meat from choice parts. On the right are element components of dry mass of meat from choice parts

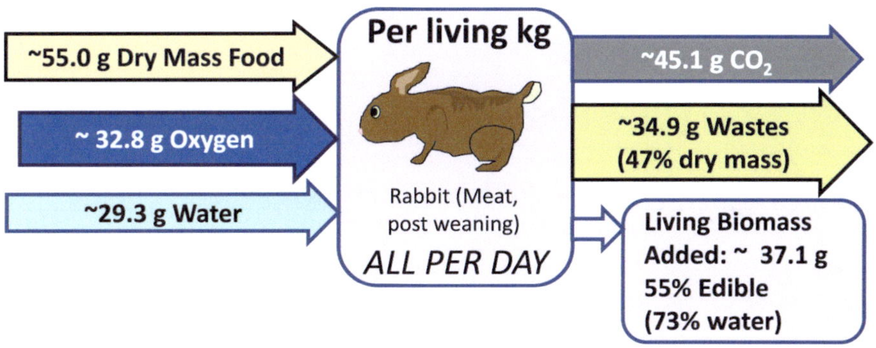

Fig. 3.36 Rough meat rabbit mass flow model. Assumes growth to 2 kg rabbit at 10 weeks of age. This diagram ignores water in and out for panting and sweat which can double or triple the water in and out

While young goats are a prime meat source for Middle Eastern, African, and Indian dishes (goat vindaloo is a favorite of your author), it may be preferable to have a smaller population of goats just to convert wastes to soil for plants, and to provide milk for various dishes, yogurt, and cheeses. The tradeoff of milk versus meat is like the tradeoff in birds of meat versus eggs, with the variable that some juvenile goats must be produced to get milk production, including time for gestation. See Fig. 3.38. Hormonal manipulation might remove the need to breed a goat to get milk.

As with other mammals, fur and skin flake sluffing may clog air filters, and effort is required to collect liquid and solid wastes, and to feed the goats. It is also a research

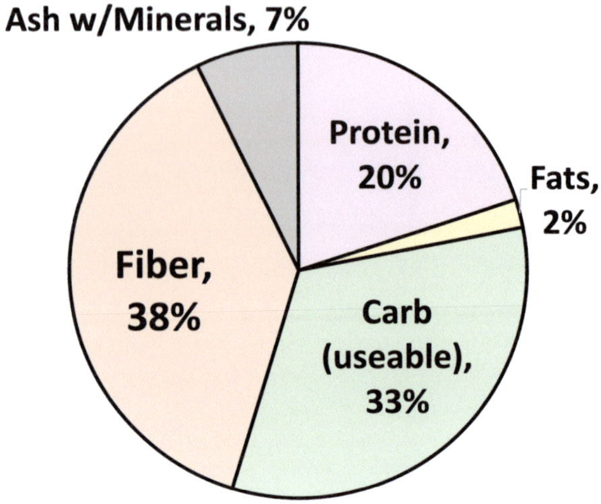

Fig. 3.37 Modeled feed for goat mass flow estimation

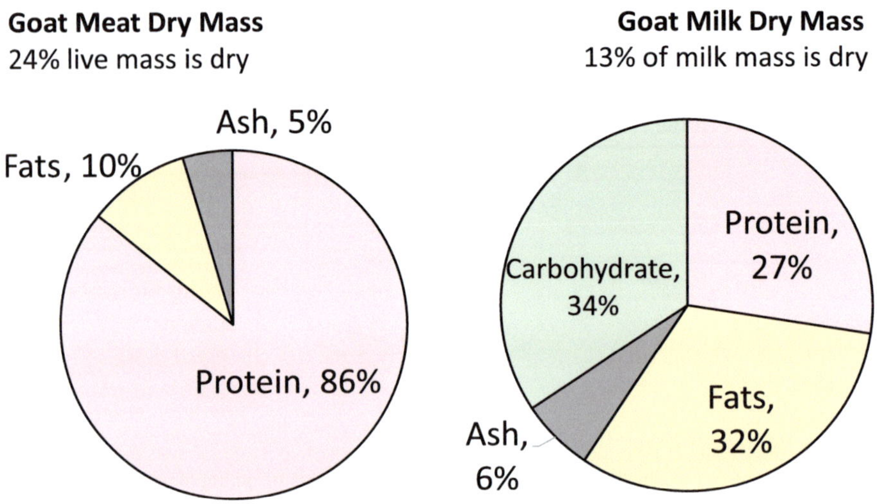

Fig. 3.38 Goat meat versus milk. Meat has more protein per kilogram dry mass, but milk uses less overall mass flow, so less energy and space

area to determine how large mammals breed and grow in lower gravity. For the goat rough model in Fig. 3.39, assume a weaned goat, that subsists only on high fiber feed and supplements, and assume weaning at 6 weeks at 13.5 kg and growth to over 35 kgs in around 160 days. Loosely calculated from above sources and from data in (AGF, 2022).

The mass flow for a milk goat is a bit different. In lieu of growth in body mass, the milk goat puts her mass into milk production for the next generation (a big nod to all mammalian mothers here). Therefore, the rough model in Fig. 3.40 is lower

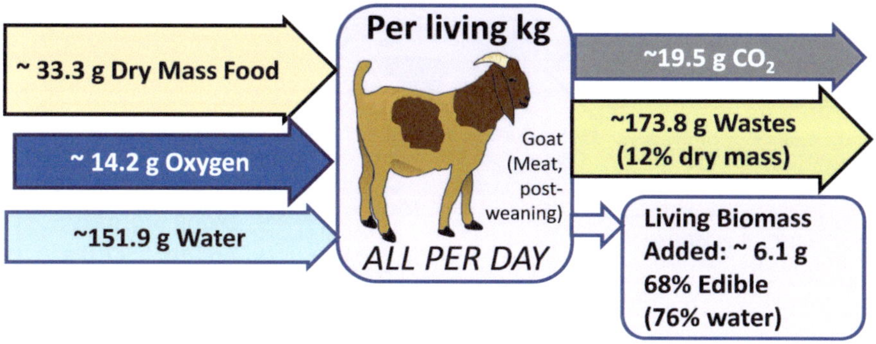

Fig. 3.39 Meat goat mass flow model. Very rough meat goat growth model per day per living kg

Fig. 3.40 Milk goat mass flow model. Milk production is a different dynamic than meat production, and lower in total mass flow per living kilogram

total flow for a larger goat, 100% edible output for the whole lactation cycle. In lieu of growing, the female goat devotes the bulk of input mass to milk production. Lactating goats are usually over 40 kg.

Flow models in Figs. 3.39 and 3.40 are super simplified and nearly notional, since there is gestation, perspiration, water in breath, and other factors involved, but these models provide a rough start for analysis. Meat is more inefficient for the slow growing goats per dry mass, even for protein production. This assumes traditional non-edible parts like bones and some organs are not also processed for production, but slow growing goats make milk a better mass conversion for these animals, unlike pigs or rabbits. Given a large grassy area, goats and sheep could provide trimming to facilitate grass growth, as in parks or terraformed areas. Further, sheep or goats can be sheared to make clothing, since wool sheep shed their coats anyhow between seasons. The wool and milk portion of the life cycle might be preferable to meat production for early settlements, though the ratio will depend on the settlement size and preferences. Martian Angora sweaters and goat cheese anyone?

For the ham-loving crowd, we should next consider hogs. Like goats, pigs are omnivorous waste processors, though from personal experience, produce astoundingly bad smelling wastes, for the gifts of pork steaks and bacon. My paternal

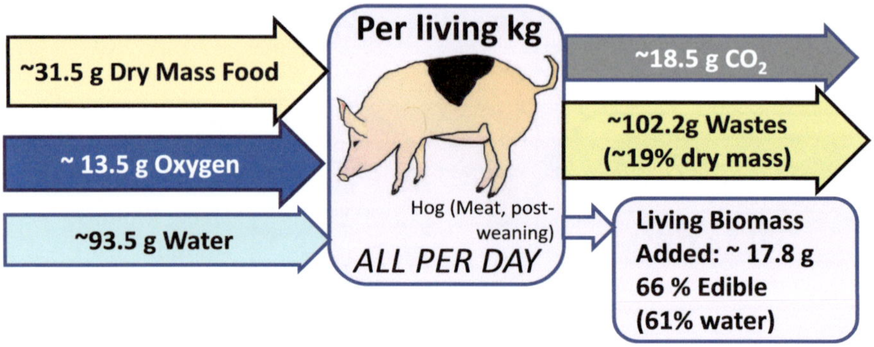

Fig. 3.41 Pig mass flow model. A very rough mass flow model for pigs based on feed and growth data averaged over 154 days (post wean)

grandparents' clothes absorbed the ammonia and other smells from their farm, and commentary by a two-year-old yours truly made my grandpa repeat the comment and laugh all his days. Dwarf varieties of hogs may be the first choice for most settlements that choose to have hogs, and given space and gravity, can be a luxury crop, or alternately, a pet. As with goats, work is required for feeding and cleaning, and pigs are notorious for disease transmission to humans. A settlement may be very well developed and decades old before using swine, either as simple waste processors, pet, or as a food crop. Work will be required to determine growth dynamics in various gravities, though the ability of pigs to grow in a variety of conditions, and in small spaces, is well documented in agriculture research, including the loose data sources for the estimates and calculations in this section: (Boler, 2014; Eikelenboom, 2004; Hendersons, 2022; USDAARS, 2022). Figure 3.41 provides a very rough model for pig mass flow, while Fig. 3.42 shows the nutritional breakout of pork meats (including organ meat).

One of the oldest domesticated mammal crops on Earth is cattle. Human history in Africa, Asia, and Europe would be very different without the decision to domesticate cattle as both food animal and beast of burden. Cattle as noted in the Earth farm examples in earlier chapters, fulfill many roles, including grass trimming, soil turn-over, and aerobic waste conversion, in addition to providing milk and meat. Ruminants like cattle also feature multi-stage stomachs that use bacteria and enzymes to partially digest fibers, such as cellulose, even if only 30% effective (see Linn et al., 2022). The initial designs for membrane based aerobic and anaerobic bioreactors were inspired by the cow's stomach and intestines, and some waste processors use rumen as a reactant. Some cheese makers use calf digestive enzymes (rennet) for cheese making. Cattle, however, are big, and require extensive space. Feedlots can hold roughly 300 to 600 cattle per hectare but do not make for happy cattle. There is a wide variance between sizes, water needs, and environmental needs, for the tremendous variety of cattle breeds. In general, a mature cow or bull is several hundred kilos, mostly gut, with under 2/3rds of that in dressed weight. It is likely only very mature settlements, in Earth-like gravity, would have cattle in any form. Getting them to the settlement

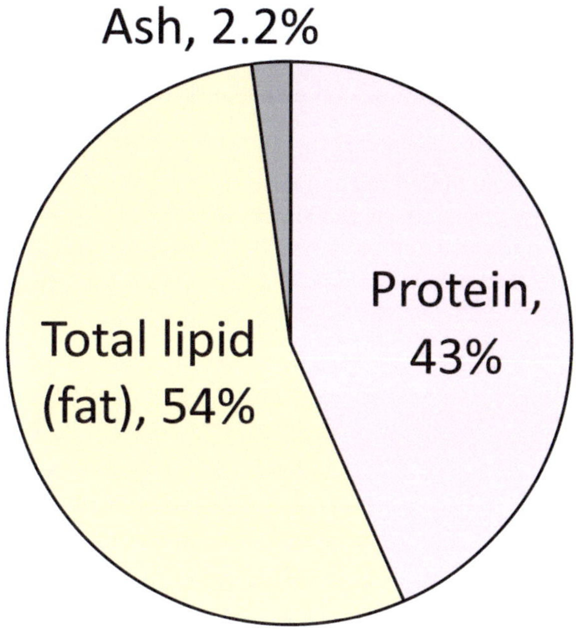

Fig. 3.42 Pork dry mass nutritional components. On average pork is 39% dry mass/61% water

would be a great challenge, as it would be for any large mammal, and may require frozen embryos and artificial wombs. Cattle (and other ruminating mammals) also pass gas lots and dealing with their methane could be a gas handling challenge. Calculations in this section loosely estimate and calculate using data from: (Jones & Heinrichs, 2017; NSWDPI, 2022; Sewell et al., 1993; Harner et al., 2021; Rounds & Herd, 2022; Ungerfeld, 2020; Bernard & Montgomery, 2002; Titgemeyer, 2007; Campbell, 2016; TBS, 2009; Turk et al., 1934; Redfearn & Bidwell, 2017; IBC, 2022; Davis, 2020; Probert, 2013; Kenzie & Williamson, 2022; USDAARS, 2022). Figure 3.43 shows a very rough mass flow model for typical beef cattle.

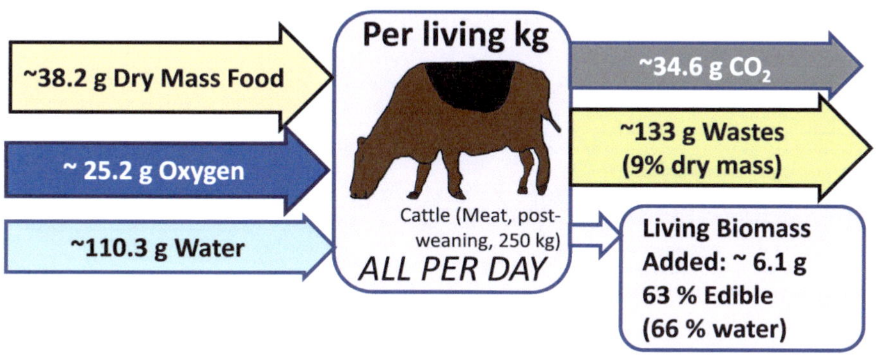

Fig. 3.43 Meat cattle very rough mass model

Beef meat is higher in fat than many meats, as in Fig. 3.44. Roughly 63% of the carcass is directly edible, 34% of that is dry mass. However, given a high degree of processing maybe all the animal mass is edible.

In contrast, dairy cows need more space, but their milk represents a daily harvest. My dairy farming uncle noted that his milk cows ran to get milked every day, since they were loaded full and needed to be emptied! Sure enough, I saw this for myself at the dairy farm near Quincy, Illinois, USA. Milk is more efficient at producing food dry mass per input per day, theoretically, but this is highly variable based on cow health and other factors. Dairy cow mass flow looks roughly like the diagram in Fig. 3.45. Data partially via Eastridge (2007).

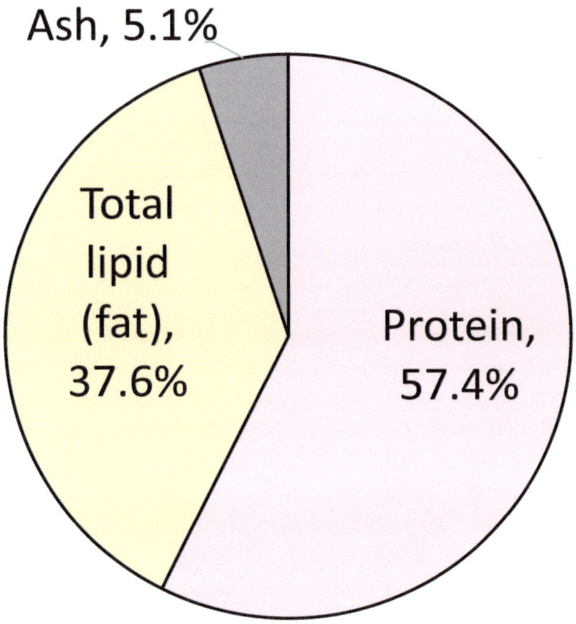

Fig. 3.44 Beef meat dry mass. Beef is roughly 34% dry mass, 66% water, and high in fat

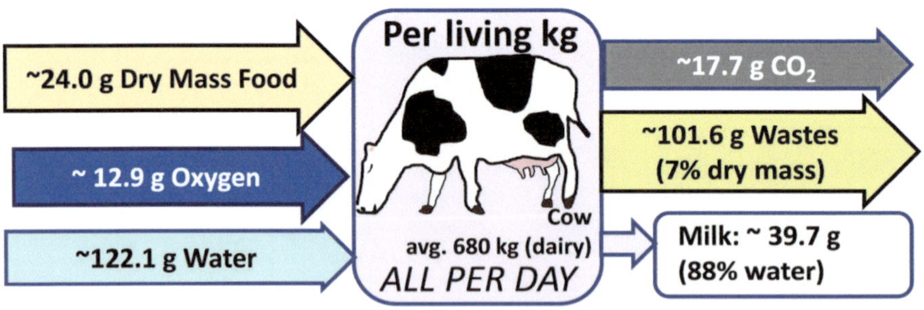

Fig. 3.45 Dairy cattle very rough model mass flow model

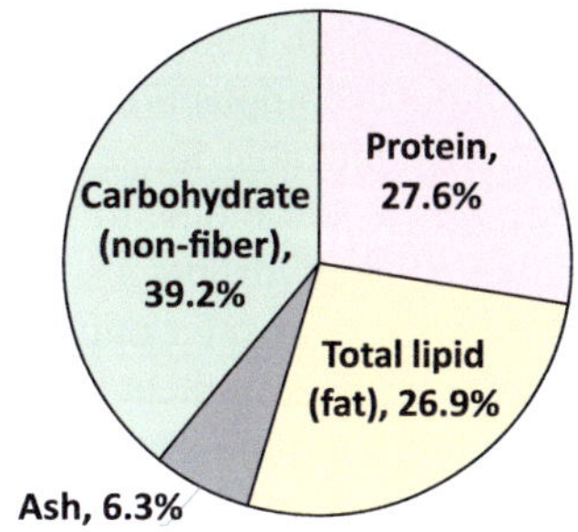

Fig. 3.46 Cow's milk nutrition. Cow's milk is 88% water, 12% dry mass, and contains carbohydrates unlike beef meat

Milk is a more complete food than meat though has more moisture than meat. However, many human genotypes cannot digest milk past early childhood and need to turn the milk into fermented products like yogurt. Milk components are shown in Fig. 3.46.

Given the mass balance of meat cattle, it is more likely settlements will favor dairy first. Given a small dairy, settlements could export milk, cheese, or meat as a very high-end food product if positioned at the right location in a system-wide economic web. For a long time to come beef may be a tissue-engineered product from cellular agriculture and printing, in lieu of from living cattle. Ditto for milk, though any cattle holding settlement will have dairy cows a very long time before meat cattle.

All the mammals in a space settlement, however, need not be either humans or crops. There is a third category: pets. Yes, space kids (or adults who are kids at heart) can have space puppies! No, they are not a crop. They are modeled however as a partial person, in the habitat, not in the farm. I would recommend short haired terriers, Chihuahuas, pugs, beagles, toy to mini sized, and the like, modeled as a tenth of a person or as a portion based on mass. As I said before, it is totally worth it. You might not want the space pup near the space rabbits or space goats, however. Kidding aside (pun intended), pups or cats would be good trial cases for mammalian reproduction in low gravity, and a boost for morale.

3.8 Fungi and Bacteria Growth and Production

Fungi and bacteria are like animals in that they take in food, water, and maybe oxygen, add mass in growth and release waste, and carbon dioxide. Some of these crops require a vat-like machine called a bioreactor.

3.8.1 Mushrooms, Morels, and Truffles

Mushrooms, forest fungi, and related morels have been a high protein food source for people for millennia. In fact, many fungi are high value delicacies and could be a low mass, high value crop for export from a settlement. Fungi are also very critical for lignin decomposition in nature. This book uses the word 'mushroom' to represent any hand pickable fungus including morels and the like. While some fungi require a complete soil biome to grow well, some can be grown on substrates like roots or structures in hydroponics. Some fungi, such as 'chicken of the woods' (*Laetiporus sulphureus*) or oyster mushrooms (*Pleurotus ostreatus*), are also expert at fostering a bacterial environment for cellulose and lignin digestion at a slow rate, in addition to adding a protein source (i.e. working as a plant waste to protein converter). Work by Dr. Jane Shevtsov (Shevtsov, 2021) under a 2021 NASA funded effort has shown how oyster mushrooms can chew up lignin and other materials to make good soil and food biomass. Figure 3.47 shows two wild examples of tree digesting fungi.

Farm candidates should be human edible, lignin digesting, grow well on crop wastes, and reproduce easily, and four candidates are in Table 3.8. Data in Table 3.8. and Fig. 3.48 is loosely calculated and estimated from values in these sources: (Imtiaj et al., 2008; Girmay et al., 2016; Kurtz, 2020; Nguyen & Ranamukhaarachchi, 2019, 2020, 2021; Hu et al., 2001; USDAARS, 2022).

As for plants, fiber, or particle substrates such as sponges or pebble beds can be fed oxygenated nutrient rich water, mixed with a sludge load from processed black water or macerated plant wastes. Alternatively, layers of ground crop wastes work as a soil like substrate. The differences between plant hydroponics and mushroom production are that the mushrooms need oxygen like an animal and use the nutrients

Fig. 3.47 Two lignin digesting wild fungi. Trametes is on the left and Coprinellus is on the right. Photo by author

Table 3.8 Sample mushroom space farm species. Note that growth estimates are highly variable and condition dependent. See Nguyen & Ranamukhaarachchi (2019, 2020, 2021) and Hu et al. (2001)

Mushroom species	Avg. days to harvest*
Lion's Mane (*Hericium erinaceus*)	52
Shitake (*Lentinula edodes*)	20
Oyster (*Pleurotus ostreatus*)	35
White Button (*Agaricus bisporus* et. al.)	42

* highly variable and rough

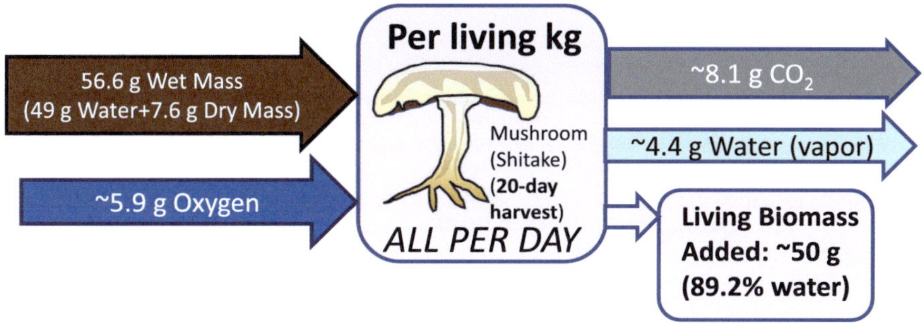

Fig. 3.48 Mass balance flow model for Shitake mushrooms (*Lentinula edodes*)

in the substrate (from bacteria that help process the nutrient rich water in the substrate bed) like food. Mushrooms also grow best in the dark, so do not need light. On Earth, mushrooms are commonly grown in porous bags with plant wastes like sawdust, or shallow mulch beds mixed with plant wastes and worm casings. As a result, the mass balance for mushrooms is just like that for animals, with the nutrients (really wastes) in the water as the food. Mushrooms take in oxygen and release carbon dioxide, but with very little in wastes out, and higher percentage than animals as growth. In the model below, fungi are aerobic, taking in a sludge or mulch bed of wet mass that roughly 13% dry mass, and oxygen gas in a roughly Earth atmospheric mix, though ideally richer in oxygen. The mushrooms release carbon dioxide, water vapor, and more mushroom as living wet biomass, with a trace of waste biomass. Respiration and growth estimated roughly from (Hu et al., 2001) and shown in Figs. 3.48, 3.49, and 3.50.

The more efficient the mechanism for getting wastes as food to the fungi, the more almost totally edible fungi mass is produced, with little waste headed out from the grow area. Most waste will be ammonia or methane in vapor or liquid form, though ideally very little of either. A healthy fungus bed usually also has a health bacteria biome which will also be aerobic and take in a very small fraction of the oxygen. This is true especially if the bed contains cellulose that is pre-digested by the bacteria or

Fig. 3.49 Wet mass breakout for oyster mushrooms. Like most mushrooms, oyster mushrooms are roughly 89% water, with a roughly 2.5:1 carbohydrate to protein ratio, and very low fats. Data from (USDAARS, 2022)

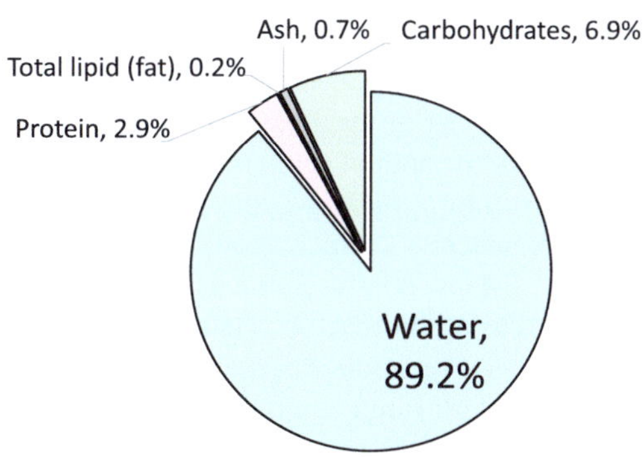

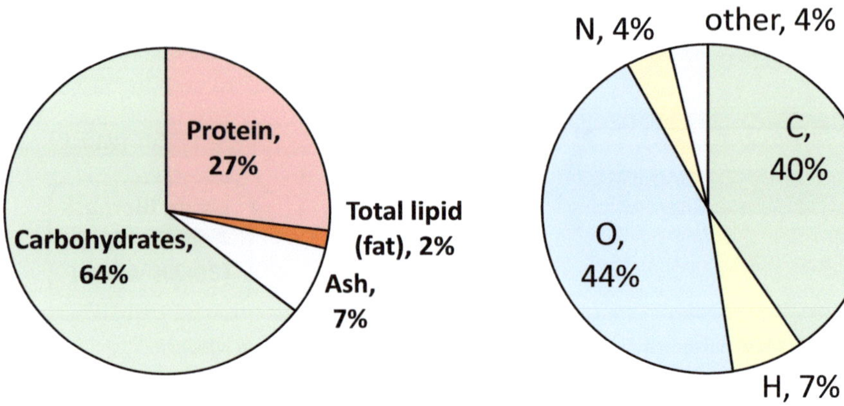

Fig. 3.50 Dry mass components for oyster mushrooms. Mushrooms are rich in minerals (seen as 'Ash'), including potassium. Dry mass by nutritional component on the left, while elemental breakout of dry mass is on the right

worms to make biomass available to the mushroom's root-like mycelium. In practice, getting the right ratio of nutrients to the mushrooms without deoxygenating, or over carbonifying, the air around the mushrooms is complex. Too much nutrient and bacteria will pollute the air and destroy the chemical balance in the substrate. Not enough nutrients and the mushrooms grow too slow or starve and stunt. Farms will require sensors and flow controls, in addition to temperature, humidity, and gas level control to make a mushroom farm stage (or mushroom bioreactor) work. It may also be possible to add the lignin digesting functions of mushrooms using genetic insertions into the genome of bacteria or yeasts, which can be grown in a bioreactor, making the aerobic bioreactor more effective. Note: Yeasts are also fungi but covered later.

3.8.2 Cellular Agriculture: Advanced Biomachinery

Bioreactors are machines that use living cells or organisms to produce a product. Nature has a variety of bioreactors; all humans are doing is making them complex and customized in machines to work for us. In this section we specifically cover bioreactors for animal cells, bacteria, and yeasts. Another name for a simple bioreactor is a culture, such as a petri dish or vat. There are also special bioreactors called 'photobioreactors', which grow algae cells or cyanobacteria, but those are in later chapters. Many pharmaceuticals and pesticides, and other chemicals, are already produced using bioreactors. Healthy soil is a very complex, though more haphazard bioreactor. As mentioned earlier, soil on Earth is a loose format aerobic and anaerobic bioreactor with fungi and bacteria. In your body, intestines are a high surface area membrane bioreactor. Membrane bioreactors are a special case of substrate bioreactors. Your intestinal walls are lined with a whole biome of many species of bacteria including *E. Coli*, that help digest your food, making complex biomolecules available to your bloodstream. Maybe best of all, fermentation vats for wine and beer are also bioreactors. (A note for the whiskey and vodka crowd: distilled spirits are distilled from a wine or beer technically). Cheeses are a bioreactor product (the cheese block is the bioreactor), as is leavened bread (the dough is the reactor, beer being liquid bread with hops). Less appetizing bioreactor examples are septic tanks or sewage treatment plants. Really complex bioreactors even grow animal cells. Therefore, it is not a shock that one of the prime ways to close the mass loop in a space farm is a bioreactor. While there are methods such as incinerators to directly process wastes to get carbon or carbon dioxide (covered in other sections), your author assumes a settlement will want the flexibility of a complex well controlled bioreactor. This bioreactor array will be a combination of reactor modules for both waste management and as a Swiss Army Knife farm balancer. Tank and membrane based aerobic and anaerobic bioreactors, assembled into a complex set of machines, can tailor the biochemistry of the farm to meet various needs. One set of reactors may be used in early settlements to produce nutritious yeast extracts, and likely another set for wines and beers. An early export from a settlement may indeed be a local vintage or distilled product served at a space tavern as proposed by Ron Kohl (a mentor to your author). Factories using bioreactors may produce specialized medical products as well. This section is focused on the biochemistry and biology of bioreactors, though much further below is the engineering of such machines.

For completeness, it is worth noting that a space settler in a small early settlement or spaceship may not have to completely forgo mammal meats or milk products. There are some exceptionally complex bioreactors married with three dimensional printers. It is possible to culture animal cells to produce additional cells for food, feeding the culture direct nutrients. A review of the processes and methods is in (Warner, 2019). The core machinery is series of membrane or substrates that hold mother cells, these cells are then transferred to growth media where they grow and are fed a solution of nutrients, water, and oxygen, again a culture membrane, substrate, or thick liquid. While very promising to vary diet using animal products, there are tradeoffs. One is that some type of processing will be required to make an appetizing result. More critically, cells are very fragile compared to whole animals. Detailed control is required to avoid contamination, and to keep the cells growing, and more critically, reproducing. Animal cells often have limits to replication cycles, so genetics are used to trick the cells into long term replication. Incredibly careful control of temperature, input chemistry, and radiation are needed to continue growth in cells that usually are enclosed in a depth of tissue.

Advances in printing, and in cell culture have allowed growth of cells in culture into muscle tissue, and it is likely the same technology could be used to have yeasts or other cells produce milk or near-milk and near-eggs. While not as culinary as the products from live animals, the result of this work will mature to a level where direct printing and bioreactors will combine to allow ground meat or meat-like products, steaks, omelets, and cheeses to appear on menus, at the cost of extra energy. It would also be possible to use yeasts or bacteria directly in lieu of animal cells to do the same artificial protein construction for meats, eggs, and milks at a future juncture, making many dishes (or near-dishes) possible from totally non-animal sources.

Further along the technological arc of bioreactors and biomachines are direct assemblers. It is possible to directly convert raw molecule inputs to make sugars and amino acids, using electricity and various chemical pathways. The two most examined are nitrogen gas, water, and carbon dioxide to amino acids, and carbon dioxide and water to sugars. There are places on Earth now that make vodka from carbon dioxide and water using electrolysis and a chemical substrate. As a side note a catalytic converter on a gas-powered car uses a similar process. The result of any chemosynthesis would have to be printed into edible materials, but at some future point these technologies might make raw materials to supplement deficiencies or to directly recycle wastes or precursors into food. I recommend if you want to head this route, to get a chemical engineering book. This text mostly uses the machines provided by nature, i.e. living organisms, but do not let that limit your imagination.

Given a chemical supply, one could of course just use a vat of enzymes to make whatever. This requires some industry to exist to make the enzymes of course. Various chemical factories rely on genetically tailored yeasts and bacteria to create a wide variety of organic chemicals, including enzymes. For example, commercial septic tank cleaners use a series of enzymes and bacteria tailored to digest cellulose from bathroom tissue and toilet paper missed by the anaerobic bacteria in underground tanks. This digestion is just like burning, and the result is either carbon dioxide and water, or methane that is then burned or used in a fuel cell to get the carbon dioxide and water. Of course, methane is rocket fuel too. Soil fungi (*Fusarium oxysporum* with *Aspergillus niger*) that make the enzyme cellulase are grown in vats in factories under tailored conditions fed by plant wastes, and the enzyme extracted from the fungi and concentrated. This same process, minus extraction, is one way to handle cellulose breakdown in the space settlement, where a bioreactor that is either a vat and/or a substrate-based reactor (including membrane based) is used to make the process of breaking down cellulose efficient. It might also be possible to engineer bacteria or fungi to produce lignin eating enzymes. Settlements need that carbon dioxide and water to close the mass flow cycle to grow dinner, and bioreactors are just a way of making natural processes in soils and marshes efficient and compact. The next section describes how and what makes a bioreactor a useful tool for the space farm.

3.8.3 Cellular Agriculture: Anaerobic and Aerobic Yeasts and Bacteria in Bioreactors

Yeasts here are an example fungus, but you can use others. Yeasts are microscopic fungi and can either take in oxygen for aerobic metabolism of sugars, or directly use sugars in anaerobic (anaerobic here means without oxygen) metabolism. Any other biomolecule must be broken into sugars for the yeasts to do their work, though that can occur too. Aerobically, yeasts operate like animals, taking in oxygen and sugars and releasing carbon dioxide and water. However, anaerobically, yeast can take the sugars and break them down to alcohol and water, as in wine and beer production. FYI, taking wheat pizza crust dough, and putting it in a plastic bin near an oven for half a day, then flipping the dough over, will almost knock over a cook with a burst of buzz inducing ethanol gas. In either case, the yeasts are like you in that metabolism releases heat. In fact, heat rejection from metabolism is a concern for the farm. In the process of metabolism, yeast also take in nutrients such as sulfur and nitrogen to make more yeast, which can themselves be consumed as a high protein food. The calculations below for yeasts very loosely use data from these sources as a starter: (White, 1952; Zakhartsev & Reuss, 2018; Haddad & Lindegren, 1953; Vieira et al., 2013; EPA-AP42, 2021) and nutrition data from (USDAARS, 2022).

By default, Baker's Yeast has the nutrition breakout as in Fig. 3.51 and by element in Fig. 3.52, though genetic manipulation may lead to a variety of results.

Looking at yeasts acting without reproduction (possible with manipulation), in the aerobic case, yeast take in sugars (biomass input) dissolved in the water around the yeasts, then take in oxygen dissolved in the same water, and make energy which

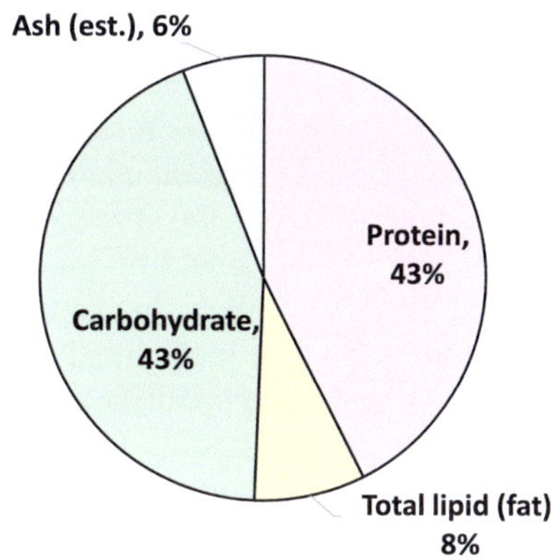

Fig. 3.51 Yeast nutrition in dry mass

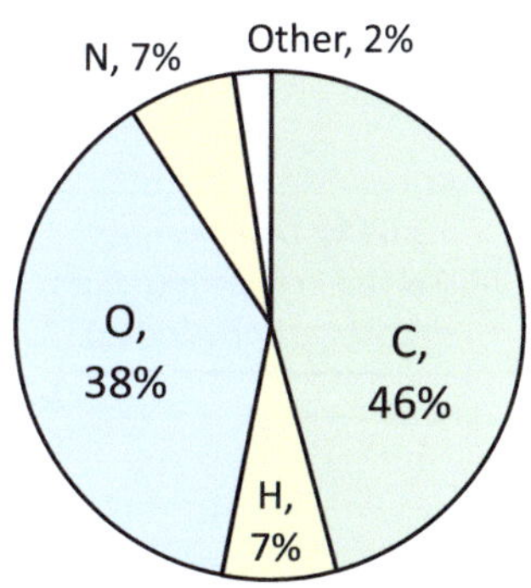

Fig. 3.52 Element breakout for yeast dry mass

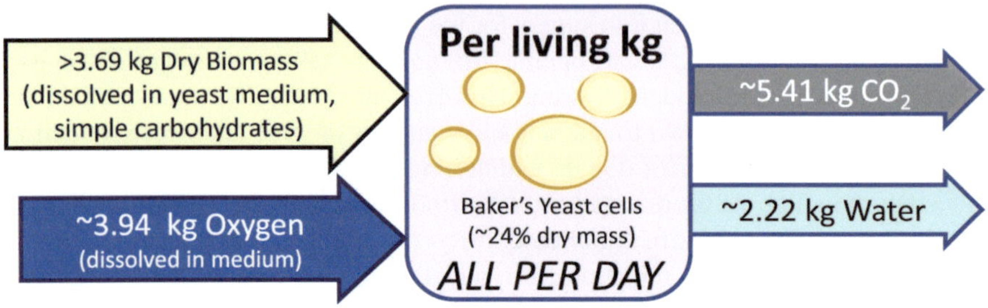

Fig. 3.53 Aerobic yeast in example conditions. Notional model

heats the batch, resulting in carbon dioxide and water as outputs. Yeast can metabolize sugars faster or slower based on how well the sugars and oxygen are mixed into the fluid around the yeast cells, and the temperature and pH of the fluids. When yeasts use oxygen, they produce carbon dioxide (often seen as bubbles) and water, in addition to heat. Carbon dioxide production is very loosely from (Vieira et al., 2013). This results in the mass flow in Fig. 3.53.

Yeasts can also use sugars without oxygen, producing alcohol and carbon dioxide, and a little bit of heat, but the alcohol eventually accumulates around the yeasts, slowing them down. In most cases, bread yeasts stop metabolizing around 11% alcohol content, while brewer's yeasts can tolerate higher alcohol levels (personally seen a batch get to 19% for mead) but metabolize much slower as the alcohol accumulates. Assuming very good conditions, and no reproduction, yeasts produce alcohol and carbon dioxide in the absence of oxygen. Eventually the alcohol suppresses the yeasts. The alcohol could be produced by design (think beer, wine) or be further burned in a fuel cell or incinerator to get more carbon dioxide and water. The anaerobic yeast case looks like Fig. 3.54.

Like yeasts, bacteria can metabolize with or without oxygen, but bacteria are more like biochemical multitools. Both natural and genetic modification can tailor bacteria to produce a complex array of products given inputs. Yeasts can be manipulated also, and yeasts are often model organisms for genetic modification. Further, like yeasts, bacteria can be grown in films bound to membranes.

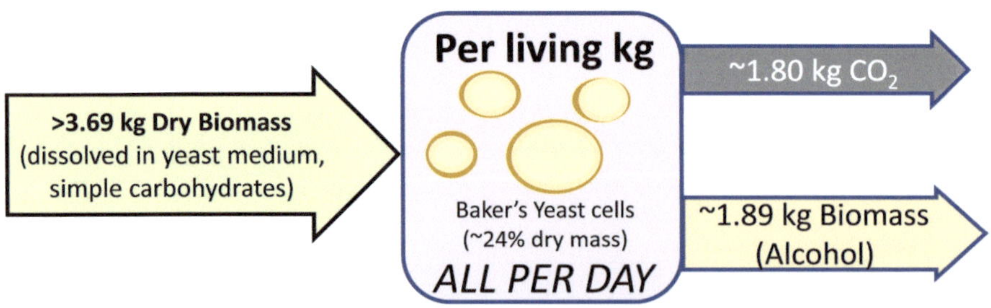

Fig. 3.54 Anaerobic yeast metabolism without reproduction. Notional model

Why membranes? Membrane confined yeasts and bacteria can be well controlled, but more importantly, membranes can be rolled up tightly to provide lots of surface area in a smaller volume and move chemicals to the cells efficiently. Nature invented membrane bioreactors, we just borrowed the method. More surface area means faster and more complete chemical reactions in the yeast and bacteria. Using drums or other tank shapes with membranes a bioreactor can have different tanks with different purposes, and different conditions, to efficiently produce different things in a small volume. One set might be optimized with bacteria to chew up cellulose to make digestible carbohydrates, while another might be optimized to take carbohydrates and oxygen to make carbon dioxide, or another to make beer. The bioreactor array will have very complex software and control mechanisms, such as pumps, heaters, and chillers, with complex sensors to measure temperatures, cell populations, and chemistries, with all the tanks, which at first will have to be made on Earth and shipped to the settlement. Back-up cultures will have to be maintained outside the main bioreactors in case of contamination, and sterilization of mass into the bioreactors, and between parts of the bioreactor, will be critical.

There are two ways to look at the mass flow for yeasts: in one way, let's call it the continuous model, as the starter culture makes more yeasts, they get flushed for harvest; but in another way, the vat model, the yeast is in a vat or container where the harvest is once a day (or periodic). The big difference is that in the continuous model, the harvested yeast doesn't get fed, so the mass flow is more straightforward, but in the vat model, the mass flow is time dependent and exponential up to harvest.

The continuous model bioreactor, see Fig. 3.55, has a continuous flow of oxygenated and feed biomass (mostly sugar) solution. The water and biomass inputs go to a mixer that then is fed to an aerator that adds dissolved oxygen. This oxygenated solution is then pumped across the membrane that holds yeast (or bacteria) cells. Any cells produced are filtered out, with either a sieve filter, settling column, or centrifuge. The remaining fluid goes to a gas separator (again a vacuum column or settling column) that extracts any dissolved gasses, for example carbon dioxide, and the water and any leftover feed biomass cycles back through the system. Ideally a second filter pulls out any alcohol, and coolers or heaters maintain the fluid at the temperature to control how the cells act. Real world membrane bioreactors use high surface area shapes like waves or spirals, but all the machinery shown in Fig. 3.55 exists in factories today.

Vats are containers with stirrers (agitators), aerators (to inject oxygen), thermal control, and input pipes for water and biomass (mostly sugars), as shown in simplified form in Fig. 3.56. Vats need to be far larger than a membrane system since they hold living multiplying cells. An interesting trait is that the living cells also need to be fed to metabolize and make more cells, so the input to a vat is far larger, based upon doubling rate of yeasts (or bacteria) in the vat. Gas bubbles up through the funnel at the top, while oxygen is injected in aerators at the bottom of the vat, stirred by an agitator to assure mixing. Periodically, the accumulated biomass of cells can settle out for collection in the funnel below (and any excess fluid is returned to the vat). In microgravity, the shape depicted in Fig. 3.56 does not work and instead the continuous case model above may work better.

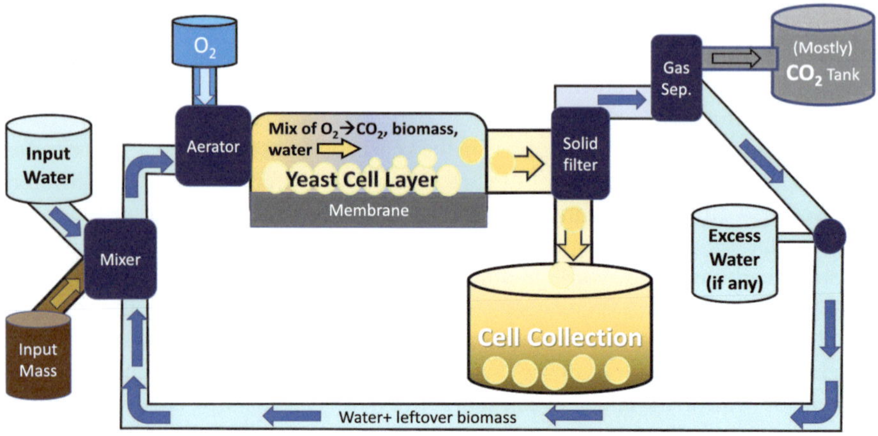

Fig. 3.55 Simplified continuous model (substrate) bioreactor

Fig. 3.56 Simple (gravity) vat fermenter

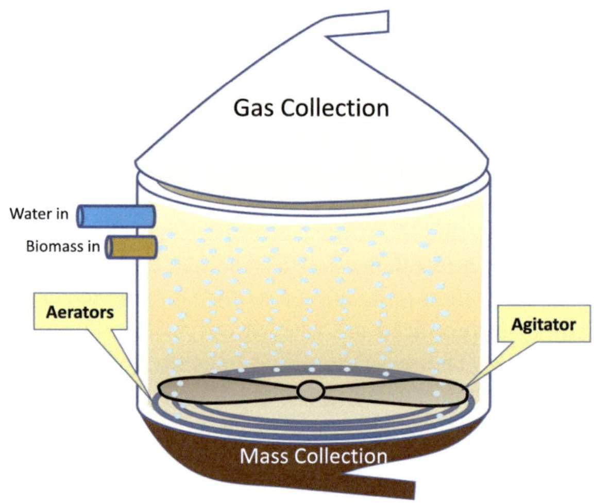

To look at the difference between models, let's assume we use yeast with a 4-h doubling rate. That means the yeast cell count doubles every 4 h, and we start both with a 1 kgwet mass of yeast. For the vat, that means for this doubling rate the container must be eight times bigger than the continuously flushed bioreactor (i.e., at least a liter for all the input water mass and wet mass produced, plus the starter wet mass of 1 kg), but the vat produces over ten times the yeast wet mass. For the continuously flushed bioreactor to produce the same wet mass, it would have to be far larger than the size of the vat. Sounds like a bad tradeoff and that you might always want a vat, right? Wrong. The tradeoff really hinges on the doubling rate, i.e. how quickly each set of cells doubles. Change the doubling rate, and the choice between vat and continuous membrane changes as well. A further consideration is gravity. Vats by default require settling to collect results, and in low gravity settling takes far longer to occur. As a result, a vat on the moon might need to be taller, and might require filters or centrifuges as part of a recirculating flow to collect solids.

For the four-hour doubling rate, the notional model, we have Figs. 3.57 and 3.58. For the first figure, the new cells are pulled out, so do not need to be fed metabolites, but the culture cells do need dry mass inputs and water to make these new cells.

In the next figure, the vat model, most of the mass input goes to feeding new cells (i.e., for metabolism for cells beyond the initial 1 kg starter), and some of the input includes water to inflate new cells, even given the fact all cells release water during metabolism.

Remember that the water made in metabolism ends up in cells in the output biomass in both cases. Note also, that for the vat case, we must feed the growing and doubling yeast cells and provide more mass for these new cells to reproduce. Most of the inputs go to the new cells to feed and reproduce. The new cells make most of the water they incorporate, but still need water inputs to inflate new cells. However, if the doubling rate is 5.45 h for this example yeast, the water made by the cells balances the water made into cells, resulting in Fig. 3.59.

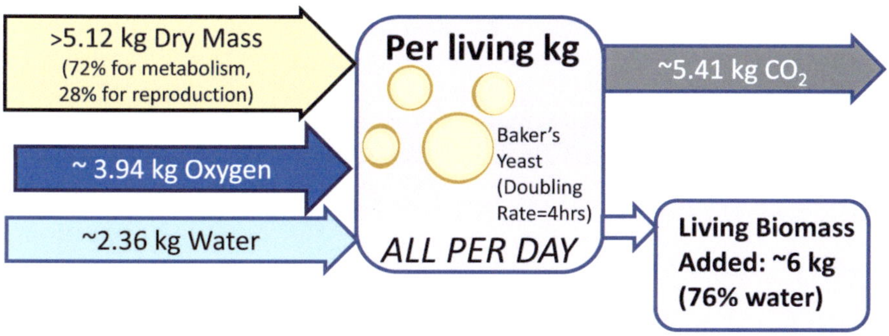

Fig. 3.57 Continuous model 4-h doubling rate. Very rough and notional

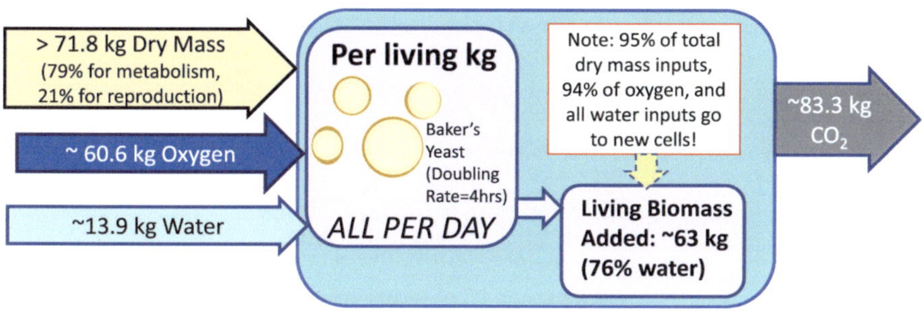

Fig. 3.58 Vat model 4-h doubling rate. Same rough notional yeast

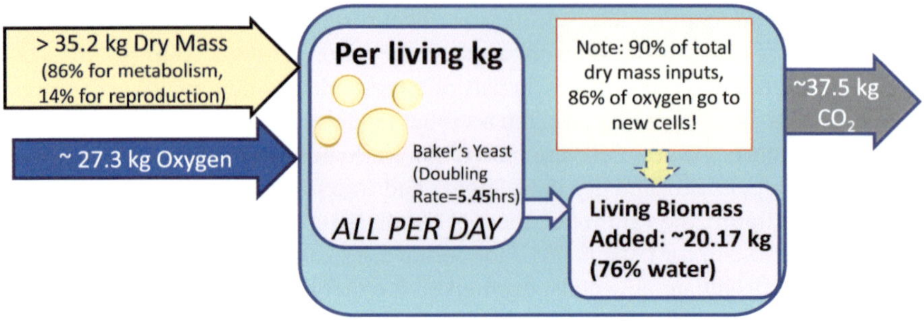

Fig. 3.59 Water balanced vat example. 5.45 h doubling rate, using same notional model

In these scenarios, the water made by the cells in the vat during metabolism balance the need for water to inflate the newly made cells. As for the 4-h vat case above, most inputs go to feed new cells beyond the initial 1 kg starter culture.

We can look at what happens for a variety of doubling rates in a graph. In Fig. 3.60, negative values indicate the need to input water for new cells, but positive values mean that the water made during metabolism of all the cells is more than the water need to inflate new cells. Note that the 'water out-V' line crosses at 5.45 h doubling rate. For the model yeast in these scenarios, the continuous model crosses at roughly 8 h doubling rate.

Now what if we just wanted to turn input wet mass into water and carbon dioxide to close the farm's mass loop? The continuously flushing model is what you need. If the yeasts are chemically bound to a membrane, they can be manipulated to put

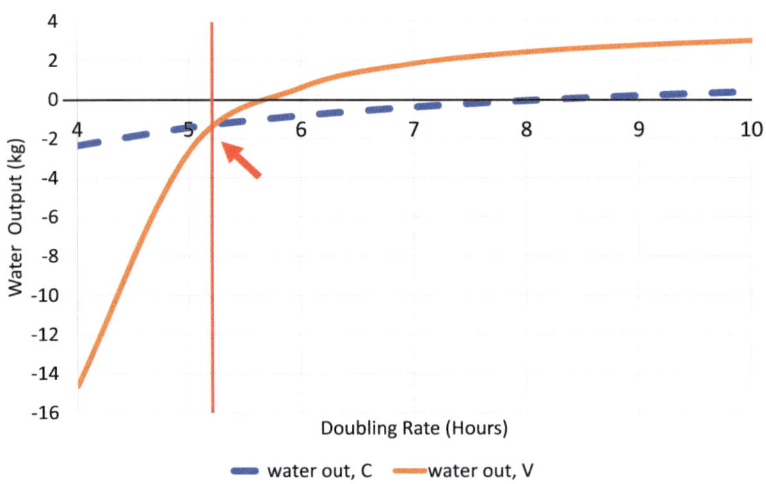

Fig. 3.60 Water balanced by doubling rate (notional model). 'C' refers to the Continuous model, 'V' for the Vat model

all their energy into metabolism, and none into reproduction. Even better, a set of vats feed a set of various membrane bound cultures of bacteria and yeasts, where the sequence and fluid flow, with other chemical and temperature conditions set in the reactors, determine desired outputs based on chosen inputs. These complex structures are very much like the chemical and pharmaceutical factories on Earth, just in miniature. We cover these far below. Such a complex bioreactor becomes the ultimate flex point for balancing the space farm and making products needed by the settlers.

Fungi are not the only thing a farmer can grow in a bioreactor. Bioreactors and be vat or substrate designs to grow bacteria, and mixed colonies of bacteria and fungi. Photobioreactors, the bioreactor for growing light using organisms, is covered in Chap. 4.

3.9 Summary

Many crops are organisms using aerobic metabolism that rely on taking in oxygen, feed, and water, and output living mass, wastes, and carbon dioxide. These can include animal crops such as fish, shrimp, mollusks, insects, birds, and mammals. This model is also followed by many larger fungi such as mushrooms, which are critical for digesting materials from plant crop waste or other wastes. Microscopic fungi, such as yeasts, and bacteria, can be used for a variety of chemical processing using aerobic and/or anaerobic metabolism, including the production of yeast biomass as a food item, or the production of alcohol. For microscopic yeasts and bacteria, growth is in a bioreactor. Bioreactors are machines that foster growth of cellular or microscopic organisms, and can be simple open vats, or membrane mediated designs. For cellular crops, doubling rate is key.

3.10 Questions

1. Assuming no harvest, if a group of animals takes in 10 kg of oxygen, 50 kg of water, and 50 kg of feed, then releases 13 kg of carbon dioxide and 47 kg of wastes, how much mass is retained in growth?
2. If you had to pick between a seasonal breeding fish and a continuous breeding fish for a food crop, which would you choose for the space farm, and what are trade-offs? What about a carnivorous fish like tuna, versus a herbivorous fish like a silver carp?
3. Why is it important to consider eating the whole fish, versus just the fillets?
4. Why is pH, water hardness, and ammonia control important for crustaceans?
5. Why are invasive species on Earth good candidates for space farms?
6. Given the mollusk table in the chapter, which would you select as a crop species: Asian clam or Octopus? Why? Which of these has alternate uses in the farm beyond food?
7. Is polyculture a likely choice for space aquaponics?

8. Beyond food, what other roles could insects play in the space farm?
9. What is the timeframe for the tradeoff of eggs versus meat from chickens? After two years, assuming the settlement has chickens, is egg laying or meat production more critical?
10. Why is water management (due to metabolism) critical for mammal crops (and humans)?
11. What are key tradeoffs when considering rodents versus crustaceans?
12. What tradeoffs occur between milk goats versus dairy cattle?
13. What is a key role of mushroom-like fungi in building soil? What is the critical waste compound that mushrooms can digest well?
14. What is a bioreactor? What is the difference between a vat bioreactor and a substrate (membrane mediated) bioreactor? Describe the role of doubling rate is selecting a vat versus other types of bioreactors? What about in low gravity?
15. Why is having a bioreactor using bacteria and/or yeasts critical for balancing the space farm?

References

AGF. (2022). Meat Goats, 2022 © The American Goat Federation, 2022. https://americangoatfederation.org/breeds-of-goats-2/meat-goats/

Anderson. (2014). Bulletin #1044, Backyard Production of Meat Rabbits in Maine, Adapted by Gary Anderson, 2014, University of Maine, Orono, Maine. https://extension.umaine.edu/publications/1044e/

Antani, K., & Burgeson, A. (2011). Blattella germanica (Online), Animal Diversity Web. Accessed March 16, 2022. https://animaldiversity.org/accounts/Blattella_germanica/

Baird, S. (2015). Can anyone stomach this orange-toothed giant rodent? JAN 14, 2015, © Modern Farmer Media, 2022. https://modernfarmer.com/2015/01/can-anyone-stomach-orange-toothed-giant-rodent/

Barkley, M. E., Knoll, K., Kime, L. F., & Harper, J. K. (2005). Meat goat production. Agricultural alternatives, UPDATED: JUNE 20, 2005, © 2022 College of Agricultural Sciences, Pennsylvania State University.

Benson, A. J., Richerson, M. M., Maynard, E., Larson, J., Fusaro, A., Bogdanoff, A. K., Neilson, M. E., & Elgin, A. (2022). Dreissena rostriformis bugensis (Andrusov, 1897): U.S. Geological Survey, Nonindigenous Aquatic Species Database, Gainesville, FL, https://nas.er.usgs.gov/queries/factsheet.aspx?speciesid=95, Revision Date: 9/13/2019, Accessed March 15, 2022.

Bernard, J. K., & Montgomery, M. J. (2002). Managing intake of lactating dairy cows, PB 1598, The University of Tennessee, Agricultural Extension Service, PB1598-500-10/97 E12-2015-00-078-098, 12/16/97, updated 2002, 2022. https://extension.tennessee.edu/publications/Documents/pb1598.pdf

Biggs. (2022a). Six milestones of chicken growth stages, patrick biggs, Ph.D., © 2022 Purina Animal Nutrition LLC. https://www.purinamills.com/chicken-feed/education/detail/six-milestones-of-chicken-growth-stages

Biggs. (2022b). How long do chickens lay eggs? Goals for Laying Hens, Patrick Biggs, Ph.D., © 2022 Purina Animal Nutrition LLC. https://www.purinamills.com/chicken-feed/education/detail/how-long-do-chickens-lay-eggs-goals-for-laying-hen

Brandon, A. M., El Abbadi, S. H., Ibekwe, U. A., Cho, Y. M., Wu, W. M., & Craig S. (2020). Criddle Fate of Hexabromocyclododecane (HBCD), a common flame retardant, in polystyrene-degrading mealworms: elevated HBCD levels in egested polymer but no bioaccumulation. *Environmental Science & Technology, 54*(1), 364–371. https://doi.org/10.1021/acs.est.9b06501. https://pubs.acs.org/doi/abs/10.1021/acs.est.9b06501

Brown-Uddenberg, R. C., Garcia, G. W., Baptiste, Q. S., Counand, T., Adogwa, A. O., & Sampson, T. (2004). The Agouti [*Dasyprocta leporina*, D. aguti] Booklet and Producers' Manual, First Edition, GWG Publications, Trinidad and Tobago, West Indies, ISBN 976-95123-0-3, Copyright © 2004 by Gary Wayne Garcia & Roxann Brown-Uddenber.

Bolla, G. (2007). Squab raising (PDF). New South Wales Department of Primary Industries. Retrieved 2024.

Boler, D. D. (2014). *Species of meat animals | Pigs, Editor(s): Michael Dikeman, Carrick Devine, Encyclopedia of Meat Sciences* (2nd ed., pp. 363–368), ISBN 9780123847348. Academic Press. https://doi.org/10.1016/B978-0-12-384731-7.00079-9. https://www.sciencedirect.com/science/article/pii/B9780123847317000799

BSD. (2022). What are the guidelines for culturing brine shrimp? Brine Shrimp Direct, P.O. Box 3044, Ogden, Utah 84409-3044, 2022. https://www.brineshrimpdirect.com/about-us/frequently-asked-questions/what-guidelines-culturing-brine-shrimp/

BSTIDNRC. (2009). Board on Science and Technology for International Development National Research Council. Micro-livestock: Little-known Small Animals with a Promising Economic Future 7—Part IV: Rodents. Appropedia. Retrieved March 17, 2022 from https://www.appropedia.org/Micro-livestock:_Little-known_Small_Animals_with_a_Promising_Economic_Future_7

Buckland, R., & Guy, G. (2002). Goose Production. FAO Animal Production and Heath Paper—154, ISSN 0254-6019, Food and Agiculture Organization of the United Nations, Rome, 2002.

Busolutely. (2021). Edible insect nutrition: silkworms, mealworms and crickets, ©2021—Bugsolutely Ltd, 2022. https://www.bugsolutely.com/edible-insect-nutrition-facts/

CAMPBELL, J. A. (2016). Understanding beef carcass yields and losses during processing. Updated: August 4, 2016, © 2022 College of Agricultural Sciences, Pennsylvania State University.

Clauer, P., & Skinner, J. (2007). Raising Waterfowl, A3311. University of Wisconsin factsheet, © 2007, University of Wisconsin. https://learningstore.extension.wisc.edu/

Craddock, F., & Steen, J. (2012). Texas Breeding Goat Guide, Texas AgriLife Extension Service, 5/21/12. http://agrilifecdn.tamu.edu/4hlivestockmentor/files/2012/05/TEXAS-BREEDING-GOAT-GUIDE.pdf

CUDRL. (2022). Hatching Duck Eggs, Cornell University Duck Research Laboratory, Eastport, NY, Cornell University, 2022. https://www.vet.cornell.edu/animal-health-diagnostic-center/programs/duck-research-lab/hatching-duck-eggs

Cummins. (2018). Americans used to eat pigeon all the time—and it could be making a comeback, By Eleanor Cummins|Published February 16, 2018, POPULAR SCIENCE, 2022. https://www.popsci.com/eat-pigeons-squab/

Davis, J. (2020). SDSU Extension Silage/Earlage Calculator, Updated August 31, 2020, South Dakota State University, Brookings, SD, © 2022 South Dakota Board of Regents. https://extension.sdstate.edu/sdsu-extension-silage/earlage-calculator

Eastridge, M. (2007). Guidelines for feeding low forage diets to dairy cows. *Buckeye Dairy News, 9*(3), Ohio Dairy Resourse Industry Center, The Ohio State University, Columbus, Ohio 43210

Eikelenboom, G., Walstra, P., Huiskes, J. H., & Klont, R. E. (2004). *Species of meat animals | Pigs, Editor(s): Werner Klinth Jensen, Encyclopedia of Meat Sciences* (pp. 1284–1291). Elsevier, ISBN 9780124649705. https://doi.org/10.1016/B0-12-464970-X/00109-4 https://www.sciencedirect.com/science/article/pii/B012464970X001094

Ellestad, L. (2020). Maximizing metabolic efficiency of commercial broiler chickens in the absence of antibiotics, UNIVERSITY OF GEORGIA, via National Institute of Food and Agriculture Grant, GEOW-2019-05980, September 1, 2020. https://portal.nifa.usda.gov/web/crisprojectpages/1022973-maximizing-metabolic-efficiency-of-commercial-broiler-chickens-in-the-absence-of-antibiotics.html

EPA-AP42. (2021). Chapter 9 AP-42: Food and Agricultural Industries. In *Compilation of air emissions factors, AP 42*(5th ed., Vol. I). U.S. Environmental Protection Agency, November 29, 2021. https://www.epa.gov/air-emissions-factors-and-quantification/ap-42-fifth-edition-volume-i-chapter-9-food-and-0

FAOAFFRIS-Tilapia. (2022). FAO Fisheries and aquaculture aquaculture feed and fertilizer resources information system, species profile, Nile tilapia. Food And Agriculture Organization of the United Nations, caputured 2022. https://www.fao.org/fishery/affris/species-profiles/nile-tilapia/nile-tilapia-home/en/

FAO-Artemia. (2022). © FAO 2022. *Artemia* spp. Cultured Aquatic Species Information Programme. Text by Van Stappen, G. Fisheries and Aquaculture Division [online]. Rome. Updated 2012-02-28 [Cited Tuesday, March 15th 2022]. https://www.fao.org/fishery/en/culturedspecies/artemia_spp/en"

FAOCAS. (2022). FAO fisheries and aquaculture "search resources factsheets, cultured aquatic species". Food And Agriculture Organization of the United Nations, captured 2022. https://www.fao.org/fishery/en/culturedspecies/search

FAO-Snails. (1986). Better farming series, farming snails, Rome 1986. Food And Agriculture Organization of the United Nations, captured 2022. https://www.fao.org/3/bp069e/bp069e.pdf and https://www.fao.org/3/bp070e/bp070e.pdf

Garcia. (2015). AGLS 6502 Lecture 15—Intensification of Agouti Production, Prof. Gary W. Garcia, Department of Food Production, University of the West Indies, St. Augustine. Updated: 18-Jan-2015. http://ostasp.brinkster.net/unit15agls6502.html

Girmay, Z., Gorems, W., & Birhanu, G., & Zewdie, S. (2016). Growth and yield performance of *Pleurotus ostreatus* (Jacq. Fr.) Kumm (oyster mushroom) on different substrates. *AMB Expr*, *6*(87). https://doi.org/10.1186/s13568-016-0265-1

Głogowski, R., & Panas, M. (2009). Efficiency and proximate composition of meat in male and female nutria (*Myocastor coypus*) in an extensive feeding system. *Meat Science*, *81*(4), 752–754. https://doi.org/10.1016/j.meatsci.2008.11.002

Grau, T., Vilcinskas, A., & Joop, G. (2017). Sustainable farming of the mealworm Tenebrio molitor for the production of food and feed. *Zeitschrift für Naturforschung C*, *72*(9–10), 337–349. https://doi.org/10.1515/znc-2017-0033

Gujral, R. (2009). Emu rearing. *Science tech entrepreneur*, Aug 2009. https://static.vikaspedia.in/media/files_en/agriculture/poultry/emu-rearing.pdf

Haddad, S. A., & Lindegren, C. C. (1953). A method for determining the weight of an individual yeast cell. *Applied Microbiology*, *1*(3), 153–156. https://doi.org/10.1128/am.1.3.153-156.1953. PMID: 13041190; PMCID: PMC1056888

Hamwi, A., & Haskin, H. H. (2008). Oxygen consumption and pumping rates in the hard clam Mercenaria mercenaria: a direct method. *Science*, *163*(3869), 823–824. https://doi.org/10.1126/science.163.3869.823. PMID: 5764478

Harlow. (2016). Homegrown turkey for the holidays. Ivory Harlow-July 8, 2016, Farm and Dairy, © Copyright 2022 - Farm and Dairy. https://www.farmanddairy.com/top-stories/homegrown-turkey-for-the-holidays/346265.html

Harner, J., Murphy, J. P., Boyer, W., Davidson, J., George H., Graber, R., Minson, S., & Harvey, M. (2021). Planning and designing cattle feedlots, MF2316 rev. December 2021, Kansas State University. https://bookstore.ksre.ksu.edu/pubs/MF2316.pdf

Hendersons. (2022). Pig growth rates and feed trough requirements, © Henderson Plastics Ltd, 2022. http://www.hendersons.co.uk/pigequip/Pig_growth_rate.html

Heppenheimer, T. A. (1977). Colonies in space, Copyright 1977, 2007 by T. A. Heppenheimer.

Hicks, C. H., & Pierce, R. A. (2022). Freshwater prawn production in Missouri, University of Missouri Extension. https://extension.missouri.edu/publications/g9471

Hu, W., Uchino, T., Yasunaga, E., Hamanaka, D., & Hori, Y. (2001). Respiration rate of shiitake mushroom in modified atmosphere packages. *IFAC Proceedings*, *34*(28), 189–196, ISSN 1474-6670. https://doi.org/10.1016/S1474-6670(17)32847-1

IBC. (2022). Calculators, Iowa Beef Center, © 1995–2022, Iowa State University of Science and Technology. https://www.iowabeefcenter.org/calculators.html

Imtiaj, A., Jayasinghe, C., Lee, G. W., Shim, M. J., Rho, H. S., Lee, H. S., Hur, H., Lee, M. W., Lee, U. Y., & Lee, T. S. (2008). Vegetative growth of four strains of hericium erinaceus collected from different habitats. *Mycobiology*, *36*(2), 88–92. https://doi.org/10.4489/MYCO.2008.36.2.088. Epub 2008 Jun 30. PMID: 23990739; PMCID: PMC3755240.

Jacob. (2022). Raising Japanese quail in small and backyard flocks. Dr. Jacquie Jacob, and Dr. Brigid McCrea, New Technologies for Agriculture Extension grant no. 2020-41595-30123 from the USDA National Institute of Food and Agriculture, 2022. https://poultry.extension.org/articles/poultry-management/raising-japanese-quail-in-small-and-backyard-flocks/

Jayathilakan, K., Sultana, K., Radhakrishna, K., & Bawa, A. S. (2012). Utilization of byproducts and waste materials from meat, poultry and fish processing industries: a review. *Journal of Food Science and Technology*, *49*(3), 278–293. https://doi.org/10.1007/s13197-011-0290-7

Jones, L. D., Cooper, R. W., & Harding, R. S. (1968). Composition of mealworm tenebrio molitor larvae. In *Proceedings of 19th American associaation of laboratory animal science*, 21–25 Oct 1968, Las Vegas NV.

Jones, C. M., & Heinrichs, J. (2017). Growth charts for dairy heifers. Updated: July 28, 2017, © 2022 College of Agricultural Sciences, Pennsylvania State University. https://extension.psu.edu/growth-charts-for-dairy-heifers

Kapetsky, J. M., & Nath, S. S. (1997). A strategic assessment of the potential for freshwater fish farming in Latin America. COPESCAL Technical Paper. No. 10. Rome, FAO (p. 128).

Kathiresan. (2013). Emu farming—Vikaspedia, © 2006–2019 C–DAC.India Development Gateway (InDG), created by: Kathiresan At 10/28/2013. https://vikaspedia.in/agriculture/poultry/emu-farming

Kayton, S. (2022). Silkworms, 2022. https://www.suekayton.com/Silkworms/whole.htm

Kenzie, O., & Williamson, K. (2022). Feedlot water system. In *Alberta feedlot management guide, elanco anial health, agromedia international, Inc.* 2022. http://www.agromedia.ca/AFMG/Section%203%20Facilities%20&%20Environment/3F1%20Feedlot%20Water%20System.pdf

Kime, L., Harper, J. K. (2008). Agricultural alternatives: rabbit production, © The Pennsylvania State University 2008, College of Agricultural Sciences, Agricultural Research and Cooperative Extension, University Park, PA. https://extension.psu.edu/rabbit-production

Kokoszyński, D. (2017). Chapter 4—Guinea Fowl, Goose, Turkey, Ostrich, and Emu Eggs. In P. Y. Hester (Ed.), *Egg innovations and strategies for improvements* (pp. 33–43), ISBN 9780128008799. Academic Press. https://doi.org/10.1016/B978-0-12-800879-9.00004-4

Kumbhar, S. (2020). Study of seasonal variations in oxygen consumption of estuarine clam, meretrix meretrix (Linnaeus, 1758) after acute exposure of cadmium chloride. *International Journal of Trend in Scientific Research and Development (ijtsrd)*, *4*(3), 525–529, ISSN: 2456-6470. https://www.ijtsrd.com/papers/ijtsrd30543.pdf

Kurtz, L. (2020). How to grow white button mushrooms. Last Updated: December 25, 2020, wikiHow, 2022. https://www.wikihow.com/Grow-White-Button-Mushrooms

Lach, J., & Schwartz, M. (2008). Helix pomatia (On-line), Animal Diversity Web. Accessed March 15, 2022. https://animaldiversity.org/accounts/Helix_pomatia/

Lanier. (2018). Nutria, nutrition and avoiding natural disasters. In K. Lanier (Ed.), *Hobby farms* (blog), © 2022 EG Media Investments LLC. https://www.hobbyfarms.com/nutria-rodent-nutrition-natural-disasters/

Liedel. (2014). The duck egg market takes off. By Emily Liedel Feb 20, 2014, © Modern Farmer Media, 2022. https://modernfarmer.com/2014/02/duck-eggs/

Linn, J., Otterby, D., Howard, W. T., Shaver, R., Hutjens, M., & Kilmer, L. (2022). The ruminant digestive system, U. of MN-Extension, © 2022 Regents of the University of Minnesota. https://extension.umn.edu/dairy-nutrition/ruminant-digestive-system

Luginbuhl, J. M. (2015). Nutritional feeding management of meat goats. NC State Extension Publications, Publication date: October 8, 2015, Revised: September 17, 2020. https://content.ces.ncsu.edu/nutritional-feeding-management-of-meat-goats

Lyn, J. C., Naikkhwah, W., Aksenov, V., & Rollo, C. D. (2011). Influence of two methods of dietary restriction on life history features and aging of the cricket *Acheta domesticus*. *Age (Dordr)*, *33*(4), 509–522. https://doi.org/10.1007/s11357-010-9195-z

Macklin, K. (2022). Nutrition for backyard chicken flocks, ANR-1317, 2022, Copyright © 2024 by the Alabama Cooperative Extension System, 2024. https://www.aces.edu/blog/topics/farming/nutrition-for-backyard-chicken-flocks/

McCune, K. (2022). How much time do you need to raise goats for meat? © 2022 Copyright Family Farm Livestock. https://familyfarmlivestock.com/how-much-time-do-you-need-to-raise-goats-for-meat/

MDC-Nutria. (2022). Field Guide: Nutria (Coypu), Missouri Department of Conservation, Jefferson City, MO, 2022. https://nature.mdc.mo.gov/discover-nature/field-guide/nutria-coypu

Meltzer, A. (1983). Thermoneutral zone and resting metabolic rate of broilers. *British Poultry Science*, *24*(4), 471–476. https://doi.org/10.1080/00071668308416763. PMID: 6667388.

Meyer. (2019). Bio-sustainability: role of aquaculture in space. In B. L. Meyer (Ed.), *National space society (NSS) international space development conference (ISDC)* 2019, presented 7 June 2019, Arlington VA.

Munywoki, G. N. (2022). Snail farming (Heliciculture): Benefits, economic viability, challenges and prospects. *International Journal of Research in Agronomy.*, *5*(2), 09–11. https://doi.org/10.33545/2618060X.2022.v5.i2a.102.ISSN2618-0618

Nguyen, M., & Ranamukhaarachchi, S. (2019). Study on the mycelium growth and primordial formation of king oyster mushroom (*Pleurotus eryngii*) on cardboard and spent coffee ground. *Research on Crops*, *20*, 835–842. https://doi.org/10.31830/2348-7542.2019.124

Nguyen, M., & Ranamukhaarachchi, S. (2020). Effect of different culture media, grain sources and alternate substrates on the mycelial growth of *Pleurotus eryngii* and *Pleurotus ostreatus*. *Pakistan Journal of Biological Sciences*. https://doi.org/10.3923/pjbs.2020

Nguyen, M., & Ranamukhaarachchi, S. (2021). Yield evaluation of king oyster mushroom (*Pleurotus eryngii*) on wheat straw mix substrates. *Research on Crops*, *22*, 224–230. https://doi.org/10.31830/2348-7542.2021.061

NSWDPI. (2022). Weaning beef calves. NSW Department of Primary Industries, 2022. https://www.dpi.nsw.gov.au/animals-and-livestock/beef-cattle/husbandry/general-management/weaning-beef-calves

NTAE. (2019). Goat Nutrition, August 14, 2019, National Cooperative Extension, New Technologies for Agriculture Extension grant no. 2020-41595-30123 from the USDA National Institute of Food and Agriculture. https://goats.extension.org/goat-nutrition/

OAH. (2019). Guinea pig life stages, November 20, 2019, © 2022 Oxbow Animal Health. https://www.oxbowanimalhealth.com/blog/guinea-pig-life-stages

Ortmann, C., & Grieshaber, M. K. (2003). Energy metabolism and valve closure behaviour in the Asian clam Corbicula fluminea. *Journal of Experimental Biology*, *206*(Pt 22), 4167–4178. https://doi.org/10.1242/jeb.00656. PMID: 14555755.

Pace, N., & Smith, A. H. (1984). Gravity, body mass and composition, and metabolic rate. In *NASA. Washington NASA space biology program, eighth annual symposium's program and abstracts*, doc id: 19840010786, February 1, 1984.

Pinchao-Pinchao, Y., Serna-Cock, L., Osorio-Mora, O., & Tirado, D. F. (2024). Guinea pig breeding and its relation to sustainable food security and sovereignty in South America: nutrition, health, and production challenges. *CyTA Journal of Food*, *22*(1). https://doi.org/10.1080/19476337.2024.2392886

Probert, T. (2013). Dairy grazing: fence and water systems, November 2013, University of Missouri, © 1993–2022 Curators of the University of Missouri. https://extension.missouri.edu/publications/m190

Quetin, L. B., & Ross, R. M. (2001). Environmental variability and its impact on the reproductive cycle of Antarctic Krill. *American Zoologist*, *41*(1), 74–89.

Randall. (2006). Raising Japanese Quail, By Maurice Randall, 17 February 2006, © 2000–2022—Global AG Media. https://thepoultrysite.com/articles/raising-japanese-quail

Rao, N. S. (2004). A study on the performance of emu (*Dromaius novaehollandiae*) in Andhra Pradesh. In *MVSc thesis submitted to the Acharya N.G. Ranga Agricultural University* (pp. 1–62), Hyderabad.

Redfearn, D. D., & Bidwell, T. G. (2017). Stocking rate: the key to successful livestock production, Published Feb. 2017, Id: PSS-2871, © 2022 Oklahoma State University, 2022. https://extension.okstate.edu/fact-sheets/stocking-rate-the-key-to-successful-livestock-production.html

Renault, D., Hervant, F., & Vernon, P. (2003). Effect of food shortage and temperature on oxygen consumption in the lesser mealworm, *Alphitobius diaperinus* (Panzer) (Coleoptera: Tenebrionidae). *Physiological Entomology*, *28*(4), 261–267, © 2022 Royal Entomological Society—Registered charity no. 213620. https://doi.org/10.1111/j.1365-3032.2003.00338.x

RIRDC. (2001). Breeding and growing snails commercially in Australia, a report for the rural industries research and development corporation. In Murphy, B., RIRDC Project No. ARH-1A, RIRDC Publication No. 00/188 (c) 2001 Rural Industries Research and Development Corporation, ISBN 0 642 58219 X, ISSN 1440-6845

Robertson, H. A. (1988). Relationships between body weight, egg weight, and clutch size in pigeons and doves (Aves: Columbiformes). *Journal of Zoology*, *215*(2):217–229, First published: June 1988. https://doi.org/10.1111/j.1469-7998.1988.tb04896.x

Rounds, W., & Herd, D. B. (2022). Cow's digestive system. Beef Skillathon, B-1575, © 2022 Texas A&M University. https://beefskillathon.tamu.edu/cows-digestive-system/

Sewell, H. B., Jacobs, V. E., & Gerrish, J. R. (1993). Backgrounding calves part 2: herd health and feeding. Reviewed October 1993, MU Extention, © 1993 to 2022 Curators of the University of Missouri, 2022. https://extension.missouri.edu/publications/g2096

Shevtsov, J. (2021). Making soil for space habitats by seeding asteroids with fungi. Trans Astronautica Corporation, NIAC Project, National Aeronautics and Space Administration (NASA), 2022. https://www.nasa.gov/directorates/spacetech/niac/2021_Phase_I/Making_Soil_for_Space_Habitats/

Soilleux, R. J. (2017). Towards closed environmental control and life support for space habitats Part II: reduced risk and increased efficiency with biological systems. *NSS Space Settlement Journal* (2), 2022. https://space.nss.org/wp-content/uploads/NSS-JOURNAL-Towards-Closed-ECLSS-for-Space-Habitats-Part-II.pdf

Taub, F. B. (1963). Some ecological aspects of space biology. *The American Biology Teacher*, *25*(6), 412–421. https://doi.org/10.2307/4440413

TBS. (2009). Beef cattle nutrient requirements, 02 June 2009, MS State Ext. Service, The Beef Site, © 2000–2020.Global Ag Media. https://www.thebeefsite.com/articles/1974/beef-cattle-nutrient-requirements/

Titgemeyer. (2007). Project report. In E. Titgemeyer (Ed.), P*rotein nutrition of beef cattle*, dir. KANSAS STATE UNIV, funded by National Institute of Food and Agriculture, to Sep 30, 2007. https://portal.nifa.usda.gov/web/crisprojectpages/0188699-protein-nutrition-of-beef-cattle.html

Tůmová, E., Chodová, D., Vlčková, J., Němeček, T., Uhlířová, L., & Skřivanová, V. (2017). Age-related changes in the carcass yield and meat quality of male and female nutrias (*Myocastor coypus*) under intensive production system. *Meat Science, 133*, 51–55. https://doi.org/10.1016/j.meatsci.2017.06.003

Turk, K. L., Morrison, F. B., & Maynard, L. A. (1934). The nutritive value of the proteins of alfalfa hay and clover hay when fed alone and in combination with the proteins of corn. *Journal of Agricultural Research*, *48*(6), Washington, D.C. Mar. 15, 1934, 2022. https://naldc.nal.usda.gov/download/IND43968520/PDF

UFIFAS-fa152 (2021) Weber, K., Hoover, E., Sturmer, L., Baker, S. The role of dissolved oxygen in hard clam aquaculture. University of Florida, Institute of Food and Agricultural Sciences, 2021. https://edis.ifas.ufl.edu/publication/fa152

Ungerfeld, E. M. (2020). Metabolic hydrogen flows in rumen fermentation: principles and possibilities of interventions. *Frontiers in Microbiology 11*: https://doi.org/10.3389/fmicb.2020.00589, ISSN=1664-302X

USDAAPHIS. (2020). Nutria. USDA animal and plant health inspection service, Last Modified: Nov 2, 2020, 2022. https://www.aphis.usda.gov/aphis/ourfocus/wildlifedamage/operational-activities/sa_nutria/ct_nutria

USDAARS. (1949–1573). Poultry—production and value 2018 summary (May 2019), ISSN: 1949-1573, United States Department of Agriculture (USDA) National Agricultural Statistics Service, director: Travis Averill. https://www.nass.usda.gov/Publications/Todays_Reports/reports/plva0519.pdf

USDAARS. (2022). FoodData Central, U.S. DEPARTMENT OF AGRICULTURE (USDA), Agricultural Research Service (ARS), 2022. https://fdc.nal.usda.gov/

Valentine, K. (2017). The amazing story of the cold war space-egg race. National Audubon Society News, December 15, 2017, 2022. https://www.audubon.org/news/the-amazing-story-cold-war-space-egg-race

Valles, S. (2021). Featured creatures common name: German cockroach, scientific name: *Blattella germanica* (Linnaeus) (Insecta: Blattodea: Blattellidae). Publication Number: EENY-2, Publication Date: September 1996. Revised: August 2008. Reviewed: December 2017. Latest Revision: May 2021. (c) University of Florida, Ganesville, FL. https://entnemdept.ufl.edu/creatures/urban/roaches/german.htm

Van Saun, R. J. (2007). Dairy goat nutrition: feeding for two. Penn State University, 4/23/07, © 2007 Penn State University. http://goatdocs.ansci.cornell.edu/Resources/GoatArticles/GoatFeeding/FeedingForTwo.pdf

Van Saun, R. J., Hill, C., Kime, L. F., & Harper, J. K. (2022). Dairy goat production. *Agricultural Alternatives*, updated: February 10, 2022, © 2022 College of Agricultural Sciences, Pennsylvania State University. https://extension.psu.edu/dairy-goat-production

Vieira, É. D., Andrietta Mda, G., & Andrietta, S. R. (2013). Yeast biomass production: a new approach in glucose-limited feeding strategy. *Brazilian Journal of Microbiology*, *44*(2), 551–558. Published October 30, 2013. https://doi.org/10.1590/S1517-83822013000200035

Waldo. (2022). DebMark rabbit education resource, raising rabbits—the basics. Dr. Mark Waldo, 2022. https://debmark.com/rabbits/basics.php

Warner, R. (2019). Review: analysis of the process and drivers for cellular meat production. *Animal*, *13*(12), 3041–3058. https://doi.org/10.1017/S1751731119001897

White, J. (1952). Variation in water content of yeast cells caused by varying temperatures of growth and by other cultural conditions. *Journal of the Institute of Brewing*, *68*, 2022. https://doi.org/10.1002/j.2050-0416.1952.tb02660.x

Xu, S., Wang, F., & Wang, Y. et al. (2019). A silkworm based silk gland bioreactor for high-efficiency production of recombinant human lactoferrin with antibacterial and anti-inflammatory activities. *Journal of Biological Engineering*, *13*, 61. https://doi.org/10.1186/s13036-019-0186-z

Zakhartsev, M., & Reuss, M. (2018). Cell size and morphological properties of yeast *Saccharomyces cerevisiae* in relation to growth temperature. *FEMS Yeast Research*, *18*(6), SN 1567-1356. https://doi.org/10.1093/femsyr/foy052. https://academic.oup.com/femsyr/article/18/6/foy052/4987208

Chapter 4
Plant and Algae Growth and Production

Abstract This chapter covers key dynamics and mass flow for crop plants, algae, and cyanobacteria. Initially the chapter explores the basic concepts for vascular plant biology, including how they grow and the difference between the root and aerial environments. Then the chapter moves to common Earth crops, including grains, fruits, vegetables, and tubers, with mass flow models and growth parameters for each type, and some examination of technologies to grow herbs, bush, and tree crops. Then the chapter moves cyanobacteria and various alga types like kelp and microalgae like chlorella, to lichens and mosses, including multiple growth models and key technologies for growth. At the end of the chapter is discussion of the maturity of space farms and the succession of crops over time.

Keywords Grains · Plants · Vegetables · Algae · Lichen · Photobioreactor

4.1 Introduction

This chapter covers key biochemical dynamics and mass flow for crop plants, algae, and cyanobacteria. Starting with common Earth crops, including grains, fruits, vegetables, and tubers, the chapter moves to various alga types like kelp and microalgae like chlorella, ending with lichens and mosses.

4.2 Vascular Plant Basics

Vascular plants are plants that have roots, stems, and leaves, where fluids flow in tubes (like in veins for animals) called xylem from roots to leaves. Nutrients and water flow from roots up through the stems to holes in the leaves called stomata. Evaporation from the stomata and capillary action act as the force that pulls the water from the roots and can cycle hundreds of times the mass of water over that used to increase mass in plant. Transpiration is how a plant pulls liquids from the

© The Author(s), under exclusive license to Springer Nature Switzerland AG 2026
B. L. Meyer, *Designing Farms in Space*, https://doi.org/10.1007/978-3-032-04395-5_4

roots up to the leaves by a combination of capillary action and evaporation from little pores in each leaf. A plant may cycle hundreds of times the mass of water it incorporates before harvest in transpiration. In other words, many tons of water are evaporated out of leaves for tens of kilograms produced in biomass. Evaporation from the pores causes a lack of water in the leaf, and this causes water to flow in by physics to the drier spot from tubes that run from the leaf to the roots. If you cut a plant stalk, you can see a cross section of these tubes. Carbon dioxide also enters the stomata, so the plant must keep them open to photosynthesize, assuming humidity is not so low as to dry out the plant. The stoma also close when there is not enough light for photosynthesis. Previous chapters covered the basics of plants and photosynthesis, so now let's look a bit more in detail. The green parts (leaves, fruit or flower, stem or shoot), use light, carbon dioxide from the air, and water and nutrients pulled from roots to grow. Roots, however, can be harmed by light, need a very high humid environment of water and nutrients, are aided by bacteria and fungi, and need some oxygen to work well. These environments are shown in Fig. 4.1.

In the leaves, lights hit parts of cells called chloroplasts where carbon dioxide and water are combined in a series of reactions to make sugars (including fructose). Some of the sugars are assembled into cellulose and other molecules to make the plant bigger, while other sugars and molecules made from sugars are sent through tubes (phloem) to roots, where the root cells make starches or other molecules to store energy. A plant must grow to survive. To use photosynthesis, the plant must incorporate the mass of the carbon atoms in carbon dioxide, and some water, at a minimum. The carbon dioxide and light taken in by plants and algae for photosynthesis are the energy sources for metabolism, and when combined with water, result in growth, minus any oxygen released, or water evaporated out to fuel plant transpiration. At night, plants use oxygen to digest sugars for energy at a low level, but the

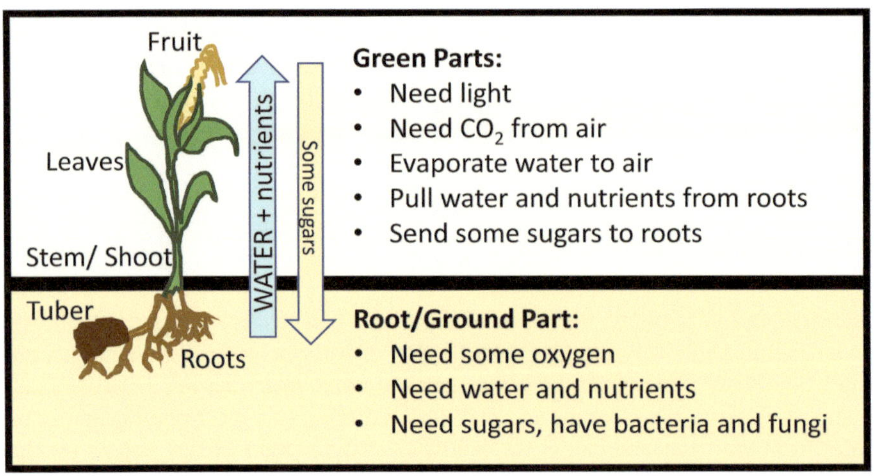

Fig. 4.1 Vascular plant environments

mass of carbons released are far lower than the mass brought in by photosynthesis. The sugars from photosynthesis make the plant grow, even if some of the sugars get used in the dark. More water is absorbed to inflate the growth, combining with sugars and other chemicals. Roots may secrete some sugars to feed the bacteria and fungi around the roots which may fix nitrogen for the plant to use or bind other nitrates for the plant to make amino acids, but these are small compared to growth. As a result, the top parts of the plant need a carbon dioxide rich atmosphere, while the roots need oxygenated water or oxygenated media. The oxygen released in photosynthesis is far more than the oxygen used in the root area (or all animals on Earth would have suffocated by now). Part of the growth of a plant might be in seeds and the stuff around the seeds (i.e. fruit).

Some plants, like grasses such as grains and others, put all their effort into the seeds, then die once the seeds are ready to drop. Others can grow indefinitely while producing fruits or other seeds periodically. Why are fruits (and yes tomatoes are a fruit) so tasty? Because the plant's genes want animals to eat the fruit to spread the seeds by pooping them out far from the plant. The animal gets a carbohydrate rich dinner, the seeds get fertilizer and room to grow.

Flowering plants start from seeds with a germ cell and nutrients to help the germ cell grow, as shown in Fig. 4.2. The seed sprouts when water is absorbed, and the first shoots and leaves push to get light and produce roots. Once the shoot and first leaves hit light, it is off to the races, as the plants use light and pull in carbon dioxide and water to grow and add overall biomass. As the plants mature, they flower, and fertilized flowers become fruits. These grow and ripen around the seeds. Eventually the fruits drop, and the seeds become the next cycle.

The part where plants merge genes with another plant via flowering is called pollination, which might require wind, a helpful insect, or in space, a helpful human or robot with a paint brush. My cousins had a tedious summer job in a seed corn

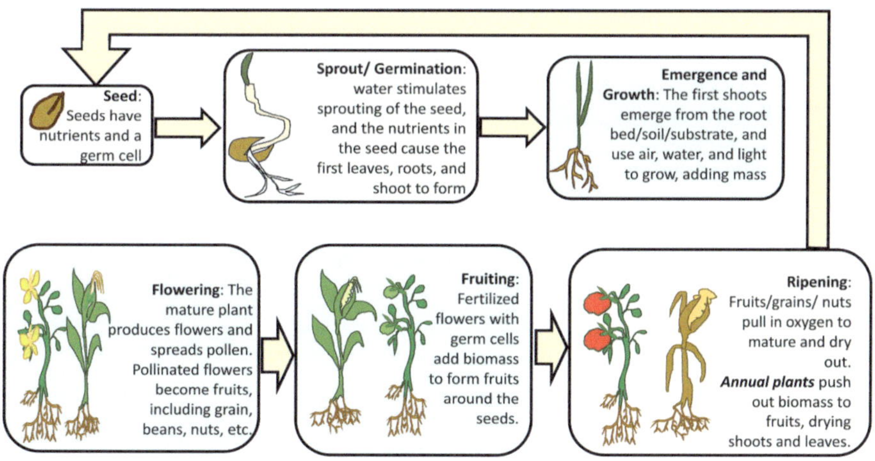

Fig. 4.2 Flowering plant growth cycle

field where they pulled the tassel of the corn plant (the male part, ouch!) to ensure correct hybridization. This tasseling job is one of the menial jobs that make character in teens to be sure. One job I 'enjoyed' was working in fast food and pizza, but I had to do some farm work for free at my grandparents' farm too. It could also be a robotic job for a variety of plants in a space-based farm. Is there a difference (in work) between a very bored teen and a robot? Up for debate, I am sure.

Fruits in this tense can also mean grains. Fruits add an extra part of the plant growth budget, and in many cases, the growing fruit needs to ripen (using oxygen) or dry before being eaten, even after the fruit stops expanding.

As space farm engineers we live by mass flow. Mass flow calculations need to assume a steady state to make the math simpler, though when starting up, the flow is not so steady. See Fig. 4.3. When starting from seed, it will take a growth season to get to steady state production. Steady state is where there are the same levels of mass inputs and outputs per day. If the plant is an annual (that means all the plant dies every season) this means a continuous replanting and harvest (covered in later chapters) each day to make sure production is steady, such as for grains and some veggies, otherwise big bang harvest means a saw tooth graph that is not steady. If the plant is perennial, this means trimming inedible mass and harvest of edible mass every day to keep steady state, as for perennial tomatoes or peppers.

Seeds are not the only way to get a plant. Many plants can reproduce using vegetative growth. In vegetative growth, a stem or root reaches away from the main plant, grows either roots on the stem, or stems on the root, and the result is a new plant, that is a clone of the original. Grass, potatoes, and many vines reproduce this way, making an army of clones. However, a gardener or farmer could cheat this same process for many more kinds of plants, including many trees and bushes. A branch can be cut from the main plant, then the branch is put in a growth solution in a tube, or a cloth. The branch grows roots, and the new rooted branch is then planted to grow a whole tree as in Fig. 4.4.

All that said, the plant as a living machine has word equations to express the mass balance:

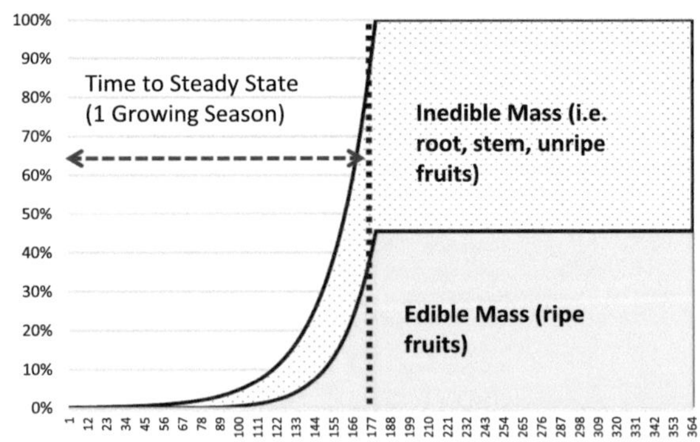

Fig. 4.3 Bell pepper growth to steady state production. Growth from first seeding to steady state

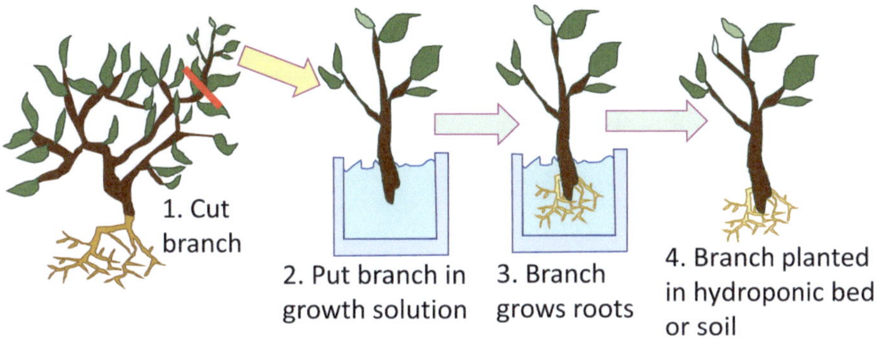

Fig. 4.4 Vegetative reproduction of plants for a farmer in four steps

- Mass in carbon dioxide + Total mass in water + nutrients yields (➔) mass added in growth + Net Mass in oxygen released.
- Mass added in growth = Mass added in inedible parts and mass added in edible parts.
- For fruiting plants, grains, and other seed crops: Mass in inedible parts = mass added in roots, stems, and leaves. Inedible mass may also include inedible coverings around the edible fruit parts, like a banana peel, apple core, nutshell, chaff, husks, etc.
- For root crops and tubers: Mass in inedible parts = Mass in leaves, other roots, and stems.
- For leaf crops: Mass in inedible parts = mass in roots and stems.

Figure 4.5 shows edible and inedible portions for various plant types. For example, humans eat the grain from rice, and the rest of the rice plant makes great fodder for cattle or shrimp, though humans could eat the rice leaves and stem if they had to (with little nutritional benefit). For some plants, like leaf crops such as lettuce, almost the whole plant mass is eaten. In some plants, the fruit component is edible, but the stems and leaves may require treatment before even animals can eat them (such as tomato leaves), and the peel may be inedible though useful (such as orange peels).

The rough mass flow modeling process can be a reverse of the modeling process for animals. To get the mass flows below, you ideally would measure all the ins and outs using a lab with lots of instruments, and this is the data we will need to build a farm on Mars or in free space in the long term. However, a cheat might use data you can find on productivity and maybe even carbon dioxide, nutrient and water usage, with growth data and chemical analysis of plant tissue nicely provided by the USDA or FDA in the food nutrition database. Armed with that, some chemistry from sections above, and some ratios of the edible versus inedible, we can get a simple mass flow roughly useful for sizing a space farm design. Figure 4.6 shows this flow. This estimation process is essentially notional, for using limited data like nutrition to back out inputs like carbon dioxide, nutrients, and water for photosynthetic crops. It is far better to have experimental data!

Fig. 4.5 Edible versus inedible for vascular plants. What is edible or inedible depends on the plant

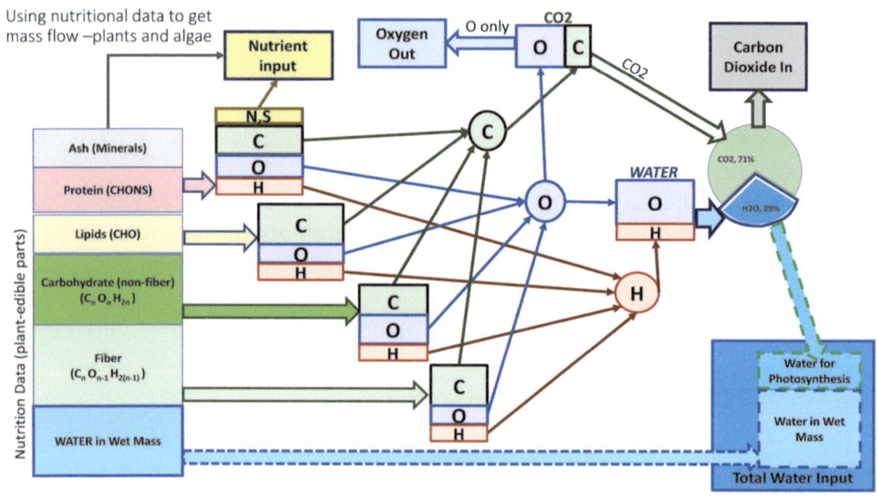

Fig. 4.6 Rough element and mass flow estimation method from nutrition data

A good way to investigate plant growth and mass flow is also in the SIMOC package (SIMOC, 2022), which is worth examining.

4.3 Vascular Plant Farm Species and Types

Technically many products called 'vegetables' are fruits, such as tomatoes or peppers, while vegetables in the stricter sense are like lettuce, radishes, carrots, and potatoes. Below are options for bushes and trees too. All mass flows here are very rough, derived from element data, and honestly need to be collected from real plants in a variety of conditions, an active research area that needs work. So use the mass flows with caution. That said, let's discuss each type and tradeoffs.

4.3.1 Fruit Vegetables

Fruit vegetables are usually very high in fiber, vitamins, and moisture, but low in calories. You would need 50 tomatoes to get a little over 1000 calories! That said, most fruits like tomatoes (as in Fig. 4.7), eggplants, chilies, and peppers are a must for a balanced diet and a good menu.

These fruiting plants grow fast, with harvestable fruits in around 90 days, and given the right cultivar, can produce fruit regularly for months or longer. Since the fruit is also the seed, collecting seeds from the fruit before ingestion can replenish the seed bank, though some of these plants can be grown from trimmings. Some vines are strictly annual, and since they like to sprawl, can be part of a polyculture like the three sisters' approach from the Americas, or used to fill up space in hallways, inside greenhouses, or in other areas. Other types of the same plant can grow like bushes.

Vines in this sprawling category include higher calorie crops like many melons, squash, cucumbers, and the like, as well as some tomatoes. Scaffolding can be used to control how the vines grow, and where runners contact substrates or soil. Sometimes small sacks can be used to allow these fruits to mature without dragging down a whole plant. Vines like these can be wrapped around columns too, providing both decoration and dinner as in Fig. 4.8, but these vines will require some human labor to manage. Around columns, root beds hold the roots, while detachable supports could hold developing fruit, and movable vine ties hold the vine to the post. In hallways, trays act as the planters, where the tray holds the root beds, lines, and developing fruits.

In practice, dedicated fruit growing chambers can have rows of suspended perennial plant vines, where fruits are collected daily, and trimming and tending keep the

Fig. 4.7 Cherry tomato plants. Cherry tomato plants produce fruit throughout a growing season, or the vines can be made perennial, with a smaller but constant crop. Photo by author

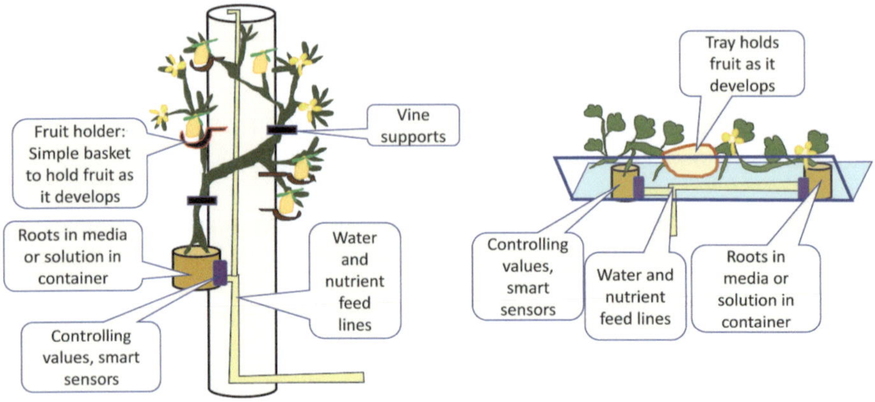

Fig. 4.8 Concepts for growing fruit plants in habitats. Left—vines can be wrapped around posts in the habitat or in a greenhouse. Right—hallways or greenhouses can hold melon vine crops in trays

vines in check. For example, in greenhouse tomato farms, strings or supports extend down to the pot containing the plant and roots, while the vine is stretches along the string or support as it grows, reaching almost all the way up to the ceiling, as in Fig. 4.9.

Fig. 4.9 An indoor tomato farm on Earth. Photo by Kai-Shu Ling, USDAARS (see USDAARS-D4685-1)

There are many, somewhat unrelated crops that fit this growth type and plan. Some are annuals only, requiring replanting and complete harvest at the end of their growth cycle, while some can be perennials that can be tended and stretched for many seasons. See Table 4.1 for a few examples.

Mass flow for a unit of these fruiting plants takes inputs of nutrients and water in solution, and carbon dioxide in the air of the farm or habitat, and produces a combination of edible fruits and inedible leaves and stems trimmed as required, and released oxygen, as the semi-notional model in Fig. 4.10. Note that this model does not include the cycling of transpired water, which is assumed to be captured by condensors and recirculated back to the roots. These flows will vary with conditions like lighting, temperature, partial pressures, and other variables, so this flow is close to being notional. If this plant is grown as an 'annual' every 180 days the whole plant gets plucked out, but in most cases, mass flow represents trimmings and growth of inedible parts like leaves and stems, growth in unripe fruits, and harvest of ripe fruits at a snapshot in time. That said, it would take some genetic tinkering to make an immortal tomato or pepper tree, and in practice tomatoes and peppers get replaced every year or so due to dropping productivity.

Nutritionally, fruiting veggies like tomatoes or bell peppers are not a primary calorie source. Due to the high-water content in most fruit vegetables like bell peppers, a person would have to eat around 6 kg per day to get enough calories to survive. While bell pepper fruit is 92% water, the dry mass component is high fiber and simple carbohydrates, and minerals (ash) as in Fig. 4.11. This breakout is like tomatoes and eggplants, though some fruit veggies could be engineered to have more calories

Table 4.1 Sampling of fruiting vegetables. All nutrition data from USDAARS (2022), harvest data is roughly estimated from experience

Plant	Time to first harvest (days) (rough)	kcal/kg edible mass	Type
Bell pepper	60–180	290	Bushy or vine (perennial or annual)
Cantaloupe	90	380	Spreading vine
Cucumber	50	150	Spreading vine
Eggplant	90	250	Bushy
Jalapeno	180	290	Bushy or vine (perennial or annual)
Okra	60	330	Bushy
pumpkin	100		Spreading vine
Tomato	75–90	180	Bushy or vine (perennial or annual)
Watermelon	90	300	Spreading vine
Winter squash	80	450	Spreading vine
Zucchini	60	170	Spreading vine

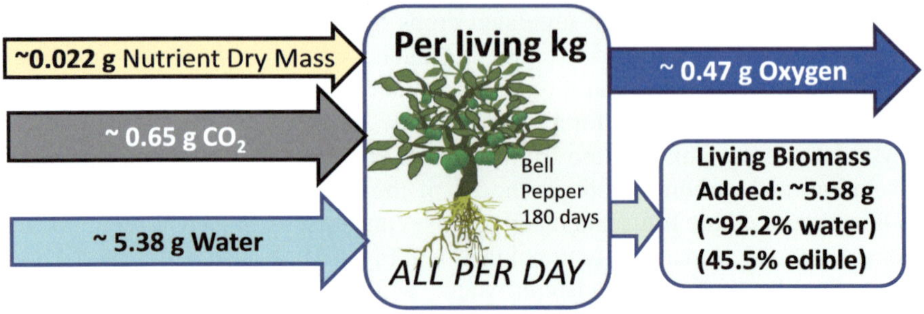

Fig. 4.10 Estimated bell pepper mass flow model. Estimated per day per living kilogram, at a very rough and conservative steady state

Fig. 4.11 Bell pepper nutrients in dry mass

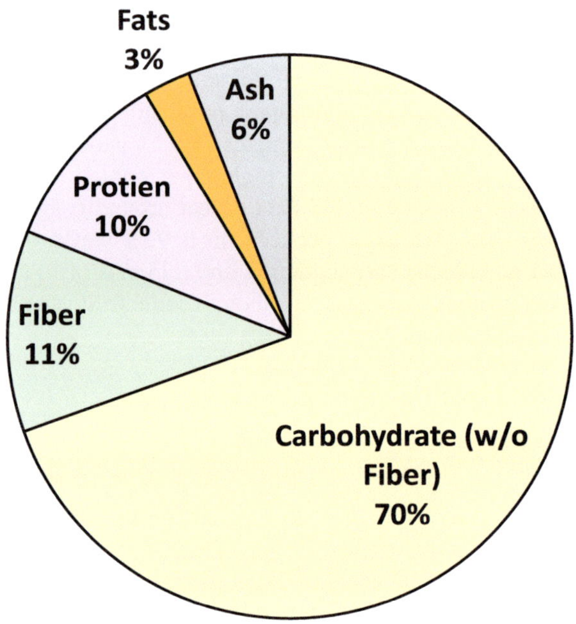

4.3.2 *Bananas and Plantains*

What about bananas and similar? Banana 'trees' fall somewhere between fruiting vegetables and trees (see "Bushes and Trees" below). They are tropical herbaceous plants that take around a year and a half to produce bunches of highly caloric fruit the first time, and are sort of perennial, making them a very possible space farm crop, most especially if brandy or mofongo is in the offing. They are not small, roughly 2–3 m in height, and may require a well-engineered greenhouse, or a good corner spot or central area in a (very warm and humid) habitat. Some varieties die after harvest, then regrow for harvest around nine months to a year later. These may be best as a soil crop, grown most of the way on the trip out, though worth research and consideration.

4.3.3 Leafy Greens

In contrast to most other crops, leafy vegetables, like cabbage, arugula, chards, rhubarb, and other lettuces, are lower in calories, and higher in fiber. However, leafy plants are critical for a balanced diet, and digestion. You are eating the leaves, not the seeds. They grow very fast, on the order of 30–45 days, but require replanting after a few harvest cycles. Harvest may simply involve trimming off leaves (as in Fig. 4.12), leaving the plant to generate more.

Many greens can be harvested after a couple of weeks if used as microgreens, i.e. harvested as sprouts. A few of the leafy plants will need to be allowed to grow until flowering maturity, making their leaves less than palatable (see Fig. 4.13), to get seeds for the next planting. Arugula for example becomes quite bitter once it flowers.

Assuming a continuous clipping of leaves, which are fully edible except for a small fraction, or full plant harvest but using a continuous model, mass flow features heavy water intake due to the mostly water nature of greens. As for most plants, the model

Fig. 4.12 Leaf lettuce growth ready for clipping

Fig. 4.13 Flowering leafy green. This leaf lettuce has grown all the way to flower and is bitter but technically still edible

below ignores transpiration, which cycles greater than 100 times the water incorporated. Transpired water can be quickly recaptured and cycled using a condenser near the plant. A living kilogram of Arugula takes over a cubic meter of volume near maturity. A loose mass model for a leafy green is shown in Fig. 4.14.

Leafy greens are high in minerals, assuming these minerals are supplied to the plants in soil or growth solution, seen as 'Ash'. Arugula (a leafy crop, see Fig. 4.15) is low in dry mass, and the dry mass is balanced between ash (minerals), protein, fiber, and carbohydrates, with roughly 250 kcal/wet mass kilograms as eaten. These leafy greens add fiber to aid in digestion, though it may come as a surprise that the dry mass in leafy greens also has amino acids in almost the same proportion as

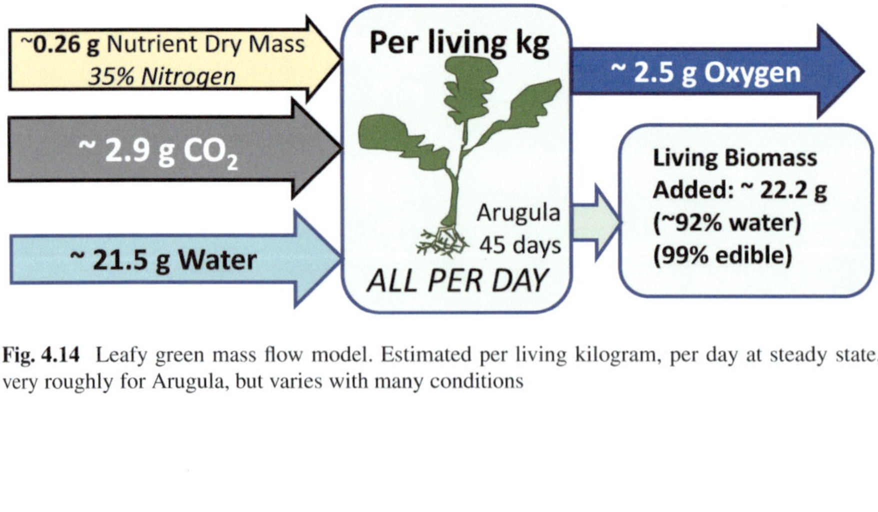

Fig. 4.14 Leafy green mass flow model. Estimated per living kilogram, per day at steady state, very roughly for Arugula, but varies with many conditions

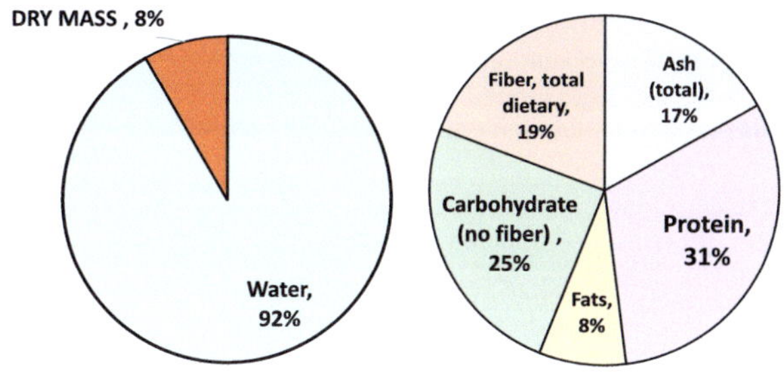

Fig. 4.15 Arugula dry mass breakout

Table 4.2 Sampling of leafy vegetables

Plant	Time to first harvest (days) (rough)	kcal/kg edible mass
Arugula	45	250
Cabbage	100	250
Cauliflower	60	250
Broccoli	80	340
Lettuce	45	140
Kale	70	350

carbohydrates. That said, it would take over two kilograms of these greens to hit the minimum daily protein needs of a person.

Table 4.2 provides a list of common leafy greens grown on Earth, and arugula is a crop that has been grown on the ISS. All nutrition data from USDAARS (2022), and harvest times roughly guessed from experience.

4.3.4 Root Vegetables (Tubers)

Root vegetables are produced when a plant stores sugars and starches, and some proteins in its root system. In most cases, these plants evolved in areas with winters or harsh climates and needed to have enough energy stored while leaves were frozen and photosynthesis impossible. Potatoes, a distributed root vegetable, for example came from the Andes. The potato growing cycle is typically perennial, and there are two options for regrowing them, from seeds, and from the tubers themselves. Seeds are genetically distinct, while plants from tubers (vegetative growth) are clones of the parent plant. Other examples of tubers like potatoes are yams and sweet potatoes. These types of tubers grow as vines, and cycle between tuber production and leaf growth. Usually, it takes around 180 days to maximize the growth cycle. Tubers like potatoes, taro, and yams can form the primary calorie source (i.e. a staple) in a diet. Nutritional components for the dry mass component of a potato is shown in Fig. 4.16. Root vegetables like potatoes are very carbohydrate heavy, though they contain all major groups including protein, fiber, fat, and minerals.

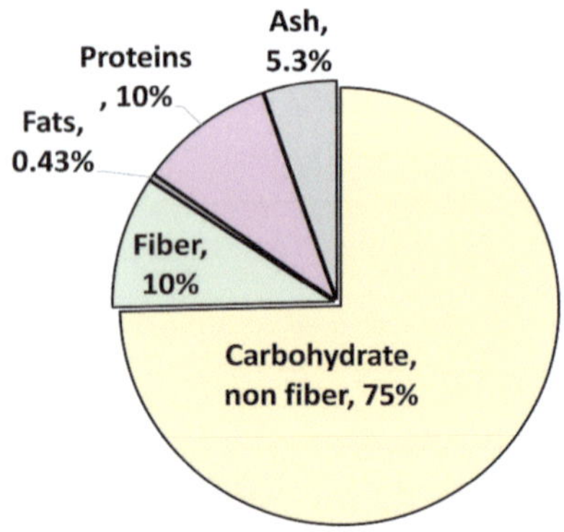

Fig. 4.16 Nutritional components in potato dry mass. Potatoes are around 79% water, the remainder is mostly starches (carbohydrates), with some fiber, proteins, and minerals, and very little fat

Distributed root vegetables like potatoes have many edible tubers scattered across the root system. In contrast, tap root vegetables, like radishes or carrots, have a single edible tuber on the main stem root (i.e. tap root). See Fig. 4.17 for a comparison of distributed tubers versus tap roots.

Tap root tubers like radishes, turnips, onions, and carrots are a bit different, in that they are usually grown as annuals (sometimes biennials), and usually grown from seeds, in lieu of vegetatively from tubers, and are higher in fiber and other nutrient compounds. The tap root (the main root) becomes the tuber. They grow quickly from seed to mature tuber in around 90 days. Typically, these types of vegetables are a side dish, in lieu of a staple, due to the high content of other compounds that add a bit of a kick, though turnips may be an exception. Others, like taro are staples. As

Fig. 4.17 Root vegetable types

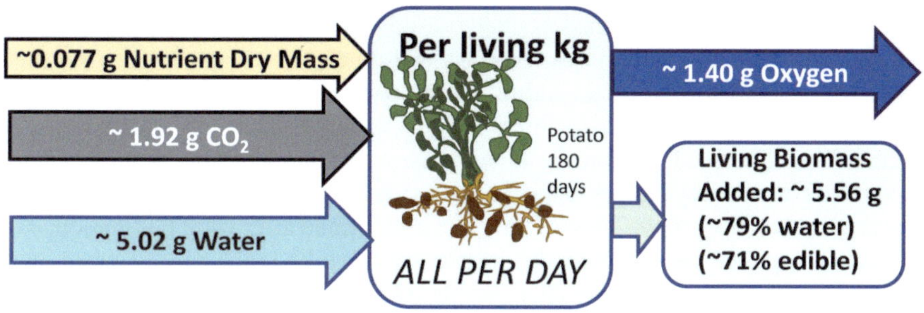

Fig. 4.18 Potato mass flow model. This is a notional potato (a distributed tuber) mass flow model per day per living kilogram steady state/harvest

Table 4.3 Sample root crops. Nutrition data from USDAARS (2022). Cassava harvest data from ACAI (2022), remainder rough experience

Plant	Time to first harvest (days) (rough)	kcal/kg edible mass	Type
Potato	110	700	Ground vine
Cassava (Yuca)	365	1600	Tall bush
Sweet potato (Yam)	110	760	Ground vine
Turnip	70	280	Tap root
Carrot	70	410	Tap root
Beet	60	430	Tap root
Radish	40	160	Tap root

is the case for other vegetables, harvest is usually before flowering and seeding, so some plants will need to be retained to produce seeds, as in leafy greens above.

There are special concerns for growing tubers, especially without soil. Tubers can rot in the root bed if there is too much moisture, and this rot can cascade to kill a crop. Further, various fungi can kill a tuber as it develops and before harvest. If that wasn't bad enough, variations in light and temperature can alter the growth budget from tuber production to vine and flower production.

Figure 4.18, like all plant mass flow models here, do not include transpiration. Water content in stems and leaves is far higher than in tubers, and tubers hold the bulk of dry mass for the whole plant at harvest. Note also that while Table 4.3 includes staple tuber crops, it does not include some tubers that may also be good inclusions for a space farm, such as ginseng or ginger.

4.3.5 Grains

Grains are staples and the primary calorie source for most diets on Earth. Corn (maize), wheat, teff, quinoa, barley, oats, and rice are the most common crops grown on Earth. Most grains are grasses, that use advanced photosynthesis to produce heavy seeds annually, dying in the process, and much like pacific salmon, all their investment is in the offspring. In addition to the capability to store grains once dried, they can be ground into flour for breads, and swelled to make various dishes, like breakfast cereals. Seeds are primarily starches, with some proteins, and once dried become useful in many recipes. The seed covering (chaff) is almost all fiber, and this covering is clustered on the shaft of the plant and formed after wind pollination. The dried shaft top (see Fig. 4.19) is also known as the spike in the case of wheat, and the components of the spike are the grains in a chaff cover.

So, growth goes from seed to shaft and leaves to flowering and seed formation, to dry out and harvest. On Earth, a wet fall or a dry spring can reduce production drastically. Too much water in the fall, and the plant becomes moldy, and not enough water in the spring and summer reduces growth. Saving a percentage of the grain for seed makes complete harvest of all plants in a field possible. Cycles are usually around 120–180 days. Since they are annuals, the whole plant gets cut, except for soil farming where the roots and retained to prevent erosion over winter. The dried shaft and leaves, called straw and hay, can become fodder for animals, especially after a little fermentation to raise protein levels. Note that there is a big difference between a modern dwarf monoculture grain, and the older heritage grain varieties. Dwarf varieties produce a far higher edible percentage versus inedible shaft and leaves the older types, at the cost of robustness and genetics. This is one of the trades when selecting grain crops for a farm. Modern plantings of the modern dwarf select crops in the midwestern USA produce far less straw and far more grain per hectare than estimated in the table at the top. In this book estimates use the high productivity of the dwarf varieties, but also assume the inedible portions of the heritage plants, as a worst case. Midwesterners are pessimistic when it comes to engineering and crops. Future space farmers will benefit from better data and less guessing. The dried-out shaft can also have a little lignin that takes energy using heat or engineered biology, or time via fungi in reactors or soil, to break down. My neck still itches from loading straw

Fig. 4.19 Wheat shaft top. Top image is a fully dried wheat spike, which is part of the shaft top. Bottom is the grain and chaff covering from the spike. Photos by author

into the loft! Grazers prefer the green shoots to avoid said lignin. Failing feeding the straw to animals which partially digest it, the straw can become soil binder, fed to a bioreactor or incinerator to pull out carbon dioxide and water, or lastly used for construction, floors, clothing, or furniture. Grains flower and need pollination like other flowering plants. In maize (see Figs. 4.20 and 4.21), silk receives pollen from the tassels due to wind, and the pollen fertilizes the developing fruit grains. The grains on the corn cob, connected to the silk, then swell with sugars and water until maturity. Near maturity, the entire plant dries out, and the fruit grains, now with starches, germ, and proteins, harden before harvest.

The grain is far drier than the rest of the plant, even near harvest, and given very efficient photosynthesis, the plant incorporates lots of carbon and nutrients, especially in wet field cultivation. The edible ratio varies highly with variety; some plants make more grain than leaf or stem (like the dwarf varieties) versus heritage varieties. We can make a loose model calculated using data found from USDAERS-RiceYB (2021), Wu et al. (2011), and USDAARS (2022) shown in Fig. 4.22.

Some grains have multiple purposes. Sorghum can be farmed for grain, or for a sweetener, depending on variety and harvest timing, as shown in Fig. 4.23.

Nutrition also varies in grain crops, some more protein-heavy due to genetic modification, while most are carbohydrate sources, especially for starches. Typical dry mass for most grains at harvest is over 90% (as opposed to stems, leaves and roots, which are wetter), and is largely carbohydrates, especially starches, with some

Fig. 4.20 Maize (corn) tassel (male part) and fruit (female part), in a field in MO, USA. Photo by author

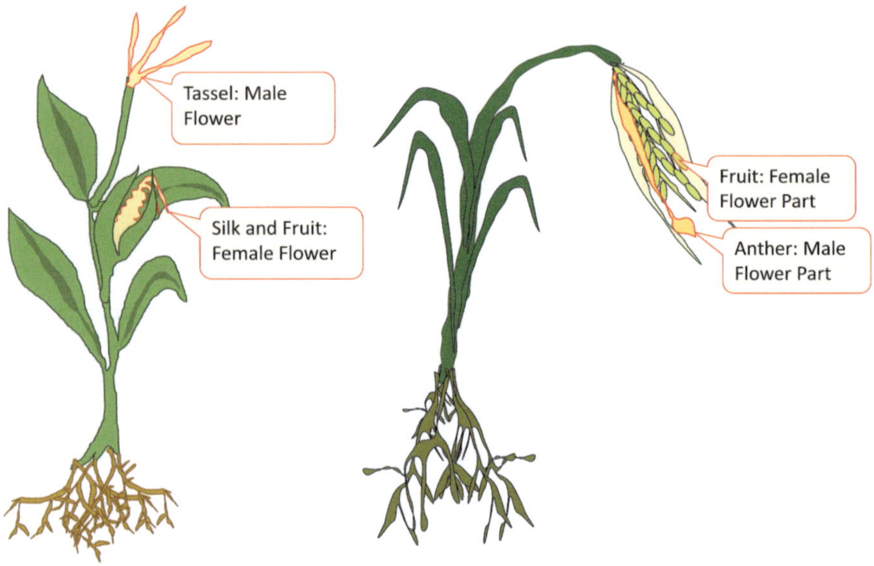

Fig. 4.21 Grain flowers. Some grains, in this case corn (maize) has a male flower, called a tassel at the top, and a female flower, the silk below. Wheat parts (left) are closer together

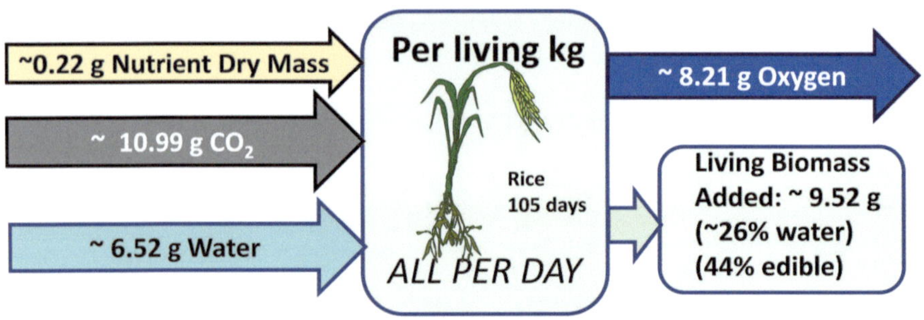

Fig. 4.22 Notional rice mass flow model. Very rough average mass flow model per living kilogram, per day at steady state for rice. Does not include transpiration

protein as in Fig. 4.24. Genetic engineering could improve, and has improved, the protein and mineral (ash) content in some grains (again, see Wu et al. (2011)).

While many cereal grains are grass plants (monocots) like rice and wheat, a few are broadleaf flowering plants (eudicots) like chia and buckwheat. Most grains in general take around four months to mature, some longer. Table 4.4 has a list of common grains, including one of the first grains domesticated, teff. Nutrition data in the table is from USDAARS (2022), sugarcane data from Babeker et al. (2020), spelt via DARBY (2024), buckwheat via Björkman (2009), sorghum via Worley and Sumner (2020), remainders via rough guesses and estimated calculations. For those with a sweet tooth, I added another grass that is eaten by processing the juice

Fig. 4.23 Sorghum in a Missouri field. Sorghum is a multi-use grass crop that can become feed, food, and sweetener. Photo by author

Fig. 4.24 Nutrition in dry rice fruit (grain) via USDAARS (2022)

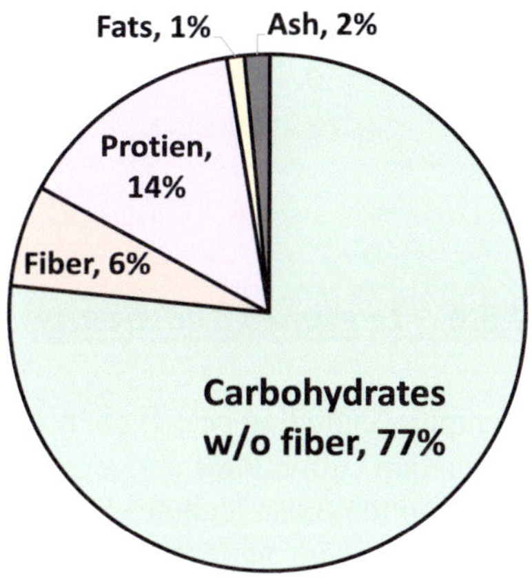

of it stalk, sugarcane. Sugarcane is a long term, high processing, and high-water consuming crop, leaving extensive residue. It is not likely sugarcane will be a space crop except in very extensive, and long developed soil farms. More likely, sugar is a product of beets or algae, maybe honey. Given grains however, there is a hack. Rice, and sorghum can be simply processed by boiling and reducing, to get a sweet juice that works for most recipes. There is a more primitive answer too. Given a bit of spit enzymes (salivary amylase), grain starches become simpler sugars, especially sugars that yeast like. Therefore, many fermented beverages all over the planet start with someone chewing and spitting into the vat. You had better really like that drink maker (preferably on kissing terms)! A more sanitary version of this process using a bioreactor could make that sweet sugar, sans spit, using some grain starch and engineered bacteria or yeasts.

Table 4.4 Typical grain crops and sugarcane

Plant	Time to first harvest (days) (rough)	kcal/kg edible mass	Type
Barley	90	3780	Monocot grain
Buckwheat	77	3430	Eudicot flower
Chia	200	4860	Eudicot flower
Corn (Maize)	100	3650	Monocot grain
Millet	90	3780	Monocot grain
Oats	88	3790	Monocot grain
Quinoa	90	3680	Eudicot flower
Rice	105	3570	Monocot grain
Rye	150	3380	Monocot grain
Sorghum	180	3290	Monocot grain
Spelt (heritage wheat)	120	3380	Monocot grain
Teff	100	3670	Monocot grain
Wheat	100	3390	Monocot grain
		kcal/kg cane	
Sugarcane	550 days to first harvest, annual thereafter	735.3	Stem monocot perennial

4.3.6 Legumes and Beans

Legumes (including beans) are flowering plants that produce high protein seeds after pollination, sometimes above ground, and sometimes below ground. In low meat consumption diets, legumes become the primary protein source. Common legumes include the many varieties of common beans including pinto beans and green beans, soybeans (in Fig. 4.25), and to some plants that you might not think are legumes, like peanuts. Most legumes are very fast growing and hardy, and while they consume lots of nitrogen from the soil, they can also work with bacteria to bind nitrogen in the air into the soil too. As a result, legumes become part of crop rotation schemes on Earth to put nitrogen into the soil before growing grains in later seasons. Most legumes are annual. There is a wide variance in growing times for legumes, from 30 days to 180 days. Since some beans are vine based, there will be a need for scaffolding to control growth. Like for grains, the shafts and leaves, and roots, can become fodder in some cases for animals, or treated as other plant wastes. Bean casings, shells of those peanuts for example, are usually treated as inedible, but immature bean pods with seeds are sometimes eaten as a salad item after early harvest, though with lower calories than the mature seeds. There is also an option to harvest the sprouts very early (after a few weeks) as a vegetable for salads and other dishes, forgoing the later harvest. Most legumes are dried and stored, then rehydrated by boiling.

For some legumes like soybeans, boiling or steaming is a must to reduce esterols. In the case of peanuts, roasting becomes a core method of both food prep and drying.

Fig. 4.25 Soybeans near Winfield, MO, USA. Soybeans are a common legume crop, high in oils and proteins. Photos by author

Like grains, beans can be made into flour, and in the case of peanuts in America, made into peanut butter. Legumes are also a key source of oils, once heated and pressed. These oils can be used to make plastics, lubricants, lotions, and most importantly in my mind, for cooking! Crops of peanuts or soybeans might be raised just to get protein powder to add to soups and sauces, and oils to fry, dress, or broil other foods. You might ask how a peanut is a legume? Simple, the peanut grows and flowers like any other legume, but then shoots the bean under the soil. Peanut farmers take the mature plant, pull it out of the dirt, and flip it upside down to dry the peanut 'nuts' which are really beans. The peanut is likely the only 'nut' most settlers will get regularly given the time to grow trees. A legume/bean mass flow model is in Fig. 4.26.

Legumes such as pinto beans are high in proteins and fiber, with more than twice the fats (oils) of most grains. Legumes like peanuts are even higher in fats. Figure 4.27 breaks out the mass flow and nutrient composition of pinto beans, a common dietary

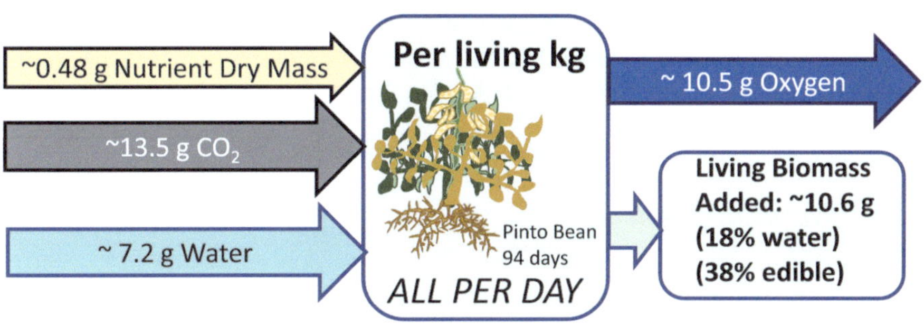

Fig. 4.26 Rough mass flow model for pinto beans. This notional model is at steady state for a living kilogram of legume near harvest. Does not include transpiration

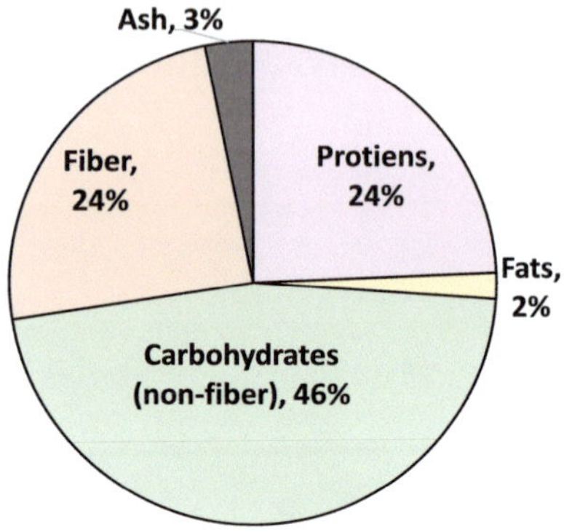

Fig. 4.27 Pinto bean nutrients by dry mass

Table 4.5 Sampling of legumes and bean crops

Plant	Time to first harvest (days) (rough)	kcal/kg edible mass
Cowpea	100	3430
Lentils	120	3520
Mung bean	60	3470
Peanuts	175	5670
Pinto beans	90	3399
Soybeans	100	4460

item, though never in Texas chili. Pinto beans before cooking are roughly 86% dry mass.

Table 4.5 covers many common legumes and bean crops. Harvest times here are very roughly from sources in Chap. 1, and McLelland (2021), Masood et al. (2009), and Dugje et al. (2009). Nutrition data is from USDAARS (2022). Most beans take two to four months for harvest, but they grow fast enough to be a Midwestern saying about adolescent kids who shoot up like bean stalks. A similar US farm saying states the relevance of a topic, i.e. what does this have to do with the price of beans in China? Beans are obviously important to both US and Chinese farmers, so if it bears on the price, the topic is relevant. Maybe a saying will develop in space, becoming something like "what does this have to do with the price of beans on Mars?"

4.3.7 Herbs and Spices

As mentioned before herbs and spice plants are typically small enough and fragrant enough to be a welcome addition to the plants grown in a habitat, and part of the grey water cycle ideally. Most are harvested as leaves or whole plant, though some like ginger are root crops, and a few are fruiting vegetables like peppers. Most of these are very resilient plants, tolerating indoor conditions and constant snipping. In the

Table 4.6 Herbs and spices. Harvest times are very roughly estimated from seed packet data and personal experience

Spice	Time to first harvest
Basil	75 days
Cilantro	30 days
Dill	90 days
Fennel	100 days
Mustard	95 days
Chives	60 days
Marigold (Candula)	70 days
Mint	90 days
Tarragon	60 days
Oregano	120 days
Cumin	120 days
Ginger	200 days
Serrano peppers*	120 days
Paprika*	150 days
Saffron	180 days
Chili peppers*	120 days
Coriander	100 days
Garlic	180 days
Turmeric	300 days
Thyme	1 year
Rosemary	1 year
Hops	2–3 Years

** = Same species as bell peppers*

close confines of spacecraft, growing a few herbs can enhance the smell environment and improve spirits. Table 4.6 is a long list of herbs and spices that might be grown in the human habitat or in the farm, and in the case of hops (the beer spice) can literally spice the spirits.

4.3.8 Water Vascular Plants

Water plants fall between crops for photobioreactors and for hydroponics. I am skipping the algae-based organisms that look like plants (i.e. kelps) in this section to focus here on the water plants that have roots and other structures. Wet rice cultivation is an example of a water based vascular plant though it is covered in grains above. Another example is taro, a tap root vegetable. Taro is a successful invasive in the continental US, evidence of its toughness. Regardless of what some say, taro, especially once made into poi, is yummy! As for other space farm candidates, invasive plants on Earth can become hardy fast-growing crops in space. Many are edible and are great foods for aquatic animals. They are also very good at cleaning water, and can take raw wastewater. The wastewater can be sewage from people or animals and the plants convert it into plant biomass, taking in carbon dioxide and producing oxygen in the process. Eurasian milfoil is invasive in the USA, but is edible, and makes a great filter. Duckweed is a floating water plant that is also edible, and again is good to clean water. The plant that is most maligned on Earth, but again might be a great space crop, is the genus *Hygrophila*. *Hygrophila* grows exceptionally fast and quickly utilizes any nutrients in the water, making it a great waste processor, and once dried and ground, food for animals including cattle, fishes, and shrimps. In Florida, *Hygrophila, Eichhornia,* and *Pontederia* (see Fig. 4.28) are used for large scale waste processing from cities and amusement parks, and the resulting plants turned into compost or fed to cattle. Like other crops and waste processing options, whether you have it in your design depends on how much area (footprint) is available. Hyacinth is a fast-growing water plant that floats well and has filtering roots. Cattails may be better option. These types of plants need more space than microalgae, but are in many cases hardier, and as a side benefit, can add greenery to water gardens or other habitat areas like parks. Many water plants can also be grown in photobioreactors with some modifications to designs. If you decide to have shrimp or fish, water plants may indeed be a very good choice since you will already be using flowing water.

A very interesting and promising water plant is duckweed (see Fig. 4.29. The duckweed (family *Lemnaceae*) is a teeny floating leafy plant, about 4 mm wide, is

Fig. 4.28 Common water hyacinth (*Pontederia* sp.). Photos by author

Fig. 4.29 Duckweed from Rabbit Run in MO, USA

about 1 mm thick on the surface, and the roots dangle about 3 mm deep. It reproduces vegetatively extremely fast, on the order of doubling every 31 h (loosely from Appenroth et al. (2018)).

It has been examined as a crop on Earth for solving hunger issues, since it grows well in stagnant water and has nutrition. There are a variety of ways to automate growth and harvest in shallow trays or dishes, and this weed may be a swap out for algae. Many fish eat it, especially carp and tilapia (personal observations). Mass flow can be modeled as in Fig. 4.30 using nutrition data and loosely from data in Appenroth (2018).

Nutritionally, duckweed is starchier then a leafy green, with more protein in dry mass. It tastes somewhere between spinach and lettuce (again, personal experience). Duckweed is 7% dry mass, of which 25% is protein, loosely again from Appenroth (2018). The graphs in Fig. 4.31 were loosely calculated from the data found in Appenroth (2018) as well. Duckweed could be grown in a type of bioreactor, as shown in later chapters, or on top of or around other aquatic crops.

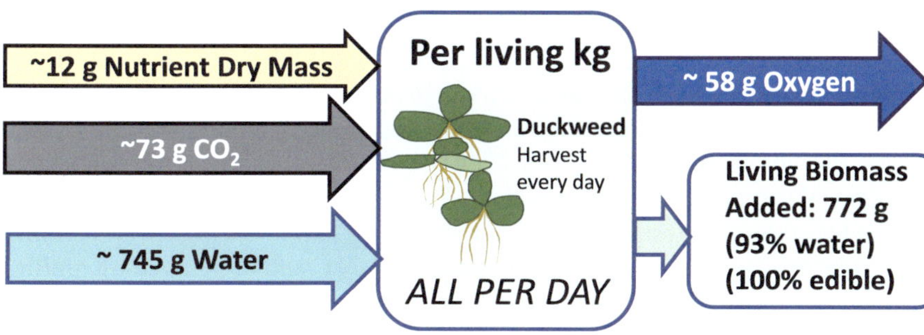

Fig. 4.30 Duckweed semi-notional mass flow model

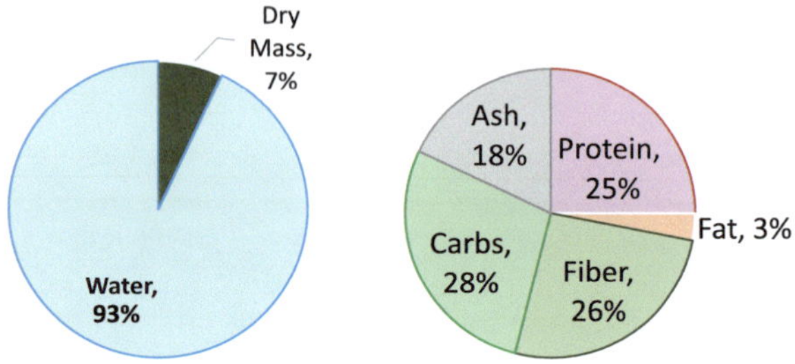

Fig. 4.31 Duckweed mass breakout. Left: dry mass to water in wet mass. Right: Nutritional components in dry mass

4.3.9 Bushes and Trees

Eventually, most settlements are going to want to grow crops that come from bushes and trees. Some bushes are well suited for greenhouses or in habitat growth in pots with hydroponic substrates or soils. Anyone who likes a caffeinated beverage in the morning will want tea or coffee, both of which are bushes. Many berries are also bushes, such as raspberries. Careful trimming and management can allow a mature bush to be transported and grown as a houseplant. They may be fed from grey water taps and grown in pots built into walls of ships or settlements. Unfortunately, most bushes take time, years, to produce, and if berries are involved as for coffee or raspberries, we will need someone with a paintbrush, or something like a robot or insect, to perform pollination. Refer to previous comparison between bored teens and robots. Woody plants, both bushes and trees, also produce lots of lignin, which takes more effort to digest than cellulose. Lignin needs either fungi, maybe including tailored species in bioreactors or mushroom grows, or fire in incinerators. Grapes can be considered bushes for this purpose for all the wine drinkers out there. Bushes would also figure highly in a larger scale soil farm, and can be used as on Earth, to control field edges and form natural fences. Some 'trees' are herbaceous fast growing plants, like bananas and plantains, and given the right planter or pot, can provide food in a few years. Several potential crops are listed in Table 4.7.

Trees are a growth up from bushes. Some trees can be grown as dwarf varieties, turning them into bushes, one example is the dwarf lime tree found in many houses. Figs can also be grown in dwarf form, often from cuttings, to make fruit in a year or two (personal experience). Some trees, however, take 5–10 years to begin production from seed or cutting, and require much root volume for leaves and branches. This time can be shortened by either transporting mature or near mature trees, or by grafting branches onto mature roots. I have grown limes in a year, starting with a 0.5-m-tall lemon sprout that grew over a meter tall in one summer. It will be a very long time before walnuts or olives show up in settlements unless mature trees are shipped in, or grafting technology is very advanced. Olives are notorious for long growth, and an Italian saying states that you plant an olive tree not for yourself, nor your children, but for your grandchildren. Other trees are quicker and smaller, like peaches or apples, though again they still need 5 years or so. Does this mean a settlement will always be an importer for its orange juice? Maybe not. It is possible using grafting (see

Table 4.7 Rough times to production and heights for a sampling of bushes and trees. Taken from experience and interviews with many farmers

Species	Height	Time to maturity (fruit) (from seed or cutting)
Blackberry	1 m–2 m	1–2 Years
Coffee	1-3 m	2–3 Years
Tea	1-2 m	2–3 Years
Blueberry	1 m–4 m, some varieties <1 m	2–4 Years
Peach	2 m–10 m	3–5 Years
Apricot	2 m–10 m	3–5 Years
Persimmon	2 m–5 m	3–5 Years
Grapes	2 m–4 m	3–5 Years
Almond	3 m–10 m	3–5 Years
Mango	3 m–10 m	3–5 Years
Fig	2 m–6 m	3–6 Years
Cocoa (Chocolate)	4 m–5 m tall	4–5 Years
Dwarf Meyer lemon	2 m	4–7 Years
Pecan	5 m–20 m	5–10 Years
Coconut	10 m	5–6 Years
Orange	2 m–6 m	5–8 Years
Apple	2 m–10 m	6–10 Years
Hazelnut	2 m–10 m	7–10 Years
Olive	3 m–10 m	7–8 Years, though some take 50 years or more
Walnut	5 m–20 m	8–10 Years

Fig. 4.32) or branch growth to cheat the time and space needed. Such methods could be extended to use mature branches alone, fed by tubes or another root base.

What is grafting? Grafting is where a branch or cutting from one plant is connected to another plant or root system. Most grape vines for example, are a combination of wild grape roots, and European grape vines. The wild grape plant is grown to maturity, then the top of the grape plant is chopped off, leaving the roots and trunk base, then a cutting of a European grape plant is taped to the root base, connecting the xylem tubes together. Eventually the roots and trunk grow together with the cutting, making a ‘Frankenstein’. Sometimes the grafting is just chopping a bit of bark away from the host tree and adding a cutting or branch from another. There are trees that produce a dozen kinds of fruit from a single trunk this way. This method might allow a quick path to fruits and berries for a settlement. In the example below the roots from the donor plant are fused with the branches and leaves of the graft planting, resulting in the grafted hybrid. Usually, the bond is a cloth or tape until the roots fuse to the branches to make a single plant.

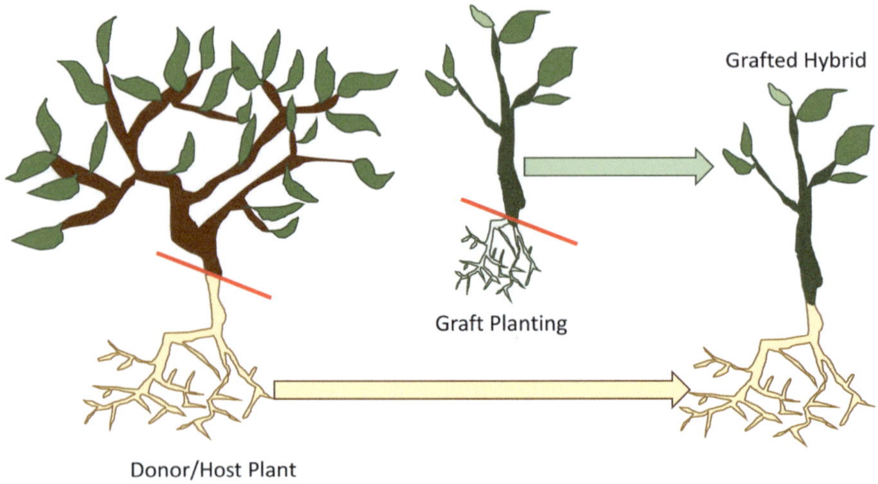

Fig. 4.32 Simple grafting. Grafting takes part from one plant and fuses it with another

4.3.10 Plants with Other Uses

There has been some discussion of crops that are not directly consumed dietarily but used for medicinal purposes. Bamboo might be grown just for materials. Marijuana for materials and medicine, though coca, poppies, eucalyptus, and others may be grown if the settlement is far from an economic web and needs to develop its own pharmaceuticals, including anesthetics and antibiotics. Aspirin for example was originally from willow bark and can be extracted using that source. Wild or domestic rose is a source of vitamin C, and a tasty food additive and summer drink in India (rosewater). There is a long list of possible plants that may be grown in small quantities for all these purposes. Some plants are just good at cleaning water as mentioned above.

4.4 Lichens and Mosses

Lichens (often called mosses, though not so) are organisms that symbiotically fuse fungi and algae (see Fig. 4.33). They have many features similar to plants, and critically, can handle very cold and high-altitude environments, and need no soil. Lichens are useful in many terraforming scenarios since they can break up rocks into soil for other growth and are very high in proteins. While they are slow-growing, they can be used in a soil-processor or simply as soft ground cover. They may line tunnel walls and provide colorful wall covers, slowly expanding the tunnel. Since lichens are both photosynthetic and fungal, they take in carbon dioxide and water and minerals and convert these into more lichen. Therefore, the lichen would have

Fig. 4.33 Lichen. Lichen are a combination of fungi and algae. Photo by author, Missouri, USA

to be eaten or digested to recover the carbon and water. If the rock is limestone or a carbonate, however, the acids from the lichen mycorrhiza (roots) might help with carbon dioxide consumption as the lichen breaks down the rock. Since most lichen grows on the order of years to grow a centimeter and require boiling to become edible, they may be reserved as an emergency ration or just a colorful scented carpet.

Mosses (Bryophytes) (see Fig. 4.34) are fern type primitive plants that thrive in high moisture and lower light areas. They grow faster than lichen, often succeeding lichen over time as an intermediate to vascular plants or larger ferns. Like lichen these can be a living carpet in tunnels or carved out areas, especially given condensation on walls and floors, and part of carbon dioxide-oxygen solution. Like other plants, the mosses will take in carbon dioxide and water and create more moss and release oxygen. That said, it is better to have lichen or moss than mold!

Reproduction in both lichen and moss is by vegetative spread (i.e. pieces growing into full growth) and spores spread in the air.

4.5 Algae and Cyanobacteria

Algae lump a wide variety of photosynthetic organisms together, from microscopic forms to slime and filaments, to structured algae like kelps. While this book does cover the details for growing the microscopic alga types in photobioreactors, other algae are interesting beyond the slime (filamentous) and cellular types. Macrophyte algae look like plants. Kelps, such as bull kelp, water lettuce, and others, are crops that add nutrients and flavor to many dishes, and thickeners for desserts and soups.

Fig. 4.34 Moss. True mosses (bryophytes) are a primitive plant related to ferns. Photo by author

Algae are also a very critical food for fish and shrimp, and it is possible a space farmer would grow algae just to feed water animals and manage carbon dioxide and waste water. In fact, the algae-shrimp or algae-fish loop are very energy efficient way to get seafood on the menu. Note that cellular algae are around 90% water, so nutrition (see Fig. 4.35) is from the remaining 10% of living mass (as if you dried it out). It is very high in protein which is why many fish and shrimp grow fast when fed cellular algae.

Some algae also form plankton in larger-scale food webs, such as diatoms, and might be something to seed ponds or streams in very large settlements or as part of terraforming. Since cellular algae can be genetically manipulated, there can be a large variance in growth rates or reproduction rates. Some alga types like to fasten

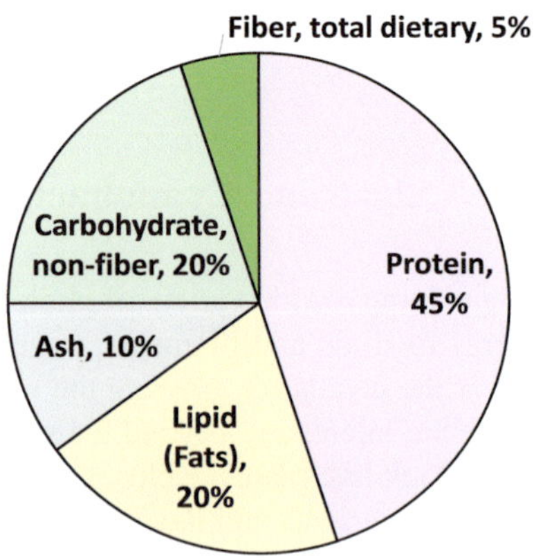

Fig. 4.35 Green algae dry mass nutritional breakout. Loosely estimated from data in Bito et al. (2020)

to substrates, but others grow better as free-floating cells. There are very large families that concern the space farmer, green algae or *Chlorophyta* which includes many freshwater alga types, brown algae or *Phaeophyceae* which includes kelps, and Red algae or *Rhodophyta* which can include cellular and seaweed forms in saltwater or extreme environments. Some photosynthetic bacteria, such as cyanobacteria, are lumped in with algae. Green filamentous algae and cyanobacteria are shown in Fig. 4.36 Cyanobacteria, often called blue green 'algae', comprises a cell type that fits many purposes in the space farm, and other photosynthetic bacteria may fit the needs of the farm. If the algae are mostly cellular, they are squarely a photobioreactor crop. If however, if the alga forms complex shapes (see Fig. 4.37), it may be a modified tank species, or part of an aquaculture structure. Fish often grow better in the presence of vegetation, and the algae of course can help balance the tank's biochemistry as noted many times in this text.

For cellular types of algae and cyanobacteria, growth is exponential, and for many species, the doubling rate is on the order of four to eight hours (see Fig. 4.38). Just like for yeast, the considerations of a vat or substrate can apply, with the same dynamics, except in this case, the inputs are photosynthetic, with carbon dioxide, water, and nutrients, with light input, and living biomass and oxygen out. Algae is 90% water, so water input is very near the living biomass out line. The living mass in the next time unit is the initial living mass from before and the living biomass produced, so it becomes like a snowball rolling downhill, gaining more mass as it gets bigger, exponentially, so that 1 kg at the beginning of the day becomes more than 8 kg at the end of the day.

The trick for maximizing cellular algae growth is balancing exposure of each cell to the right solution concentrations of carbon dioxide and nutrients in water, with the right circulation to carry away oxygen produced, with the right light levels in the right spectrum, at the right pH and temperature. Just a touch off, and the algae cells can either go into suspended animation or die. Lots of work in biodiesel and biofuels has benefited algae culture, and as a result many tank shapes have emerged to balance the

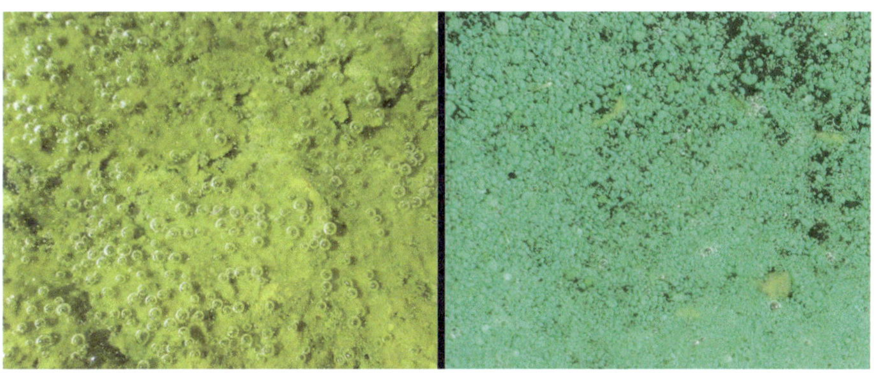

Fig. 4.36 Green algae and cyanobacteria. Filamentous green freshwater algae (left) and cyanobacteria also called blue-green 'algae' (right). Photos by author

Fig. 4.37 Four macroalgae. Kelp brown algae (upper left), sargassum brown algae (upper right), red macrophyte saltwater algae (bottom left), and green macrophyte saltwater algae (bottom right)

chemistry and growing conditions of algae. Note also that many species of algae need a dark time with lower light to allow them to reproduce and balance their internal chemistry (and build proteins, etc.). This means a space farmer that uses algae needs to be well acquainted with cellular biology, in addition to chemical engineering. Since algae grows so very fast, it can become a serious engine for processing carbon dioxide and wastewater. Algae is also size efficient, so that a glop producing farm might be the first version of the farm in a starter settlement (Fig. 4.39).

Type of algae is also a tradeoff. For example, Spirulina (cyanobacteria) (see Fig. 4.40) are very high in protein but grow far slower than green algae.

Getting past the cellular side, more structured algae, i.e. filamentous or macrophyte types, are nice additions to menus though grow slower. Kelps and Sargassum, both saltwater brown algae (again in Fig. 4.37), are very good candidates for a space farm, and good in salads and soups, especially combined in flows with saltwater aquatic animals. Elodea, a freshwater green alga, is good waste processing and animal food that should be added to fish or shrimp tanks whether as an effluent catcher, or in the tank itself as part of polyculture. Elodea could also be grown in its own tanks, then fed to poultry, insects, or cattle. Macrophyte algae, such as the kelp, sargassum, or elodea, can be mass modeled like a vascular plant due to their slower non-exponential

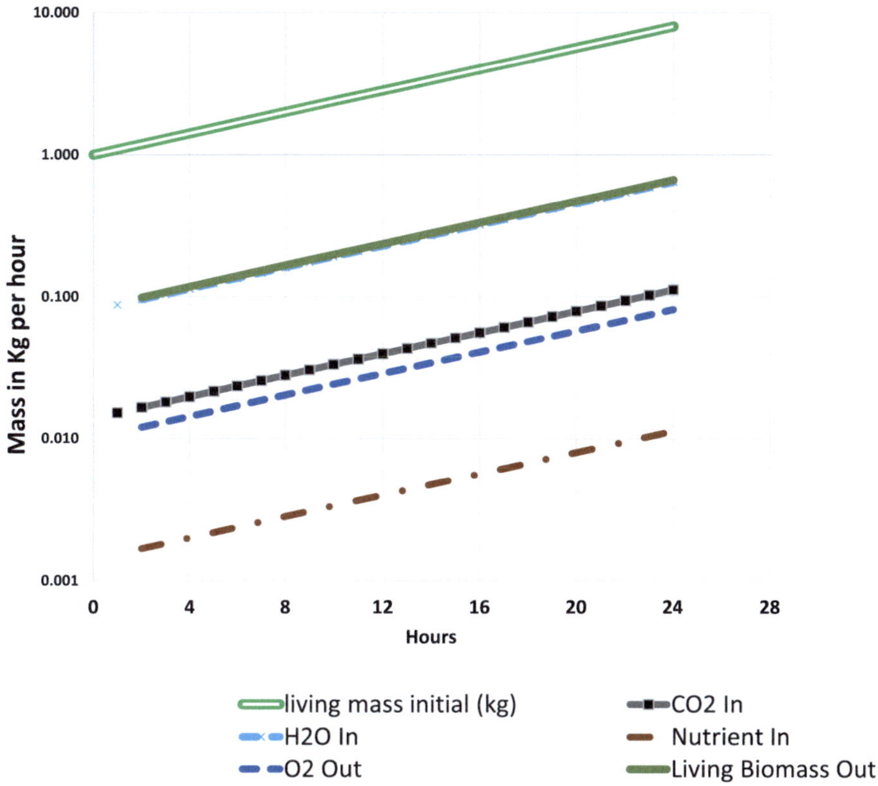

Fig. 4.38 Mass flow model (semi-notional) for an initial 1 kg culture of cellular green algae (*Oocystis* sp. or similar), with 8 hour doubling rate. Loosely calculated from data in Csavina et al. (2011), Sorokin and Krauss (1959)

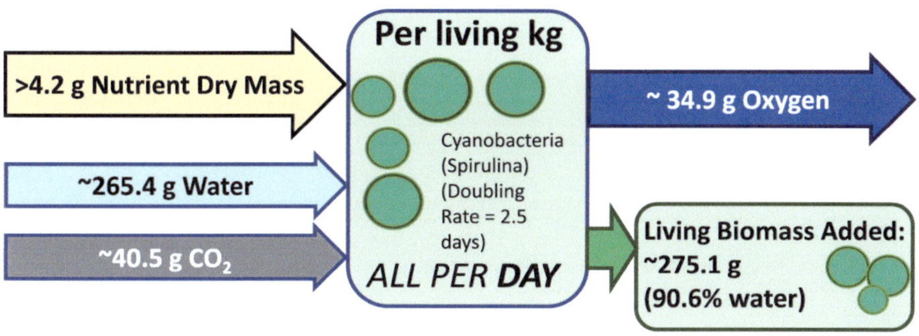

Fig. 4.39 Rough mass flow for Cyanobacteria ('blue-green algae', *Spirulina* sp.). Loosely calculated from data in Rajasekaran et al. (2015) and USDAARS (2022)

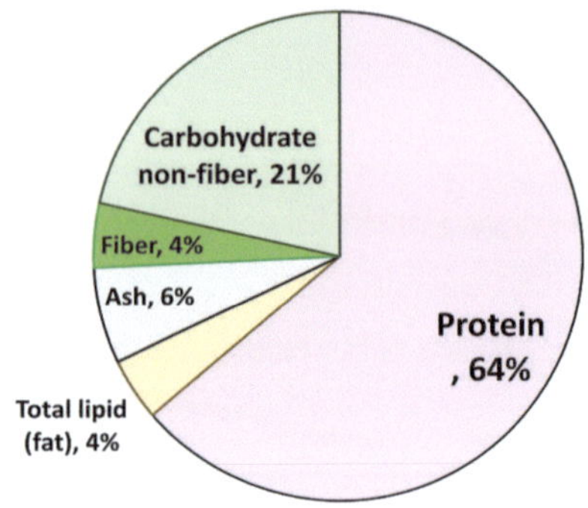

Fig. 4.40 Spirulina dry mass dietary nutrients. Using data from USDAARS (2022)

growth, like leafy vegetables such as lettuce, assuming daily harvest (see Figs. 4.41 and 4.42). Just like lettuce, the whole plant is edible, though some parts are crunchier than others. Kelp grows roughly 2% per day derived from data from Reed et al. (2008), which is in line with many terrestrial plants.

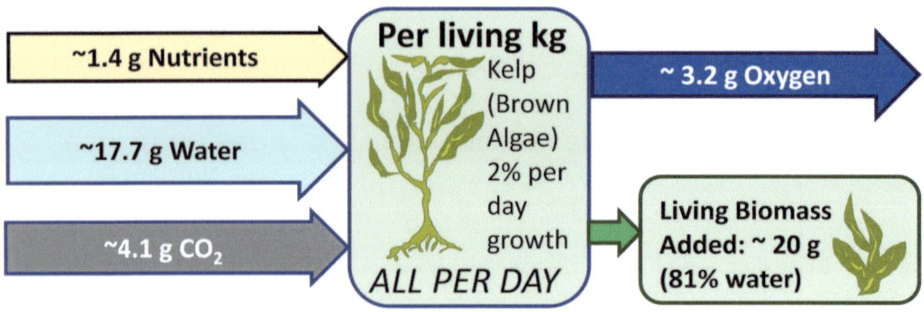

Fig. 4.41 Kelp rough mass flow model. Loosely calculated from data in Reed et al. (2008) and USDAARS (2022)

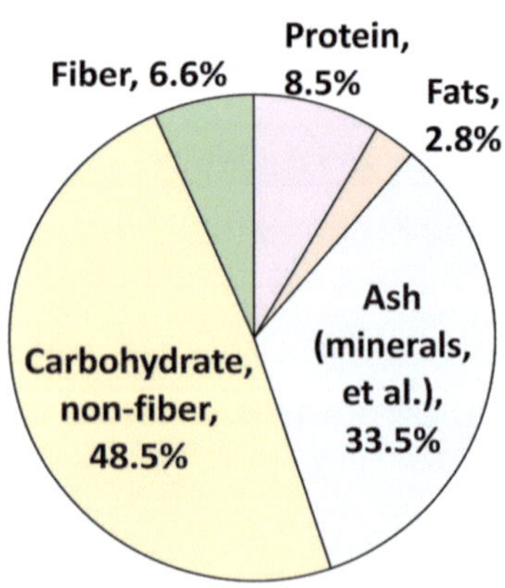

Fig. 4.42 Kelp dry mass nutrition. Loosely calculated from USDAAR (2022)

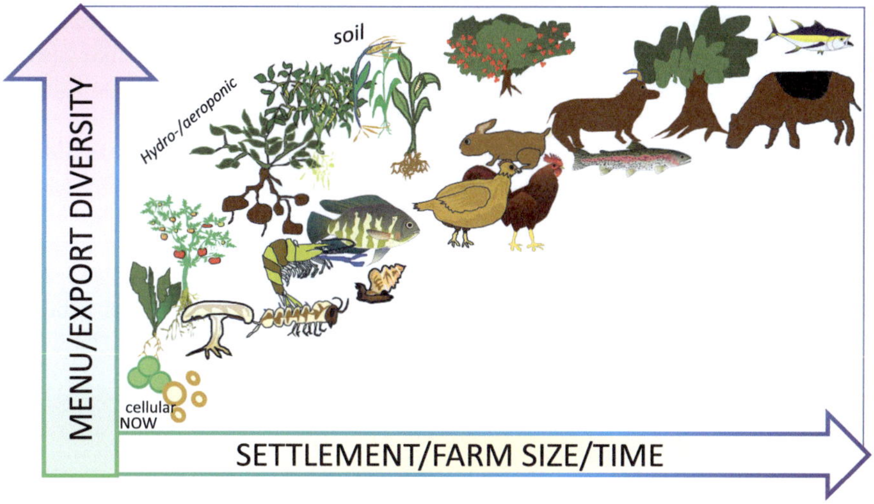

Fig. 4.43 Crop option evolution over time. Starting with cellular crops, the settlement grows and crop options increase with enclosed volume, mass available, and time

4.6 Crop Species Over Time

In the succeeding chapters, we cover how to take all these living things and link them together to grow dinner and process wastes, especially all the machines to get to a full mass flow. In Fig. 4.43 is a spectrum that shows the evolution of crops in a space farm over time. As a settlement grows, with more volume enclosed, and more time, a greater variety of crop options are possible. From initial salad plants, yeast, and algae options for an early settlement, growth in time and size will allow a succession of species including larger plants and more diverse sets of plants such as traditional grains, and maybe bushes, with aquaponic systems adding animals. In more time, with more room will allow tree crops, and maybe terrestrial animal crops, to a distant time with large growing areas that may allow larger livestock, many large plants and trees, and maybe even carnivorous fishes.

4.7 Summary

This chapter covered the growth dynamics, life cycles, and mass flows of several candidate photosynthetic organisms. Vascular plants, which have roots, stems, leaves, and circulation from roots to leaves and back, make up most of Earth's farm produce. Among these plant crops, most of which have growing seasons aligned to spring, summer, and fall, are flowing plants. Grasses and some other flowering plants make up grain crops, while legumes and beans add amino acid rich seeds, and fruiting vegetables such as tomatoes add fiber and vitamins. Other crops are root based, including tubers that can be dispersed like potatoes, or in a tap root like carrots. Salad greens like lettuce are crops where we eat the leaves in lieu of the fruits or roots. Other photosynthetic crops can be algae and cyanobacteria. Algae can be filamentous and microscopic, or macroscopic like kelp forming plant-like shapes. Microscopic alga and cyanobacteria can be grown in machines called bioreactors that can be a simple vat tank, or more complex membrane (substrate) structures, each with growth dynamics tied to doubling rate. Eventually, as space settlements and their space farms evolve, early crops that are more cellular based will be replaced (succeeded) by crops that are more complex such as salad greens and grains, to bushes and trees, and smaller animals with larger animals. This farm evolution is similar to natural succession in forests and prairies on Earth.

4.8 Questions

1. What is the difference between a vascular plant and kelp? What about water hyacinth and kelp?
2. What is the difference between a fruiting plant, a tuber vegetable, and a leaf vegetable? Which of these foods are technically fruits: tomatoes, wheat, potatoes, corn, onions, bananas?
3. What is the difference between the needs of roots versus the needs of leaves and stems in a vascular plant like corn?
4. What is the genetic relationship difference between a plant from seed via flowering, versus a vegetatively sprouted plant, or a grafted plant?
5. What types of plants are good to grow along hallways, along columns or in corners of habitats?
6. What kinds of plant crops should I select to get lots of protein?
7. What are good plants to filter wastewater?
8. Why is a lichen like a plant?
9. What is the similarity between harvest of a leafy green plant and kelp?
10. Which crops are more likely in an early settlement versus a later mature settlement.

References

ACAI. (2022). Scheduled planting and harvest recommendations for Cassava. ACAI Project & Akalimo, National Root Crops Research Institute, Nigeria. Found 2022 at: https://acai-project.org/wp-content/uploads/2020/05/Scheduled-Planting-and-Harvest-Cassava-Nigeria-Acre-latest.pdf

Appenroth, K.-J., Sree, K.S., Bog, M., Ecker, J., Seeliger, C., Böhm, V., Lorkowski, S., Sommer, K., Vetter, W., Tolzin-Banasch, K., Kirmse, R., Leiterer, M., Dawczynski, C., Liebisch, G., &

Jahreis, G. (2018). Nutritional value of the duckweed species of the Genus Wolffia (Lemnaceae) as human food. *Frontiers in Chemistry, 6*. ISSN 2296-2646, https://www.frontiersin.org/article/10.3389/fchem.2018.00483.

Babeker, A. M., Ahmed, A. R., & Mastafa, G. A. (2020, February). Evaluation the quality parameters of sugar cane and raw sugar samples at season 2017 with reference to (SASTA, 2009) and (ICUMSA,1994) standards. *European Journal of Food Science and Technology, 8*(1), 55–71. Published by ECRTD UK, Print ISSN: ISSN 2056-5798(Print), Online ISSN: ISSN 2056-5801(online) found at: https://www.eajournals.org/wp-content/uploads/Evaluation-the-Quality-parameters-of-sugar-cane-and-raw-sugar-samples-at-season-2017.pdf

Bito, T., Okumura, E., Fujishima, M., & Watanabe, F. (2020). Potential of Chlorella as a dietary supplement to promote human health. *Nutrients 12*(9), 2524. https://doi.org/10.3390/nu12092524. PMID: 32825362; PMCID: PMC7551956.

Björkman, T. (2009). Information for buckwheat growers: When to harvest buckwheat. ©2009 Cornell University, Cornell Cooperative Extension found at: http://www.hort.cornell.edu/bjorkman/lab/buck/guide/harvestwhen.php

Csavina, J. L., Stuart, B. J., Riefler, R. G., & Vis, M. L. (2011, August). Growth optimization of algae for biodiesel production. *Journal of Applied Microbiology 111*(2):312–318. https://doi.org/10.1111/j.1365-2672.2011.05064.x. Epub 2011 June 23 PMID: 21624020.

Darby, H. (2024, January). 2023 Spelt variety trial. University of Vermont Extension, Burlington, VT 05401, found at: https://www.uvm.edu/sites/default/files/Northwest-Crops-and-Soils-Program/2023%20Research%20Rpts/2023_Spelt_Variety_Trial_Final.pdf

Dugje, I. Y., Omoigui, L. O., Ekeleme, F., Kamara, A. Y., & Ajeigbe, H. (2009). Farmers' guide to Cowpea production in West Africa. International Institute of Tropical Agriculture. Ibadan, Nigeria.

Masood, A., Masood, A., Singh, K. K., Pramanik, S. C., & Ali, M. O. (2009). Cropping systems and production agronomy. In *CABI* (pp. 213–228). https://doi.org/10.1079/9781845934873.0213

McLelland, J. (2021, December 4). The growth stages of Mung Beans. Hunker. Retrieved 5 December 2021.

Rajasekaran, C., Ajeesh, C. P. M., Balaji, S., Shalini, M., Siva, R., Das, R., Fulzele D. P., & Kalaivani, T. (2015). Effect of Modified Zarrouk's Medium on Growth of Different Spirulina Strains. *Walailak Journal of Science and Technology (WJST), 13*(1), 67–75. Retrieved from https://wjst.wu.ac.th/index.php/wjst/article/view/1416

Reed, D., Rassweiler, A., & Arkema, K. (2008). Biomass rather than growth rate determines variation in net primary production by giant kelp. *Ecology, 89*, 2493–2505. https://doi.org/10.1890/07-1106.1

SIMOC. (2022). Scalable, interactive model of an off-world community (SIMOC). Arizona State University, School of Earth and Space Exploration, Interplanetary Initiative Pilot Project, and National Geographic Society, found at: https://simoc.space/

Sorokin, C., & Krauss, R. W. (1959). Maximum growth rates of chlorella in steady-state and in synchronized cultures. Department of Botany, University of Maryland communicated by Sterling B. Hendricks, October 21, 1959. *Journal of Botany, 45*.

USDAARS-D4685-1. Image Number D4685-1. Photo by Kai-Shu Ling, USDA Agricultural Research Service, (USDA-ARS) found at: https://www.ars.usda.gov/oc/images/photos/aug21/d4685-1/ (use notice: https://www.ars.usda.gov/oc/images/copyright/).

USDAARS, FoodData Central. (2022). U.S. Department of Agriculture (USDA), Agricultural Research Service (ARS), found at: https://fdc.nal.usda.gov/

USDAERS-RiceYB, Rice Yearbook, U.S. Acreage, Production, Yield, and Farm Price. (2021). Economic research service. U.S. Department of Agriculture, updated 3/31/2021, found at: https://www.ers.usda.gov/webdocs/DataFiles/53146/U_S_and_State_acreage_yield_and_production_all_rice_and_by_class_marketing_year.xls?v=9656.4

Worley, J. W., & Sumner, P. E. (2020). Grain sorghum: harvesting, drying and storing. Circular 1017, The University of Georgia © 2020, found at:https://extension.uga.edu/publications/detail.html?number=C1017&%20Drying%20and%20Storing

Wu, C. Y., Feng, Y., Shohag, M. J., Lu, L.-l., Wei, Y.-Y., Gao, C., & Yang, X.-E. (2011). Characterization of (68)Zn uptake, translocation, and accumulation into developing grains and young leaves of high Zn-density rice genotype. *Journal of Zhejiang University, Science B, 12*(5), 408–418. https://doi.org/10.1631/jzus.B1000291

Chapter 5
Basic Machines for the Farm

Abstract This chapter covers the various devices, mechanisms, and structures that are needed to control and contain space farms. Starting with devices to move mass, such as pumps, fans, then contain mass in vats, tanks, and columns. Other machines are required to separate components of mass, such as filters, settling tanks, distillation columns, and other separation technologies. Condensers and coolers are key to controlling humidity. Injecting gases via aeration is via combinations of pumps and various structures are important for many purposes. Lighting has many purposes and shapes in the farm, such as sterilization, plant growth, and safety. Thermal control and heat rejection are critical to control growth environments. All these machines use electricity, which the chapter examines as energy that must be modeled for design. Finally, the chapter covers control software and systems, which can use artificial intelligence in various roles.

Keywords Valves · SCADA · Artificial intelligence · Machine learning · Electricity · Lighting · Pumps · Filters

5.1 Introduction

A space farm is composed of more than just living things. In this chapter, we cover some of the basic machines needed to move mass and create the environment needed to grow crops and recycle wastes from humans and crops. Some of these machines are simple containers, while some high technology devices pulled from various industries. Important areas for devices are fluid separation, containment, pushing and pulling, and circulating. Lighting is also critical, especially to grow photosynthetic organisms, to allow animals to orient, and for human safety when working in the farm. All these machines create heat, and living things release humidity, which both need to be handled. Finally, controlling these devices, assembled into systems and stages, called Supervisory Control and Data Acquisition (SCADA), can be borrowed from chemical factories and refineries, with added artificial intelligence (AI).

© The Author(s), under exclusive license to Springer Nature Switzerland AG 2026
B. L. Meyer, *Designing Farms in Space*, https://doi.org/10.1007/978-3-032-04395-5_5

5.2 Pumps, Fans, Stirrers, and Pump Types

In this book a pump is anything that uses energy to increase pressure in fluids. Pumps use electrical energy to push a mass of liquid at a pressure. Since a fluid can be a gas or liquid, fans are technically pumps too. Pumps are used heavily in any space farm design, from a simple hydroponic garden sharing air with a habitat, to circulating water in tanks, to a more complex mass balancing farm.

5.2.1 Liquid Pumps

Much of the energy in a farm will be moving liquids around using pumps. There are many current technologies, and few that could become useful in future.

5.2.1.1 Impeller Pumps

The most common type of pump on Earth now is the impeller pump. An impeller is like the propeller of a plane or prop on a boat, or turbine in a jet. Impeller pumps use blades to pull in fluid from the center, then push it at higher pressure to the edge for output. Blades grab and push liquid from an input pipe to an output pipe, increasing the pressure between in and out. Many impeller pumps also have valves to make sure the liquid always flows from in to out. If the impeller blades are made to chop when spinning, the pump is called a macerating pump, since it takes input liquid with chunks of solids and chops up the solids as it pushes out the liquid. Macerating pumps are all over in daily life on earth, from in sink garbage disposals, to sewer plants, to bathrooms on small boats. That is why some reefs offshore have extra fish activity as the fish feed on the macerated output of boat toilet heads. I wouldn't keep those fish in that case. If instead you push fluid through an impeller (see Fig. 5.1), it becomes a turbine for generating power, such as the turbines in hydroelectric dams. Impeller pumps are very common for liquids, but a version can be used to move gases. An impeller pump for moving gases like air at lower pressures is called a blower.

5.2.1.2 Stirrers

Stirrers are simple shapes that circulate liquid by spinning (see Fig. 5.2). Home aquarium waterfall type pumps use a stirrer that is magnetically connected to a coil in a shaftless design. Another shaftless example is the stirrer bars dropped into a beaker in chemistry labs. Scaled up, a shaftless stirrer might be used in bioreactors, or in fish tanks. Other types have a shaft that passes from the liquid part to a motor. Anytime you make a hole for a shaft, you risk leaks. Seals around these shafts must

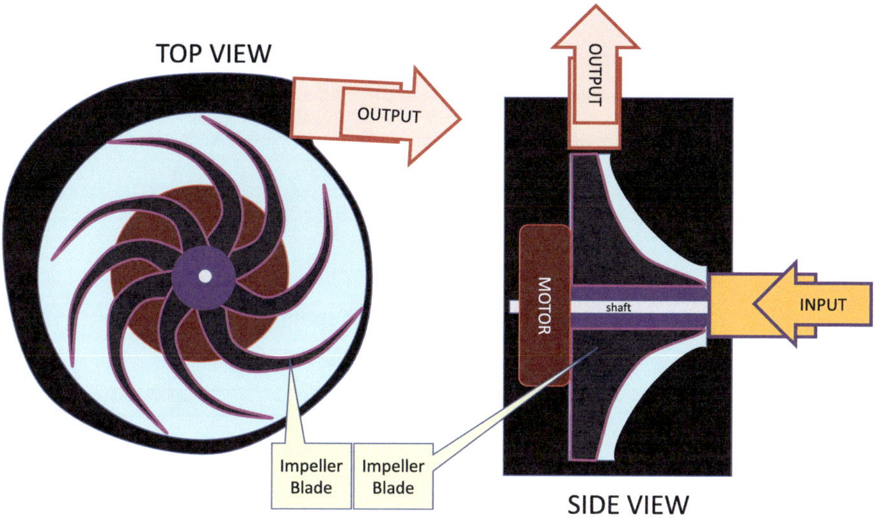

Fig. 5.1 Impeller pump schematic

be maintained and replaced often. That said, shafted stirrers are more power efficient than shaftless designs. Shafts can come in from above or below and likely will see use in open format initial settling tanks for waste handling parts of the aerobic yeast-bacteria bioreactors.

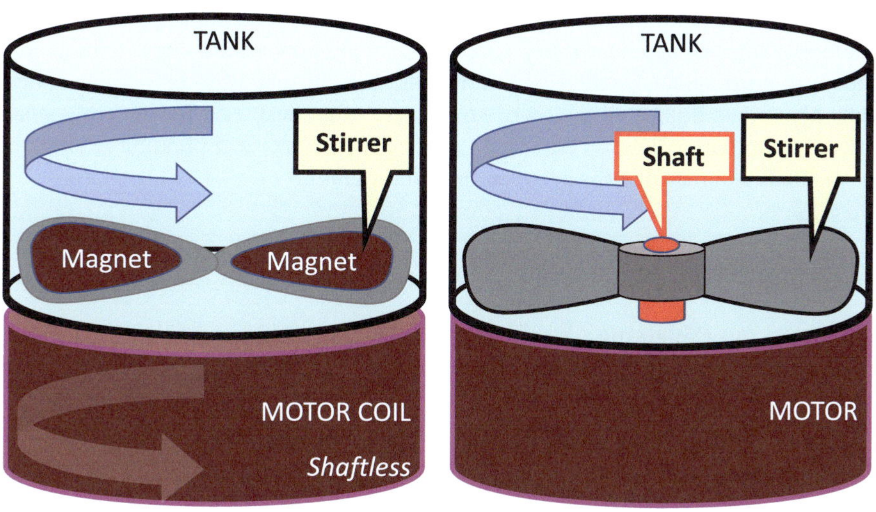

Fig. 5.2 Stirrer designs. Stirrers can be shaftless or have a shaft. Shaftless designs rely on magnetic fields made by the coil to spin the stirrer. Shaft designs simply have a motor that spins the shaft and hence the stirrer

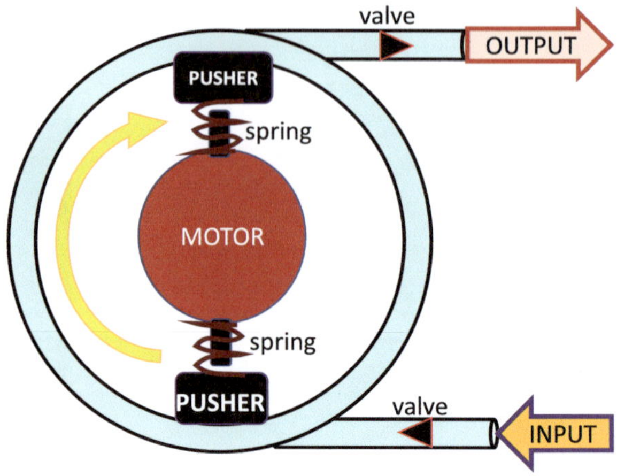

Fig. 5.3 Capillary pump schematic. Capillary pumps use tubes that are compressed by rotating pushers to move liquids. These pumps are common on dialysis machines and drug dispensers in hospitals

5.2.1.3 Capillary Pumps

Capillary pumps are similar to those used in medical devices and are used in the farm in places where crushing cells is a problem. Some designs might use these pumps in the photobioreactor or open tank yeast or bacteria bioreactors, in lieu of using aerator bubbles to circulate, or in lieu of using a slow fan type open impeller. In the capillary pump (see Fig. 5.3), a tube is compressed, like how toothpaste or lotion is squeezed from a tube, by a slow turning motor arm. Each end of the tube has a simple one-way valve that keeps the liquid moving forward (see diagram). Future pumps along these lines might work the way peristalsis moves material in your intestines. The down sides of capillary pumps are that they are power inefficient, tend to clog once stuff accumulates on the edges of the tube, and the tube itself fatigues and pops. They also move liquids very slowly, so they are often best used for injecting small amounts into larger flows. They might also be used to re-seed photobioreactors or move small volumes of cells between bioreactor tanks.

5.2.1.4 Magnetohydrodynamic Pumps

One of the most interesting future technologies will be using magnetic energy to move liquids without any moving parts. They work by creating a powerful magnetic field that acts on the molecules of the liquid (see Fig. 5.4). Currently, magnetohydrodynamic pumps are found in factories to move liquid metals or highly concentrated ionic solutions, but one day could see use in a space farm if room temperature superconductors are found and developed. (Note: in a certain fictitious movie/novel a certain Soviet submarine used the same technology.) Since these pumps have no moving parts they could make pumps that need little or no maintenance.

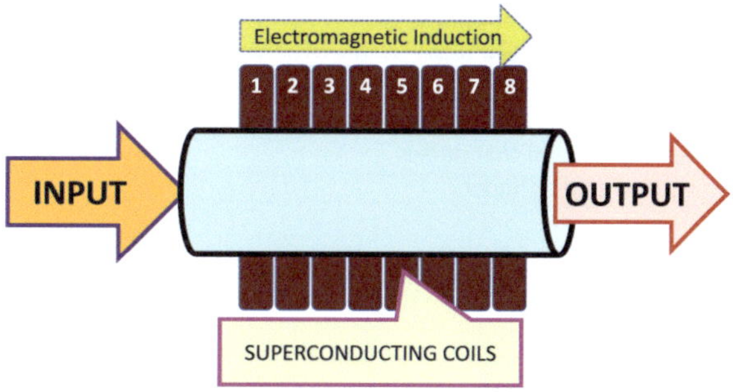

Fig. 5.4 Magnetohydrodynamic pump schematic. Pumps of this type use strong magnetic fields to pull the molecules of the liquid itself

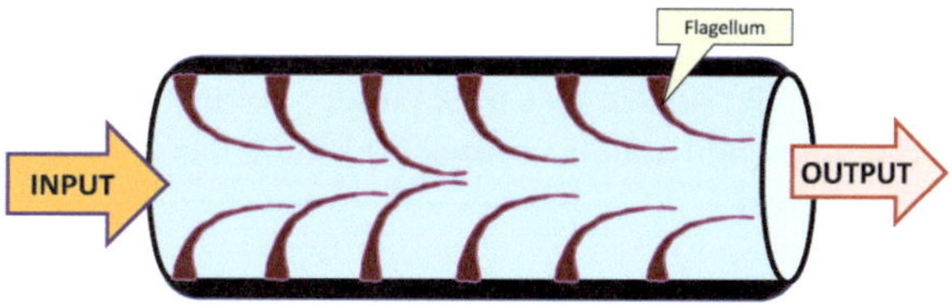

Fig. 5.5 Flagellar pump schematic. In flagellar pumps, the flagellum (whips) ripple in motion to move liquids. These work best for thicker fluids or small pipes

5.2.1.5 Flagellar Pumps

Another future technology might use a method borrowed from nature: mats of tiny hair-like whips called flagella that push liquids (See Fig. 5.5). This is how many small animals like protozoa move, down to paramecium cells that move the same way, and how water is moved inside some very small sea animals. Given advances in muscle metals from robotics, larger scale flagella could line pipes and push a variety of liquids.

5.2.2 Pumps for Gases

Gases are fluids, like liquids, but unlike many liquids, they can be compressed. There are many ways to move gases through the farm by increasing or decreasing pressure, using valves, and many other methods. Almost all require electricity via motors or coils.

5.2.2.1 Diaphragm Pumps

Small aerators for home aquariums or even larger high-pressure aerators for ponds on Earth often use diaphragm pumps. A blacksmith's bellows is a close relative. In this type of pump, a cylinder has a solid or rubber (or other material) flat piece that is pulled and pushed by either a rotary electric motor or linear coil (see Fig. 5.6). A valve lets in gas as the vacuum forms in the cylinder, then another valve lets

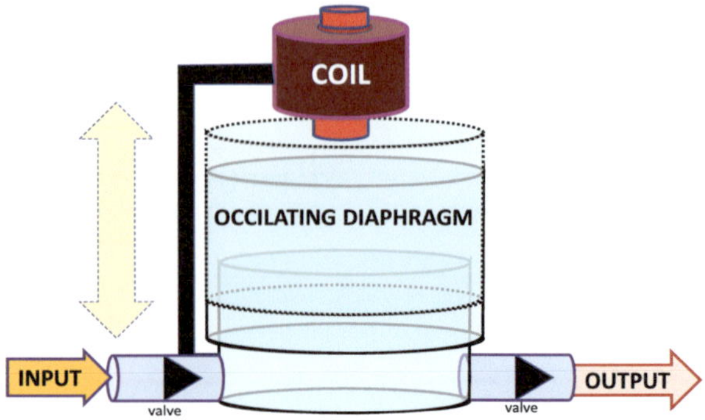

Fig. 5.6 Diaphragm pump schematic. The diaphragm can be a flexible metal or plastic, or the diaphragm can be a solid element that pulses on a lubricated cylinder

gas out under pressure as the cylinder volume is compressed. Bioreactors, root water circulation, and aquatic animal tanks will use these or similar pumps for gas injection and aeration.

5.2.2.2 Fans

Just like the stirrer for liquids, a fan has blades that move fast or slow, in either a duct like a box fan or vent fan in a car or in the open like a ceiling fan (see Fig. 5.7), to move large volumes of gases at lower pressures. Hydroponic areas have fans to move carbon dioxide rich air to plants and pull oxygenated air way from the plants. The fans move air for cooling as well in many places in the farm, especially for gardens, but also for terrestrial animal areas, in the human habitat and around machinery in all parts of the farm for heat rejection. Fins on pipes that contain cooling liquids are blown by fans to send heat into the liquid and send the heat away from the machinery, tanks, or air around plants and animals. Like an impeller, pushing a fan to turn it makes the fan a power generator as for windmills.

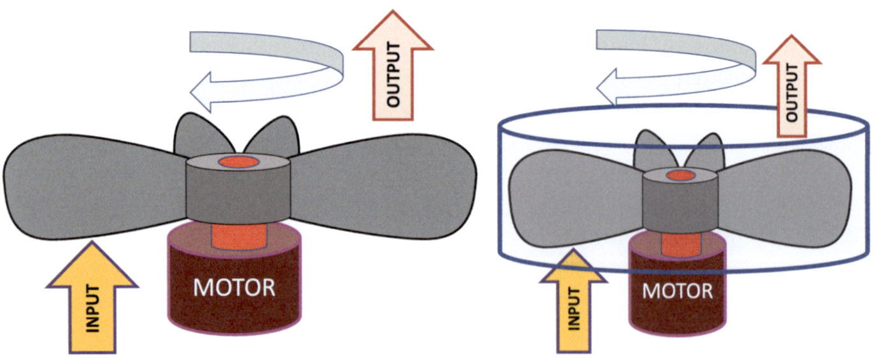

Fig. 5.7 Open and ducted fan schematics

5.2.2.3 Blowers

Blowers are between pumps and fans. A blower is essentially an impeller pump for gases, though can also be a ducted fan, depending on pressure needed. Very high-pressure blowers are turbines like impeller pumps above, and one version on Earth is a turbocharger on a car. Another common blower impeller design is for leaf blowers. Lower pressure versions, such as ducted fans, simply have rows of fans in a tube, usually found on earth for ground cooling ducts for airplanes parked at a gate. Impeller designs provide higher pressure but for lower volume, versus ducted fan designs which are lower pressure but provide high volume output.

5.2.2.4 Others

There are other technologies for moving gases, such as heat, sound pulses, peristaltic pumps, and other methods yet to be determined. The trades are volumes and masses moved versus energy required to do the movement, equipment service life, and design complexity. Remember energy is a critical resource for a space farm.

5.3 Settling Tanks, Filters, and Filter Types

The basic function of settling tank or filter is to separate mass in a flow. The mass might be dust from clean air, or oils from water. A basic commandment in fluid engineering is thou shalt always use a filter before pumping, though macerating pumps are the exception. There are many methods to make the separation work, and there are some examples below.

5.3.1 Settling Tanks and Centrifugal Methods for Liquids

A simple filter type is a settling tank. Settling tanks are basic means to use gravity to extract solids or denser liquids from liquids. In most uses the input liquid has suspended solids that gravity pulls to the bottom of the tank, and cleaner liquid gets pulled out minus solids (see Fig. 5.8). Liquid enters from the top, solids settle to the bottom (assuming gravity, lower the gravity, the slower it settles), and an output pipe enters at some level above the bottom of the tank to pull out cleaner fluid. A third pipe at the bottom of the tank pulls out the accumulated solids or sludge. The bottom might be a funnel shape to make it easier to pull the sludge out. This same method works for liquids of different densities. In some scenarios the tank (usually stretched into a column) can be used also to separate denser liquids from less dense liquids. There are many places in the farm that will use a settling tank, mostly to

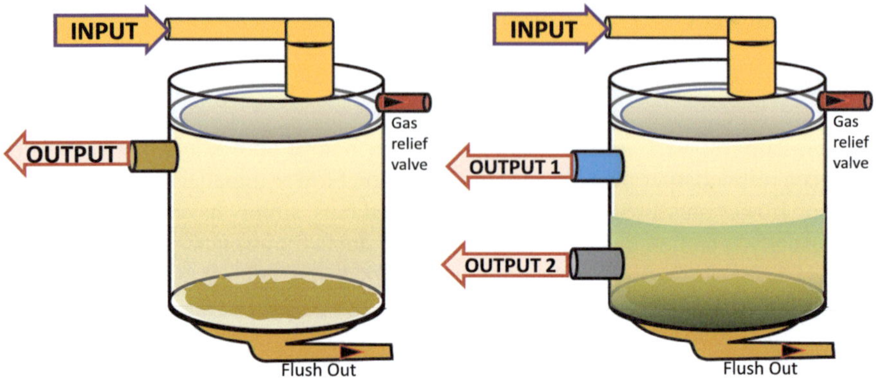

Fig. 5.8 Settling tank schematics. The tank on the left is for separating solids from liquids, the tank on the right is for separating liquids of different densities and separating solids

prevent clogging of later parts. Good designs also include a gas relief valve to prevent pressure from bubbling gasses popping the tank!

In lower gravities like the moon or Mars, the settling tank takes more time to settle solids from liquids, or liquids of different density. A Mars tank might take at least 3 times longer to settle than the same tank on Earth. Assuming microgravity, a centrifugal separator (see Fig. 5.9 can spin to pull out and separate solids from liquids (e.g. algae cells from solutions), or liquids of various densities. Centrifugal flow separators are common on Earth already and are available from many companies. This adds some complexity, but if there is little gravity to use, you make do.

5.3.2 Settling Areas for Gases

Similar designs to settling tanks can use gravity for air or gases loaded with dust. Air in vents can be pulled into slower flowing areas to settle out particulates (see Fig. 5.10), like how dust collects in corners and bends in vent pipes in heating systems on Earth.

Static electricity can be used to aid the dust pull out in lieu of just gravity, and the static settling concept will be key for places like the Moon with charged dust. A canister vacuum cleaner combines a settling area (the canister) with a screen. This is applied in a less good way in vent pipe bends, edges, or corners that grab dust. This principle will apply to many methods that follow.

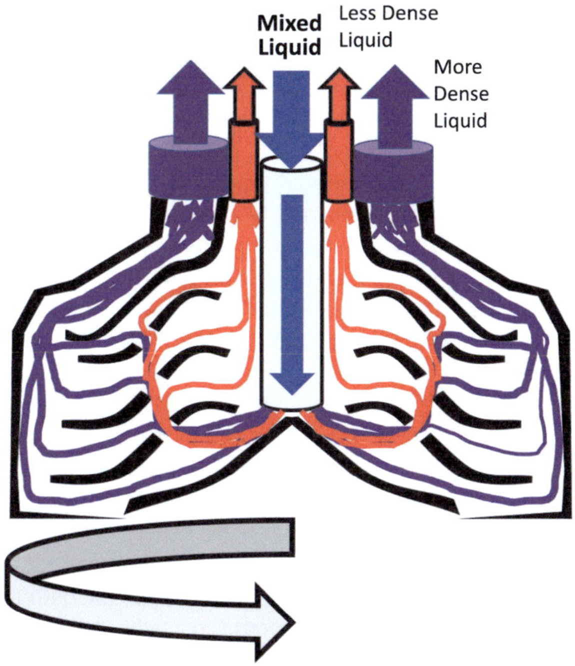

Fig. 5.9 Centrifugal separator schematic. Rotation forces denser liquids or solids to the outer edge, while lighter liquids (fluids) are pushed to the center

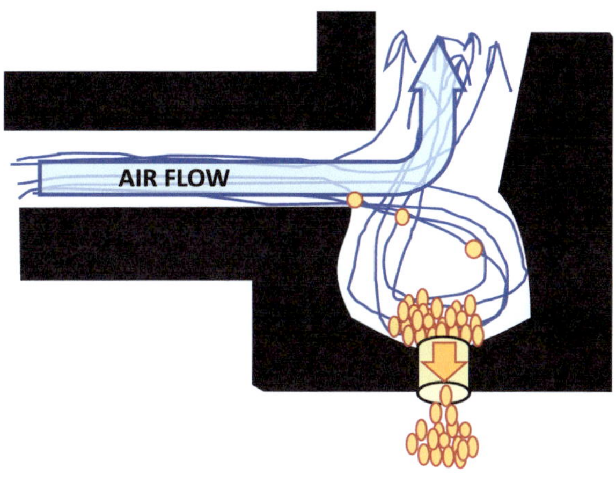

Fig. 5.10 Using air flow to settle dust

5.3.3 Simple Screens and Canister Filters

A screen is a simple filter where liquid or gases (both of which are fluids) are pushed through a mesh. The mesh catches particulates down to the mesh size, and cleaner fluid goes through. This type of filter works well as long as the filter gets cleaned and replaced. A drain in a dishwasher has a screen, as does the filter for an air conditioner or heater, and exiting the canister in a vacuum cleaner. Some filters use fiber mats in lieu of metal or plastic meshes, and some meshes are electrically charged to capture extra particles. The finer the holes in the mesh, the more force is required to push fluid through it, the quicker it will clog, but the cleaner the fluid result. The larger the surface area exposed to the flow, the slower it clogs, and greater the throughput. Often folds are used in mat-type filers to increase the time between replacements, and lower the force needed to push through the fluid. Building and home HVAC (heating, ventilation, and air conditioning) filters are commonly many centimeters thick folds of a very thin paper and cloth mat. Dust accumulates first in the folds. Some arrangements for liquids crank the screen mesh to allow material to settle on the bottom, where a pipe pulls out the material to keep the screen area cleaner, as in Fig. 5.11

A filter might also have a round mesh cylinder in a canister, combining a settling tank with a screen. Fluid flows into the larger canister, clean fluid pulled from inside the cylinder. For combustion engines, the same method applies to oil and fuel filter. If you open that canister vacuum or oil filter, you will see this arrangement, likely using a paper fan folded fiber mesh. There is another type called a bag filter, where the high-pressure input goes into a looser bag filter that is inflated by the pressure, the fluid pushed through the filter becomes the output. This filter type is common in small vacuum cleaners and pool screens but requires constant cleaning.

In some liquid versions, a third pipe with a valve, called a'flush out., is periodically used to pull out the thicker material that accumulates in the bottom of the canister. The canister-cylinder screen plus settling tank approach delays screen clogging and provides more surface area than a flat screen in a pipe would.

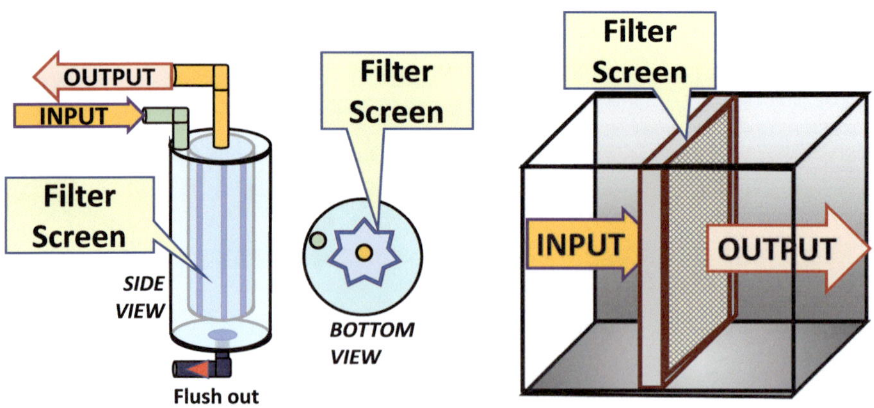

Fig. 5.11 Canister type rigid filter and simple box screen filter schematics

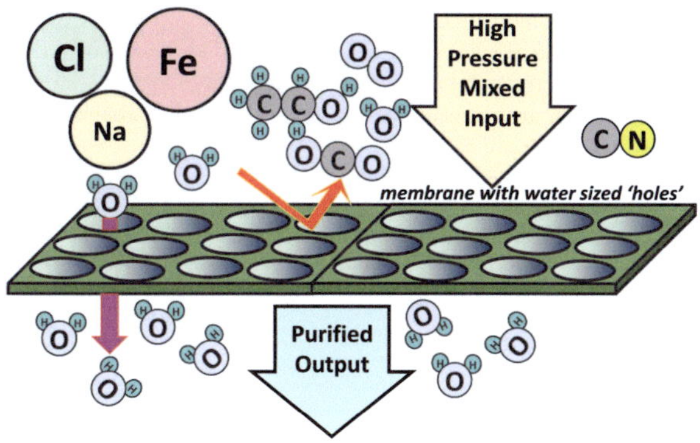

Fig. 5.12 Reverse osmosis membrane basics. Reverse osmosis membranes have molecular holes that selectively allow water molecule sized compounds through, but block molecules larger than water

5.3.3.1 Reverse Osmosis Filters

An extreme form of 'mesh' is a semi-permeable membrane. Cell walls are made of membranes using lipids, but a larger artificial version can be made from cellulose-based materials. The mesh is at the molecular level, and molecules of only a certain size or smaller make it through the membrane. See Fig. 5.12. In most cases, the membrane would have flow in favor of the osmotic balance, so that if a membrane had saltwater inside for example, and pure water outside, the pure water would flow in to balance out the salt water, popping the membrane. Reverse osmosis uses pressure to reverse the flow, i.e. push through cleaned liquid (usually water), and keep back anything that has a bigger molecule than water like salt, dirt, sugars, etc. Given the hyper-fine nature of this screen, lots of pressure is required to push through liquids, and careful designs to prevent clogging. Early designs used a bag filter approach, though more modern designs use the rigid filter structure. Usually, a reverse osmosis filter is used after another filter that catches the bigger stuff before the liquid goes to the reverse osmosis filter. Like in other examples the reverse osmosis filter uses folded membranes held in place by plastics, usually in a canister tank design. Desalinization and water recycling use reverse osmosis filters on Earth.

Future versions might take advantage of nanotechnology or bioengineering. In a nanotechnology version, special molecule machines selectively push water molecules (or whatever molecules we want) across but block everything else, very close to what some cells do. This is like molecular gates in cell walls that selectively pull in molecules needed by the cell. In the bioengineered version, membrane-based cells do the same thing, passing through the stuff we want, and preventing passage of the stuff we don't. These filters may be sandwiches of membranes in lieu of canisters.

5.3.3.2 Particulate and Fiber Media Filters

There are many forms of particulate or media filters, the most common devices people see are on home swimming pools and fish tanks. There are very large-scale examples

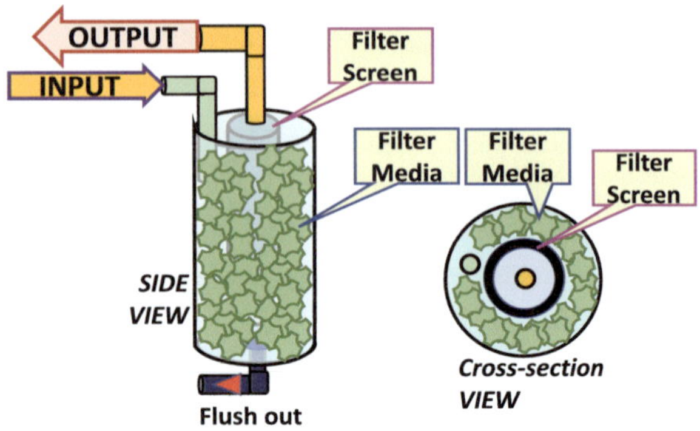

Fig. 5.13 Canister media filter schematic. This type is very common for fish tanks, and the media are usually seeded with aerobic bacteria

in nature too, such as marshes and gravel streams. Commonly, a canister filter is filled with filter media (see Fig. 5.13). This media can be synthetic fibers (like aquarium floss), rocks, sand, gravel, lava rocks, porcelain shapes, charcoal, plant fibers like straw or coconut husks, or even hollow beads. The coarser the material, the less likely the filter is to clog. Debris or particulates are captured by the media, and a simple cylinder shape in the center, usually with a simple coarse mesh, is used to pull out the cleaner fluid. These media filters when used for liquids can double as bioreactors if the media is seeded with bacteria that help chemically balance the liquid. The bioreactor version is called a biofilter. A form of this filter that will be core to most space farms is the use of plant roots in a hydroponic tray as a biofilter. Charcoal adds an additional benefit as media since it removes contaminants and toxins. Zeolite is a mineral that is good for ammonia control in water or carbon dioxide control in air. For gases, the same principle applies except that typically you do not want the media colonized by mold or bacteria, so media filters for gases may include a sterilizer. For an air filter that uses media it is critical to sterilize the air, and dehumidify the air, before it enters the filter media.

5.3.3.3 Electrostatic and Magnetic Attractor Filters

Assuming the stuff you want to filter out can be attracted by static or magnetism, charging or magnetizing a screen or media can pull out particles or sort dry flows. Dust filters in some air purifiers use statically charged meshes to catch dust. Magnetic filters can be used to pull out metal particles shed from manufacturing or metal parts in use. For in-situ mining, magnetic fingers or fins can sort iron ores out of regolith. Static filters can also be used on the moon to catch dust (FYI: moon dust is charged) before it enters the farm or habitat.

5.4 Condensers, Coils, Fins, and Columns

Various methods are used to either heat or cool air. Sometimes cooling can be used to collect any vaporized liquids, or even condense gases into liquids, for collection. Like the reverse of a settling tank, a column can collect condensed liquids from a mixed gas using temperature and gravity.

5.4.1 Condenser Coils and Fins

Coils and fins are shapes used to cool or heat fluids including air. They work by passing a cooling fluid through a pipe, which then cools the gas or liquid near it. The more surface area in the shape, the better the heat transfer from the cooling fluid to the air or liquid. A nice effect is that with cooling coils or fins in humid air, pure water condenses on the fin or coil and drips off to be collected to reuse. Combining a fan with a coil or fin array makes a dehumidifier, which is critical in gardens to collect transpired water, or in habitats to collect water in breath and evaporated sweat and cool the air. Combine a fan with the same coil or fin array, but pumping in a heated fluid, and the machine becomes a heater. Directly exposing a coil in a tank can control the temperature in the tank based on heated or cooled liquid inside the coil. Finned mechanisms are common in heat rejection schemes where one set of fins are in the higher heat area, a second set of fins is exposed to a lower heat area or free space. Fluid is pumped inside the pipes that connect the fins to move the heat from the higher heat area to the lower heat area or radiated to free space. In the Fig. 5.14, in the coil (left) a cooling fluid is pumped through, pulling heat from the gases or liquids around it. In the fin array (right) the same principle applies, where cooling fluid is pumped through, and the fins improve heat transfer from the surrounding gases or liquids. Given warmer fluids in the coil or fin in lieu of cold liquid, the coils or fins become a heater.

In Fig. 5.15 is a dehumidifier. This machine combines a coil (pictured) or fin array (not pictured) and fan to cool and dehumidify air. Cooling fluid is pumped through the coil and the ducted fan blows moist air over the coil. The coil condenses out water vapor to liquid that can be collected and reused.

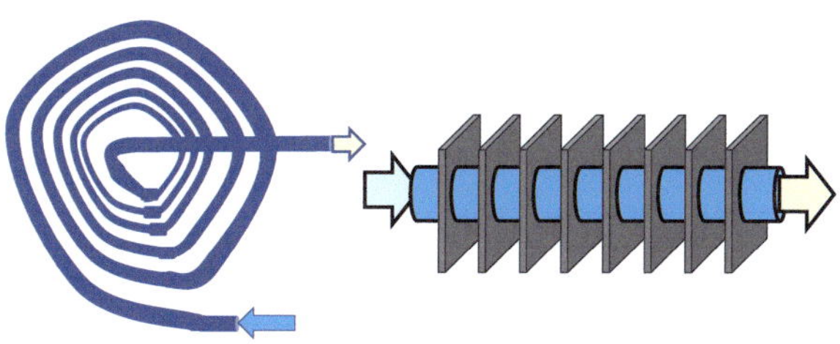

Fig. 5.14 Coil and fin cooling heat transfer methods

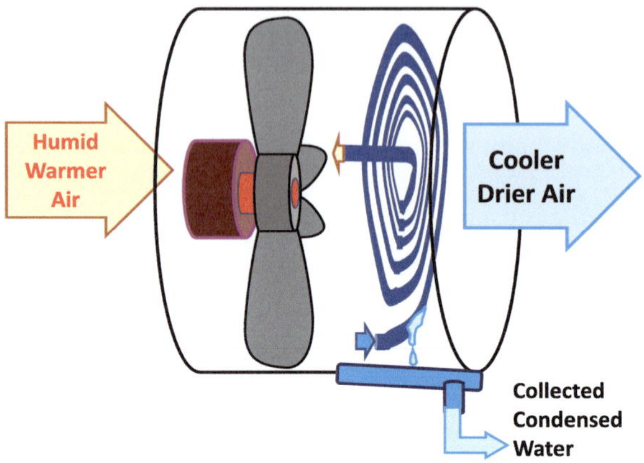

Fig. 5.15 Simple dehumidifier (condenser) schematic

A simple fan-less fin condenser might work for pulling water from exhaust gas from an incinerator. In this event the fins extend into a pipe or box with a liquid collector at the bottom, though a condenser column as below might be better yet. Fins may be of many shapes to steer gas flow, such as spirals, corkscrews, or waves to make the flow and heat transfer more efficient. Note that the same design for a dehumidifier works for cooling fluids or air, with fins or coils. Usually, the fluid in a dehumidifier is passive cooling fluid that went underground or through cooling fins somewhere, but sometimes you need to use energy to make cold fluid. The machine that makes fluid cold is a chiller.

5.4.2 Chillers

Chillers are devices that cool fluids for use (see Fig. 5.16). Most people have a chiller in their air conditioners and in their refrigerators/freezers. Many chillers use energy intensive thermodynamic processes of various types, including compression then expansion of fluids like freon or ammonia with compressing pumps.

There are some solid-state designs using thermoelectric cooling, such as Peltier coolers, as well, and it is likely space settlements may use other types too. In Fig. 5.17 is such a cooler. A thermocouple with two semiconductors of different resistances is powered with electricity, the result is that one side gets cooler, the other get hotter. The hot side of the cooler usually has fins and maybe a fan to pull away the heat.

Ideally, cooling is part of the overall heat rejection strategy and uses less chillers, and more movement of heat around in the farm. Unless you want to get deep into thermodynamics and heat transfer, we will leave it at the fact you will need chillers usually to generate cool fluid to condense liquids, or air condition habitats, and those chillers use lots of electricity.

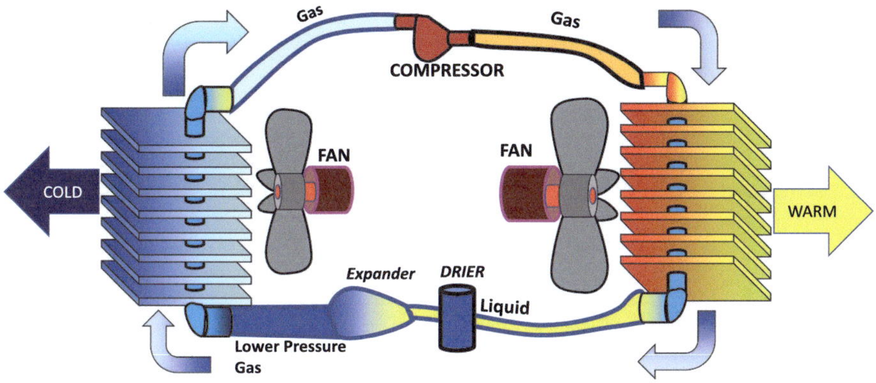

Fig. 5.16 Classic state change chiller schematic. A fluid like freon or ammonia is compressed and cooled to a liquid, then expanded to a cold gas

Fig. 5.17 A rough Peltier thermoelectric effect chiller schematic

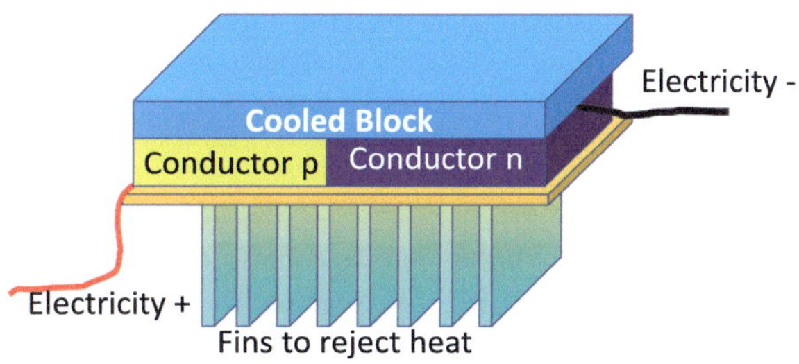

5.4.3 Distillation and Condensation Columns

One way to separate mixtures of liquids is boiling the stuff then condensing the various vapors. Distillation columns work by boiling liquids to vapors then using the differences in densities of the vapors and temperatures of condensation to pull the component liquids apart. A basic distiller cools vapors to get purified liquids. See Fig. 5.18. Anyone who has used or drank alcohol is consuming a distilled product. Moonshine is made using a backwoods distiller. A smart liquor distiller understands that methanol (the stuff that makes you blind) vapors first, then as temperature climbs, ethanol (the good stuff) comes off next. Go hotter yet, water boils and you get just purified water (which is good but doesn't have the kick). Some desalination uses distillation, especially on older boats or older city water plants. Take the distiller and point it vertically, adding take off points at key locations in the column, and you get a distillation column. Petroleum refineries use distillation columns to separate the chemicals in raw crude oil. As the oil is boiled, asphalt and sludge stay on the bottom, jelly comes out next, kerosene as you go further up, gasoline near the top, and the rest get flamed off, or condensed using a cooling tube for other uses. Other chemical processes also use distillation columns. Using control systems can make

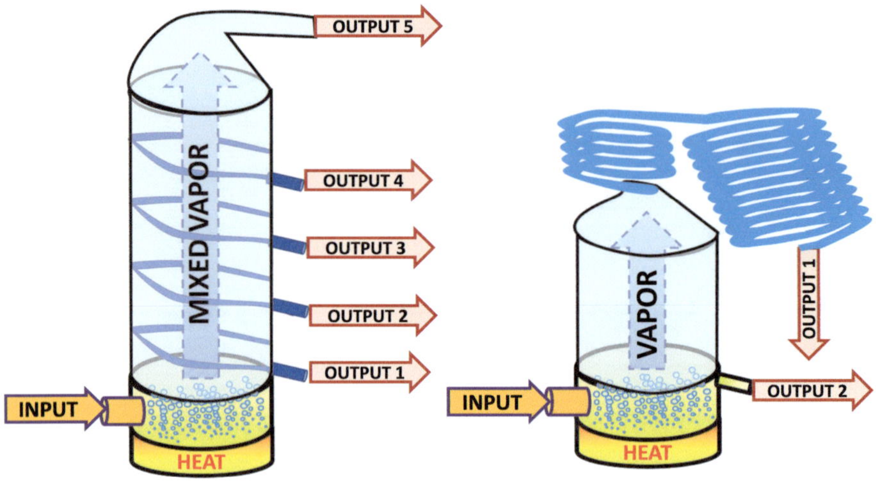

Fig. 5.18 Distillation column schematics. The column on the left is like those used in petroleum refineries, where the most volatile compounds are extracted higher up the column. The column on the right is like those used in alcohol production

distillation columns very useful to the space farmer and in situ miner, at the cost of heat and heat rejection. That said, there is a cheat. Vacuum makes many liquids boil at low temperatures. So, if you are willing to lose a bit of mass to space, vacuum can replace the heat in your column for free space, moon, or asteroid settlements.

A condensation column or cryopump works in reverse. A vapor is passed through a tube, surrounded by a container (jacket) of cooler liquid. The tube is cooled, and the vapor condenses to liquid. Gravity affects this behavior too. See Fig. 5.19. You will need taller columns with gravity on the moon or Mars. In the next figure, you see three example columns. All three columns have the same volume but must be shaped differently to have the same settling behavior when put on Mars or the Moon. The column on Mars is narrower and longer due to the 1/3rd gravity of Mars. The Moon has 1/6th Earth's gravity, so its column is twice as tall as the Mars column, and narrower yet.

An interesting combination of condenser column and a vacuum, or simple tall column can perform a critical function in the farm: take carbon dioxide rich air and water vapor, and pull out the water, leaving the carbon dioxide mix. Since humid carbon dioxide rich gas is the result of aerobic respiration or burning, as after an aerobic bioreactor or fuel cell, getting the water out for reuse is very important. The remaining gas is usually useful by itself, but if you have to further pull-out carbon dioxide from a mix of oxygen and nitrogen, you will need to use the gas separation methods farther below.

In Fig. 5.20, the left drawing is the basic condensation column. Cool liquid is pumped into the cooling jacket where a coil of tube cools vapor to liquid. On the right, the same column is connected to another simple tube column (a simple gas extractor) that allows dissolved gases to bubble out from the condensed liquid.

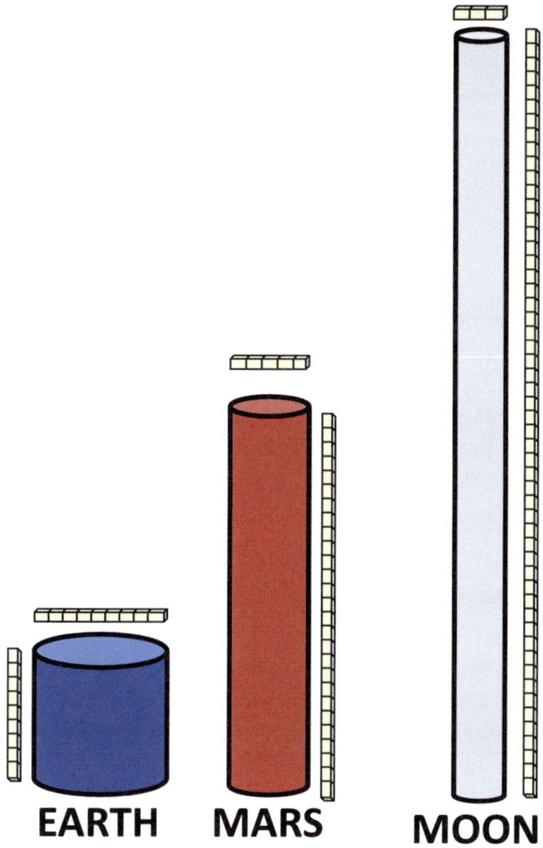

Fig. 5.19 Column with different gravities

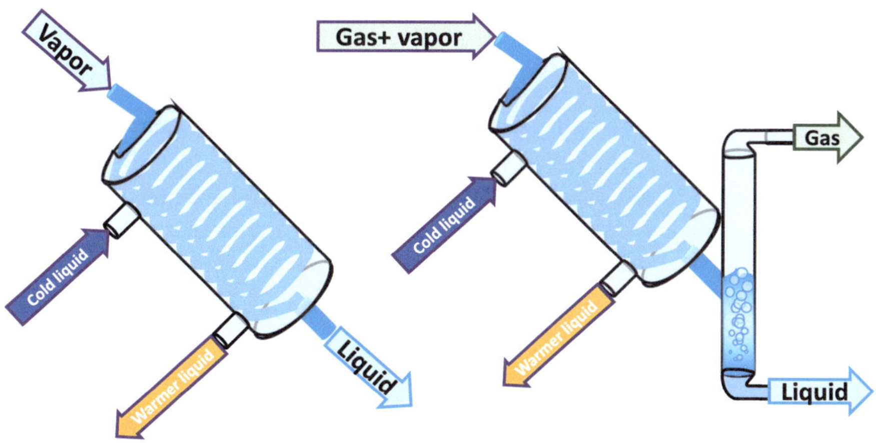

Fig. 5.20 Condensation column schematics

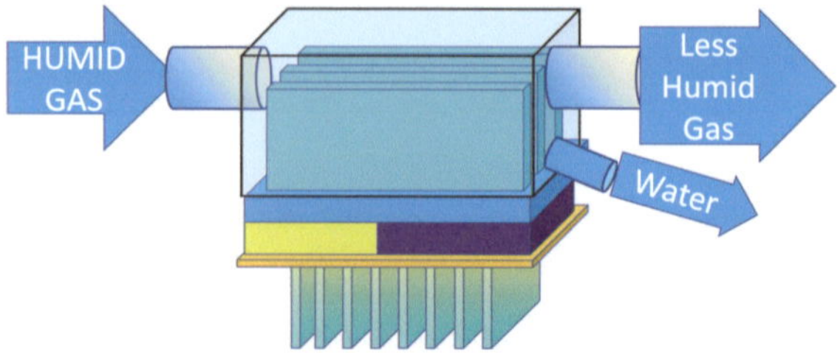

Fig. 5.21 Solid state condenser and cooler

A more solid-state design combines the electric solids state cooler with fins in a box or container, and if the gas has a very high relative humidity, might be more space efficient, as in Fig. 5.21. Air (gas) with a high-water content enters the box that has cooling fins on top of a semi-conductor cooler (wires to connect electricity not shown). Some of the water condenses as the air cools and is extracted, while the now less humid air exits.

5.5 Infusers and Aerators

Infusers inject gases into liquids. Aerators are a type of infuser. One of the most common and simplest types of aerators on Earth is used in aquariums, where a pump pushes air through a tube to a bubble stone to bubble through the water in the tank. Sometimes the bubbles are confined to a tube to improve mixing. Add a few more spirals in the tube, the mixing improves more, making for happy fish and shrimp. Screens can also slow the bubbles and increase dissolved gas concentration. In an aerobic bioreactor, an aerator adds oxygen to the nutrient solution for yeasts or bacteria. In lieu of air and oxygen, carbon dioxide can be injected into water for algae in a photobioreactor. Oxygen also needs to be injected into the liquid around roots of hydroponic plants, especially in deep water hydroponics. It may be easier in low gravity to dissolve gases in liquid due to increased time for bubbles to reach the top of a column.

In Fig. 5.22 the leftmost schematic is for a simple tube and bubbler ring design, where high pressure gas is forced through small holes in the rings at the bottom, forcing the liquid to flow up, and dissolving the gas into the liquid. The middle schematic has a spiral shape in it to slow the bubbles and increase dissolved gas concentration in the liquid. The rightmost schematic uses a set of screens to, again, slow the bubbles to increase gas concentration.

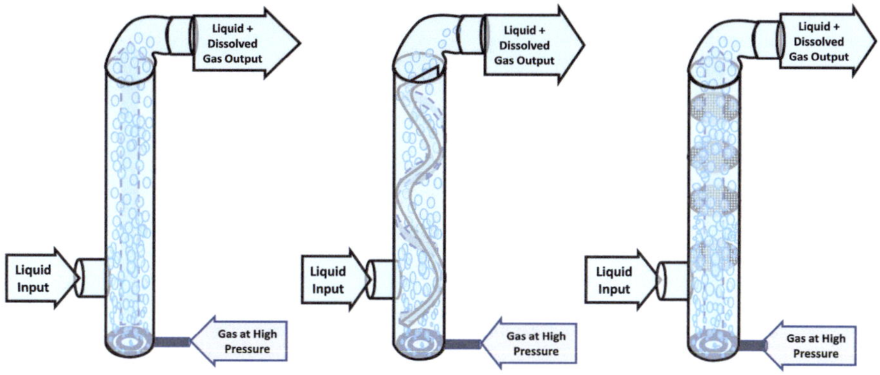

Fig. 5.22 Three examples of aerator schematics

5.6 Sterilizers

Many microorganisms are either not good for crops or people, or in some cases, are good but need to be contained in a location. Ways to kill microorganisms include ultraviolet (UV) light, low dose radiation, intense light, or heat. Knowing that, sterilizers are parts that kill almost everything in a flow of air or water. A commonly used UV sterilizer is in a home air purifier, killing pesky cold viruses, mold spores, or bacteria. The same technology is also common for fish tanks and fishponds, to keep fish and shrimp healthy, or to keep diseases from one tank working into another tank. See Fig. 5.23. Inputs flow into a chamber that has an ultraviolet light, and typically the interior of the chamber is coated to reflect UV light to enhance light levels and sterilization effectiveness.

Another sterilizer design uses low dose radiation (see Fig. 5.24). Low dose radiation is used to preserve packaged food and could be used effectively in settlements in high exposure environments like free space, asteroids, or the moon. Microwaves can also be used to sterilize liquids, though they add significant heat. In essence, daylight in space or on the moon kills just about everything, so clear pipes running through sunlight would make a low-cost sterilizer. The daylight can be concentrated using mirrors to increase dosage.

Heat itself is a sterilizer too. The canning process uses above boiling point heat to sterilize food going into cans, and the same process (pasteurization) is used for

Fig. 5.23 Ultraviolet (UV) sterilizer schematic. The bulb could be any lighting technology

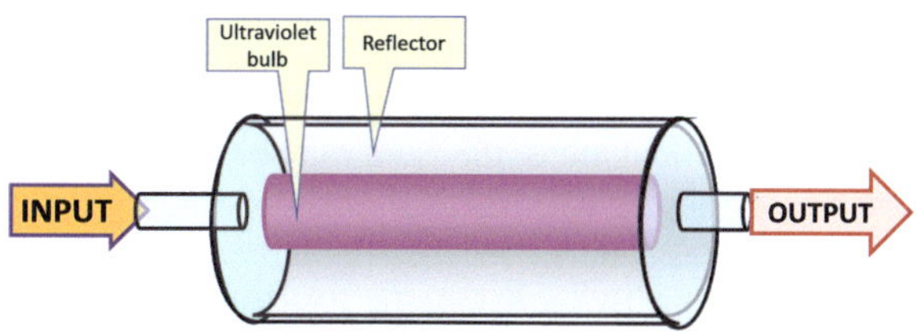

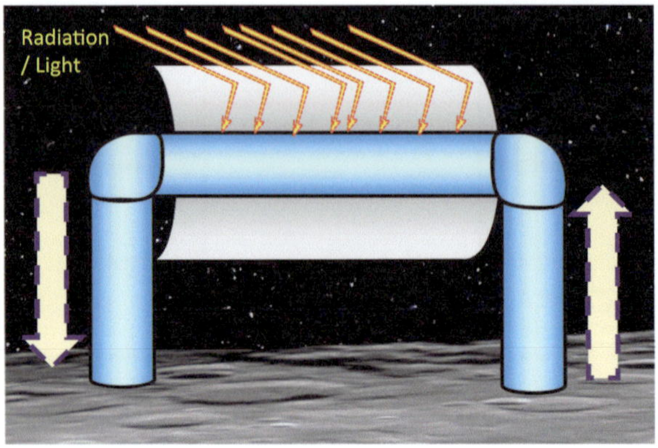

Fig. 5.24 Passive radiation sterilizer. Using bright sunlight on the Moon or other locations to sterilize fluids

liquids. Incinerators also kill organisms. Heated pipes or canisters therefore work as sterilizers too (though at a higher energy cost, and cost in heat rejection).

5.7 Lights

There are many ways to generate controlled artificial light, and likely many more ways for future space settlements. The most common method in most US households in the 2020s is via Light Emitting Diode (LED). LEDs are commonly used for overhead lights, side lights, and backlights for screens on various computing devices and displays. Other methods could include florescent lighting or incandescent lighting theoretically, but not likely. Some light will be routed from natural sources, using light pipes with mirrors or fiber optics, which will save energy. Future methods might use bioluminescence or to be determined methods.

Regardless of method, lighting control is critical for both growing photosynthetic crops, but also for work lighting and psychology of settlers and workers. Further some animals like fish orient to light sources, and many animals and plants need light of a certain spectrum and for certain times. A daily cycle is a must for most animals, and many plants, though not all plants. Dark cycles are critical for some algae as well.

Work lighting, usually white light in farm areas will need to be bright enough, and dispersed enough, to allow people to fix components, tend to plants and animals, and walk without tripping. The U.S. Occupational Health and Safety Administration (OSHA) recommends white (or near white) lighting in standard number 1915 (OSHA-1915.82, 2011), of at least 108 lux in most working areas. In places like offices or medical areas, the need may be three times brighter, though more focused. Light loses power with surface area covered, so that the further the light is from a surface, the less energy reaches the target per unit area. A simple example is with a directional light like a flashlight. The closer the flashlight is to a wall, the brighter

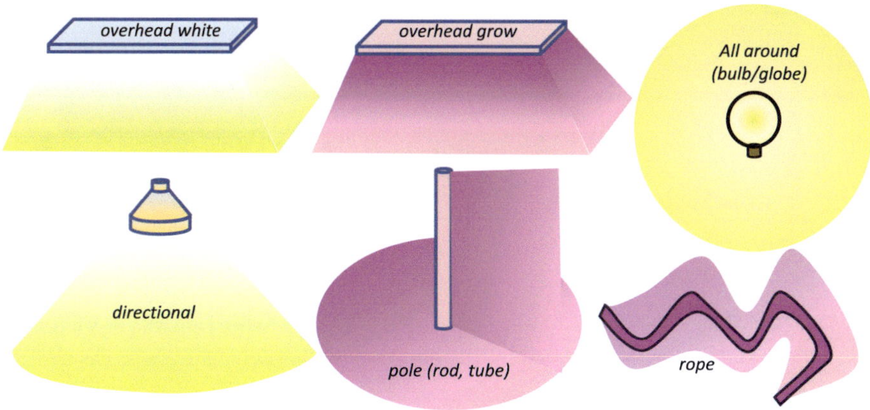

Fig. 5.25 Common lighting types

the light on the wall, but the smaller the lit area. The further the flashlight is from the wall, the more area is lit, but the dimmer the light. Since lights also emit heat with light, where lights are located are important, as to not kill the organisms lit with too much light, or too much heat from the light. Heat is also generated by the light source itself, so air flow and heat rejection for lighting is a core design item. In Fig. 5.25 are six lighting types. Overhead broad area lights are made to put light over a large area either above, below, or from the sides. All around spherical lights illuminate a broad volume. Directional lights focus brighter light over a more-narrow angle. Pole lights illuminate a cylindrical volume. Rope lights provide a region (volume) of light, or highlight edges, such as emergency lighting to show walkways or other hazards.

Lighting for photosynthesis is usually optimized for certain parts of the spectrum. See Fig. 5.26. Photosynthesis largely uses the sections of the spectrum from red to yellow (low energy) and again from cyan to ultraviolet (high energy). Most plants algae reflect green but absorb the region from near-ultraviolet to blue-purple, and orange through red. Lights that focus on growth spectrums are called grow lights and look a bit magenta pink to the human eye. There also are full spectrum grow lights that add a bit of the reflected light spectrum and look mostly cool white. Picking the right light depends on plant species and timing. That does not mean all light for the plants or algae need to be grow lights! Some plants need a bit of white light during fruiting times, or other spectra at other times, and experimentation is needed to establish these parameters.

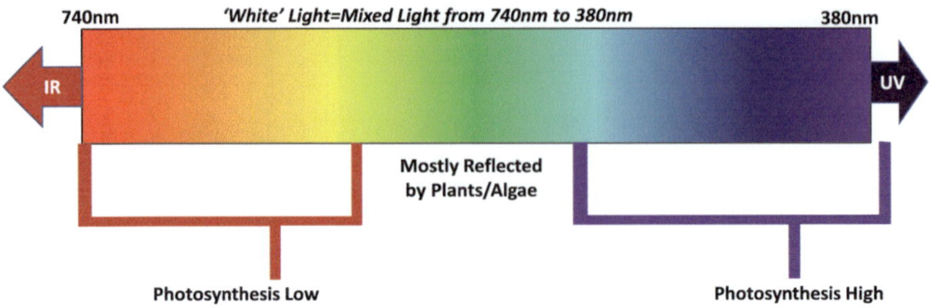

Fig. 5.26 Visible light contains wavelengths from 380 to 740 nm, viewed together as white light. Data drawn loosely from (NASA-Vlight, 2010)

5.8 Electric Heaters

Thermal control includes heating liquids, air, or organisms so that everything works together optimally. Most common heating involves using the coils and fins where a warmer liquid is pumped through the coil or fins. However, warming up fluids can also use an electric coil that uses resistive heating. In a (fish) tank heater as in Fig. 5.27, the electric coil is surrounded by an insulator to prevent a shock. The coil uses resistive heating to warm the inside of the shield, then the shield (the one that prevents that shock) conducts the heat to the surrounding water or gas.

Another type of electric heating is more indirect, and very familiar to a midwestern hog farmer or cook at a fast-food restaurant: the infrared heating lamp. The infrared heating lamp is like the lighting concepts above, except that the spectrum for the light is concentrated in the infrared spectrum, so that it heats the area illuminated. Most infrared heating lamps are incandescent, some gas filled. Anyone who had a kid's toy oven cooked cakes using the heating of an incandescent lamp. The infrared light version might be key for livestock, but also for warming people after a trip into the Martian cold, or even plants in cold sections of a settlement.

Fig. 5.27 Electric heater schematic

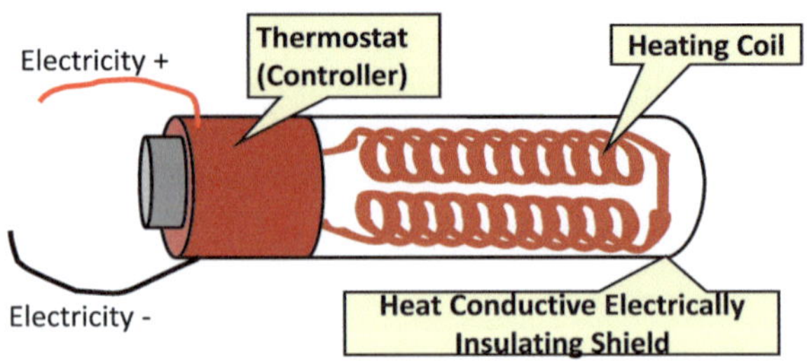

5.9 Sensors

Sensors are devices that report a numeric measurement or array of data measurements for a criteria. Meshes of sensors of many types are very critical to determine the state of the farm, and regulating the various parts of the farm to ensure safety and optimal production. They will be married to supervisory control and data acquisition (SCADA) systems and machine learning/response systems for control. Smart valves are the marriage of local processors with rulesets, sensors, and control values.

5.9.1 Light and Optical Sensors

Sensors to detect light levels in a particular spectrum range can be CdS (Cadmium Sulfide) cells or transistors, though can be camera sensors like CMOS (Complementary Metal-Oxide Semiconductor) or CCDs (Charge-Coupled Devices). Simple optical sensors are critical parts of many of the other sensors below. Using a camera image sensor is needed if it is more important to determine a pattern of colors and shapes, or to detect motion, in lieu of a simple light level in a spectrum range.

5.9.2 Chemical Sensors

Chemical sensors essentially 'taste' or 'sniff' the air, media, or a liquid to determine chemical composition. Liquids and soils, salinity, hardness, various ions, oxygen levels, carbon dioxide levels, dissolved solids, and pH are the biggest needs. In air, oxygen level, carbon dioxide levels and the levels of toxins like carbon monoxide, volatiles, and other dangerous compounds are critical. Many of these sensors use conductance, capacitance, or optical sensors.

5.9.3 Hygrometers

Hygrometers measure humidity in soils and media or in air or in other gases. The oldest variety uses a human hair which expands in higher humidity and contracts in lower humidity. These hair using dials have long since been replaced in digital devices by capacitance, resistance, or optical solid-state sensors.

5.9.4 Pressure Sensors

Pressure sensors measure the relative 'push' or 'pull' of fluids including gases and liquids. These can be electromechanical with a diaphragm and spring, or use Piezo electric circuits, or fiber optics with an optical sensor, or other technologies. Pressure sensors are key for regulating air flow in the farm especially into bioreactors, and for making sure there are no clogs or overloads in pipes, pumps, and tanks. They are also deadly important for detecting leaks to vacuum or breaches in bulkheads as from meteorites, quakes, popped locks, or other hazards.

5.10 Incinerators

Incinerators are machines designed to use heat, and oxygen, with high temperatures, to convert wastes, mostly thick fluids or solids, into nutrient ash, carbon dioxide, and water. In essence, burning is a hot version of aerobic respiration, though unlike the wildfire example, a space settlement burning solution would use precisely controlled vessels that would inject fuel (i.e. wastes including fiber and lignin) and oxygen under controlled conditions to maximize carbon dioxide and water production and minimize byproducts like carbon monoxide. Since the wastes will also contain nitrogen and mineral compounds, other byproducts might emerge and need to be handled. The resulting ash will contain the ash from food assays, i.e. minerals, salts, and metals. A downside of this ash is that it might need additional processing for plants or algae to use it. Heat rejection will be critical, though burning wastes can also generate energy using a turbine. All vessel materials would be high temperature tolerant, and the combustor could be compact. In a simple incinerator as depicted in Fig. 5.28, biological wastes are burned with oxygen to produce solid ash, and a gaseous mix of carbon dioxide and water vapor. The gaseous mix is cooled in a condenser to liquify out the water, leaving carbon dioxide gas if the burn is nearly ideal.

In many current designs for larger scales, a rotating bin (see Fig. 5.29) is maintained at temperatures of over 1000 °C, and waste is either sprayed in liquids, or dumped if solids. The heat source can be simply gases from the incinerator itself, or methane from settling tanks and other sources. To start, the heat could also be generated using solar or nuclear sources but since the incinerator is itself a heat producer, usually burning wastes in the incinerator itself provides the heat needed. Oxygen is added, then the waste is combusted. Controlling the waste to oxygen ratio is key to making sure undesirable gases are not produced, such as carbon monoxide. The gases are collected and cooled. Water is condensed out and then sent for use in the farm or habitat, while remaining gasses pushed through a carbon or similar filter to remove toxins, then are sent to either photobioreactor or plant crops for photosynthesis. Since burning produces heat, the incinerator is usually a net heat producer, and the heat can be either radiated out as part of the heat rejection scheme, or even better used to push a turbine to generate electricity for use by the settlement

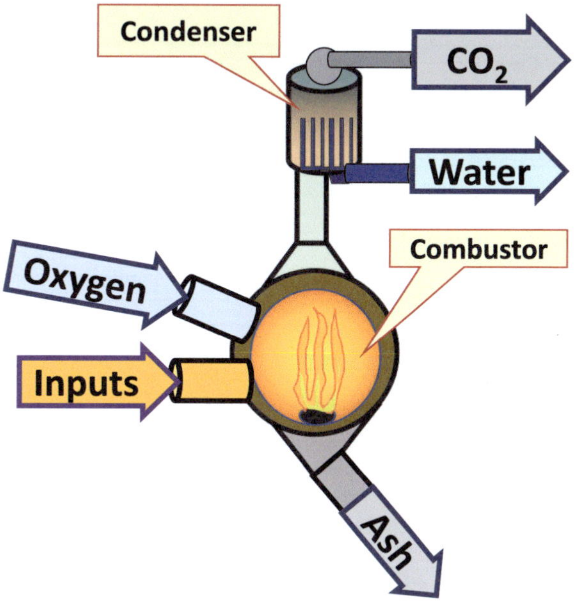

Fig. 5.28 Simple incinerator schematic

or colony. Incinerators are quicker processors for given volume than bioreactors, and great for hard to digest compounds like lignin, but have the cost of tight control, and excess heat generation. Having flames in a tight colony might give small settlements pause, but in a very large settlement, an incinerator can be used as an add on to the yeast-bacteria bioreactor array, providing a buffer for excess mass flows, or to sterilize infected or toxic wastes.

Leave out the oxygen, and instead of general waste, focus on plant wastes rich in cellulose and other fibers, and an incinerator produces charcoal. As noted above charcoal has many uses in filters, and as a plant soil. Charcoal can also be used to spin fibers for graphite composites, or to create ceramics.

5.10.1 Biocarbon and Vacuum Burning

An alternative to the traditional incinerator uses pressure and high temperatures to create a combination of carbon and ash, known as biocarbon, that can be used as soil or for building materials. Biocarbon production is a variation of burning, using high pressures and low oxygen, in a device that looks and works like an incinerator (see above), though starved for oxygen. Instead of a lower volume of ash and a high volume of carbon dioxide and water, biocarbon processes use higher heats to produce carbon rich ash, like charcoal production on Earth. The biocarbon then can be used for planting or used as stock for materials like nanotubes, graphite composites, and

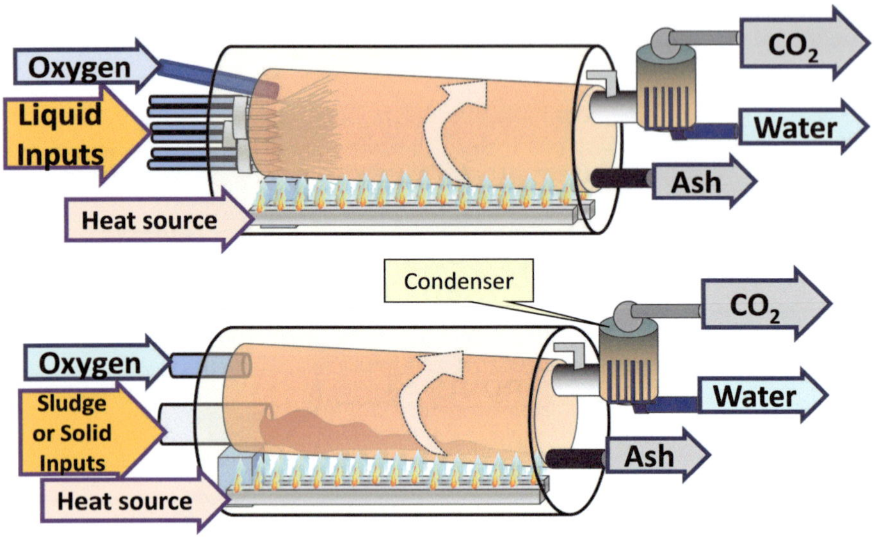

Fig. 5.29 Rotary incinerator schematics

plastics. This practice assumes regular in-situ or re-supply of carbon to replace the mass lost by the farm to biocarbon.

5.11 Fuel Cells

Fuel cells are machines that take organic gases like methane and oxygen use metal plates to generate heat, electricity, carbon dioxide, and steam. See Fig. 5.30. The input gases are passed over the plates which are folded and rolled to maximize surface area. The surface pulls on the hydrogen groups of the molecules in the gas, breaking the carbon-hydrogen shapes and combining them with oxygen. The result is chemically almost the same as an aerobic bioreactor or incinerator, and similar to a catalytic converter, producing warm carbon dioxide rich water vapor. The water can be condensed out and supplied to the rest of the farm (usually after a charcoal filter) as potable water, and the carbon dioxide to photobioreactors or hydroponic gardens.

In a large settlement, a fuel cell could receive gasses from first stage buffer or settling tanks that take black water. Since bacteria may create methane, or similar gases, the fuel cell could capture both energy and carbon dioxide from these biproducts of fermentation. Additionally, if pigs are used as livestock, air vented from pens could pass through a fuel cell to break down the fragrant output. Trust me, you want to clean pig air! My grandpa, who worked on the family pig farm, used to remind me that as a toddler I noted the smell that had soaked into his clothes and skin using a four-letter word that peppered his vocabulary. Again, before using the fuel cell

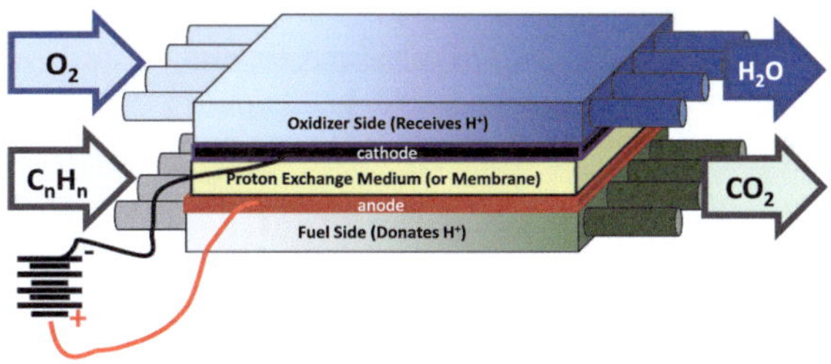

Fig. 5.30 Fuel cell schematic. Fuel cells generate electricity during proton exchange from fuel to oxidizer

products, it might be wise to add a charcoal filter or bubble tank to make sure toxins don't make it into the food supply. Sensor arrays are key to monitoring the ins and outs of the fuel cells and controlling their use.

5.12 Valves

The lowly simple valve is an often-overlooked component of the farm and plays critical roles in all aspects where gases or liquids get moved. There are many kinds of valves of various scales, some simply control the direction of flow, while some are throttles that might have intelligent controls and sensors. Scale is important in valves, as small valves for liquids for example become chokepoints and sites for accumulation of stuff that clogs tubes and pipes. Even air valves, such as those in flues or on vent pipes, can accumulate lint and dust to become sources of mold. A good space farm engineer seeks to minimize the count of valves. Valves are often inside, or paired with, pumps and tanks. Given complexity, the more intricate the valve, the more likely to break, and more likely it is shipped from Earth.

5.12.1 Simple Directional Valves

Every human heart contains simple directional valves that make sure each contraction of the heart pumps blood in the right direction. The same is true for a smart aquarium owner who puts a check valve on hoses to make sure their tank doesn't drain out on the floor in a power failure. Likewise, anyone with a drain in the basement that goes to a sewer has a check valve to make sure water goes down into the sewer and the neighbor's wastes doesn't flood back into the basement floor (ugh!). A check valve is a simple one-way valve, most used for liquids have a ball inside a pipe. When flowing in the right direction the ball is pushed into a larger section that lets the liquid flow. If the flow reverses the ball pushes back against a smaller section of pipe to block flow. In Fig. 5.31 on the left, the ball and pipe valve stop flowing one way when the ball hits the smaller pipe but allows flow in the other direction. In the same figure on the right, a duckbill valve closes one way but opens with flow from the other direction.

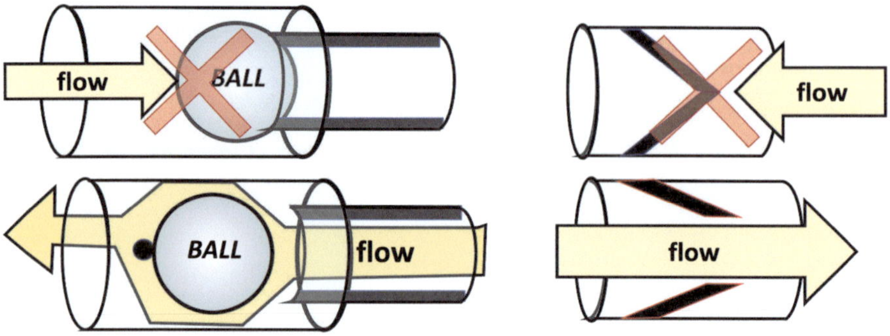

Fig. 5.31 Two simple directional valves

While these valves work great for preventing accidents, they often become inefficient if the liquid has anything that can clog, stopping the ball from closing, or making the ball a clogging source. In some forms a spring pushes the ball to make sure a minimum flow opens it, and there is positive pressure regardless of orientation to close flow. A further modification is a full flow design that uses a bend where the ball is push away from the bend enough to permit full flow. Another type of check valve for liquids is a cone or duck-bill valve. The cone or duck-bill is a silicone or similar flexible material, that opens to admit flow in the right direction of either gasses or liquids, but as soon as pressure drops, closes and prevents backflow. Most aquarium check valves are like this. Tradeoff for this simple design is that the flexible part eventually fatigues or is damaged, and that a minimum flow pressure is required to open it. Another even simpler version is a simple flap. The flap is either hinged or is made from a flexible material. Your heart valves are close to being simple flap values. A type of valve familiar to most brewers, vintners, and plumbers, is the S-valve. In this gravity-dependent valve, a horizontal s-shaped pipe (like an inverted 'N') contains water or another liquid. For the bubbler type, gasses push out and are bubbled through the liquid out of the container attached, but gasses from outside the container can't push back. This assumes that the pressure in the container is higher then the pressure outside of course. For a bioreactor like a fermenter, the CO_2 escapes through the S, but the air outside doesn't get into the fermenter to contaminate the batch. In the plumbing case, the S-pipe is in a toilet, and only on a flush does it let waste flow around, keeping stinky gasses from the sewer out of the toilet. Guess what happens however if the sewer pressure builds too high? The S-bend reverses and out comes a wonderful aroma and effluent all over that bathroom floor! As noted the S-pipe only works if the pressure between the two sides has a gradient, and only if the liquid at the bottom of the S does not dry out.

5.12.2 Throttling Valves

There are many types of valves that regulate flow and can control flow rates. The ball and cock valve is one of the most common valves, found in faucets, gas pipes, and other spots. A ball with a hole drilled through it, is in a pipe, and a lever outside the pipe controls how much of the hole is exposed to the flow. Turn the lever 90 degrees and close off the flow, turn to 0 degrees, full open, and any position in between regulates how much gets through as in Fig. 5.32.

A variant of this valve uses a screw into a pipe that lets flow through all the way after several knob turns or closes it off completely by many turns in the opposite direction. See Fig. 5.33. Your shower faucets use this method most likely since it allows a fine degree of control (to keep from scalding you). A fancier version for controlling two flows into one, like the single knob shower or faucet, has a screw that raises a cylinder and pipe or screw and pipe on one side while opening the opposite side simultaneously.

Sphincters, of which your body has many, are valves where a flexible material or camera shutter-type iris obstructs a circular pipe or opening, to various degrees,

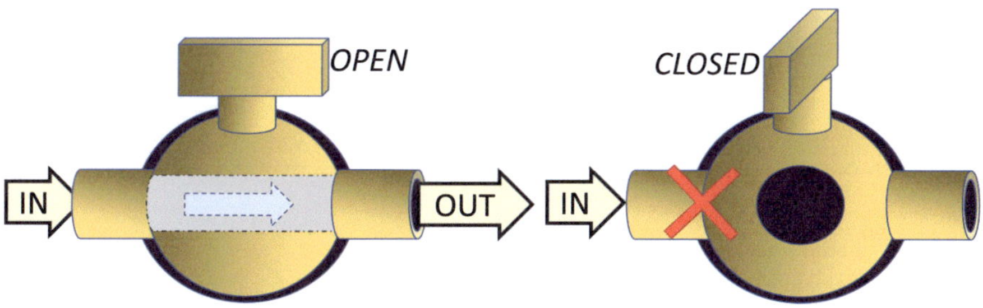

Fig. 5.32 Ballcock valve schematic

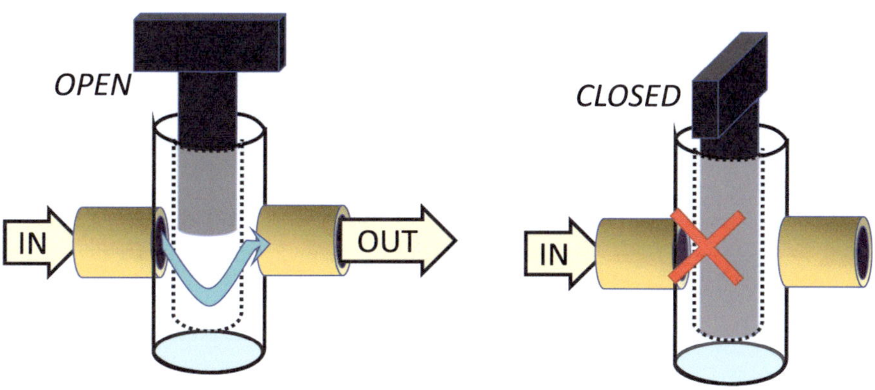

Fig. 5.33 Screw valve schematic

obstructing from the edge all at once. Muscles and flexible proteins make your own irises work, opening or closing to control light to your retinas. Similarly, arraignments of wires, strings, memory-metals (metal muscle wire), or even fluid sacks (like plant stomata), close or open the hole. In a camera shutter type arrangement, plates are set to hinge inward in a spiral. See Fig. 5.34.

Throttle bodies for turbochargers or carburetors on combustion engines sometimes have another type of controllable valve. A disk (usually metal or ceramic) had a rod through the edge bisecting the disk as in Fig. 5.35. The disk is slightly bigger than one end of the pipe. A lever turns the rod and twists the disk to open the valve. Sometimes a spring is used to keep the disk open only as far as the lever controls. This type of valve, usually for course control, usually assumes pressure is much stronger on one side of the valve than the other, though these valves can respond swiftly to changes.

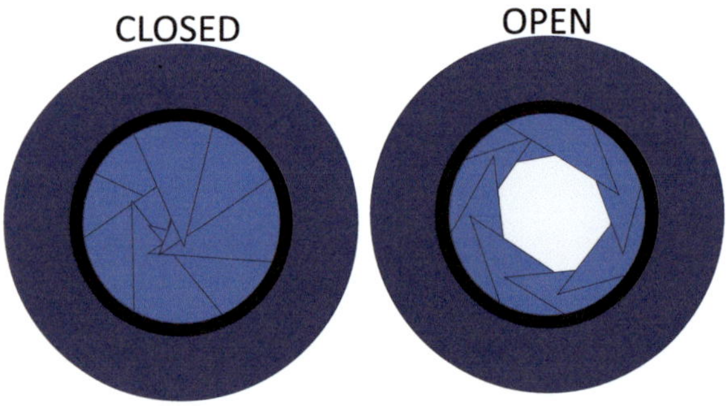

Fig. 5.34 Simplified notional shutter valve. These tend to be more fragile than other valve designs

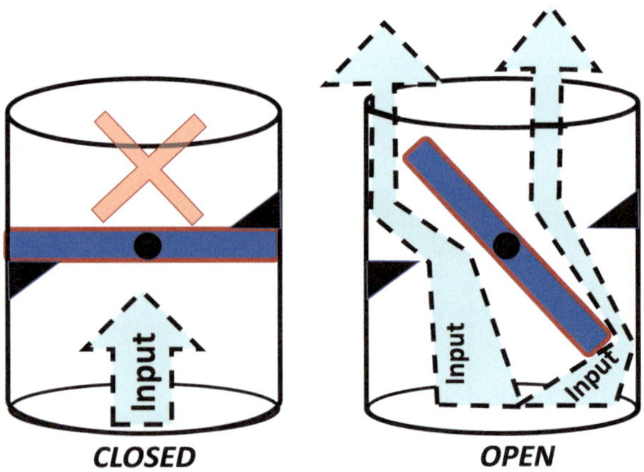

Fig. 5.35 Throttle body (disk and rod) valve. These can be implemented with two half pipes or a set of seals. Very common in combustion engines or similar devices

5.12.3 Smart Valves

An old natural example of a smart valve is the plant stomata. The stomata opens when it receives light, letting water out for evaporation, and air in for photosynthesis. Local, quick, response to a change in conditions. The stomata an example of a throttling valve paired with a (light) sensor and control law (chemically implemented). Smart valves are the same concept, pairing sensors and simple processors with control motors or electrically actuated parts of valve types above, to keep flow as planned as in Fig. 5.36.

Smartness can vary from the simple table of responses to machine learning artificial intelligence. Even in the absence of sensors, the actuator and valve should be tied to a control network that keeps flow at the rate as needed. The concept of a smart valve is to respond locally to changes in conditions quickly, with pre-programmed rules. These smart vales push data up to larger scale control systems for the farm and receive rule changes from the higher-level systems. An arrangement might be used in case of a rupture to prevent mass loss and implement backup flows for example. The sensor-control-valve works in dishwashers or laundry machines, and in chemical factories and breweries all over, in that tense called a Programmable Logic Unit (PLU). All the sensors and PLUs already exist on Earth and will get smarter over time. These smart valves also already find use in current space efforts on spacecraft and space stations. The farm should have many of them, in a variety of places. Put a series or mesh of these smart valves together to make an array, and they could implement complex rules (or artificial intelligence (AI) responses) to control complex parts of the farm, as in the flowchart in Fig. 5.37.

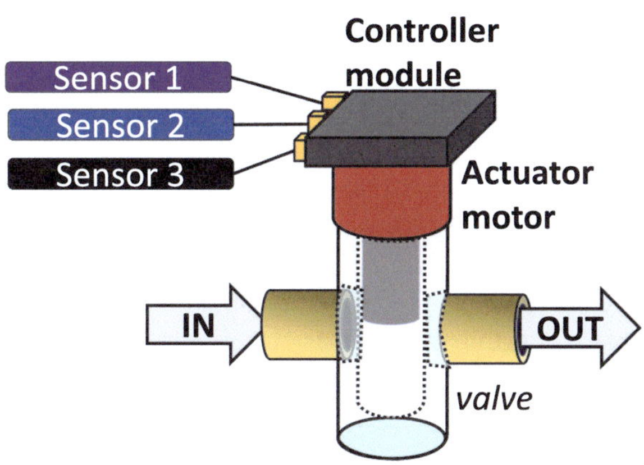

Fig. 5.36 Smart valve schematic. Combination of sensor, processor, actuator, and screw valve

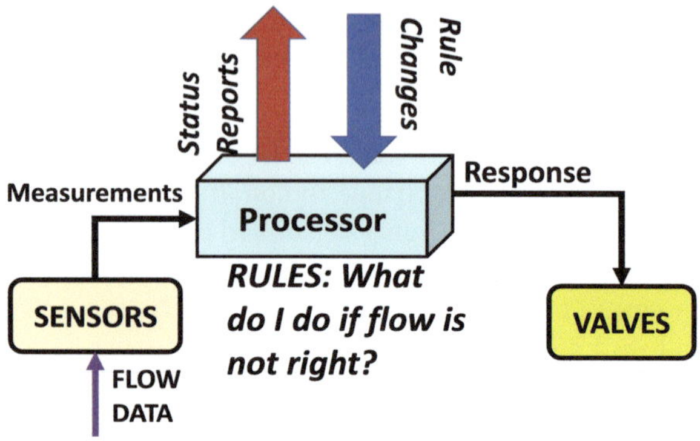

Fig. 5.37 Smart valve array flowchart of operation

5.13 Gas Separation Machines

Separating mixed flows of gases is important to making sure the various parts of the farm, and the organisms in these parts, stay alive and balanced. There are many devices that can deal with the tricky nature of separating gases, though these devices ideally supplement organic methods like blowing air through plants or algae.

5.13.1 Zeolite Filters

For every space mission from the beginning to now, controlling the level of carbon dioxide in habitats is critical. All the way into the 2020.s scrubbing carbon dioxide and monoxide from human air was performed by a simple zeolite filter. Zeolite is a naturally occurring mineral, that is now made artificially to improve performance. Air was initially pushed through a canister filter full of zeolite and charcoal, though more modern versions bond zeolite and charcoal to a series of fiber filters. The carbon dioxide can be extracted from the filter using heat. While great as a backup for photosynthesis in animal and human habitats, the zeolite filter method wastes carbon dioxide mass, or at least locks it up for a time. Zeolite is also very useful as a backup method to scrub ammonia from water, again the zeolite will have to be heated to get the ammonia back.

5.13.2 Cryopumps

Cryopumps (and cryotraps) are also critical to gas separation, until replaced by semi-permeable membranes or other less energy intensive technologies. Where you see cryopump/cryotrap in this book, you can swap in the replacement technologies as they mature. Cryopumps require very cold temperatures and strong temperature control.

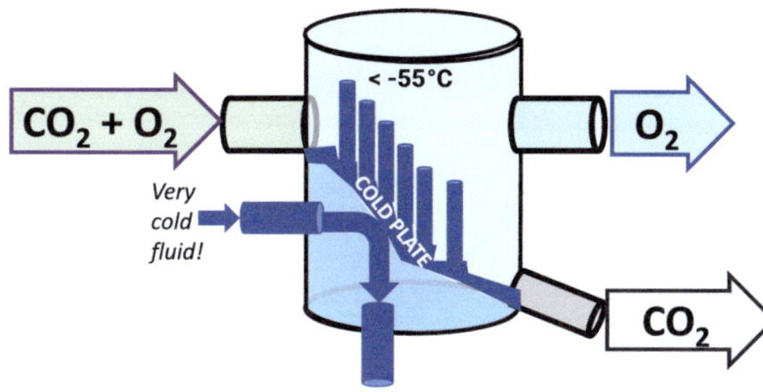

Fig. 5.38 Carbon dioxide and oxygen gas cryotrap (cryopump) separator schematic

These pumps are common in carbon capture schemes on Earth. In deep space the cold supply may not be an issue since cold is a freebee in dark areas of vacuum. The concept is loosely the cold version of a distillation column. Mixed gases are pushed by pressure (pulled by vacuum) though a column with different temperatures and pressures to condense out gases for collection as in Fig. 5.38. For carbon dioxide and oxygen separation, the mix is first dehydrated by a condenser, then the carbon dioxide and oxygen mix is pumped into chamber where a cold plate is dropped to below − 55 °C, condensing the carbon dioxide to liquid for collection, but leaving the oxygen as gas which condenses at a far colder temp. Oxygen gas condenses at a three times colder temperature than carbon dioxide, allowing the carbon dioxide to condense on the cold plate (with fingers or fins to improve surface area) and is collected below, leaving oxygen gas (and nitrogen or other colder condensing gases). The process works very well but requires very careful pressure and temperature control.

5.13.3 Catalytic Converters

Commonly used to clean car exhaust, a catalytic converter uses plates or high surface area forms of platinum group metals to convert gaseous hydrocarbons into carbon dioxide and water. See Fig. 5.39. These can be adapted for other types of filtering too. Sandwiches of metal with gaps for gases make the converter. These converters may be critical for cleaning up incinerator outputs, due to incomplete burning, then chained with a condenser to recover the water, and leave the carbon dioxide for photosynthesis.

5.13.4 Electric Hydrolysis

Sometimes, a settlement might want to split water into hydrogen gas and oxygen gas, especially if there is an easy supply of water, or the settlement needs hydrogen for fuel or for other processes. The process itself is simple and a common chemistry experiment for elementary school kids. See Fig. 5.40. A cathode and anode are put into a

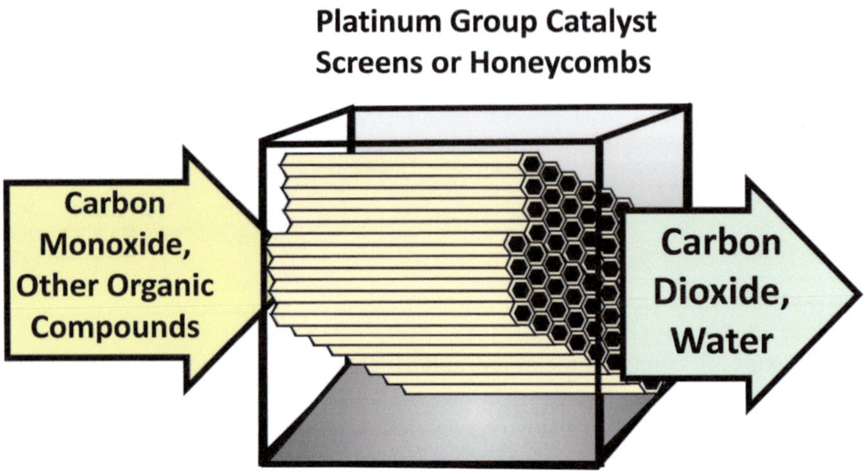

Fig. 5.39 Catalytic converter schematic

container of liquid water, and current applied to the electrodes, producing hydrogen gas and oxygen gas if power is applied, and the electrodes are not too corroded or covered with debris. Electric current breaks water molecules into hydrogen and oxygen ions. The negative side collects the hydrogen ions which join into hydrogen gas. The positive side collects oxygen ions which bubble into oxygen gas. The less pure the water, the more likely the electrodes will need to be cleaned, and there are many designs using membranes in the commercial world that make the process more efficient.

Fig. 5.40 Simple hydrolysis schematic

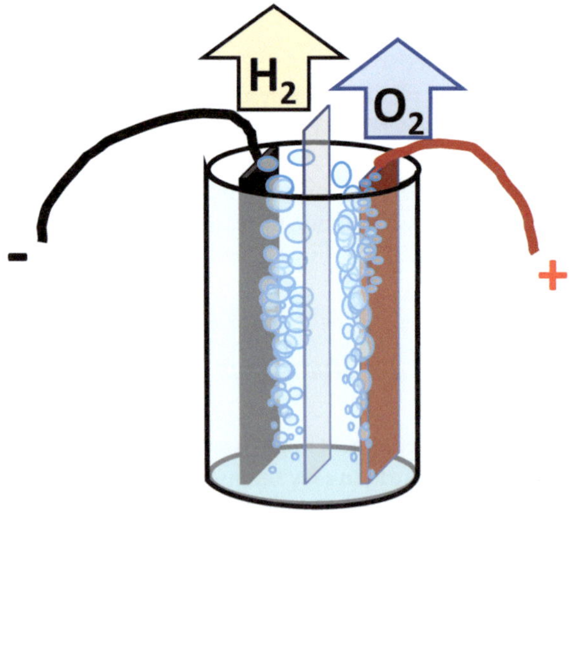

5.13.5 Other Methods Including Selectively Permeable Membranes

Biology and nanotechnology provide a wide variety of methods to separate gases and liquids, and gases from a mixed flow, using selectively permeable membranes to active molecular selection as for living cells. These low energy methods will replace cryotraps or zeolite filters for gas separation. Many are already in development, and nature has many instances to copy. They will require a pressure gradient or use passive means. In design, these membrane filters would essentially be designed like reverse osmosis membrane filters in canisters or sandwiches.

5.14 Chaining Machines

In chemical engineering it is common to chain machines together to get various products. Chaining various machines together can clean gases from mixed flows or capture critical parts like carbon dioxide. For example (see Fig. 5.41), a fuel cell or catalytic converter can be stuck in front of a dehumidifier/condenser, then go to a cold trap to take methane infused, carbon dioxide and water rich air as from animals or fungi as in the following schematic. Classic flowcharts are a starter, but to calculate and program these processes, there are commercial chemical engineering software packages available now. These software packages can design exceptionally complex flows graphically, have models and plug-ins for common commercial valves and machines, then run real time simulations before employment and use.

Other options might be to bubble air through water to capture gasses like ammonia for algae, or use a zeolite or charcoal filter to capture ammonia, then use the condenser

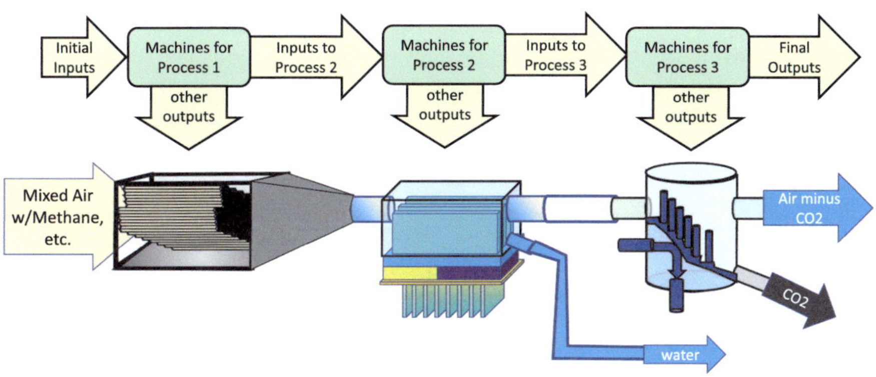

Fig. 5.41 Notional chain of machines. In this case, a notional chain of catalytic converter, condenser, and cryotrap is used to transform gases into carbon dioxide and water vapor, the condenser separates out the water, then the cryotrap removes the carbon dioxide, leaving gases like oxygen and nitrogen

and cold trap. These chains of machines will be key to later design concepts for various parts of the farm.

5.15 Energy (Electricity) for the Stage

A stage is the sum of all growth chambers, tanks, human work areas, and machines, to grow a crop in the farm, as in Fig. 5.42. All the machines above need electricity to function.

Energy (electricity) use in the stage comes from these large types: lighting, recirculation and condensation, input/output including injection and separation, heat rejection and electrical temperature control. Heat rejection is moving heat away from places in the farm to open space, planetary atmosphere, or regolith. Since every machine, circuit, and light generates heat based on the electricity used, plant areas have too much heat due to lights and pumps, and this heat can be moved using finned pipes to either cold regolith or open space using more fins. Animals and yeasts also generate heat via aerobic and anaerobic respiration, though this heat is minor compared to lights and pumps. Note, there is a whole paper on this topic, with math in (Meyer, 2023). A balanced farm with a diverse mix of plants, algae, yeast, and aquatic animals roughly uses at least 0.11 kW/m^3. Energy is mostly in lighting all the diverse plant crops, then in moving fluids around (this includes heat rejection), either in recirculation, or in bringing in inputs or outputs. In Table 5.1 are a few examples, note that dietary diversity drives both farm size and energy use.

5.15.1 Energy Used in Lighting

Lighting comes in three forms: work lighting, light for photosynthesis, lights for sterilization, using lights just like the devices in Sect. 5.7 Lights above. Getting the

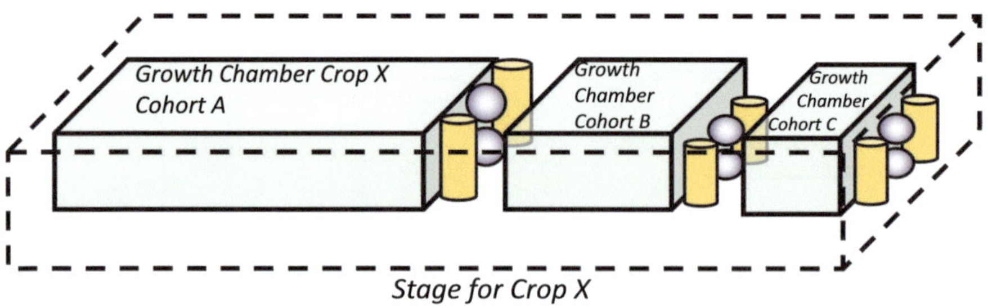

Fig. 5.42 Stage and growth chamber concept. Stages must incorporate all the growth areas (chambers), tanks, work areas, and plumbing in footprint, mass flow, total mass, and energy calculations

Table 5.1 Electrical energy requirements for three types of space farms for a habitat with 100 people. Loosely from (Meyer, 2023)

Farm (Diet) type	Photosynthetic light (kW)	Fluid movement (kW)	Work light (kW)	Heat rejection (kW)	Total (kW)
Glop Diet of algae and yeasts, with some salad	117	1.34	0.23	8.06	127
Veggies, beans, potatoes, and some fish	1945	6.46	0.86	30.55	1983
Highly diverse set of plant and animal crops	11,470	77.73	3.89	139.71	11,691

energy needed for all that light depends on crops, footprint for the stages and work areas, and for sterilization, flow rates.

5.15.1.1 Energy for Work Lighting

Work or animal lighting is a soft white light to white light, in a similar spectrum to the sun as seen on Earth, needed to allow humans to work in an area, or set diurnal rhythms and orientation (i.e. sun is assumed to be above) for animals. As a result, part of the energy required in the farm will be for work lighting for humans and animals. Energy for work lighting here is assumed to be in a linear relationship with the footprint of a stage (i.e. enclosed area), the light intensity required in lumens per unit area, and the lighting efficiency of commercially available LED bulbs or light rods or strips in watts per lumen. Work lighting is assumed to be of the same spectrum as in most homes and for most animals and working people, as in (OSHA-1915.82, 2011) or similar standard. Usually, intensity is around 10 lumens per square meter at floor. This is generally lower than required for plants and algae, so is an add-on for those stages, especially since photosynthetic lights will be in spectra that may cause eye strain to human color vision (personal experience), and some plants need white light to trigger life events. Note that light for animals will need to be balanced with dark, as for day-night cycles on Earth, so 12 h out of a day could be used, though for worst case power, 24 h is a rule of thumb.

To get the energy for the stage for work lighting, find the wattage for your sample lights for the intensity at around 3 m (a standard light height), and multiple by the footprint of the whole stage, i.e. the growing areas, plus an added 10–25% to allow people to work. Most commercial bulbs in 2021 provide around 90 lumens of light where needed per watt of electricity used, so a good guess is around a watt for every 8 square meters for work lighting.

5.15.1.2 Energy for Photosynthetic Lighting

As discussed earlier in this chapter, for vascular plants, Photosynthetically Active Radiation (PAR) is the technical light measurement used by botanists, but there is a simple rule of thumb to figure out energy requirements. See Fig. 5.43. In 2024, an LED grow light uses about 180 watts for each square meter of plants lit, assuming the light is about 1 m above the top of plant. Light shapes are in a wide variety of arrangements, sizes, and spectra. If you stack plants, or arrange them in other patterns, the calculation gets a bit more complex.

In an algae or cyanobacteria reactor, light is a delicate balance. Too much will kill all the cells, and too little will make them dormant. So, cells need to move around to get equal exposure in a tank, or if on a substrate, the lights need to be arranged to be at the right distance. A worst case is around 120 watts per square meter of algae lit, though in a tank, this may mean calculating the exposed algae at any one time. Another way to guess the light for a photobioreactor at worst is around 60 watts per day per cubic meter volume in the tanks or membranes, but this very, very rough just for sizing power systems, and in reality, may be far lower.

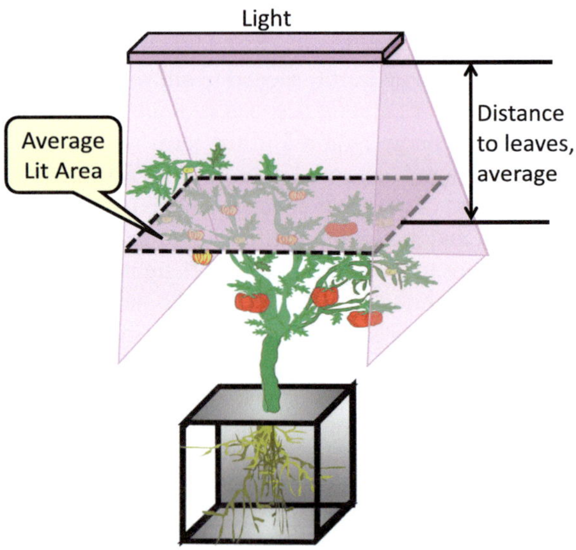

Fig. 5.43 Light energy estimation. One way to estimate light energy at plant leaves using distance to light to find lit area

5.15.1.3 Sterilization Lighting or Radiation

There are many options for sterilization, heat is one, using ultraviolet or low dose radiation is another. Commercial ultraviolet liquid sterilizers usually can get to at least 99% sterility using around a watt per 2 L (kilograms) per minute assuming a pump is pushing the fluid at the right rate. In air or a gas, ultraviolet sterilizers in HVAC, with or without ionization, follow a fiber filter, and can kill around 90% to 99% of germs in air by using very bright light. The bulbs as of 2024 (using a very cursory survey of commercial designs) are usually around 75 W for 200 L per second of sea level pressure air, or around 375 watts per cubic meter per minute. In mass terms, around 290 watts per kilogram of a gas per minute.

5.15.2 Energy for Recirculation and Condensation

Recirculation is the movement of gases or liquids around inside the stage itself, either between growing chambers in the stage, or inside the grow chambers. For example, in a greenhouse, fans need to move air to allow the plants to breathe and transpire, and to keep leaves cool. Some of these fans will blow across cooling fins or coils to pull excess humidity from the air to allow transpiration. The condensed water is pumped for reuse by the roots. If the plants are grown in a shallow water hydroponic system, the liquid in the troughs is recovered, filtered, aerated, nutrients added, and recirculated back to the roots. In a deep-water system, aeration in the deep water usually does the recirculation, though stirrers can add more water movement too.

In an open volume, especially in a greenhouse, all the air in the chamber should circulate at least 12–15 times per day. This is similar air over fish tanks, though for land animals crowded together, this is closer to 30 to 50 times a day. If this was at sea-level pressure on Earth at standard temperature and pressure, density is 1.23 kg per cubic meter. A good fan needs around 2 watts per kilogram. So for the volume of the growth chambers together, watts (per day) is roughly 1.23 * 12 * 2 per cubic meter for plants, and a factor of 2 or 3 higher for land animals. Note that you can't forget the work areas around the chambers, that need circulation too, about 12 times a day or more.

Liquid recirculation is needed in several places in the farm. Liquid pumps, including macerating pumps that move suspended solids in liquids, take around 1.2 watts per day per kilogram in 2021. In a shallow water or deep-water hydroponic system, root water needs to be recirculated, very worst case, around 130 times per day (i.e. around 10 times an hour) shallow water, though less in deep water systems if an aerator is used in a bucket system. This is one of the trades in design since the amount of fluid in the root system is determined by architecture of the root bed and need for root room. This rate varies highly with each plant, and in the diaper or bucket scenario, the inputs are trickled in in lieu of being recirculated. In a fish or shrimp tank, the tank water is recirculated around 40 times per day, again depending on species.

For temperature, there are passive systems that just put a coil or fan in the way of the air, acting as a dehumidifier is addition to cooling the air, though how much liquid to push through coils or fins is a more complex calculation, relying on the dark art of heat transfer. There will be liquid pumps involved, and possibly additional fans if the lights, animals, or other pumps generate too much heat. Overall the heat rejection needs for the stage or the farm as a whole can be estimated as a percentage of the total electricity used, usually around 2 to 7% of the total energy (electricity) used in the farm from all other sources.

5.15.3 Energy to Drive Inputs, Outputs, Injection, Separation

Many gases, liquids, and solids go into and out of a stage. Some of these are lower pressure inputs, as from blowers, versus high pressure inputs from gas pumps. Since lower pressure pumps take around 5% of the power that high pressure gas pumps do for each unit mass and time, knowing when you need each is a big decision for power consumption. Blowers use around 7 watts per day per kilogram, while high pressure pumps use around 180 watts per day per kilogram. Any time a gas is injected into a liquid, a liquid pump or stirrer is somewhere pushing the fluid through, while the high-pressure gas pump shoots in the gas. Liquid pumps are also power user, and all liquid stages use recirculation, as does any hydroponic design in the root bed. That said, aerators can be very efficient if the pressure of the expanding bubbles does the liquid movement, as in fish tanks or tank style bioreactors, and in this case the energy is the high-pressure gas injection alone. Gasses entering a stage will not go in as fast jets that can blast off hair and leaves, but will be instead slowed down to a gentle breeze, so blower energy rates should work for gasses into and out of greenhouses, animal areas, or air above and around tanks, while high pressure pump energy is a good estimate for gas separation, injection like aerator or into incinerators, fuel cells, or other complex machinery (Table 5.2).

5.16 Control Systems and Software

Fortunately for future settlers, sensor array monitoring and Supervisory Control and Data Acquisition (SCADA) is already a high art on Earth, with complex software and systems engineering processes There are examples for each stage type in later chapters in this book. For the whole farm, a network of servers, switches, controllers tie to sensor arrays, smart valves, and pump and light controls in every stage and in the systems between stages as in Fig. 5.44. Coordinating both machine and human work are workflow/ticketing software systems already in common use, using a combination preprogrammed rules and AI adapted processes.

The story of linking all those sensors is as follows. Inputs from sensors are aggregated together, each signal coordinated by a shared clock, and tagged with a time

Table 5.2 Summary table of rough rules of thumb for electricity (Energy) Use in a Farm. These numbers are semi-notional, given real world draw of these devices. Data from (Meyer, 2023)

Energy type	Device	Input estimate	Rough multiplier (to get Watts per day)
High pressure gas	Gas pump	Mass of gases	0.45 W/day/kg
Low pressure gas	Blower	Mass of gases	0.01 W/day/kg
Fan	Fan	Mass of gases * times per day * volume in chamber	0.0031 W/day/m^3 for plants, 0.01 W/day/m^3 for land animals
Liquid	Liquid pump	Mass of liquids and rate of recirculation	0.0015 W/day/kg inputs and outputs (Note the real draw may vary)
Photosynthetic light	Grow lights	Lit area	180 W/day/m^2 lit (can be > 500 W/day/m^2 in some designs)
Photosynthetic light (Photobioreactor)	Grow lights	Lit volume	60 W/day/m^3 L
Work light	Overhead white light	Lit area	> 0.125 W/day/m^2 L
Heat balancing and rejection	Blower or fan	Volume in stage and energy in stage	2 to 7% added to total energy used in the farm

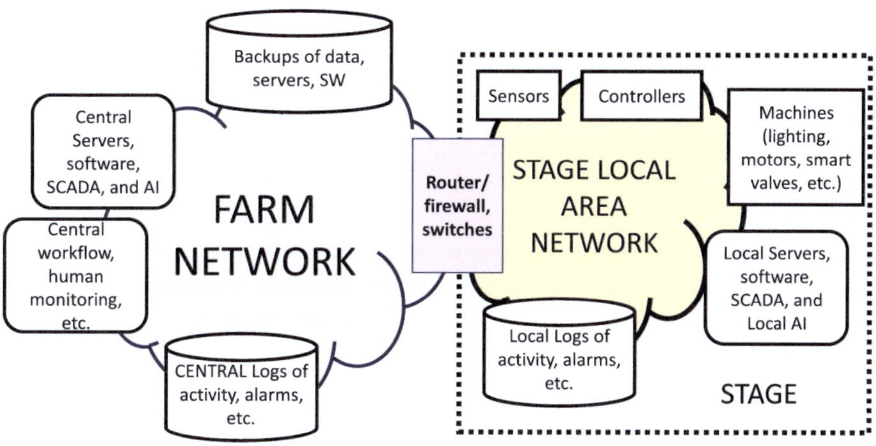

Fig. 5.44 Hierarchy of servers, control systems, and AI for the farm and stage

(i.e. time stamp) from the common clock. The signals go over a network, similar to the Internet, and get pooled together in each area by local computers. The inputs, regardless of sensor type, are put into a common format by a networked node, with the time stamp, so for example temperature data and pressure data can be compared. All the common input messages are feed to a log, which is a raw list of the messages, tagged not just by time, but by source, type, and other important tags. The log is read by a learning array, usually something between a search algorithm and neural network that looks at the log against a list of rules. Any data or combination that violates the rules triggers an alarm, to a human operator, and to automatic response routines. Severity, based again on the rules, determines what happens. An icon on a screen might turn yellow or red, a part shut down, a replacement ordered from inventory of 3D printed, or a human readable message might be sent to list of people or robots. Hopefully the trouble system does not become self-aware and determine the humans are a risk to the mission ala a certain 1960.s space movie. The alarm usually triggers an incident report of some type, and over a longer term, reports are examined to prevent future issues. Complex software usually data mines the log in the long term to look for trends and concerns.

Far from theory, the story above goes on every day, now, in chemical factories, refineries, pipelines, electrical grids, city manager offices, vertical farms, cloud data centers, and Internet networks with hundreds of thousands to millions of nodes. The practices, systems, and software have strong economic motives to continue to improve and will be even better as time progresses. Time is money, and response time doubly so.

Typically a control systems engineer designs the system and programs the control logic using real time control software and specialized algorithmic languages and plans the network layout to connect all sensors, valves, and other controls for each part of the space farm and settlement. On Earth, technicians (usually paid well) install all the networks and hardware. Operators (paid even better yet) sit in control centers to babysit the whole system to make sure the chemical plant or factory runs smoothly and efficiently. There is also an ample role for AI and Machine Learning (ML), together called AI/ML. In Fig. 5.45 is a simplified arrangement of AI and control systems.

AI is already in use on Earth for many robotic control systems and back-ups. Fuzzy logic with a rule engine is an AI strategy, combined with neural networks for sensor processing. The combination of software, SCADA and AI feature in various levels of the control systems and operation systems hierarchy due to the chance of local communications or equipment failure or software corruption, and feed to troubleshooting systems that will again use AI to prioritize issues, diagnose across networks and devices, and even fix or route around problems to keep the critical systems running. Neural networks are usually the method for reading issues and fining fixes to send to the maintainers and operators.

In space it pays to control local, then send data back to more centralized components and controllers. Since connections can fail, networks or controls break, each part of the farm should have controls that can run absent guidance from the human habitat or other farm elements. Solar storms, meteors, or just bad luck can be mitigated by decentralization and redundancy. It is very likely farms will be set up and run remotely on the moon or Mars, with occasional control commands sent via high latency connections, and AI performs fixes while waiting for the humans to arrive. Therefore, control systems in addition to decentralization, will need some degree of

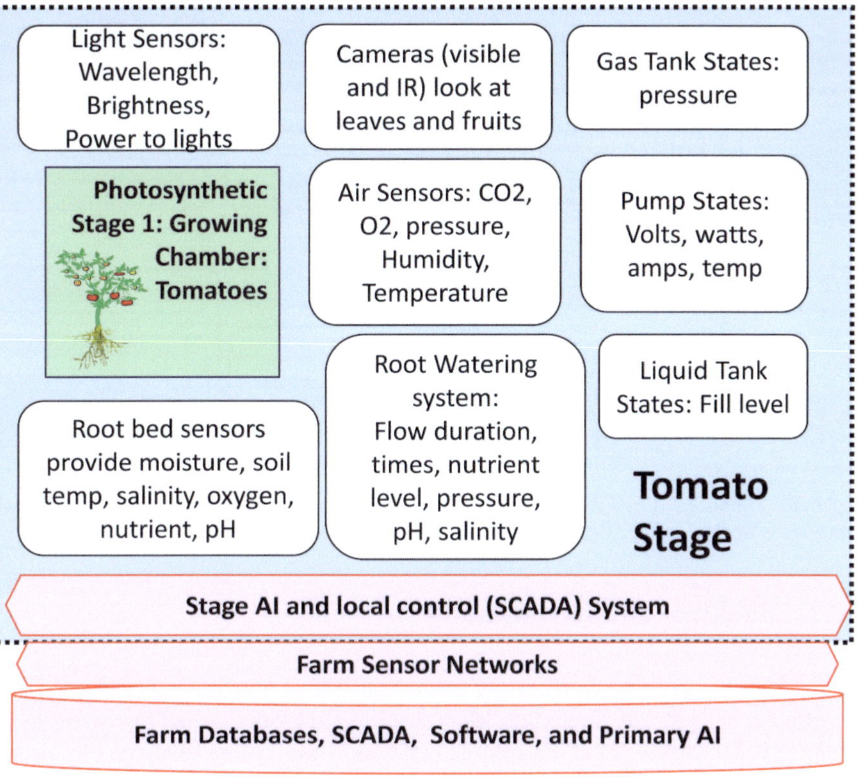

Fig. 5.45 Simplified schematic example of local and centralized SCADA, monitoring, and AI

autonomous operation. Setting up the farm early means fresh food can be ready for settlers when they arrive, avoiding reliance on preserved foods or glop.

Scenarios are important for figuring out how AI, SCADA, and humans will operate synergistically to keep a space farm working. In Fig. 5.46 is a sample scenario where an AI/ML system helps fix tomato growth for a space farm.

As shown, the AI and SCADA system run through a series of checks and controls but must default to a human technician using the workflow system who is aided by augmented reality (AR) displays, customized procedures, and real time checks. Once fixed, the AI systems close out the workflow ticket and continue to monitor and store logs for the status of the problematic parts of the stage.

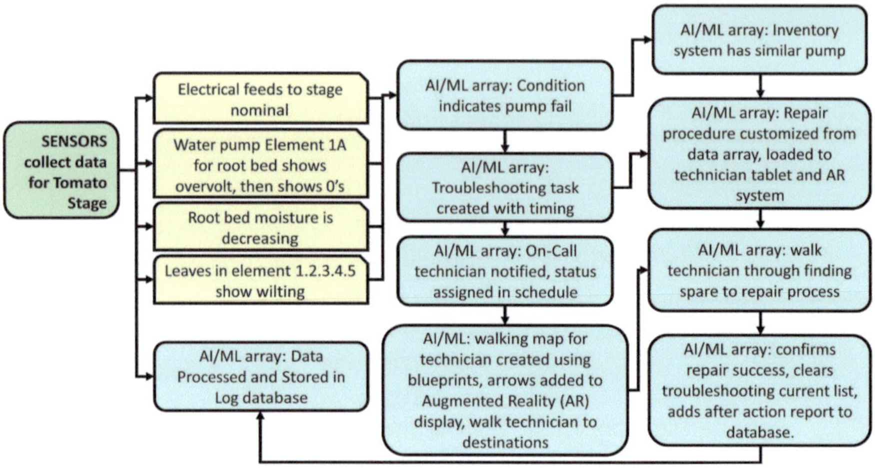

Fig. 5.46 Simplified use of AI and humans to troubleshoot. In this case, a growth problem with tomatoes is worked by the AI-human team to find, then repair the issue

5.17 Backups and Redundancy

Given time to resupply or even get basic help for any space mission, especially those on distant locales including Mars, having back-up systems ready for switch over are key. For biological organisms, mayhem is even more likely due to disease, radiation, failure of support systems or environment, hybrid breakdown, and many other reasons. Time to grow crops to maturity means that food needs to be stored to cover farm failures, and crops need to be kept at the ready, and ideally in separated diverse volumes. Using continuous harvest and oversizing growth areas will ensure short term losses and provide material for storage. For a settlement beyond a few individuals, multiple domes or rooms separated by airlocks and sterilizers would ensure that vacuum, disease, or other catastrophe doesn't claim the whole farm. Further, banks of seeds and for animals, frozen eggs, and sperm, provide back-up and genetic reserves, and take little room. Another backup involves in habitat growth of some plants in corners and nooks.

For machinery, ready to switch hot standbys come at an energy and mass cost but may be worth it for critical systems. One way is to oversize critical systems, such as making multiple pipes, tanks, and pumps. This strategy has the advantages of planning for growth and reducing wear and tear on machinery. As noted earlier, clogging is always a problem in waste and farm systems, and parallel piping allows for shutdown of one route for cleaning, while another carries the fluids. Other parallel items might include the heat rejection/heating equipment, which are also critical for human life support.

Computing systems will also need redundancy and backup. In addition to managing mass, farms generate lots of data, and need constant monitoring and management. The farms network should be redundant, with sensors that overlap coverage, and diverse in location and box, for arrays of storage and processors. A mesh of processing nodes and storage with pre-designed continuous backup and update, and long term buried or off-site intermittent backup might be a architecture concern. Long reach data communication to satellite or back to other settlements might provide potential reboot options, at the cost of introducing cybersecurity concerns via the connection. Machine Learning (ML) is one way to combine and sort signals from the sensors and correlate these signals in to human intelligible notices and alarms (as in the AI examples in this chapter). Several networking and control systems on Earth already use software to take the traffic in data from sensors and reduce it to critical results and create additional reports.

Getting data between spots will use multiple methods. Wireless can be problematic in the high interference environment of space, but of course may be necessary over long distances. Current in space wireless uses unreliable data protocols (UDP) that assume data might not get there, then software assembles the data packets in whatever order received, then asks for the missing parts. On top of this, compression makes big data sets small enough to send in realistic timeframes. Given the long distances in space, latency, i.e. the delay between sender and receiver is a big concern. For example, it will take years to get all the flyby data from the New Horizons mission back to Earth and process it. In our farmer sense, shorter distances between farm parts are more likely, even if the signal back to home base is longer. Wire might be used if well shielded for short runs. That said, fiber optics are favored over wire for the simple reason that wire cables collect signals, and in space even shielded wires will get some static, slowing bandwidth. Fiber optics will need to be toughened for the temperature swings of the off-Earth environment (therefore likely be thicker) but pick up less interference and can use dense wave division multiplexing to send parallel signals. Further, silica and doping agents used to make fiber optics are plentiful in space, especially on the moon. A further long-range method of communication can be free space lasers, which work well in vacuum, but require careful alignment. Likely for distances to ten meters, the connections will be wire and fiber. For distances of tens of meters to kilometers, you will use fiber or free space laser. For kilometers too far off, communication is by wireless of some type, possibly relayed by satellites.

5.18 Summary

This chapter covers the devices and machines required to move mass, separate mass and various chemicals, provide light, and regulate temperature. Most, or almost all, of these devices are already in use on Earth in factories, fish farms, vertical farms, greenhouses, and other places. The devices for the space farm will be similar, except possibly hardened against conditions of the settlement, and made more resilient to minimize the need for maintenance. These device are building blocks that implement and link the parts of the growth chambers, stages, and between stages.

5.19 Questions

1. What are a few types of devices for moving liquids, and which might be best for: recirculating algae cells that shouldn't be crushed, pumping liquids that contain solids that should be chopped, pumping a liquid in a place where maintenance will be very difficult?
2. What types of pumps can be used for injecting gas at pressure into liquids? What devices are used to enable injection of gases into liquids?
3. What device without moving parts is used to settle out liquids of different densities? What concerns would be applied to this device in lower gravity environments?
4. What are the benefits of a canister filter versus a flat screen filter?
5. What is a use for a reverse osmosis filter?
6. Why would you need a UV sterilizer with a canister media filter?
7. What device (or devices) could be used to reduce the humidity in a room?
8. What device might be used to radiate heat away into space?
9. What are components of a state change chiller?
10. What is a device useful for separating a mixed liquid that contains liquids of different vaporization and condensation temperatures? (Clue: used on stills for moonshine, and in oil refineries too).
11. What method could I use if I wanted to sterilize large volumes of liquids, but did not want to use the energy for a UV lightbulb, and my settlement was below the surface of the Moon? What if instead by settlement was on Mars, but had lots of waste heat available from a nuclear reactor?
12. Why would I need to allocate electrical power for work lights? Where in the farm are they used?
13. Why would lights for growing some plants look purplish in lieu of white?
14. Assuming my lights and heat rejection from other stages is insufficient to warm my banana and orange trees in a park in a lava tube on the moon, what devices would help warm my trees and park visitors?
15. What are three kinds of sensors and what do they measure?
16. What is a device I could use to recover carbon dioxide if my farm produced so much lignin that I could not use a digestor?
17. How could I resupply my canister carbon filters using farm wastes?
18. What device could be used to control the flow of water to my fish stage? The flow oscillates often so the device should not be fragile.
19. What are they key difference between a zeolite or carbon filter and a cryopump?
20. Assuming my settlement is positioned next to a very large water ice deposit, and my settlement needs to ship farm produce to market via chemical rocket, what method could I use to produce rocket fuel?
21. How does what I can have for breakfast, lunch, and dinner affect energy use by the space farm?
22. Why does the space environment justify back-up systems and stage level logging and control?

References

Meyer. (2023). Electrical requirements for a spectrum of multi-stage space farms. *National Space Society (NSS) Space Settlement Journal*. Published by the NSS, (11/2024). https://nss.org/wp-content/uploads/2023/06/NSS-JOURNAL-Space-Farm-Electrical-Requirements-updated-2023.pdf

NASA-Vlight National Aeronautics and Space Administration, Science Mission Directorate. (2010). Visible Light. Retrieved April 05, 2022, from NASA Science http://science.nasa.gov/ems/09_visiblelight

OSHA-1915.82. (2011). 1915.82—Lighting, U.S. Occupational Health and Safety Administration (OSHA) Standard 1915.82, [76 FR 24701, May 2, 2011].

Chapter 6
Harvest Dynamics and Stages as Mass Flow

Abstract This chapter examines how crops in space farm stages are harvested, examining two harvest methods: big bang harvest (i.e. traditional seasonal harvest) and continuous harvest. Big bang harvest is how field crops are harvested at the end of the growing season on Earth. In the traditional method, crops are harvested, the field is cleared and then replanted. Plants are then grown to maturity months later, and all plants of the same crop are at the same level of maturity. Continuous harvest is where a portion of a crop is harvested each day and replaced by planting every day, as is done for vertical farms and some tropical crops. Both harvest methods are compared using planted area (footprint), for plants and animal crops such as fish. The resultant comparison informs stage size and mass flow, and therefore space farm size.

Keywords Harvest · Footprint · Area · Timing · Planting

6.1 Introduction

This chapter examines how crops in space farm stages are harvested, examining two harvest methods: big bang harvest and continuous harvest. Big bang (traditional) harvest is how field crops are harvested at the end of the growing season on Earth. Continuous harvest is where a portion of a crop is harvested each day, as is done for vertical farms and some tropical crops. For big bang harvest, fields need to be big enough to grow enough food to last until the next growing season, and big enough for the maximum footprint of the mature crop, and mass flow ramps up exponentially to harvest. Continuous harvest is sized so that food is produced each day, fields are sized smaller to take advantage of the organism's growth model, and mass flow is the same every day.

© The Author(s), under exclusive license to Springer Nature Switzerland AG 2026
B. L. Meyer, *Designing Farms in Space*, https://doi.org/10.1007/978-3-032-04395-5_6

6.2 Stages and the Harvest Problem

While there is a good definition of a stage above, here it is again, to allow discussion of why how we will harvest is important. A stage, as defined in this book, includes all the living biomass of a species in the farm, the enclosures (growth chambers) to grow this species at optimum conditions, with all the machines, tanks, lights, pipes, and other equipment, with control systems and sensors for this species. Biomass of the species, tracked as living kilograms, includes all generations in the farm. We need to know how many living kilograms we will need at its largest, to figure out how to engineer and size the stage. If we just harvest every time all the organisms hit harvest, harvest time is the biggest the stage will be. If we harvest crops on Earth such as wheat, we plant with just a few living kilograms of seeds in a volume and with all the equipment needed for the eventually mature crop at harvest roughly 110 days later. I call this big bang harvest, where you wipe out almost all the living mass at harvest, then replant starting near zero living mass. For most of the 110 days while the wheat grows, the mass flow is very low, and most of the growth chamber is nearly empty. This is fine on Earth where the area is already there, with copious water and free carbon dioxide and sunlight, but having a giant volume sitting idle in space farm for most of the growth cycle is very expensive and inefficient. It is more like renting a giant warehouse for four months, but for three of the four months, it is mostly empty, but you must still pay for electricity, security, and rent for the entire space.

To illustrate this problem, you will have to pardon me here for a bit of elementary school math using Table 6.1.

At roughly averaged steady state near maturity, roughly 40% of the wet mass is in the edible part (yes, at harvest the percentage is even higher). So if you need around 1.1 m^2 for a living kilogram of wheat, getting you 0.40 kg of dinner, and you want enough calories to feed 50 settlers, each settler needs 2000 kcal per day or about 0.54 kg of wheat grain per person per day, or around 27 (= 0.54 kg * 50 people) kilograms per day for the settlement. These 27 kg of food means you need 68 (=

Table 6.1 Assumptions and data for wheat footprint example

Basic data table: assumptions for people and Wheat	
50	People, assumed
2000	kcal per settler per day, assumed
3700	kcal/kg wheat grain (USDAARS, 2022)
0.541	kg wheat grain per person per day (assuming a 2000 kcal diet)
110	Days to harvest wheat (assumed)
1.10	Footprint in square meters per living kg (measured)
1.25	meters, height of wheat top of stalk to bottom of roots, personal measurement
2.52	Ratio living mass Whole plant to grain mass (from calculations in prior chapters)
40%	Ratio grain mass to living mass Whole plant (assumed)

Measured the wheat footprint myself

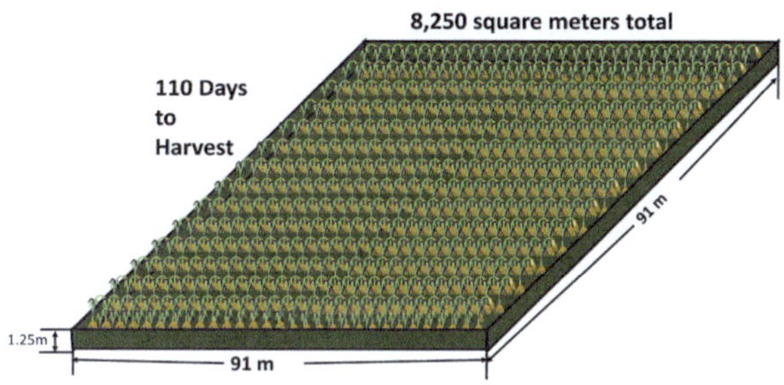

Fig. 6.1 Wheat growth footprint. Estimated footprint (area) to grow 110 days of wheat using traditional (big bang) harvest

27 kg * 2.52) living kilograms of wheat per day, the other 41 kgs of the 68 kgs is in straw, chaff, roots, etc. Since it takes 110 days to grow wheat, you are going to need around 7.5 (= 68 kg * 110 days) tons of living biomass at harvest every 110 days. 7.5 tons of living wheat plant need around 8250 m^2 or 83% of a hectare (1 hectare = 10,000 m^2), or 2 acres. Most people can't visualize that area, so let's say about the size of eight USA residential lawns or roughly 40% of a city block in Chicago or New York City. Wheat plants grow to around a meter or so tall and use about a quarter meter of root depth, so a box of 10,312 (= 8250 m^2 * 1.25 m) cubic meters at least, assuming the crop never fails, and you don't need any extra room for lights, or machinery to make it all work, or extra human workspace for the wheat. This layout is shown in Fig. 6.1.

6.3 Continuous Harvest Concept and Periodic Problems

Since most of us are familiar with the concept where we go out to a garden each day and pick a few things for dinner that day, or out to a pond and catch dinner for just the day, we already understand a bit about continuous harvest. Continuous harvest is where we always just collect what we need for today, and tomorrow that harvest will be replaced by the organisms left growing. This works for grains and annuals too. In our wheat example, instead of allocating the whole 8400 m^3 and leaving it mostly empty, we calculate how much we need to make sure that each day, we get 68 living kilograms of wheat plant, giving me 27 kg of grain for food each day.

First let's look at how wheat grows. If we get a living kilogram in 110 days and start from just seed mass (essentially 0), the chart looks like a classic exponential graph as in Fig. 6.2.

The graph in Fig. 6.2 is for exponential growth. Over time the graph looks something like the sawtooth graph in Fig. 6.3, though I added the case where the organism grows linearly.

Since footprint and volume are just a multiple of the living kilograms, assuming one harvest every 110 days to feed my 50 settlers to produce 7480 kg of wheat plant to get about 3 tons of grain food. However, due to the growth chart above, I only use a portion of the total footprint area as the crop grows assuming the growth chart

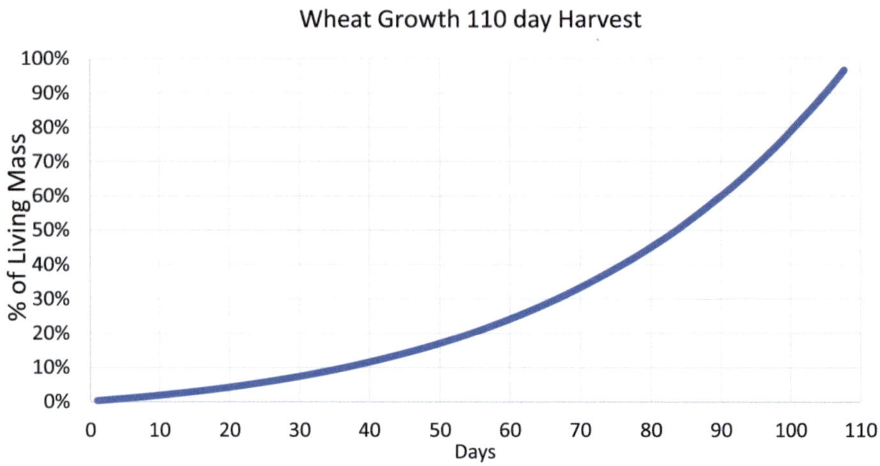

Fig. 6.2 Basic growth curve (estimated and rough) for wheat. Graph assumes exponential growth. Dry out might affect this graph

Fig. 6.3 Multiple big bang harvest growth models, linear and exponential, over time (days)

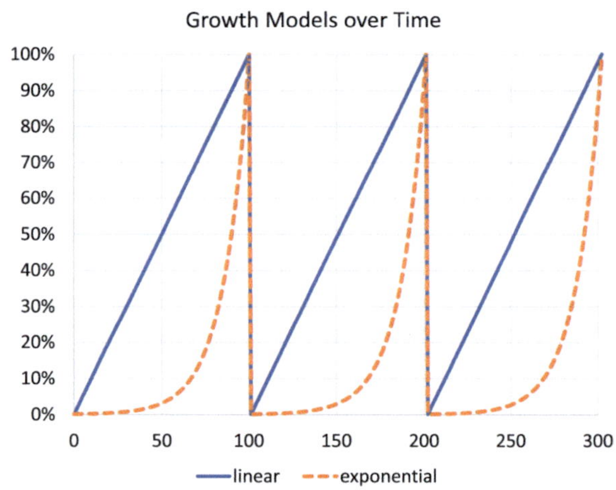

above and the footprint per mass from Table 6.1. In practice a soil farm just looks empty until the plants grow tall, since the wheat crop doesn't get transplanted.

There is an old joke in the wheat growing Midwest USA, since wheat looks like any other grass for most of its life. A guy knocks on the farmer's house. Farmer answers. Guy says "Hey, I saw that all the grass in the field looked overgrown, so I mowed it for you." Farmer faints.

If the plants are contained individually, they can be easily moved. In a hydroponic scenario, or pot or diaper scenario, the plants might be shuffled around. Table 6.2 takes the percentages from the growth graph above and multiplies it by the total

Table 6.2 Wheat growth and area used for 7.5 tons of living mass at 110-day harvest

Footprint Area by Day			
Day	% living mass (%)	Living mass (kg)	Area (m^2)
1	0	Small	Very small
30	13	986	1087
60	35	2590	2856
90	68	5103	5628
110	100	7480	8250
Living mass = % living mass (from growth graph above) *7480 kg			
Area = kg living mass *1.1 m^2/kg			

living mass size at harvest (7480 kg), then multiplies that by the footprint area per living kilogram (1.1 m^2/kg). Again, this table assumes plants are ideally spaced.

I must allocate 8250 m^2 for the wheat to grow into, though, for up to 30 days I would only use around 1087 m^2 if all the plants were laid out ideally (see Fig. 6.4) since the plants are still under 13% of their total size.

At 60 days, the plants are 35% of their final size, but I still have the same 8250 m^2, now using 2856 m^2 of the 8250 m^2 total footprint, so around two-thirds empty (see Fig. 6.5).

At 90 days, the plants now use around 5628 m^2, using around 68% of the final total (see Fig. 6.6).

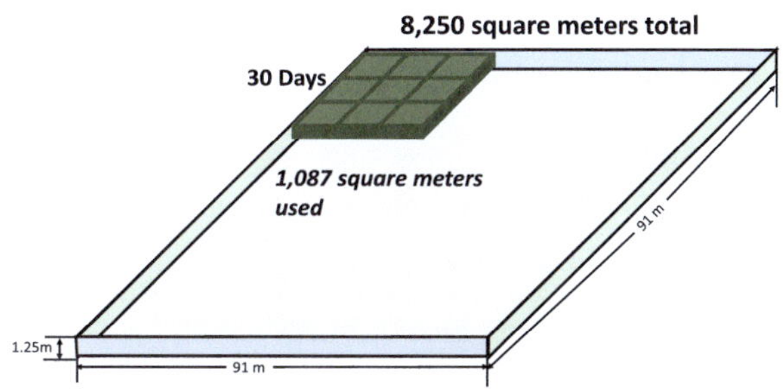

Fig. 6.4 30-day footprint. The area occupied by wheat plants for the first 30 days based on the footprint for the size of each plant at 30 days

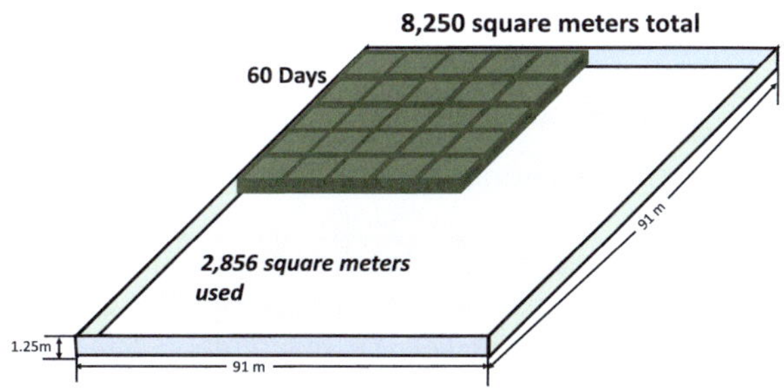

Fig. 6.5 60-day footprint. Area occupied by wheat up to day 60

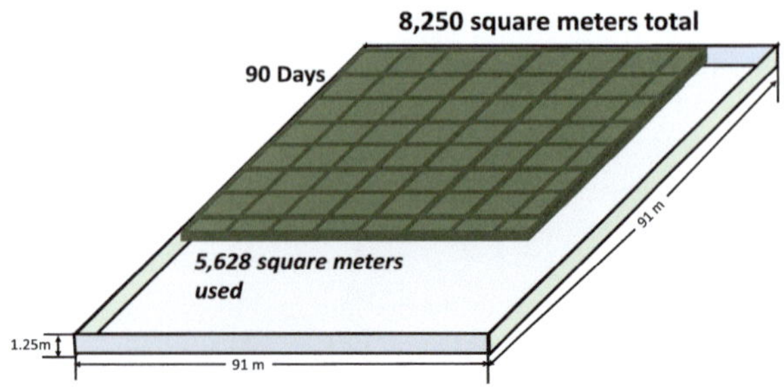

Fig. 6.6 90-day footprint. Footprint occupied for wheat to 90 days

Using the 120-day area for 90-day, or 30-day, or 60-day plants is not a good use of space in space, especially since every bit of volume and footprint must be sealed against vacuum or near vacuum, protected from radiation, and heated or cooled and filled with air. Believe it or not, since things grow exponentially, the volume the crop will then need is far less if we do continuous (daily) harvest. Therefore, we should use continuous harvesting, even for annual crops like grains or meat animals. For daily crops like eggs or perennial plants like tomatoes, continuous harvest is less about replanting, and more about simply harvesting, replacing organisms that age out, and maybe trimming.

While ideally, we would plan our pots or hydroponic trays for our wheat plants sized for each day to maximize area and volume, this is not feasible in practice, especially for hydroponic trays or pots, due to the work and risk involved in moving the wheat plants. We will have to actually shift pots that may be oversized or use a root container to make it easy to move the plants without killing them. Choices in root bed design, and species, will affect if continuous harvest can be used. Root and shoot damage is always a risk in moving plants, but cannabis and vertical farmers have this figured out on Earth. For animals like fish or shrimp or even chickens, you could move them often since they are more portable, and on Earth most animal farmers do exactly that. Chicks start in an incubator, then move to larger pens as they incrementally grow. Ditto for fish, especially for predatory fish such as trout. For wheat, planning for space in monthly groups until harvest might be realistic, so we will be setting up our grow for intervals at day 30, 60, 90, and harvest day 110 (See Fig. 6.7). A cohort is a group of organisms at roughly the same growth stage. One of the strong upsides of the replant concept is that conditions for each cohort can be optimized. For example, in the 90–110-day cohort, conditions can be drier with a shorter daylight interval than during the previous cohorts to trigger grain maturity and dry out.

Instead of ~ 7.5 tons of live mass, each day's harvest is 68 kg of plant (27 kg edible), since that is all, I need each day to feed my 50 settlers. Granted, in both big bang and daily harvest scenarios, I must wait over 110 days to get my first crop.

The difference is that while I store the grain from the big bang harvest for 110 days till the next harvest, I instead just get my daily crop every day if I plant the wheat in my settlement daily as well (i.e. indefinitely). See Fig. 6.8. I pluck up today's harvest, then move in the most mature of the plants in the 60–90-day group in the now empty

Fig. 6.7 Movement of wheat assuming 30-day cohorts

pots or trays. Then I replace the group I moved from the 60–90-day group with a group from the 30–60-day group, then replace the moved plants in the 30–60-day group from the plants in the 0–30-day group. In the 0–30-day group, I plant seeds in the now empty pots or trays. This means some real work for the gardeners (or robots), but that is life. Tables 6.3 and 6.4 together are sample schedule that repeats every 30 days. So on Day 1: Cohort 4, Position A (i.e. 4(A)) is Harvested; then Cohort 3 Position A is moved to Cohort 4 Position A (i.e. 3(A) → 4(A)); then Cohort 2 Position A is moved to Cohort 3 Position A (i.e. 2(A) → 3(A)); then Cohort 1 Position A is moved to Cohort 2 Position A (i.e. 1(A) → 2(A)); then Cohort 1 Position A is seeded (i.e. Seed → 1(A)). You would need a schedule like this for each continuously harvested crop.

In practice also note that we would have to have backup and a bit of excess species biomass due to accidental losses. With daily harvest, we could isolate each month's cohort in its own volume, therefore limiting damage due to disease or accidents.

To get the graph in Fig. 6.9, each day is a sum of all the percentages from the growth plot from day 0 to the day in the graph. So, for day 4 for example, the multiplier is 0.3% + 0.7% + 1.0% + 1.4% = 3.4% (or 0.034) of the daily living mass need. This means at day 4, that will eventually be fully grown for harvest in 110 days to 106 days from now, together have 13.3 kg mass, and take up at least 14.7 m^2. This is the pessimistic way of this calculation, because optimistically I would subtract the previous field sizes from each prior cohort, but for plants, this doesn't quite work out, so pessimism is used here. The group of plants for multiple days grouped together are called a cohort. Using the graph above, we can pick the minimum multiplier of the daily living mass needed at 30, 60, 90, and 110 days so that we move plants only once a month. This means that plants sit in the cohort until they are old enough to

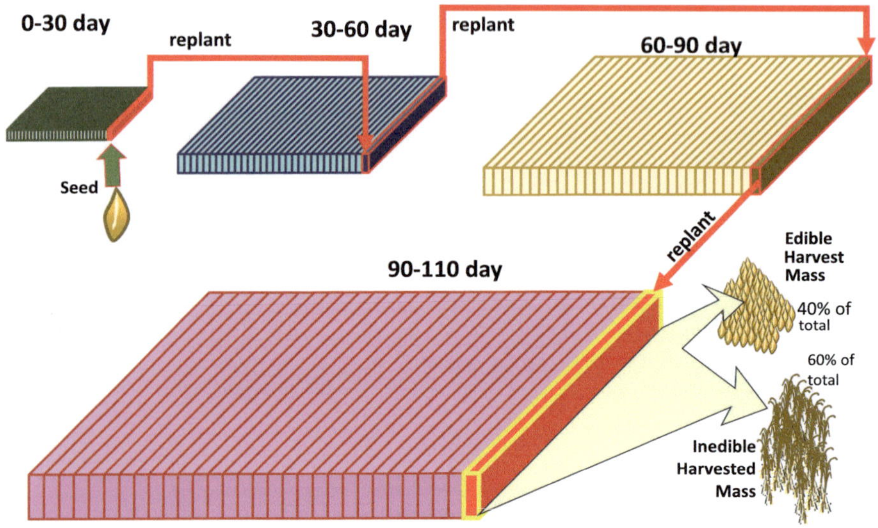

Fig. 6.8 Daily harvest and cohort seed, replant, and harvest

be moved up to the bigger growing space. Daily mass for each month in the cycle is shown in Table 6.5.

Given this method, you will note that the sum of 5691 m^2 is lower than the big bang number of 8250 m^2. See Fig. 6.10. Let's say I did a daily move instead of every 30 days; the sum would be a much lower at 3072 m^2 (= 110-day multiplier from graph which is ~ 41 * 68 living kg/day * 1.1 m^2/kg). Sticking with the monthly cohort idea, a few drawings might be useful. The drawing below shows how the area is used. In the 0–30 day area, seeds are planted and sprout for 30 days. Then grow out for 30 days in the 30–60-day area, then in 30 days move to the 60–90-day area, then in another 30 days to the 90–110-day area, then harvest in 20 more days.

This continuous harvest concept looks like an exotic concept but is in use right now, on Earth. Greenhouse crops, especially cannabis and salad greens, are already grown using this exact method to maximize space and production, and in space we can use this method for many other crops too.

6.4 Growth Models, Mass Flow, and Continuous Harvest

Mass flow gets very interesting given either big bang harvest or continuous harvest. Graph shape becomes very critical. If a dry-out time is involved or a more complex growth graph, we may have to use the same logic as above or sizing but adjusted based on graph shape. For example, if we look at various models for pinto beans, we see that picking the right graph makes a difference. In this case, the closest to

Table 6.3 Sample work schedule for wheat continuous harvest example for Cohort 3 and Cohort 4

DAY	Cohort 4 (90–110 day)			Cohort 3 (60–90 day)		
	Age	Position	Action	Age	Position	Action
1	110	A	4(A)→Harvest	90	A	3(A)→4(A)
2	110	B	4(B)→Harvest	90	B	3(B)→4(B)
3	110	C	4(C)→Harvest	90	C	3(C)→4(C)
4	110	D	4(D)→Harvest	90	D	3(D)→4(D)
5	110	E	4(E)→Harvest	90	E	3(E)→4(E)
6	110	F	4(F)→Harvest	90	F	3(F)→4(F)
7	110	G	4(G)→Harvest	90	G	3(G)→4(G)
8	110	H	4(H)→Harvest	90	H	3(H)→4(H)
9	110	I	4(I)→Harvest	90	I	3(I)→4(I)
10	110	J	4(J)→Harvest	90	J	3(J)→4(J)
11	110	K	4(K)→Harvest	90	K	3(K)→4(K)
12	110	L	4(L)→Harvest	90	L	3(L)→4(L)
13	110	M	4(M)→Harvest	90	M	3(M)→4(M)
14	110	N	4(N)→Harvest	90	N	3(N)→4(N)
15	110	O	4(O)→Harvest	90	O	3(O)→4(O)
16	110	P	4(P)→Harvest	90	P	3(P)→4(P)
17	110	Q	4(Q)→Harvest	90	Q	3(Q)→4(Q)
18	110	R	4(R)→Harvest	90	R	3(R)→4(R)
19	110	S	4(S)→Harvest	90	S	3(S)→4(S)
20	110	T	4(T)→Harvest	90	T	3(T)→4(T)
21	110	A	4(A)→Harvest	90	U	3(U)→4(A)
22	110	B	4(B)→Harvest	90	V	3(V)→4(B)
23	110	C	4(C)→Harvest	90	W	3(W)→4(C)
24	110	D	4(D)→Harvest	90	X	3(X)→4(D)
25	110	E	4(E)→Harvest	90	Y	3(Y)→4(E)
26	110	F	4(F)→Harvest	90	Z	3(Z)→4(F)
27	110	G	4(G)→Harvest	90	AA	3(AA)→4(G)
28	110	H	4(H)→Harvest	90	BB	3(BB)→4(H)
29	110	I	4(I)→Harvest	90	CC	3(CC)→4(I)
30	110	J	4(J)→Harvest	90	DD	3(DD)→4(J)

real-world growth is a combined graph that begins as an exponential graph, becomes a linear graph, and then flattens during dry out, from Fig. 6.11.

The key is to pick a model that is closest to your chosen crop. First, we must consider if the crop is an annual or perennial. Is the crop is a periodic whole organism harvest, i.e. an annual, like wheat, or tilapia? If a crop is where only a part of the

Table 6.4 Sample work schedule for wheat continuous harvest example for seed, Cohort 1, and Cohort 2

DAY	Cohort 2 (30–60 day)			Cohort 1 (0–30 day)			SEED	
	Age	Position	Action	Age	Position	Action	Age	Seed
1	60	A	2(A)→3(A)	30	A	1(A)→2(A)	1	Seed→1(A)
2	60	B	2(B)→3(B)	30	B	1(B)→2(B)	1	Seed→1(B)
3	60	C	2(C)→3(C)	30	C	1(C)→2(C)	1	Seed→1(C)
4	60	D	2(D)→3(D)	30	D	1(D)→2(D)	1	Seed→1(D)
5	60	E	2(E)→3(E)	30	E	1(E)→2(E)	1	Seed→1(E)
6	60	F	2(F)→3(F)	30	F	1(F)→2(F)	1	Seed→1(F)
7	60	G	2(G)→3(G)	30	G	1(G)→2(G)	1	Seed→1(G)
8	60	H	2(H)→3(H)	30	H	1(H)→2(H)	1	Seed→1(H)
9	60	I	2(I)→3(I)	30	I	1(I)→2(I)	1	Seed→1(I)
10	60	J	2(J)→3(J)	30	J	1(J)→2(J)	1	Seed→1(J)
11	60	K	2(K)→3(K)	30	K	1(K)→2(K)	1	Seed→1(K)
12	60	L	2(L)→3(L)	30	L	1(L)→2(L)	1	Seed→1(L)
13	60	M	2(M)→3(M)	30	M	1(M)→2(M)	1	Seed→1(M)
14	60	N	2(N)→3(N)	30	N	1(N)→2(N)	1	Seed→1(N)
15	60	O	2(O)→3(O)	30	O	1(O)→2(O)	1	Seed→1(O)
16	60	P	2(P)→3(P)	30	P	1(P)→2(P)	1	Seed→1(P)
17	60	Q	2(Q)→3(Q)	30	Q	1(Q)→2(Q)	1	Seed→1(Q)
18	60	R	2(R)→3(R)	30	R	1(R)→2(R)	1	Seed→1(R)
19	60	S	2(S)→3(S)	30	S	1(S)→2(S)	1	Seed→1(S)
20	60	T	2(T)→3(T)	30	T	1(T)→2(T)	1	Seed→1(T)
21	60	U	2(U)→3(U)	30	U	1(U)→2(U)	1	Seed→1(U)
22	60	V	2(V)→3(V)	30	V	1(V)→2(V)	1	Seed→1(V)
23	60	W	2(W)→3(W)	30	W	1(W)→2(W)	1	Seed→1(W)
24	60	X	2(X)→3(X)	30	X	1(X)→2(X)	1	Seed→1(X)
25	60	Y	2(Y)→3(Y)	30	Y	1(Y)→2(Y)	1	Seed→1(Y)
26	60	Z	2(Z)→3(Z)	30	Z	1(Z)→2(Z)	1	Seed→1(Z)
27	60	AA	2(AA)→3(AA)	30	AA	1(AA)→2(AA)	1	Seed→1(AA)
28	60	BB	2(BB)→3(BB)	30	BB	1(BB)→2(BB)	1	Seed→1(BB)
29	60	CC	2(CC)→3(CC)	30	CC	1(CC)→2(CC)	1	Seed→1(CC)
30	60	DD	2(DD)→3(DD)	30	DD	1(DD)→2(DD)	1	Seed→1(DD)

organism, including seeds (like apples or oranges), eggs, or leaves (like tea bushes), is harvested every so often but the organism (tree, hen, bush) still lives to produce more, that is for our purposes called perennial. Annual crops could be any of the shapes in Fig. 6.10. That means mass flow would shadow this graph in a big bang harvest. Perennials are essentially a flat line, because any mass input is taken out in

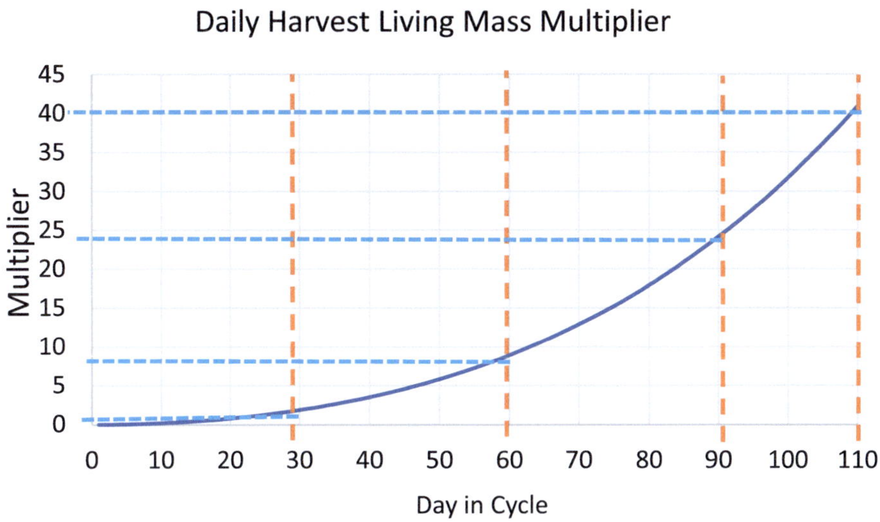

Fig. 6.9 Graph for daily harvest living mass multiplier

Table 6.5 Multipliers for cohort example for wheat

Area by Day				
Day	Multiplier	Total living mass for cohort (kg)	Area (m^2)	% of total footprint (%)
1	0.00	0.23	0.25	0
30	1.87	127	141	2
60	8.90	605	668	12
90	24.13	1641	1810	32
110	40.97	2786	3073	54
SUM			5691	100
Area = kg living mass *1.1 m^2/kg				

almost equal amount by harvest or waste. What continuous harvest does in practice is make a whole stage of annuals look like a perennial.

Let's quickly review how things grow and mass flow (see Fig. 6.12). Starting with a seed, egg, or growth culture, inputs are added as needed to make the organism or culture growth. The amount of these inputs depends on how much the living thing will grow and is a multiple of how big it is that day. The bigger it is, and the more it grows, the more mass it needs as input. Outputs are then the wet mass that becomes growth, plus anything that gets released like waste wet mass or gases. For photosynthetic crops, like plants and algae the inputs are carbon dioxide, water, and nutrients, outputs are more crop living wet mass and oxygen. For aerobic crops, the inputs are food wet mass, water, and oxygen, and the outputs are more crop living

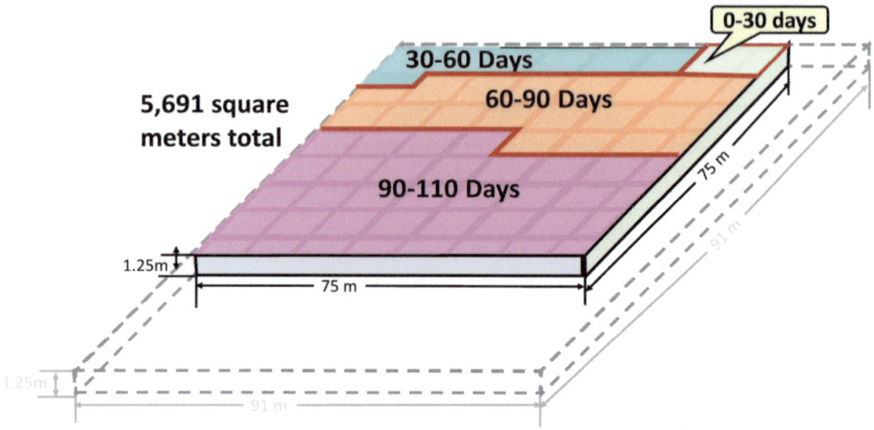

Fig. 6.10 Layout of cohort planting footprint for daily harvest and 30-day cohort example

Fig. 6.11 Growth models for pinto beans. Various growth models for an annual, in this case a legume

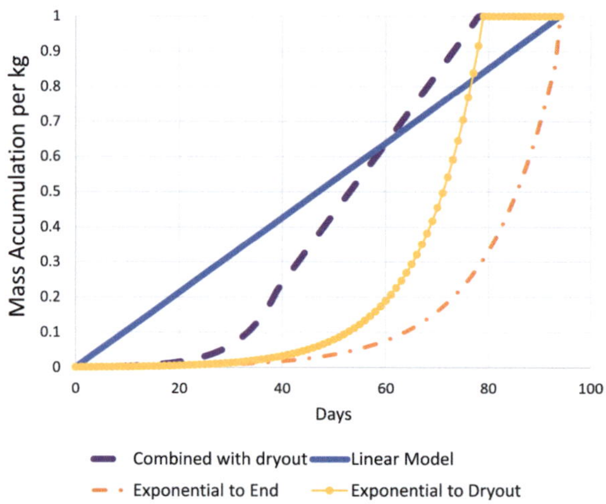

wet mass and carbon dioxide, water, and waste wet mass. In this tense, eggs and milk are waste wet mass too.

Many plants like bushes, vines, and trees fit the perennial model, like fruit trees and grape plants. Only excess inedible wet mass (leaves, branches pruned) and mature fruits (edible) are harvested. In this case, harvest time may be longer from seed, but the mature plant remains alive. In this case, harvest time is daily as fruits mature, making for a very small harvest per day. This concept is like the biomass in excretions of an animal, or eggs from a chicken or milk from a cow, with the hen or cow as the 'tree'.

Many crop plants however are annuals, as are meat animals/insects, and mushrooms. Grains, legumes, some salad greens like cabbage, and many others are called

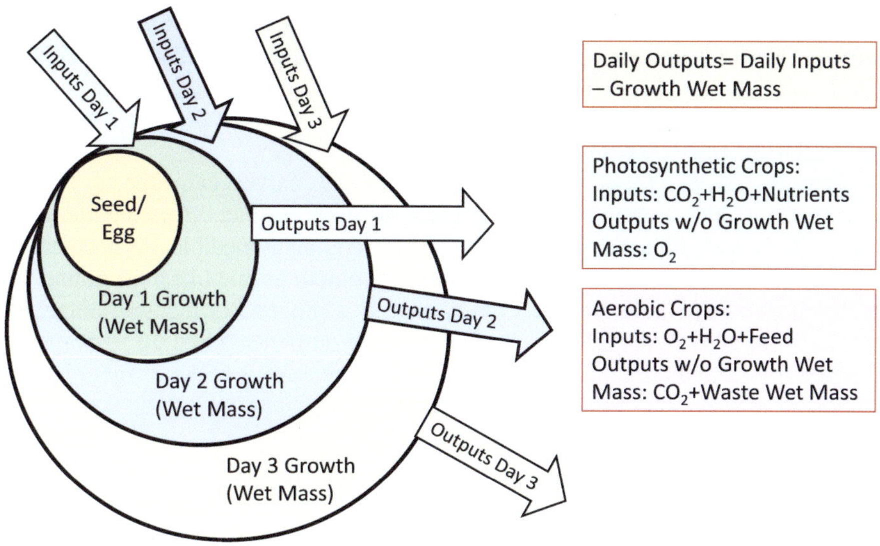

Fig. 6.12 Daily growth by addition concept

annuals here. Animals (fish, shrimp, and mollusks) follow a similar dynamic to the wheat example, since the entire crop organism is harvested, but with straight exponential growth, and an overgrowth that includes a breeder population that has individuals much larger and older than the crop. For many species of shrimp and fish, breeders are twice as massive as the crop, though very small in number (less than 1% to 0.1% of a large population of fish or for shrimp, represented by fecundity of the particular species, and brooding behavior). There is also mortality of organisms, ideally added < 1% by population count, but insignificant by total mass, that die before maturity, reflected as inedible biomass output. This mortality should consider the fact that mortality mostly occurs when the mass of each organism is exceedingly small. All this results in overgrowth of roughly 1% for fish or shrimp, derived from references, and multiplied in the growth.

For algae, bacteria, and yeasts, the doubling rate affects the mass in the stage (i.e. total living biomass), making them more like annuals than perennials. Living mass literally doubles on the order of a few hours to days, and continuous harvest includes the entire growth mass beyond that confined to the growth membranes or as basic mass in the tank. For the purposes of each farm here, productivity is time for the mass in the bioreactor to double in days, assuming full use. A small backup culture of minimal mass is maintained to restock the reactors, though insignificant when compared to mass and volume of the stage. As a result, the differences are essentially the doubling time plus one.

We already know that sizing for volume and footprint are different for perennials and annuals. Perennials need enough room, and therefore living wet mass to provide the steady-state continuous flow of daily harvests. Inputs are then balanced by the harvested wet mass, only some of which may be edible, and other outputs like carbon dioxide for photosynthetic crops. Mass flow for that whole crop is then pretty simple. Daily inputs are a multiplier of living mass in the whole crop. Daily outputs are the living mass that gets harvested, plus other outputs including any wet wastes and gases. Harvest and outputs balance the inputs.

Annuals present a bit of a problem that continuous harvest can address. To look at this, we can use a notional rice-like grain to simplify things. Assume this grain takes 100 days to harvest, produces as much edible grain as inedible wet mass called silage (i.e. roots, stems, leaves), and has an input/output profile like most grains: for every kilogram of edible grain wet mass, there is a kilogram of inedible silage wet mass and takes in 2.3 kg carbon dioxide, 0.02 kg nutrients, and 1.36 kg water. In addition to wet mass in grain and silage, this grain releases 1.68 kg oxygen.

6.4.1 Big Bang Versus Continuous Redux: Linear and Exponential Growth Models

Given a big bang harvest, the crop growth looks like Fig. 6.13.

Zooming in on a few days, the graphs look like Fig. 6.14.

In an exponential model, the inputs and growth get bigger as the plants/animals/etc. get bigger. In a linear model, the growth and inputs are the same every day, even as the plant or animal or fungus gets bigger. In the exponential case, as the mass of the crop increases, the inputs and outputs increase, which means a sawtooth graph of inputs, starting as a trickle as in the beginning turning into a flood right before harvest. That would mean sizing supplies, tanks, pipes, and pumps for the biggest

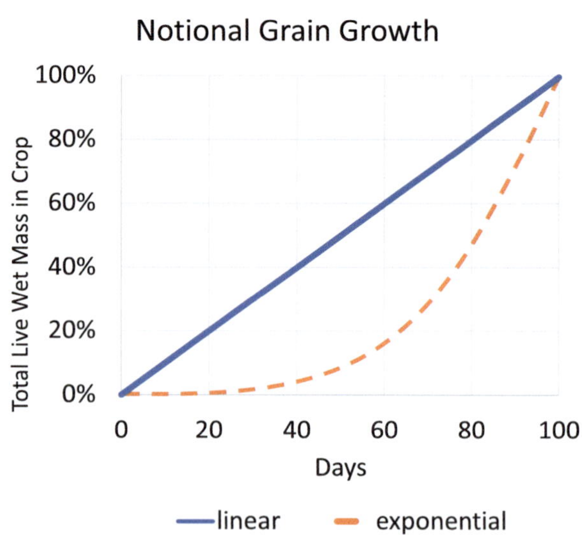

Fig. 6.13 Graph of linear growth versus exponential growth for one harvest cycle

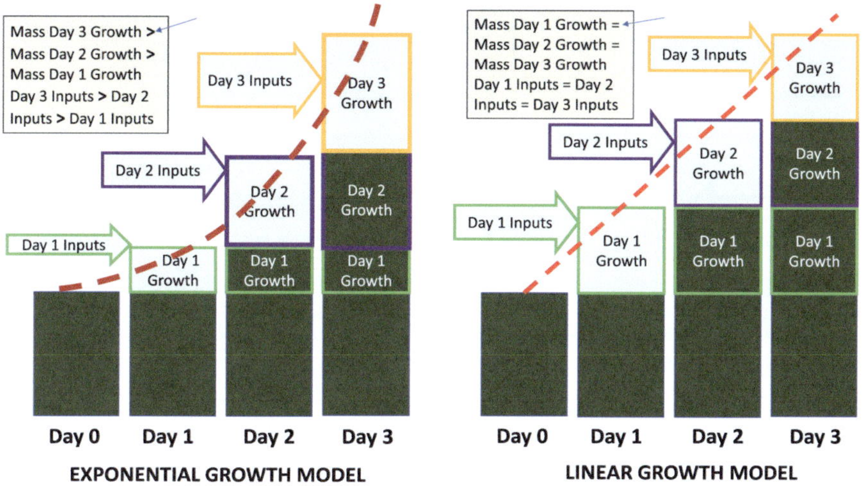

Fig. 6.14 Comparison of linear and exponential growth by day

time (right before harvest), and then barely using anything right after each harvest, making for a very large crop area, and lots of storage area. This problem does exist before steady state no matter what, but no need to exacerbate it!

Usually, to find out how much of a crop you need, you pick the outputs first like how many kilograms of a food, how many kilograms of oxygen, etc. Then you back into the size of the crop needed to make those outputs, then figure out the growth and harvest type, big bang or continuous, then reverse out the living mass at any time in the overall crop area. If you know the living mass and growth model, you can link it to the organism's mass flow model to get the inputs and outputs for the whole crop mass at any time. The goal is to make that mass flow as even as possible to shrink footprint and volume and to help that crop fit into the overall farm. Also let's ignore transpiration.

For our example, we use grain (see Fig. 6.15): to get 1 kg of grain in 100 days, it takes an input of 3.68 kg over 100 days to get 2 kg of plant (which leads to 1 kg grain) at harvest, and 1.68 kg of oxygen released over 100 days. The growth model determines how much each day the crop takes in, how much living mass is in the total crop, and how much oxygen it releases at any point in the growth cycle.

If we assume big bang harvest like on Earth, the crop plants are all the same age in the field, and the mass flow goes from near zero to maximal right at harvest.

Assuming a linear model for this big bang case, the plants will take in the same amount each of the 100 days, and release and growth the same amount each day, roughly whatever mass divided by 100. That means for this linear, big bang case, each day the grain takes in 2.3/100 kg carbon dioxide from the air, 1.36/100 kg water from the ground, and 0.02/100 nutrients from the ground, resulting in 2/100 kg of plant growth and releasing 1.68/100 kg oxygen to the air. On any day, the total mass

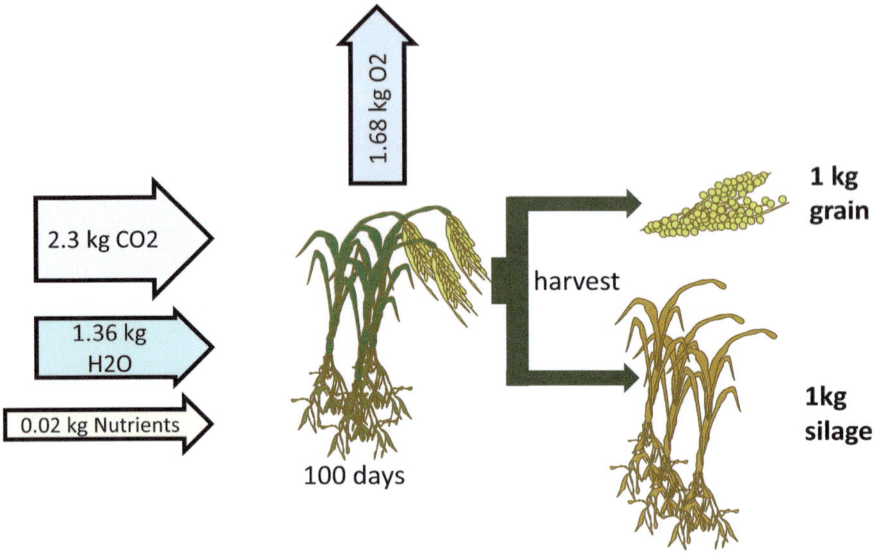

Fig. 6.15 Wheat plant mass flow to get 1 kg food, notional grain

of plant is 2 * day/100 kg, e.g. at the end of day 10, the plant is 2 * 10/100 kg = 0.2 kg, at harvest 2 * 100/100 kg = 2 kg.

Assuming an exponential growth model, and big bang harvest, the mass flow and total mass are more complex and depend on total mass to date and vary by day. Let's assume 7.15% growth per day; then on any day, mass at the end of the day is mass the day before * (1 + 7.15%). The same applies to all inputs and outputs, so at the beginning of day 65 plant mass is 0.1654 kg, carbon dioxide input is 0.0137 kg, water input is 0.0081 kg, nutrient input is 0.000119 kg, resulting in 0.0119 added wet mass and 0.00999 kg oxygen released, and beginning on day 66 the plant living mass is 0.1773 kg.

In both linear models and exponential models, for big bang harvest, your harvest at day 100 is 1 kg of grain, and 1 kg of silage, with no harvest until day 100. This means the grain must be stored for 100 days, hopefully with no losses to rot or rats. This is exactly why the ancient Egyptians had lots of cats to eat the rats that ate their stored grain! If you need 1 kg of grain per day as food, you must harvest 100 * 2 = 200 kg of plant every 100 days, pulling 1 kg of grain out of storage every day for use until you get the next harvest.

So how do we level the mass flow out using continuous harvest, getting a harvest every day instead of waiting for every 100 days? Ideally, we stage the plants to pull 2 kg of mature plants, for 1 kg of grain, every day, replacing them with seed that will mature in 100 days. First, we assume we already spent 100 days to get to steady state for the first harvested crop of 1 kg grain. For each model, the total mass of the crop being grown is the sum of the living plant mass at each day, so for the linear model, it is something like in Fig. 6.16.

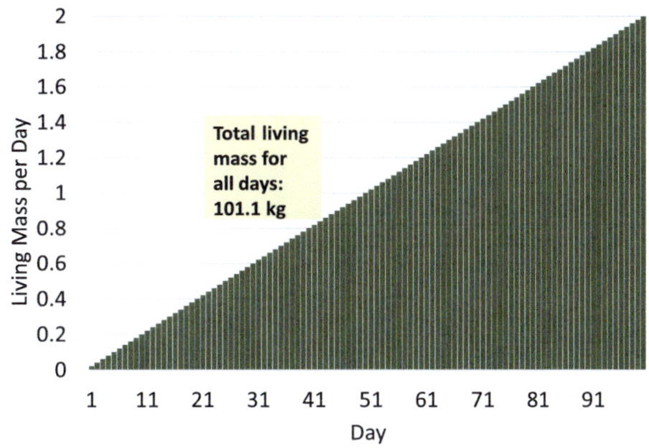

Fig. 6.16 Linear model mass (in kilograms) daily cohort for continuous model

That means that we pull 2 kg every day for each set of plants that have grown for 100 days, and replant plants that will be harvested in 100 days, keeping a continuous 101.1 kg in the fields, pulling 1 kg of grain every day. This also means the inputs and outputs are consistent the whole time, added exactly like the living mass for each day.

This principle is magnified for the exponential case, assuming the same 7.15% growth rate. The figure shows the living masses for the exponential case in Fig. 6.17.

This means assuming I get this kind of growth, I only need roughly 30 kg of living mass in my fields to harvest 2 kg of plant every day, yielding 1 kg of grain daily. This is very optimistic, but that is what you may get with continuous harvest.

In both linear and exponential growth models, the mass flow is consistently the same for every day, and the same for both models. Don't believe me, let's just look at CO_2 input for linear model in Fig. 6.18.

Now the same type of graph for the exponential case, in Fig. 6.19.

In both linear and exponential models, the daily mass flow is just like in Table 6.6.

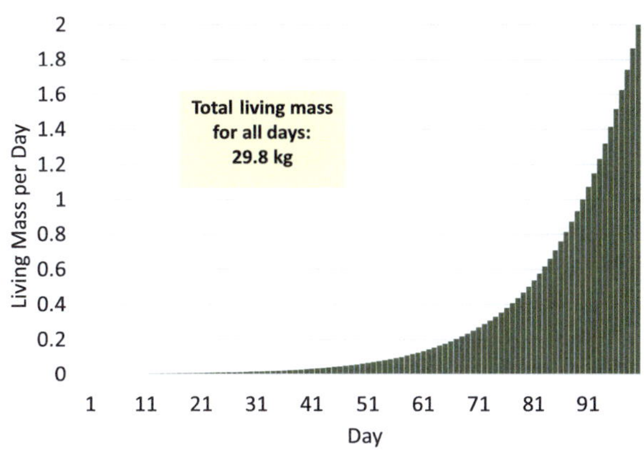

Fig. 6.17 Continuous harvest exponential model mass (in kilograms) by daily cohort

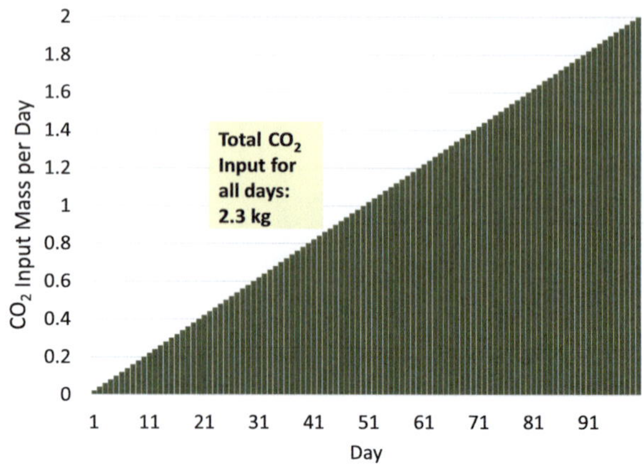

Fig. 6.18 Linear model continuous harvest CO_2 input by daily cohort

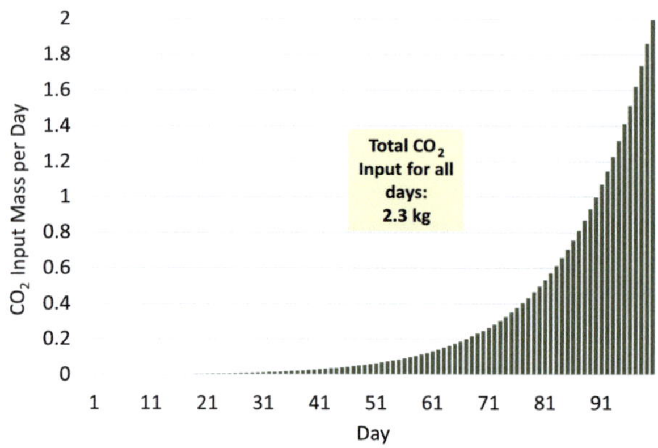

Fig. 6.19 Continuous harvest exponential model CO_2 input per day for each daily cohort

Table 6.6 Simple mass model for example grain

Inputs		Outputs	
CO_2	2.3	Grain	1
H_2O	1.36	Silage	1
Nutrients	0.02	O_2	1.68

The difference is in how much living mass in each field (stage) is needed to get there, or 101 kg for a linear model versus 30 kg for the exponential model. Since most real-world crops are at least partially exponential, continuous harvest eliminates the need for storing crops, and shrinks the field, providing the same mass flow each day.

Figure 6.20 illustrates the exponential increase in carbon dioxide for a big bang harvest model versus the even flows in the continuous harvest model. So that is the trick, mapping the real-world growth model to the continuous harvest concept for the sake of evening out mass flow, and getting a fresh daily harvest. Since the real-world

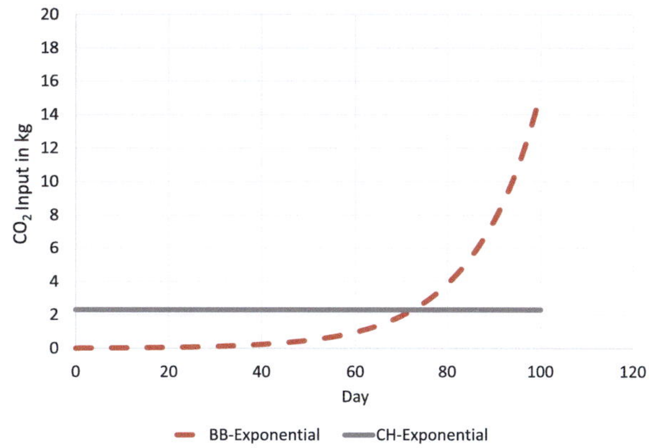

Fig. 6.20 CO_2 input for each growth model and harvest type. CH = continuous harvest, BB = big bang harvest

features more exponential growth than linear growth, we need to think more about getting closer to the real stuff.

6.4.2 *Combined Growth Models and Mass Flow*

So now, let's see what the notional grain would look like, assuming the same 100-day harvest, but with a more realistic growth curve with linear growth, without dry out. In this case, the plants grow exponentially for the first 50 days to 75%, then at slower linear rate till harvest, as in the plot in Fig. 6.21.

The mass flow by day changes, so that day 30 has more input mass and growth than on day 29, but on the linear end, day 60 has the same input mass and growth as on day 59 as in Fig. 6.22.

'Day' here is a set of plants at that age in the field. Note that using the continuous harvest idea, the total living mass of around 100 kg falls in between the strict exponential case (30 kg), and the strict linear case (101 kg) above.

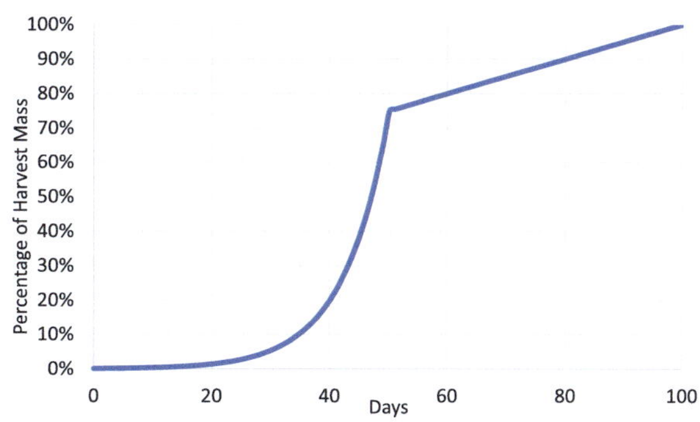

Fig. 6.21 Combined growth model for notional grain

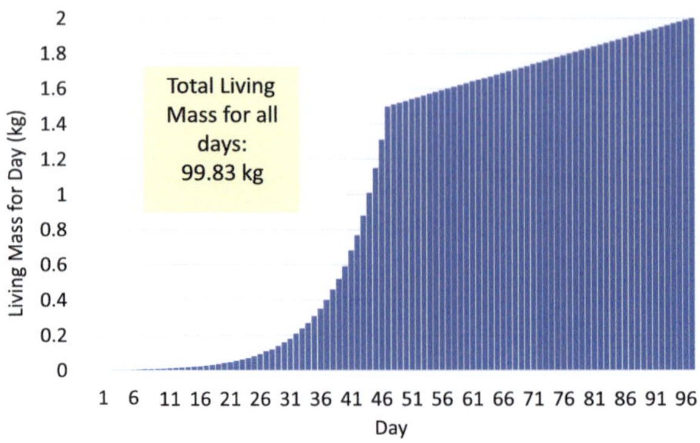

Fig. 6.22 Combined growth model daily cohort continuous harvest masses

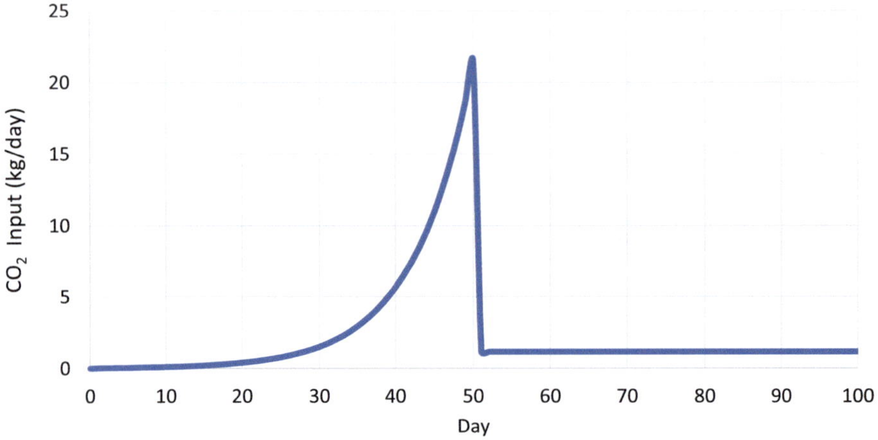

Fig. 6.23 Big bang harvest, combined growth model, CO_2 inputs

For big bang harvest of the combined graph, mass flow rates change all the way to day 50, then levels off, as for the linear cases earlier in the chapter, and in Fig. 6.23.

This is reflected in the CO_2 input schedule graph in Fig. 6.24.

Again, continuous harvest levels off the mass flow, even with this combined graph.

6.4.3 *Adding Dry Out for Mass Flow*

So now, let's add in a dry out period. Essentially in a dry out period, mass flow is 0, but the chemistry of the plant changes. In some cases, the mass even decreases, but that is harder to model. For now, assume the combined graph with the change that

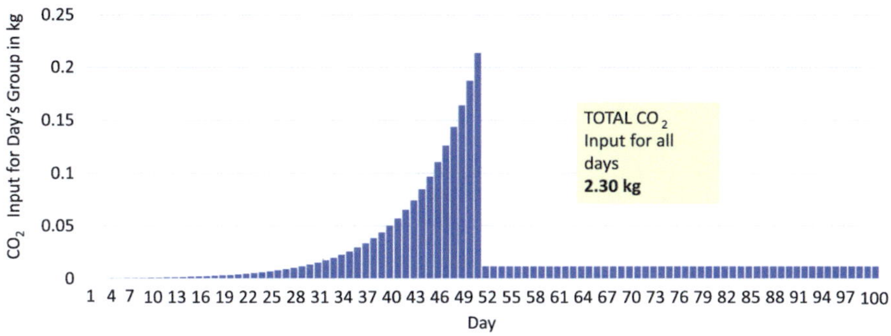

Fig. 6.24 Combined model CO_2 inputs by daily cohort continuous harvest

the last 7 days are a dry out time, so that mass hits 100% at day 93, is exponential to day 50, and linear from day 51 to day 93 as in Fig. 6.25.

Given this curve, the mass accumulation schedule looks like the graph in Fig. 6.26.

Yes, the total mass is if bigger than the linear case. For dry out, you would have to allocate a growth chamber or area that is drier than the other day classes of the crop, similar to the difference in weather from June in the Midwest which has some rain and warm temperatures to late August which is very dry (no rain) and very hot.

Looking at the mass flow, the big bang case looks like the combined model big bang chart in Fig. 6.26, with a zero-input time from day 93 to day 100 as in Fig. 6.27.

Yet again, continuous harvest levels out inputs by day like in the graph in Fig. 6.28.

Overall, mass flow is the same regardless of model if you use continuous harvest, but model dictates footprint and volume differences. You really need to know your

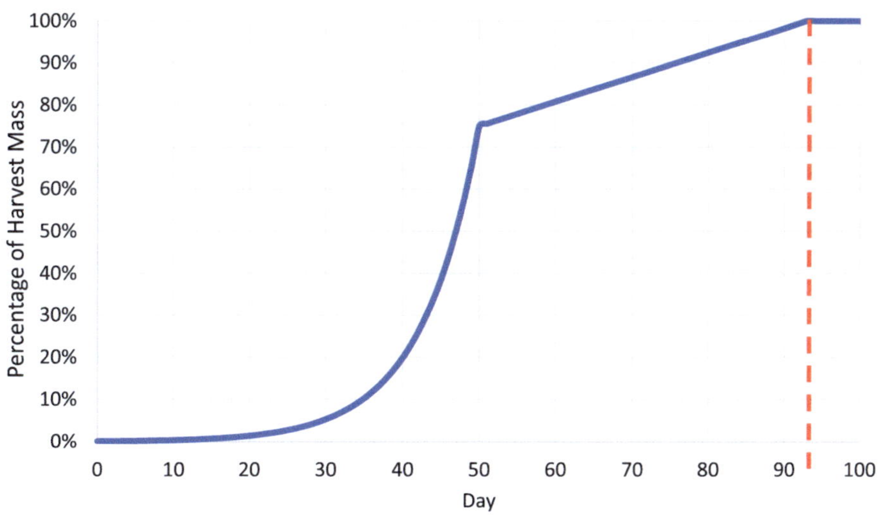

Fig. 6.25 Notional grain combined growth model with dry out

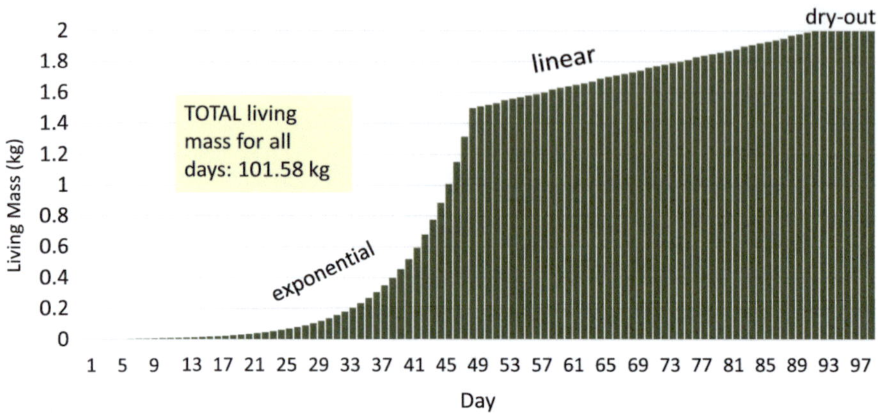

Fig. 6.26 Living mass by daily cohort for combined growth model with 7 day dry out

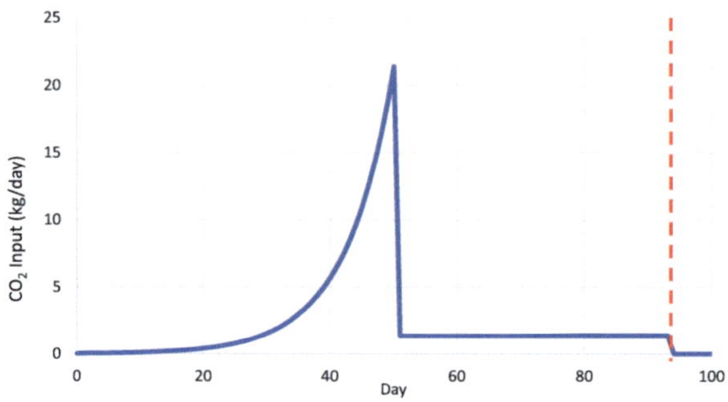

Fig. 6.27 CO_2 inputs for big bang harvest, combined growth model with dry out

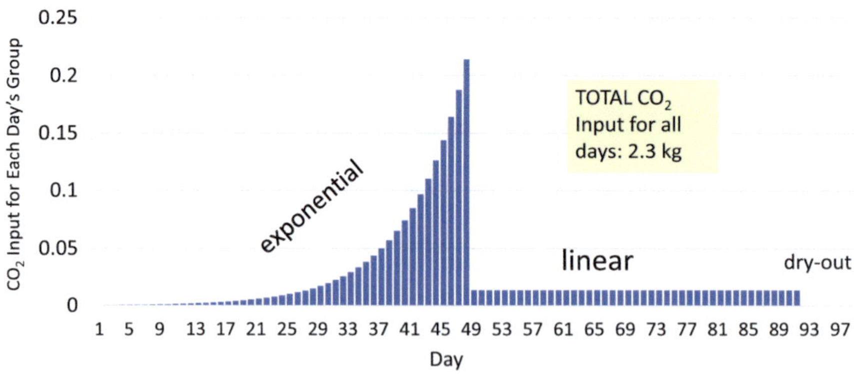

Fig. 6.28 Combined model with dry out CO_2 input daily cohorts

crop to get the footprint right, and knowing the crop lets the space farm engineer lay out the growth areas for each crop.

6.5 Volume and Rules of Thumb for Stage Sizing

A few rough fudge factors are good to know for taking a first guess at the size for each stage. These will essentially require getting the data for a living kilogram per day, then multiplying as in continuous harvest above.

- You will need to add volume for at least 1 to 2 days total inputs in tanks assuming worst-case timing. For that volume, you will need to add an additional 20% for the structures of the tanks themselves. For gases, assume roughly 180 kPa (like a car tire). For a big bang harvest, the tank size is 1 to 2 days at the day of harvest (which is worst case).
- You will need 1 m around the footprint all around the top of the total volume and footprint for access, and at least a meter between chambers.
- Hydroponics or aeroponic systems that use trays will need a meter between rows for access. Assume a human can only reach about a half a meter with arms. Robots however are more versatile. A rough factor is to add an extra 20%.
- Bucket deep hydroponic systems or plant diapers will cost an extra 10–15% volume for access and spacing.
- Most plants per living kilogram need two meters in height for greenery and from leaf top to lights, and a square meter footprint for growth. Root space depends on growing media. Aeroponics and soil are similar in root volume. Roughly a quarter meter volume will work for most plants for a living kilogram.
- Tank walls need to be a few centimeters thick in all directions for aquatic animals, and for safety allow two centimeters for all bioreactor or storage tanks for walls. Roughly add 10% volume added to tank volume.

6.6 Getting the Farm Started Before Steady State

The mass flow for the first crop, even in continuous harvest, shadows the growth curve of that first crop for each species. In the case of continuous harvest, the mass flow curve also includes future harvests too. As a result, a farm will need mass input for at least a full harvest time (or a year ideally) for every crop, and about a year's food for the humans. These inputs will need to be stored in tanks and slowly used for the first crops until the farm reaches a stable level of inputs and outputs, i.e. steady state.

6.7 Summary

This chapter covered two models for planting and harvest: big bang (i.e. traditional) and continuous. The big bang model requires reserving a large footprint for enough crop to last for the duration of growth (and supply), so that I have to reserve enough area and volume for a 120-day crop that provides enough food for at least 150 days, harvesting supply for 150 days at harvest, then replanting the whole area. The continuous harvest model essentially treats a seasonal crop as a perennial, so that for a 120-day crop, I harvest every day 1 days' worth, and plant new plants to eventually replace the day's harvest, occupying far less area, especially since most crops grow exponentially. Big bang harvest requires storage, while continuous harvest provides fresh harvests every day. Continuous harvest also levels out mass flow calculations and associated management systems, at the cost of effort to move organisms, and risk of damage to organisms during movements.

6.8 Questions

1. What is the difference between a big bang harvest and continuous harvest?
2. If I have a crop of orange trees, and only remove fruit and dead leaves each day, what type of harvest is that?
3. Given the same orange trees, assuming the only growth mass outputs are trimmed leaves and dead leaves, and ripe oranges, would the graph of mass flow for carbon dioxide and water input be: (a) flat, (b) linear growing, or (c) exponential?
4. Traditional 110-day wheat on Earth is harvested by which harvest model: big bang or continuous?
5. For traditional grain crops and big bang harvest, what shape does the mass flow graph over time have?
6. Using four cohorts for the wheat example, and assuming the original big bang footprint is 8250 m^2, how much area do I save using continuous harvest and mass models? What shape does the mass flow for the continuous harvest case have? (Hint: see Figures).
7. Given the assumptions for the first table in the chapter, and continuous harvest, what is your estimate for volume and footprint for the total wheat stage?

Reference

USDAARS. (2022). FoodData Central, U.S. Department of Agriculture (USDA), Agricultural Research Service (ARS), 2022. https://fdc.nal.usda.gov/

Chapter 7
Vascular Plant Stages and Associated Technologies

Abstract This chapter examines the machinery and methods to grow plant crops inside the space farm, where each species of plant is grown in a stage. We will look at strategies and tradeoffs for different types of plants, and different methods and technologies to grow them, from aeroponics, through hydroponics, to soil. This chapter also covers connecting the lighting, root areas, inputs, outputs, and condition management in plant growth chambers and assembling these into a stage.

Keywords Crop plants · Aeroponics · Hydroponics · Soil · Lights · AI · SCADA

7.1 Introduction

Armed with growth dynamics, machinery, lights, controls, and plant crop choices with their associated growth models, we can now discuss how to pick growth methods and build chambers and stages. Design of a plant stage is all about tradeoffs. If I pick hydroponics, how does this affect tray layout and lighting? If I pick soil, how do I manage the soil and extensive growth area? Given my choices on growth media and layout, what chambers will I need? How many storage tanks and pumps will I need, where? This chapter will first explore the growth methods for plants such as hydroponics, aeroponics, and soil. Then we explore how to layout lights and choices for lighting. Discussion then follow for carbon dioxide enrichment, conditions in growth chambers to control harvest, making soil from regolith, and whether one species or many species should be in the growth chambers. Machinery is also critical; how will we keep conditions optimal for plants? What about using robots? Can we grow edible plants in human used areas in the habitat? What about growing tree crops without the tree? Finally, the chapter covers plant species selection, and how little plants like duckweed can be grown in a photobioreactor-like machine.

© The Author(s), under exclusive license to Springer Nature Switzerland AG 2026

B. L. Meyer, *Designing Farms in Space*, https://doi.org/10.1007/978-3-032-04395-5_7

7.2 Vascular Plant Stage Basics

As noted in prior chapters, a stage is all the equipment, volume and area, and growth chambers needed to grow a crop to harvest in a repeatable way. Vascular plants are plants that have roots, stems, and leaves, and the roots need a different biochemical environment than the green part. Growth chambers might be self-contained and sealed but hold a growth stage (cohort) of the crop in an environment best for the crop at that period. Each growth chamber will have some way of containing the roots for a set of plants in a dark, oxygenated, and moist container. The chamber has a way to support stems and leaves and maintains temperature, air, and light for that set of plants. In root and green parts, there will be recirculation of liquids and gases. In the aerial part, there will be dehumidifiers, heat rejection, and circulation fans. The chamber also must have room for humans or robots to access the plants, and sensors to monitor the chamber. The chambers together will need tanks for gases and liquids, bins for solids and tools, control systems that keep each chamber and the plants in the chambers at optimal conditions and send alerts to the humans running the farm if anything goes wrong. This means the areas in the stage at some point need work lighting, room, and air circulation or other means to allow humans to work. It is likely vascular plat stages will have at least 2 chambers, one for nursery that seeds or grows young plants till they are old enough for transplant with a backup set of breeding plants, and a second chamber for grow out of the crop. To optimize space in continuous harvest, there will be three or more chambers for annual crops like grains. Your farm, your call. Adding all this together makes a stage as the book means it. At a top level, Figs. 7.1, 7.2, and 7.3 show what the plant growing chamber (and stage) flows might look like. Plants need light, water, and carbon dioxide to grow, but that is not the whole story. Condensers to capture transpiration, fans or blowers circulate air, cooling coils pull out excess heat from the lights, aerators and pumps circulate water to the roots, and valves and separators pull out excess oxygen for use in the rest of the farm or the human habitat.

Adding these functions together require machinery as in Fig. 7.2. Basic functions are as in Fig. 7.1, laid out here as simple schematics. Aeration would occur for water inputs using air in stage or from oxygen outputs, especially required for deep water hydroponics or carbon dioxide enriched air. There would also be more heat rejection for the areas outside of chambers due to pumps.

As you can see here, there is more than just light, water, and carbon dioxide. The green parts of a vascular plant (i.e. with roots, leaves, and circulation) take in carbon dioxide from the surrounding air, and, take in light energy from a star or artificial light, and pull up water with nutrients from the roots. The pull comes from transpiration, which is evaporation from the leaves. The roots need a little oxygen, however, to survive, and process nutrients.

Adding all these chambers and all the equipment together make a stage. In the diagram, Fig. 7.3 is just one simplified way to lay out a stage and growth chambers. The stage could be scattered, and growth chambers can be different than depicted. There are two growth chambers in the example in Fig. 7.3, and that figure is not to

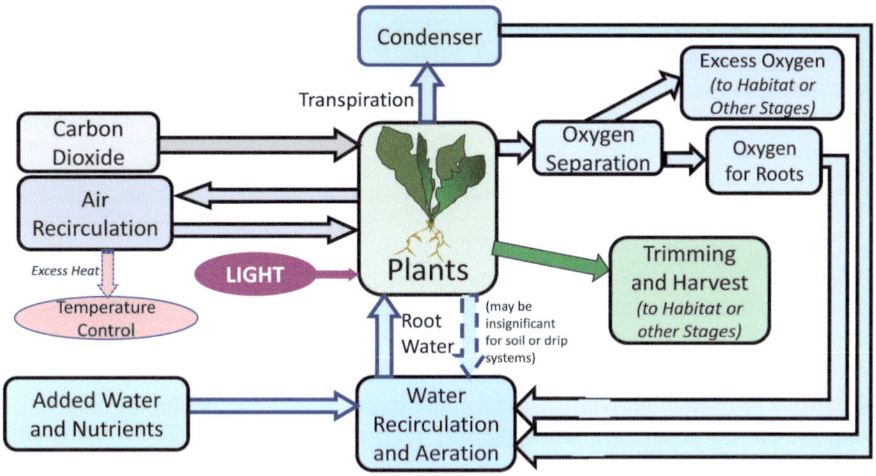

Fig. 7.1 Plant growth chamber and stage flows

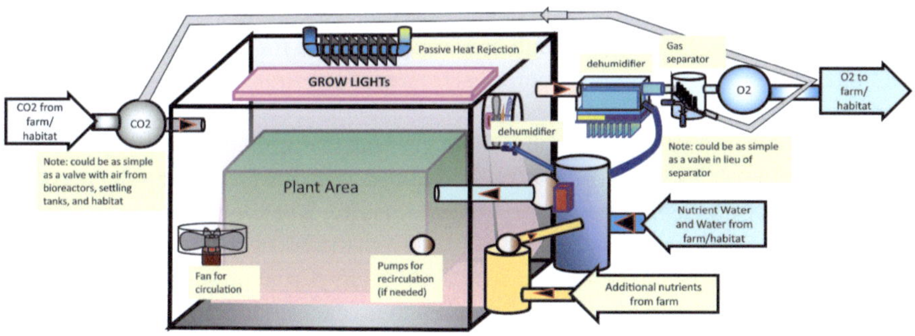

Fig. 7.2 Basic plant growth chamber rough schematic

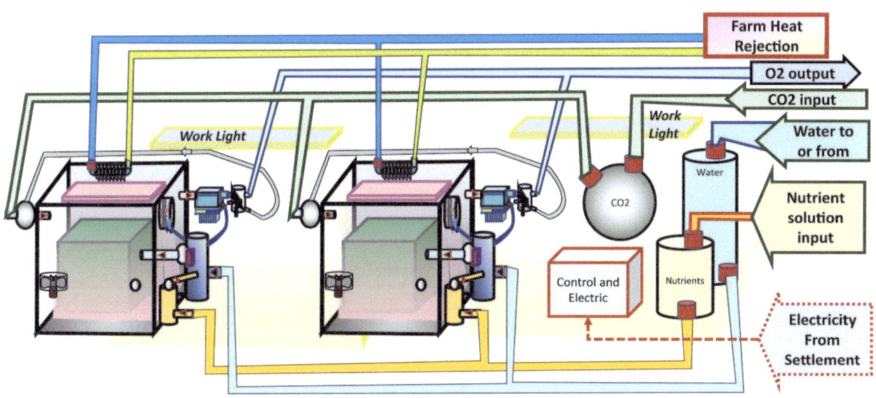

Fig. 7.3 Vascular plant stage rough schematic

scale. Every tank has at least a controlled pump, and plant lighting can be in racks and rows, or other arrangements as needed. Inputs to the stage come from the rest of the farm or habitat, buffered by more tanks and pumps. Outputs the reverse. Not shown in the diagram: Biomass (harvest) output which is a task that might be automated or manual.

7.3 Soilless Technologies

Decisions on how to grow plant crops affect the design of the aerial portion and root portions of the plant, and how each plant will occupy the overall volume for each plant (see Fig. 7.4) in each growth chamber in the stage. Soilless technologies mechanize plant growth by separating what happens with the roots from what happens with the shoot, leaves, and green parts in a highly controlled way. An engineer needs to make a choice early on whether to use soil or soilless technologies. Aeroponics uses a simple spray of nutrient water onto roots, but hydroponics puts the roots into a liquid. Substrates are materials that we can put around roots. Substrates can span from sand or fibers that behave like soil, through bigger pebbles and rocks or ceramic shapes. It is a trade to decide if your hydroponic system needs substrates or not, covered more later.

Pumps and aerators make sure the roots get water, nutrients, and oxygen. Light sources generate heat that needs to be pulled out. Extra water in the air needs to be pulled out by condensers to let the leaves transpire. Remember that transpiration is often over 100 times the water that is used by the plant to grow. Air needs to move around to make sure carbon dioxide at the right partial pressure ends up in the leaves. Regulating gas levels, heat, light and light cycles, chemistry of the roots, and moving fluids takes energy and equipment. All this movement, regulation, lighting, and cooling imples lots of machinery that outdoor farmers on Earth do not need.

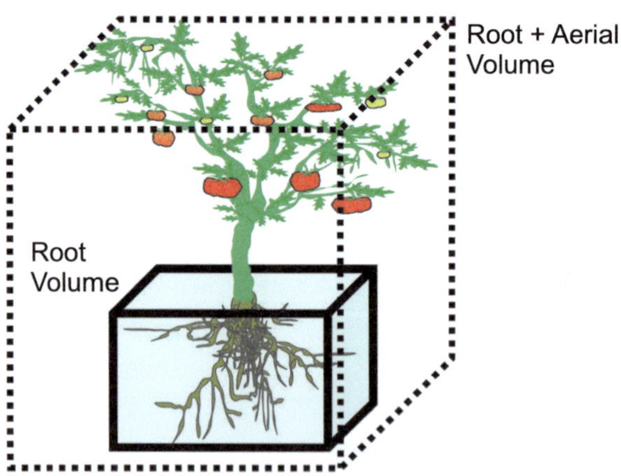

Fig. 7.4 Volume needs for a plant

When we go to space, we need to use technology to do what nature provides on Earth.

7.3.1 Substrate-Less Farming

It is amazing to me how many ways life adapts to the conditions we make. It is indeed possible to grow plants with nothing around the roots but moist air, or a slow flow of nutrient-infused water. If there is nothing around the roots but a container (bag or sleeve) and a mist spray of a mix of nutrients and water, then we have aeroponics, or 'air growth' as in Fig. 7.5. In microgravity, plant roots are in a dark box or tube, where fans pull in oxygenated air, then sprayers atomize the liquid mix (grey water would work here) ahead of the roots, the fans pull the hydrated air over the roots, and the roots send and process the moisture to the rest of the plant. The leaves and stems (the green parts of the plant) poke out of the tube or box through a hole big enough to accommodate the stem, but tight enough to maintain the pressure in the root box. Again, fans facilitate transpiration, moving the moistened air away and carbon dioxide-rich air in. In gravity, a set of scaffolds provide support to the leaves and stems of plants like tomatoes or beans. Drier air is circulated over the leaves, with light shining from top and sides. The drier air allows the plant to transpire, and excess moisture can be condensed and used, since the plant essentially purified it. A drawback of the aeroponic box approach is that there is little surface area for bacteria or fungi to help the plant, beyond the roots themselves, but the aeroponic approach works for microgravity and requires little mass to set up.

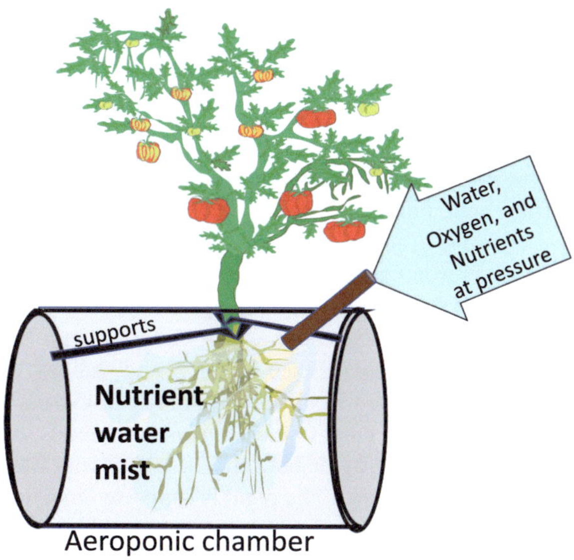

Fig. 7.5 Aeroponic schematic

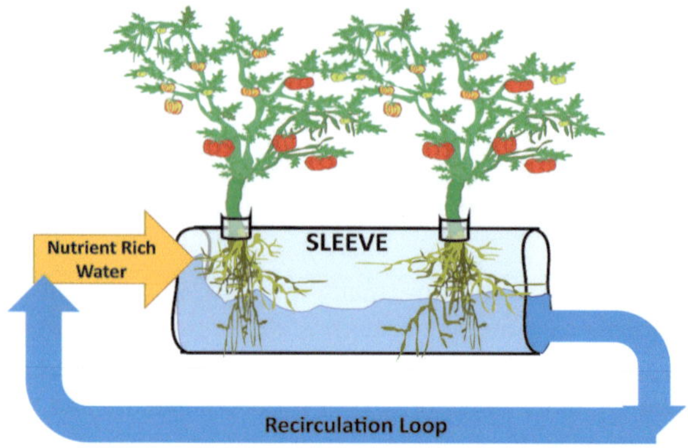

Fig. 7.6 Sleeve system simplified schematic

A similar approach, proposed and used at the University of Arizona for eventual use in space (see Kacira et al., 2012), involves using sleeve tubes that hold plant roots, and scaffolds to support the plant, similar to the simplified schematic in Fig. 7.6. Liquid is pumped around the roots in the sleeve as a skim, leaving an air gap to oxygenate the roots. Most of the sleeves are gaps, and capillary action pulls the liquid into the plant. The fluid is again a mix of nutrients and water, likely grey water, or wastewater. The sleeve is a soft plastic membrane with supports for plants. If the nutrient water injected is at the same rate as consumed by the plants, then the recirculation loop may not be needed, and in that case the sleeve end is sealed.

This sleeve approach may be less microgravity-friendly but would work in places with gravity like the moon, Mars, or free-floating bases and is again low in mass to set up. Since most long-term settlements will have gravity of some type, this method may be chosen due to the low mass to set up. The stage is shipped out compacted together, then extended to make the plant stage. Very little initial mass is required for this architecture. A downside or the sleeve is like the box approach above, because again there is little surface area for bacteria and fungi beyond the roots and the walls of the sleeve. Rigid tubes can be used in lieu of sleeves if the tubes also serve as structure for the planting area. In all tube and sleeve methods, ribs and bumps, maybe even 'hairs' can be added to the sleeves to aid in bacteria growth by increasing surface area, but they are in less contact with the roots than soil or other media, so less effective at making a rich root biome.

A step further along the soilless curve, like above, the roots are in tubes, trays, or sleeves, but the liquid is all encompassing of the roots; the liquid is however heavily oxygenated and pumped over or around the roots. There are two major approaches, deep water hydroponics (see Fig. 7.7) and shallow water hydroponics. Deep water hydroponics uses a well of water that fully immerses the roots in a net or holder that floats on the water tank. In the deep water methods, floating root balls sit well above the bottom of the tray or bucket. The plants root ball is enclosed in a net or mesh, and the aerial portion of the plant is supported in a float. In some versions in vertical farms on Earth, the floating plants are currently pushed along the long set of trays as they grow, then harvested at the end, in a continuous harvest.

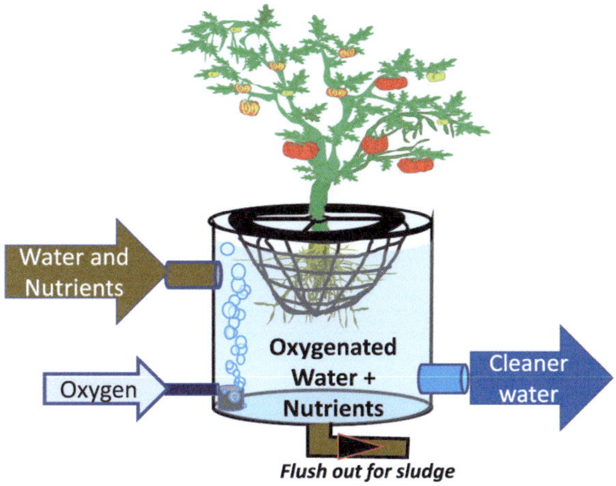

Fig. 7.7 Deep water hydroponics in a bucket

In shallow water hydroponics, the roots are partially, to fully, immersed in slow-flowing water, with or without media (see Fig. 7.8). Rice paddy agriculture works like shallow water hydroponics, in an unpowered setting, but given the right liquid chemistry and the right crop plant, a submerged root base can provide a good growing method, with the same benefits and drawbacks as above. The tube will need more mass since they hold a full volume of fluid, but the cross section might be smaller since the roots will be efficiently fed.

Before the water enters the root bed, an aerator (a column or snail spiral shape) will add oxygen (see Fig. 7.9). This oxygen infuser is needed to make sure to have at least 5 ppm dissolved oxygen dependent on temperature (via Patil et al., 2020) in the root liquid to keep from suffocating the roots. Bubble control in the bed itself will be key to maximizing the oxygen absorption of the roots. The pumped fluid will need a settling tank after the root bed to catch any solids, and this tank may provide bacteria to aid the chemistry of the root liquid before it is circulated back to the roots. Great care will be needed to ensure water molds or other unbeneficial infections do not occur in the root bed, since the liquid can carry diseases from plant to plant, so

Fig. 7.8 Simple shallow water hydroponic tray, no media

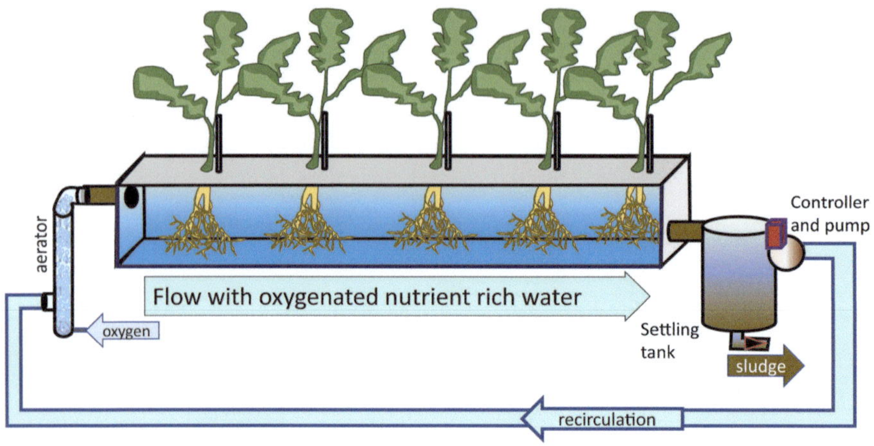

Fig. 7.9 Substrate-less hydroponic tray with added machinery

a sterilizer may be required at some points in the circulation loop between trays of plants.

7.3.2 *Substrate Concepts*

Substrate for plants is any material or medium that supports roots and holds fluid. There are engineering choices and tradeoffs for deciding whether to use or not use substrates, and then what granularity or type of substrate to use. Figure 7.10 shows some of these tradeoffs.

On Earth, the soil is a mix of particles like sand, silt, or clay, and organic material such as decaying leaves, feces, carcasses, or other wastes. The particles are the substrate of the soil, as they give the soil structure and surface area for bacteria and fungi to colonize. On Earth, worms and insects aid aeration and might be something

Fig. 7.10 Substrate spectrum with tradeoffs

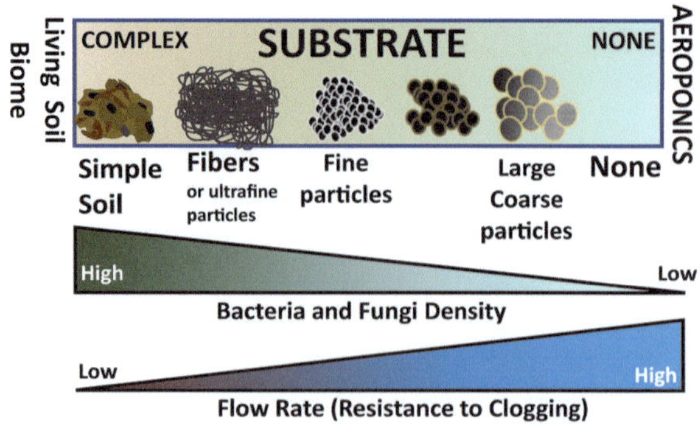

to add to soil in space. The soil biome overall interacts with plants as shown in Fig. 7.11. However, soil can have very slow flow for water and oxygen and can become unbalanced in smaller areas, leaving dead plants, fungal outbreaks, or other issues. Fine soils like clay and silt move water and oxygen slowly but have more area for minerals to leach in and surface area for bacteria. Gravel and sand move water better and can be well oxygenated, but have fewer chances for nutrients to be absorbed, and less surface area for bacteria.

For high-density growing, a balance between having a healthy soil biome, and volumes, may favor using substrate, such as hollow artificial spheres (see Fig. 7.12) and marbles, fibers, sponges, or in situ obtained gravels and pebbles, with machines to add nutrients and minerals. The substrate can be partially submerged and pumped or use capillary action with drip feeds. Great care is required to make sure gravel does not contain metals or chemicals that can harm plants or the people that eat the plants. So, this method may favor insoluble minerals such as silicates. The substrate is first conditioned by adding bacteria and fungi that aid in plant growth, then the plants are added so that the roots infiltrate the spaces in the substrate. A tray, like a roof gutter, could hold the substrate and roots and be capped with a lid that has holes and scaffolds to support the growing plants.

Liquid, again sterilized grey water, or filtered wastewater, is oxygenated and pushed through the tray, then circulated. The stems and leaves again are lit and in a drier environment with fans to circulate the air. Given some set-up time, a rich root bed biome emerges and helps achieve better plant growth, just like in gravelly soil, but more controlled.

In microgravity as on the ISS or even in lunar or Martian gravity, plant roots may be in 'diapers' or 'pillows' of fiber that is drip irrigated. Like soil, the substrate allows the microorganisms chosen to aid plant growth, but unlike soil, the substrate allows faster, controlled, flows. In Fig. 7.13, a fiber matrix of rock wool or similar is the media in the diaper, with a mesh or permeable membrane cover, and a drip hose near the roots to provide nutrient mixed into water as required by the plant. In some

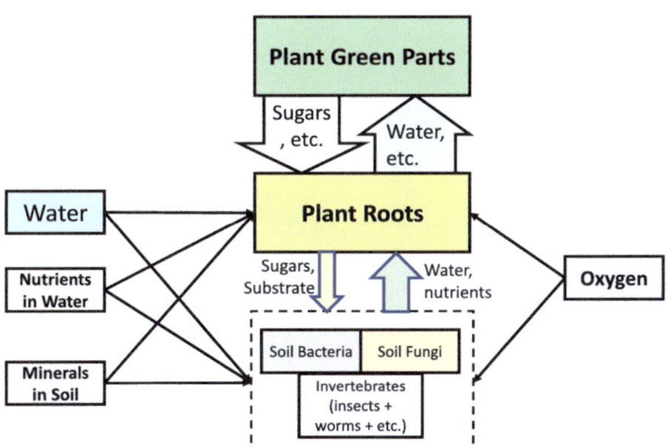

Fig. 7.11 Plant to soil biome interaction

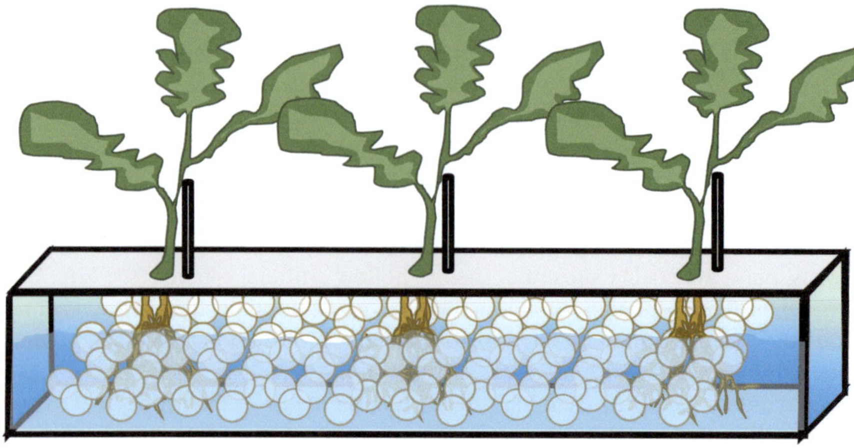

Fig. 7.12 Simple substrate for hydroponics

systems, the nutrients are in a packet that sits in the fiber and releases nutrients to the plant. Sensors determine the drip rate, and fans circulate air around the plant to allow transpiration.

If you were to use the diaper method in a stage, it may be arranged somewhat like Fig. 7.14. Diaper plants can be enclosed in a chamber with lights, gas controlled for the right partial pressures of carbon dioxide and oxygen, dehumidifier with fan, a tank with controller and pump to drip liquid into the plants as needed.

The obvious drawback to using a substrate is the need to make, bring, or find the substrate, which means mass. Research into the tradeoff of mass versus productivity

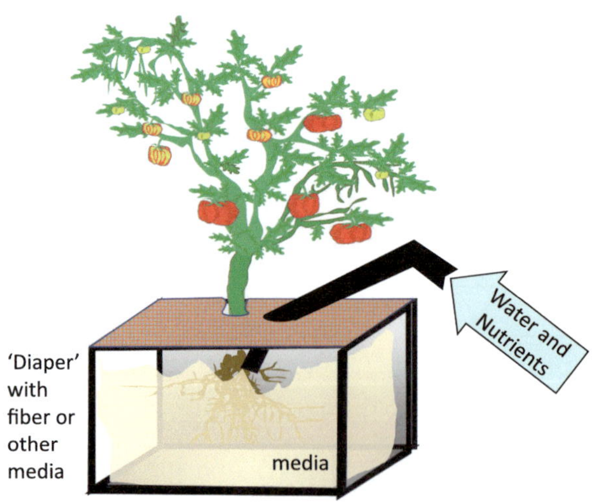

Fig. 7.13 Diaper method schematic

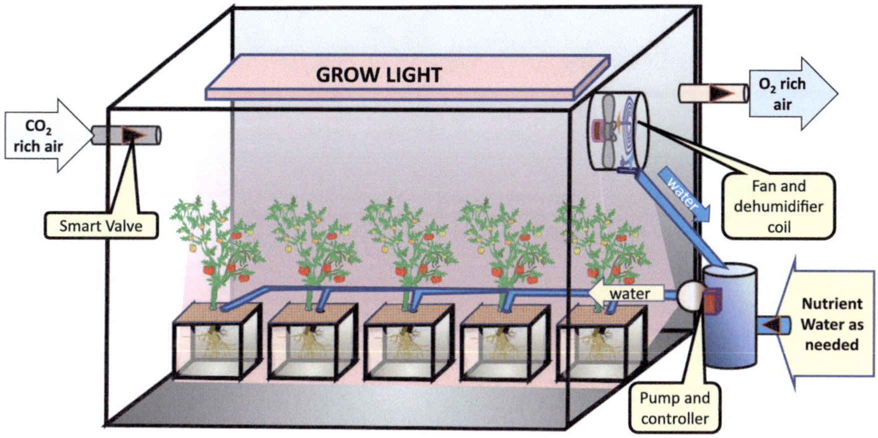

Fig. 7.14 Diaper concept in a growth chamber (of a stage)

is still ongoing, but each settlement and crop set will have different factors favoring different solutions.

Tray length and shape are a tradeoff. Tray length, from a simple holder for a single plant's roots, to long trays of dozens of plants, is a tradeoff, with mass of the farm in hoses and pumps and tray walls, traded against ability to manage flow of fluids around the roots of individual plants. A long tray might have too much oxygen or nutrients in the front (the place where pumps bring in the nutrient- and oxygen-rich water) and not enough at the back (where the remaining fluid is pulled for recycling). A short tray or bucket will have better flow per plant but will require more pumps and hoses and tray material (see Figs. 7.15 and 7.16).

Each species might have a different solution. Tomatoes or peppers might engineered to be long-lived perennials, with very tall vines, that continuously produce fruits, so managing the root bed of each individual plant may be very important. Contrast a lettuce that is small and grows to maturity in weeks to be yanked out and replaced with fresh seedlings, where the tradeoff favors a longer or wider tray with

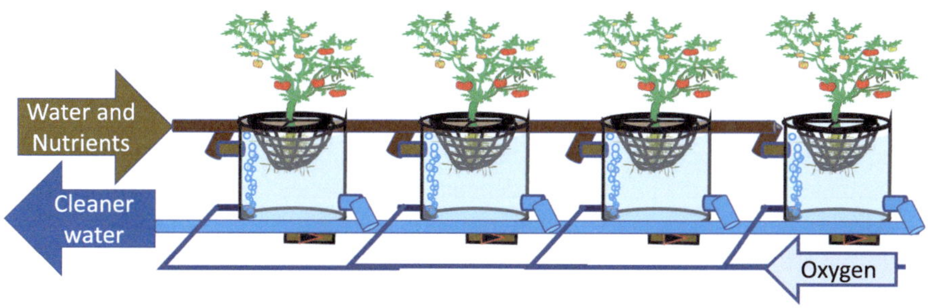

Fig. 7.15 Multiple containers schematic

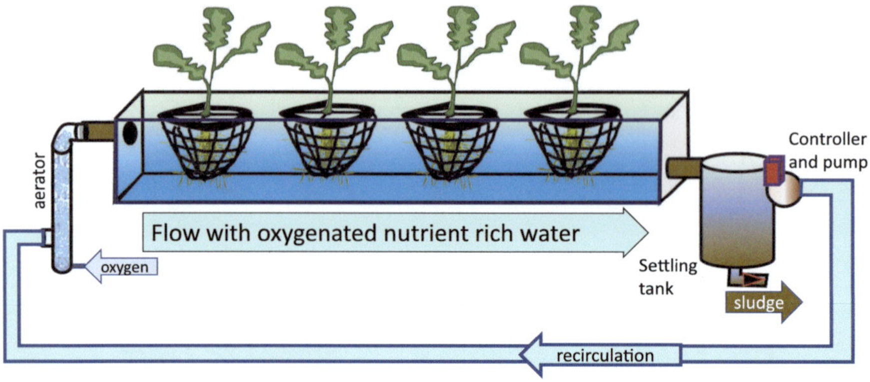

Fig. 7.16 Single tray system with baskets

a simpler set of pumps and hoses. Note that some plants devote more mass to roots in aeroponics, versus substrate hydroponics (see Li et al., 2018).

7.3.3 Putting Soilless Methods into a Growing Volume: Beds, Trays, Columns, Rows

After picking a growth method and crop plant, all the plants of at least one type are arranged to make the best use of available area and volume, while still allowing room for humans or robots to maintain and harvest the plants and produce. An introductory resource for simple on Earth soilless growing if you want to explore further is in (Johnson, 1998). The basic unit in most scenarios is a tray (sleeve) with or without substrates that hold the roots of the plant, a separator that has a hole or holes for stems, supports to hold the plant and branches as it grows, light elements for photosynthesis above and/or around the plant, and sensors to monitor the plant. This plant module will have pumps, filters, tanks, and fans that are connected to each module either in central form or isolated to each module. The tray is usually smaller than the aerial (i.e. green) part of the plant, depending on species. In the soil, the root ball can spread as deep as a plant is tall. However, space farm engineers should count on around 15% to 35% (my estimate) of the aerial volume as tray volume as a very rough rule of thumb for a very well-managed, oxygenated, root bed. For many crops, this means a meter or two between the top of the tray to the light source, and a square meter around the plant, again depending on species. Some vines might need more height, less area, or are wrapped into spring shapes. As a result, rows of plants can be set so that the root trays or troughs and lights form shelves, with walking areas, or rows of trays on shelves that rotate for access stacked to fill the available volume as in Fig. 7.17.

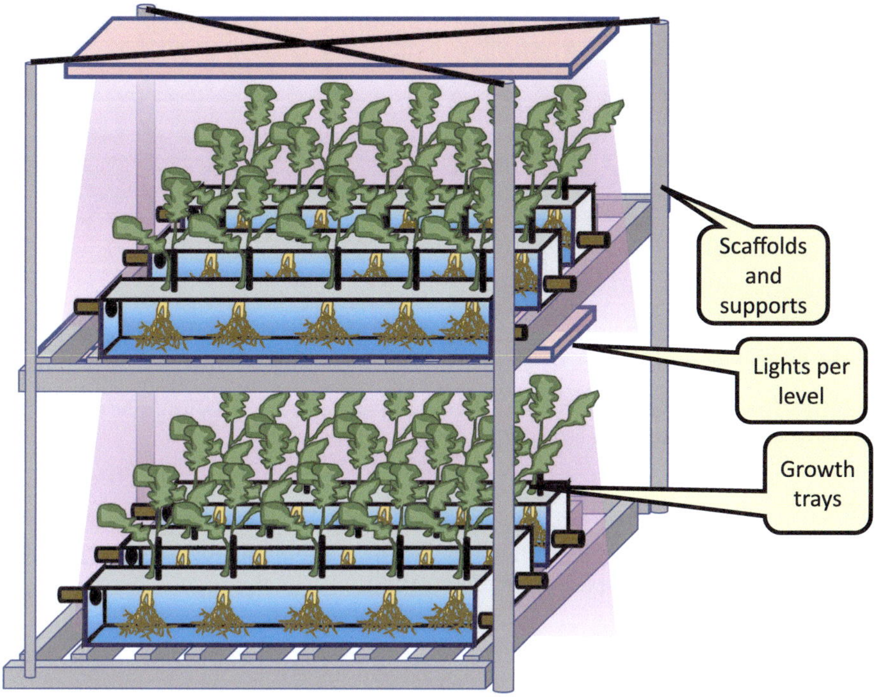

Fig. 7.17 Multiple shelf hydroponics

Airflow and temperature management are part of shelf arrangement. Productivity depends on keeping ideal flows of CO_2 to the leaves for photosynthesis, and keeping humidity and temperature managed to avoid burning out the leaves from the lighting. From personal experience, I know how easy it is to over-light a plant or dry a plant out by keeping humidity too low, or molding plants by keeping humidity too high. Remember water moves in a plant from root to leaf by the evaporation of water from the leaves. Ideally, sensors including cameras are located near each plant to make sure fans, pumps, and light sources are also synchronized to optimize growth for each plant.

Since the plant stages are but one part of the total farm, the stage concept applies to all the other things growing in the farm as well. With these basic concepts, we now can go a bit deeper in the sections below. Figure 7.18 shows two types of green walls. In the left, the plant roots are in a single tube per row, versus on the right each plant's roots are in their own container fed by a drip line. Lighting can simply be room lights or dedicated grow lights in front of (and above or parallel to) the wall. A second design puts embedded lights behind the plants or in tubes above each row.

Current commercial grow walls are mostly for air cleaning and aesthetics, though some are for simple vegetables too. Many are managed as a single unit with automatic liquid management, some with included lights. Green walls could be made as a series of parallel partitions in a dedicated plant stage, or part of the walls and parks of the

Fig. 7.18 Two types of green walls. Left design uses a single tray per row, while the right design uses taps that feed root vases for each plant

human areas and hallways. Grow walls could also line hallways of spacecraft or be added to animal stages for air purification. Picking your plant growing strategy is but one of many tradeoffs in space farm engineering.

7.4 Additional Plant Stage Design Trades

This section covers designs to develop various growth strategies for plants using soil from regolith, using lights, carbon dioxide enrichment, controlling harvest timing, and polyculture versus monoculture.

7.4.1 Soil Ecosystem and Using Regolith

On Earth, productive soil is a combination of non-living and living things. In the Great Plains of North America for example, the soil has particles of various sizes from Ice Age glacial moraine to clays and silts either flooded in or blown in, to broken bedrock from freezing-thaw cycles. Plant root action breaks down particles further. Added to these particles are decomposing parts of bodies or wastes of plants and animals, roots of plants, and a fauna of bacteria, fungi, insects, and worms. Altogether, living soil is far more productive than simple dead dirt, and as noted above, as a support for roots (see Fig. 7.19), as a buffer for the root biome chemistry, and aerobic and anaerobic ecosystem of its own.

The worms and insects break down formerly living material and aerate the soil. Bacteria and fungi break down material even more and incorporate nitrogen and aid in water movement, and minerals in the particles are consumed by the plant as they become soluble.

Fig. 7.19 Roots in living soil. Living soil has many particles that support roots. In this case, the soil is supporting my tomato plants. Photo by author

In Fig. 7.20, there are worms, particulates, bacteria, fungi, and decaying plant parts from previous growth. Looking even further in Fig. 7.21, you can see the aggregation of clays, sand, and material, dug into by worms and insects.

In space, it can be assumed the regolith from planetary bodies that are just particles of rock of one type or another. To use soil in a space farm, the soil will have to be made. Soil can of course be made without regolith by just using wastes and composting it, but the regolith adds room for the biome and roots. Making a living soil ecosystem or biome takes time and processing as in Fig. 7.22. This can be accelerated using a bioreactor/compost tank, worms (Williams, 2024), mushrooms, and plants like legumes that are tilled under to feed the soil. The soil at all phases of development will need oxygen and water to develop, so calculations make soil a stage unto itself.

Fig. 7.20 Soil root biome. This soil is from a grass field in Missouri, USA. Photo taken by author

Fig. 7.21 Zoomed-in soil root biome. From a grass field in Missouri, USA. Photo taken by author

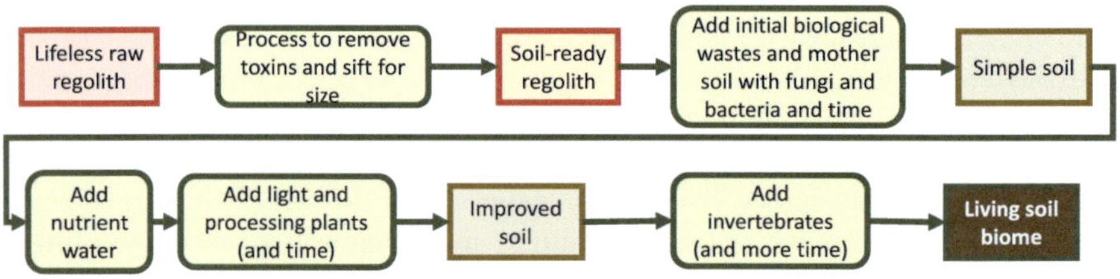

Fig. 7.22 Simplified process from regolith to soil

NASA ISS Expeditions 63 and 64 have indeed explored using and making soil (see Lehmann & Irons, 2020). A few technical resources for the details of soil-making can be found in Irons and Irons (2021), and Shevtsov (2021). For each mass of regolith, a mixture of organic wastes can be added, then 'seeded' with at least bacteria, such as Handy et al. (2021), Fackrell and Schroeder (2019), and many other projects and papers going back decades. For the astute, you will note that your gut fauna can be used, so yes, raw human waste makes a starter for making soil (as in a certain Hollywood Mars epic). However, the first set of bacteria may be better sourced from a mother batch taken and cultured from Earth, then customized, using sterilized wastes and regolith to start the recipe. The recipe works best if we also add fungi, such as mycorrhiza. Both bacteria and fungi can be cultured and grown in-route to a settlement, then stored to make starter, or the mother batch. Really good soil also has earthworms and other crawlies. These are harder to ship, but a small bed of worms can be kept in transport, then grown to reproduce in soil as it is made. When making soil, water and oxygen need to be used by the soil biome, and the soil will generate heat that must be moved away and dealt with. Living soil exhales carbon dioxide that will need to be mass balanced in the farm.

At this point, it is important that the regolith have the right chemical composition. Perchlorates and other contaminants require pre-processing the regolith before use in soil making. Perchlorates are hard to eliminate, though some methods may evolve using salts, water, and electrolysis. There are many papers on the topic, one is Archer et al. (2019).

Soil development takes time. Even given all the regolith you would ever want, it will take years to get a fully developed soil biome of hectare sizes, though as the soil develops, crops can still be grown, just at lower productivity rates, using careful irrigation (drip to the roots or other methods) and management of nutrients (fertilizers). Early soil is essentially substrate hydroponics. As soil is developed in a long-term colony, it will become a large-scale oxygen consumer mostly from the oxygen released by the crop plants. But the soil is also a producer of carbon dioxide and water for the same plants. The carbon fixed by the plants and oxygen produced by the plants in the soil will far outstrip the soil contribution to the farm's balance, but in soil using farms, calculations must include the soil as a stage (like the fungi or aerobic animal stages), or the mass will not balance in the long term. The soil at some point becomes the cellulose processor as well, breaking fiber molecules to release carbon dioxide and water, taking in oxygen.

Every space farm will start simple and small, and some will become more complex. Soil-based open air parks and farms feed the human mind as the space and time allow, but soil-less methods are quicker out of the starting gate and more volume efficient. Given a large-scale lunar, Martian, or large O'Neil style free space settlement, and time, soilless stages may be supplemented by soil-based farms, then replaced for some crops, by soil farming over years or decades. At some point, it may be that even small settlements with gravity get a small park with a few plants and maybe a lawn, with dirt for babies to eat and train their immune systems, and adults to feel a connection to life and the Earth they left behind. A minute of toes wiggled in grass beats a minute of psychotherapy any day.

7.4.2 Light from Above and Around: Mapping Light Energy to Mass Flow

In Chap. 5, we covered light types and light spectra, and here we explore using those lights. While in most vertical farms, and outdoor farms, light comes from roughly overhead, plants can use light from almost any direction. Care needs to be taken, however, to keep too much light from burning out the plants. Light onto plants heats the plants, and this heat needs to be balanced with humidity and circulating air. Plants typically take in light, on average, 12 h a day of varying intensity outdoors, but plants can grow with different light periods. The light timing and duration depend on species and growth phase. Some plants need lots of bluer light when maturing, then redder light for shorter hours when flowering. Crops on Earth like maize in far northern or far southern latitudes get long days with 16 h of daylight in the summer growing season, while the same maize in equatorial areas like Africa or India get even 12-h light year-round. Maize grows in both places, so plants can be accommodating. Sunlight is a mix of wavelengths, many of which are not always used by the plant but add heat to the leaves and stems. Too much heat will either break down and kill leaf cells directly or indirectly dehydrate the leaves faster than the stems and roots can feed in water. So, it is a careful balance to optimize growing light in spectrums the leaves can use, to maximize growth, without over-lighting. Put the lights too close, and the lights themselves add heat to the air around the plant. Lighting needs to be placed at a distance that optimizes the light coverage of the plant, at the right level, and allows room for growth. Since not every species has the same need for light, trays in a greenhouse might need to have different light amounts, or greenhouses (i.e. rooms with trays of plants). An extra twist is that some plants need an interval of light in the whole white light spectrum to flower and then make fruits and seeds. Again, this is where a good set of sensor arrays and software with artificial intelligence come in handy. In general, an array of lights overhead, and some lights or reflectors around the plants, should make sure the whole plant gets the right dose of light for its time of life and species.

The day timing of the dose of light is another matter. Plants key in on light direction, intensity, duration, and spectrum to trigger life events. Some plants use dark periods to build biomolecules or use duration in addition to spectrum to key flowing, leaf shedding, and dry outs. One of the key reasons to monoculture stages of the plants in parts of the farm is to optimize each stage for the species inside. What is great for rice might not be great for lettuce or tomatoes. Light dynamics combined with carbon dioxide levels in the air around the plants should be optimized for the species involved. Further down is a discussion of carbon dioxide, but here is a rough rule for now: leafy plants (C3 photosynthesis for the biologists reading this) might be able to use extra carbon dioxide to grow faster, but some grasses (like grains, read C4 photosynthesis) may not be able to use extra carbon dioxide. It is an active research area to determine the ideal combinations of light, carbon dioxide, and nutrients to optimize crop production for most crop species, so there will be some trial and error for each strain and species grown. Gravity variations might change all the data that was collected on Earth or the ISS. For settlements with access to natural light, periodicity might prove interesting as well. For an Earth orbit of 300 km, sunrise and sunset come 45 min apart. Determining if the natural light can be used given such a short 'day' is yet to be determined fully.

Light source will determine the efficiency of the conversion of electrical power to light used by the plants. Light is measured by a few methods for growers, such as PAR, lumens, and watts per area. The goal is to map energy consumption by lights to mass flow input and output and eventually calories from the stage, limiting heat, burning of leaves, and optimizing food growth.

Photosynthetically Active Radiation (PAR) is the light spectrum most used by plants during photosynthesis. No matter how much light hits the leaves of a plant, production of dinner depends on how the plant can convert all that light energy into photosynthetic chemical energy thence into food calories for us. Roughly, it depends on the combination of effective energy in the usable spectrum, over the right effective area. Some methods to find the area include guessing the leaf area of the plant and working from there. Honestly, as a farmer, I can be a bit more empirical early in design and use rules of thumb to estimate effective area based on the geometric shape of the light from above and sides.

For a very rough guess in a paper to map light energy, I initially used a multiplier of 3, so there is three times as much area covered by light as the area of the array itself, assuming a 3-m-high row of plants. Therefore, if I had a square of lights a meter wide and long, it would illuminate 3 square meters of plant where the plant can use it. Then I multiplied by the footprint of the stage, assuming the rating of roughly 170 watts per square meter of footprint area would be good enough. Rule of thumb then became 3*170*footprint of plant stage in square meters. This is a real 'wag' (or 'wild (pick a word, mine has 3 letters) guess') at best. This approach and rule of thumb are also covered in the machines chapter above.

Another trick is to use Earth farm data for a similar species. Most places have good records and data for crop productivity for a location over a known growing season for a unit area, like acres or hectares, as from each U.S. state's Department of Agriculture or U.N. Food and Agriculture Organization (FAO). Given the heavy use of solar panels, there is also good data on how much sun energy hits the ground in the same locale, usually in kilowatt*hours per day or month, for example from the U.S. Department of Energy. If you notice, we bridged two big government bureaucracies here, just like magic. This number above is in the first chapters of this book, 10 MW per hectare, 12 h a day. Since the plants in a field can only use light after they sprout (a.k.a. after emergence) find the light energy from emergence to harvest and combine with the mass produced at a known harvest date, and we have a worst-case energy to mass over time and area calculation. Usually, productivity is for the part you eat, measured in bushels. Bushels are a fuzzy thing, meaning different masses for different crops, but the mass of a bushel per crop can be looked up easy enough. Then the trick is to find out how much mass in silage or residue remains after that bushel is harvested from the field.

What's residue or silage? Those reading this book and skipped a few chapters and who have never driven a combine or gotten up at 4 A.M. to drop a salt block likely don't know. Farmers like to use plant leftovers as feed for cattle or other animals. The leftovers in a field after the human foods are grabbed can be cut down to near ground level, then piled together, dried out a bit, then left to partially ferment to feed to animals. Hay and straw are examples of the stalks of grasses cutoff then tied together in either round- or cube-shaped bales for movement and storage. The very unfun part of making straw or hay bales, then putting them in the top part of a barn or loft is memorable. Think of very hot summer days, confined spaces, heavy work, and a blizzard of little pieces of dried grass in every crevasse of your sweating body, under every bit of clothing. I itch the back of my neck compulsively every time I think about it! Again, agricultural data to the rescue. There is data for the mass of silage or residue per acre for a region. With that data, we got the edible mass, plus the above ground part of the inedible mass, if we assume moisture content in the leftovers. Roots pose a problem however, but for many crops it is a roughly solvable one. Data on mass distribution for some plants can be found in biology journals, and with that, we can now bridge a worst-case rough guess at energy per mass per area. Knowing edible mass produced, a run to the government nutrition database provides food calories per mass as in the sections way above on vascular plant mass flows. For wheat, this combination of calculations to rough estimate is the one in example at the beginning of this book. This is how to map light energy required for the mass flow in the chapters above for each living kilogram. Then multiply for stage size and harvest dynamics to get the mass flow and light energy for the whole stage. We want a simple multiplier that says this much light energy in this design takes this much carbon dioxide and water to make this much food.

For a better estimate of mass flow with light energy, it would be better to measure the change in dry mass of the plant (given plant size and species and strain) with a known light source, in a sealed growth area. This change could be used to find the real function of plant mass to CO_2 used to light energy provided to crop mass produced. Usually, this experiment means freeze-drying the plant. Then feed the data function into a model of the space farm to find the power needed to get the mass flow from the plant stage to get the calories the humans need. In short, a good coefficient for watts of light at a set distance and growing volume takes in x carbon dioxide per time unit for plant mass. These numbers may be out there, or more likely need to be found by experiment for the various conditions in a space settlement. Modeling for mass flow per species is covered extensively in the sections above but is an area that needs lots of new research to get good numbers beyond the rules of thumb in this book for the myriads of species and conditions we will encounter, then used to pick the best lighting.

7.4.3 Carbon Dioxide-Rich Air and Oxygenated Roots: Optimizing Mass Flow

While carbon dioxide-rich air is not so good for animals, it is good for plants, to a point. Combining enriched air around the green parts of the plant, while oxygenating the roots, and maximizing light, is likely the key to supercharging growth, and minimizing growth volume for plants. Some indoor operations have used up to 1000 ppm for leafy greens (as per conversations) or 0.1%, roughly double the atmospheric level in rural areas, or just a tad more than a very polluted city in 2024. Carbon dioxide levels become dangerous to humans above 0.5% if exposed over 8 h as per US OSHA (OSHA-183). Some work by various greenhouses (Mortensen, 1987; Poudel & Dunn, 2017; Schwarze, 1993) shows increased production by using carbon dioxide levels over 0.1% for some species. This means a smaller plant stage can produce more food and process more carbon dioxide to oxygen. Some species of plants such as broadleaf plants like tomatoes can use the extra carbon dioxide to boost output, and some grasses can too. Some plants can only use extra carbon dioxide with lights on but need less during dark periods. Some plants need more carbon dioxide during early growth, but extra oxygen just before harvest. It is a research area to find the optimal function of partial pressure of carbon dioxide in air with oxygen in roots, so super high enrichment might require human workers to wear breathers or limit work time. Note that ripening fruits (fruit in the technical sense, so including tomatoes and peppers and the like) need oxygen to ripen. Commercially, farms on Earth pick fruits while still immature, then ripen them on the way to, or at, the store. Space farms likely will not need such trickery since production for a long while will be near point of use, so daily harvested vegetables and fruits will be at their full plant ripened flavor.

7.4.4 Controlling Harvest by Conditions

Light is not the only cue to control harvest in plants. Ideally, the plant stages are sized using the growth models to allow continuous fresh harvest. Given the fact that the settlement is an artificial environment, we can manipulate the plants environment and biochemistry to make each set of plants grow and produce as needed. As mentioned, light is the most controllable of factors, and for many species, simply timing light levels will be enough to control harvest timing. In some, temperature or spectrum are important, as for a certain popular medicinel. In other plants, water flow will do the trick. All that said, the use of pheromones or hormones may be required to both time harvest and get desired yields. Hormones are commonly used on Earth to control fruit production and ripening. Methods include hormone regulation, epigenomic regulation, light, and temperature as in these references: Klee and Giovannoni (2011), Buckley (2013), Liu et al. (2015). Many plants also require a dark cycle to grow effectively, and light control will be used to maximize the light–dark periods for each species.

7.4.5 Polyculture or Monoculture

One of the trades when designing growth chambers in stages is to grow a single species, monoculture, or to combine species that can grow together, polyculture. In a very small settlement, it is likely there is only one growth chamber where pots or trays of plants of different species reside next to each other. There are other options, such as the Milpa system where squash, corn, and beans are grown together in the same pot or tray. There are also aquaponics systems where deep water or shallow water hydroponics is combined with fish or shrimp. The EPA even has a business guide for aquaponics, see (EPA 560-K-16-004). If deep water hydroponics is used, and the plants are in baskets, the same bucket that holds crop plants could hold shrimp, fish, or mollusks if the tank biochemistry favors plants and animals, even if the animals are not a crop per se. Animals in this tense might just be nutrient engines for the plants or cleaners for the roots (small shrimp and snails make good root and tank cleaners). Figure 7.23 shows two polyculture examples. The left is an aquaponics

Fig. 7.23 Two examples of polyculture

deep water hydroponic system, where the tomatoes share the root water with snails and fish. The right is a classic Milpa, though potted, with corn, beans, and squash.

Rice paddies, mentioned in previous chapters, are also a classic aquaponic polyculture, with shrimp, fish, frogs, snails, and rice. The trades when choosing to use polyculture should be examined like so:

- Can all the organisms share the same environment for roots? Can the plants, and or animals, grow well in the same salinity, pH, nitrate levels, temperatures, bacterial and fungal cultures? Will animals eat the plants? Will plants of different species try to crowd out each other's roots? Will their bacteria or fungi in the roots compete?
- Can the plants prosper in the same aerial volume? If they can grow well in the same humidity, carbon dioxide mix, and sunlight, with the same circulation, then they might be able to be grown together.
- Will the plants fight each other using pheromonal warfare or crowding? Yes, plants fight. Plants will crowd out other plants and strangle them, or emit chemicals on contact with each other, or via chemicals in the air, to suppress or kill another competing plant. Leaves in contact will brown out the enemy leaves. I have seen tomatoes and other veggies compete. In this case, plants of each species will need to be spaced far from each other.
- Are the plants cooperative in nature? Milpa is an example of mutual benefit between corn, beans, and squash. The beans use the stalk of the corn as support, while the beans fix nitrogen for the corn roots. The squash outcompetes other

ground plants, and benefits from the nitrogen fixing of the beans, and the squash plants hold moisture in the soil by shading the ground for corn and beans.

After you answer these questions then decide if your plant of choice needs a private chamber or not.

7.5 Using Machinery and Controls for Plants

Indoor agriculture on Earth, especially for high value crops such as cannabis, relies on complex networks of sensors, control systems, and software to optimize production and detect and fix issues. Even outdoor fields use satellite and Uncrewed Aerial Vehicle (UAV) surveillance to track chlorophyll and moisture levels. Given the vast importance of crop health to the survival and well-being of the human settlers, it is expected that sensors and control systems will be deeply sophisticated and meshed, likely with neural networks and some forms of intelligent arrays and 24 × 7 human monitoring. Even if they use current off-the-shelf models of sensors and control systems, as those for chemical factories, refineries, and indoor farms, the sensor network will have a very critical role in the very survival of colonists. Food security is a must when help is a long way off. Figure 7.24 is a schematic of how control systems enable growth. Sensors feed controllers in the stage, which includes the grow chamber and all equipment. In the Water to Roots box are pumps, valves, and tanks. The Nutrients to Water box also includes pumps, valves, and tanks. Each box is in itself a system of machines and tanks, all controlled as part of a farm's overall systems. Several of these systems will have AI in some degree.

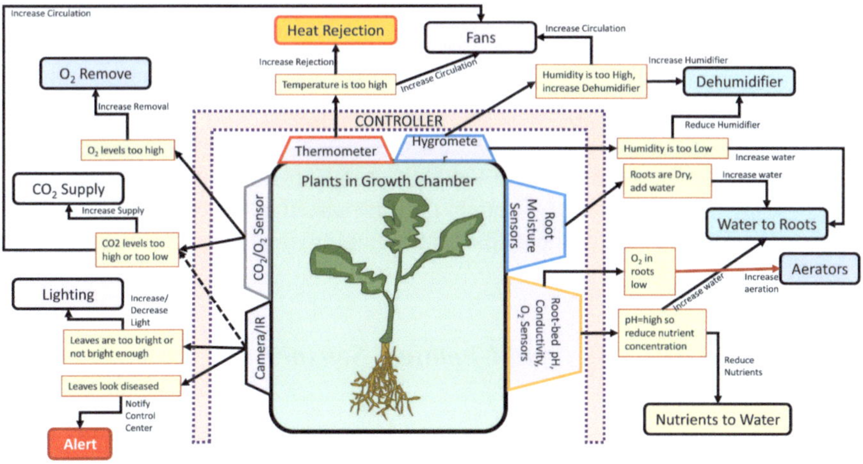

Fig. 7.24 Sensor to controller to equipment sample responses

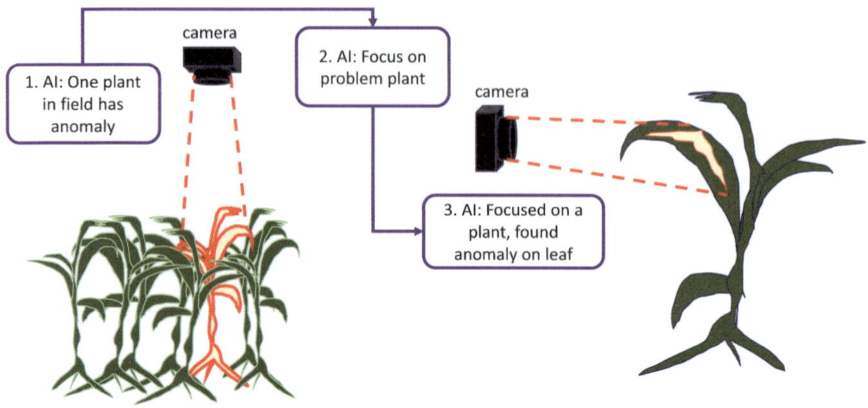

Fig. 7.25 Cameras and AI to diagnose plant issues

7.5.1 Use of Light Sensors and Cameras

There are a vast variety of sensors in the current market for plant growth. Visual sensors range from cameras in infrared, ultraviolet, and visible spectrums, to simple photosensors that feed light intensity at a given location.

Usually, the camera-type sensors are placed to observe leaves from a few angles (see Fig. 7.25), which can be used by humans as a visual check, then AI software filters the images and correlates them in various spectra to find chlorophyll activity and plant health against a knowledge array. This technique is already in use at macroscales for field crops from aircraft and space, and microscales for indoor crops. Issues found by camera examination using AI software are then used to send alerts, and filtered data is logged continuously for later use.

Simple sensors can be placed at numerous points around the plant growth chamber to measure light levels from light sources, and like all sensor data is correlated with plant health and productivity. A series of sensors picking up low light in normally well-lit spots may indicate either overgrowth or bad light. Light level can be correlated to chlorophyll activity. If there is enough light, but the plant is still not producing, or the leaves are not green enough, it might indicate a need to change conditions in the stage or check against other sensor data from humidity and carbon dioxide.

7.5.2 Use of Humidity and Chemical Sensors

Humidity and chemical sensors are also key. Hygrometers measure humidity levels in each part of every greenhouse. Humidity determines the effectiveness of photosynthesis, and the health of leaves and stems. Every species has different requirements, so in a polyculture humidity control may require a deeper network of sensors, versus a stage with a single species. Humidity and gas sensors will need to be placed after some rudimentary aerodynamic analysis of the stage's interior, matched with fan locations, to find dead spots and high airflow areas. Misters may be needed to balance humidity with dehumidifiers for some crops. Matching infrared or camera sensors with humidity sensors can be used to troubleshoot mold or algae growth and find dry pockets. If land snails or insects are added to the stage (to control algae and add a secondary crop), they can be used as an indicator of humidity as well, especially if Radio Frequency Identification Devices (RFIDs) are attached to the creepy crawlies.

Levels of carbon dioxide, oxygen, and various other biochemicals must be measured as well. Carbon dioxide and oxygen sensors for both the aerial and wet portions of the hydroponic garden control and improve growth and health of the crop, paired with the rest of the sensor arrays. These sensors are currently available and used for a variety of purposes and come with various levels of intelligence. Also currently available are sensors for a variety of biochemicals, some of which indicate fruit ripeness, plant health, and plant cycles.

Added to the chemical sensors above are sensors for acidity (pH), salinity, turbidity, and hardness. All are currently available and in use on Earth. The root fluids and media need to be maintained at levels to maximize growth and maintain the root biome for bacteria and fungi. These sensors should feed automatic chemical dispensers to balance the root fluids, and sensors will need to be in many places in longer tubes or trays, though at a minimum at input and output ends and in tanks. Turbidity sensors are essentially infrared emitter receiver pairs (though can use other means) and measure clarity and particle content of a fluid. Deviations can indicate blockages, contamination, diseases, or other problems in the hydroponic system.

Pressure sensors are also critical for aerial and root/fluid parts of the chamber. Plants need certain atmospheric pressures to grow well, and combined with gas sensors and chemical sensors, can describe the partial pressures in the air part of the chamber. A big drop in pressure might also be a leak to vacuum, thus ending your garden if not fixed quickly! Some actions like vacuum decompression may require a series of swift automated responses, such as sealing the chamber to prevent further loss. In the liquid portion, pressure sensors can prevent overflows, underflows, troubleshoot, and detect clogs or leaks. Liquid pressure frequencies can also be a sign of a failing valve, filter, or pump.

7.5.3 Fluid Manipulation for Plant Stages

As already mentioned, liquid pumps are needed for all non-soil scenarios. Pumps are matched to tanks to buffer flow (i.e. account for too much or too little). If the root beds (in gravity canted to have gravity assist flow) are fully immersed with a good seal, liquid can hit the pump at a regular flow, but the flow is more likely to surge and wane, so a tank will first catch the flow from the root beds. The pumps then pull fluid from above a settling level, so any substrate or root pieces don't end up clogging the pumps. The post-root fluid is then filtered and sterilized, then enters a feeder tank, likely above the root beds, though in lower gravity any level requires a pump. New sterilized water from other stages is added as needed from a tank in front of the stage. Water condensed by dehumidifiers is added to the feeder tank as well. A dehumidifier uses fans and heat exchangers (with pumps) to pull excess humidity from the air around the plants. Sensors check the fluid for chemical composition and oxygenation, and issues alarms if the fluid is not chemically sound for the plants in the stage, possibly adding salts or minerals. An oxygenator adds oxygen, using an air pump to push oxygen gas into the liquid using a bubbler or other infuser. The oxygenated fluid is then pumped into the root bed, closing the cycle. In the air above the root bed, fans (essentially ‘open air pumps’) circulate the air to cool the plants and move carbon dioxide and oxygen. An array of gas pumps will extract excess oxygen from the stage in optimum farms, especially if the air around the plants is enriched in carbon dioxide to accelerate growth. Other gas pumps will add carbon dioxide to the air as sensors and controls dictate to maximize growth. Carbon dioxide can come from the yeast-bacteria stage, a burner, or concentrated from the habitat or animal stages. In early settlements, a simple ventilation system from the habitat or aerobic stages can send air to the plants and return air after the plant stage to the habitat or animal stages. Likewise, an early settlement can use a simplified flow from the top of a septic tank through the root bed. Cleaner water after passing through the root bed can be pumped back to a charcoal filter and sterilizer tanks for the habitat. More about how to join stages together is in later sections.

7.5.4 Manual Harvest and Robotic Tending

As mentioned previously, room is required inside the hydroponic plant stage for tending the crops. At a minimum, all crops will need: planting, transplanting, pruning, harvest, and other tasks. Plant mass from pruning is key to maintaining mass flow in the farm and is required to keep each plant within growth parameters and remove dead material. Fruiting crops like tomatoes or coffee will need to be trimmed to stay inside a growth volume. Harvest of fruiting crops can be continuous. Dead plants and roots might need to be removed. Flow maintained by unclogging sleeves, trays, and tubes. Pump filters changed and replaced. Lights fixed. All tasks are familiar to an indoor gardener or farmer. Sensor arrays and software will indicate tasks for each plant or group of plants. In some crops, like tubers, entire sleeves or trays may need to be cleaned out and plant debris (silage) removed.

A meter between trays would allow for human workers to walk and work (see OSHA-1926.451), though the stage could use a rotating means to bring trays to light and to workers, or robotic devices could perform some tasks. Robot multiarms, as in Fig. 7.26, like those already used in factories, could be mounted on tracks to perform simpler tasks, and if tied to an AI SCADA system, could pluck contaminated plants or trim plant parts. An application may be to build out, then operate, the hydroponic stage by remote operation and AI before humans get to the settlement. When humans

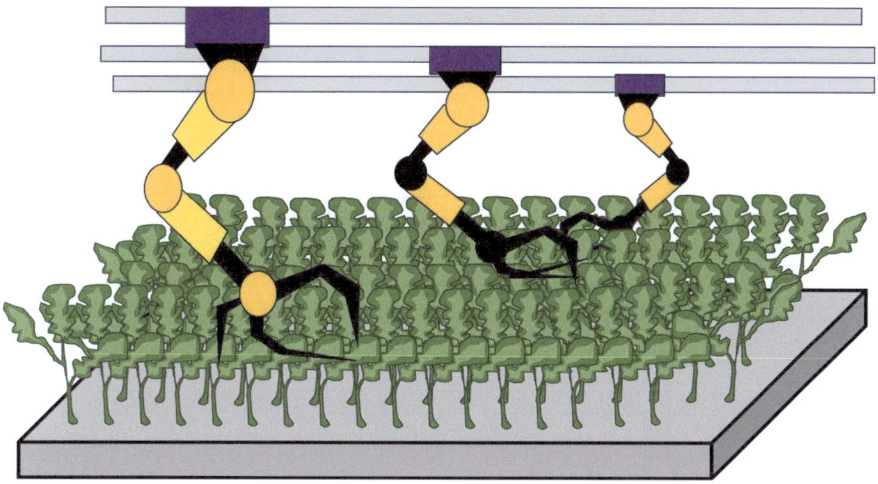

Fig. 7.26 Robotic arms on tracks

arrive, the hydroponic stages could then grow larger, and food stock from robotic harvests could make the initial supply for the new settlers.

On Earth, there are many types of automated harvesters, as described in Valle and Kienzie (2020), that supplement human labor, and smaller versions of these can be moved, or built on site, for space farms.

7.6 In Habitat Growth: Spices and Vines

As mentioned in other sections, grey water is a resource that can be directly cycled inside the habitat to grow plants in corners, around columns and supports, along hallways, and in parks. People need to see greenery often, especially in the sterile to near-sterile landscapes and open volumes of space itself. Houseplants are great at filtering the air of toxins, converting carbon dioxide to oxygen and buffering the gas balance inside living areas. Ions released by growing plants help mood as well. So why not have some plants where we will live? Corners and room edges are a great place to start. If each occupied area has a few taps of grey water (drip lines in particular), a cup of gravel or a fiber-sponge can be a root medium for herbs and spice plants, many of which grow in small spaces quickly, and smell good. Figure 7.27 shows two examples of ways to put plants in habitats to optimize footprint. Simple cups with root media or solution can accept grey water and be put in corners, or along columns in the habitat to process water, and to improve the air inside the habitat. If these plants are herbs or spices, the air will smell better, and the food will taste better too. One of the most common complaints of astronauts on the ISS is the smell, and this is a quick fix.

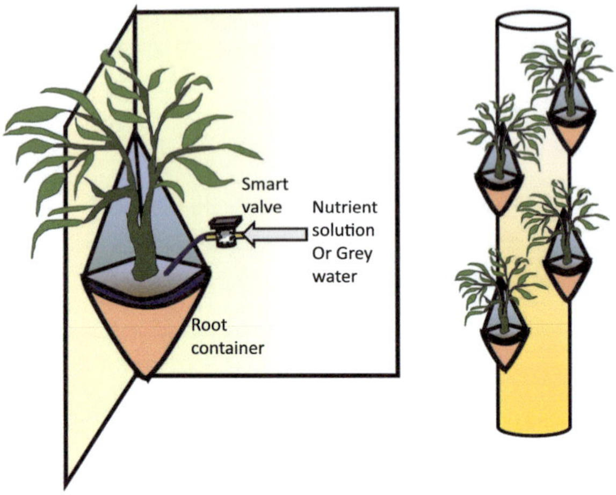

Fig. 7.27 Corner and column plant schematics

Residents can tend these plants for mentally therapeutic activity, and swap spices to improve their diet. Others may grow medicinal items. Such indoor gardening could become a common topic of lunch conversation. Note there is a common medicinal grown commercially on Earth in hydroponics closed cycle environments and is a well-trod path we can copy to be sure. Around ceiling edges, along posts, and along hallways (loosely discussed as in prior chapters), trays plumed with the same grey water taps, matched with lighting, and supports, can grow a wide variety of vines. Larger trays (see Fig. 7.28) may hold gourd or melon crops, though more likely other vines will be used first, again cycling air, and providing additional crops. Space grapes? Why not?

Given a tube in lieu of a tray, you can do a further grow type discussed in the next section.

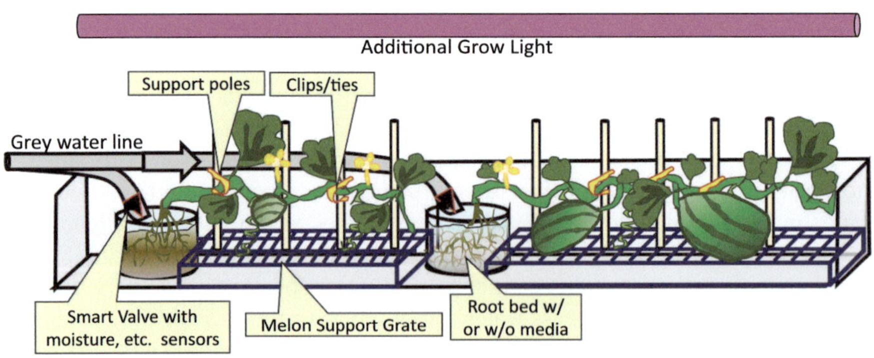

Fig. 7.28 Hallway hydroponic schematic with vine melon crop

7.7 Chopped Plants: Tree Products from Branches

What if I wanted yummy tree crops, like oranges, apples, or cocoa fruits, but lack the time and room to grow the whole tree? We could just grow the branch with fruit, theoretically as in Fig. 7.29. Each branch of a fruit tree is fed water and minerals from the trunk, and in a small branch, sugars made by photosynthesis transit from other branches. If I matched leaf count to each fruit, I just need the water and mineral part, with some hormones to regulate growth. Grafting is another option. In grafting on Earth, farmers match the branch from some fruit they want to a trunk and root of another plant to make it grow. In Missouri, USA, wine grape branches are grafted to wild grape roots and trunks to allow disease-free growth. In many places, trees with many kinds of fruits are grown from the same root and trunk using grafting as in Fig. 7.30.

So, in space, we can use a similar set of methods. For the tube in lieu of root option, the branch is attached to a set of tubes that replace the feeds from the trunk, and a support to keep the branch stable. Add light and carbon dioxide, and some hormones to trick the branch into fruiting, use a paintbrush to pollinate the flowers, and like magic, the branch fruits. While dwarf fruit trees can be grown in the hydroponic stage or in habitat, full-sized fruits can be grown this way, likely at scale in a hydroponic-like stage. Nothing is free in space, and the biochemistry of the branch feeding fluids is very tricky, but likely not impossible. The right ratio of leaves to fruit is key. It is possible that taps from other plants can be paired with the fruit branches with some intermediary controls, making an extended 'Frankenstein' graft. More likely,

Fig. 7.29 Chopped branch schematic. Chopped tree branches can be connected to tubes that provide the same inputs as the xylem of a whole tree

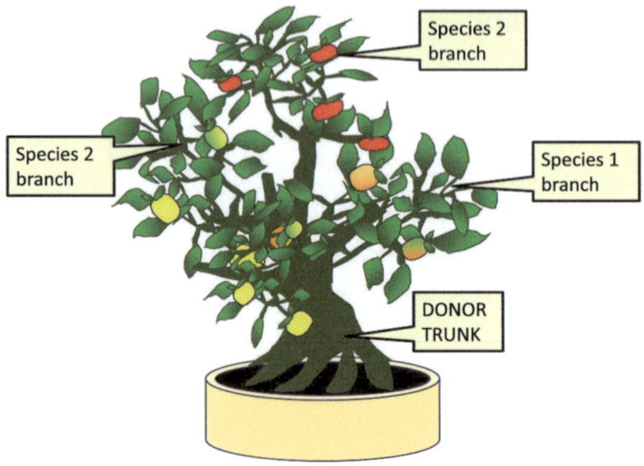

Fig. 7.30 Poly-species graft. Here a donor trunk is grafted to three species of fruits

space farmers will use conventional grafts using roots grown for the purpose, maybe imported from Earth, like Fig. 7.30.

7.8 Plant Species Selection: Time, Space, Nutrition

As with other crops, vascular plants should be selected for nutrition, ability to grow in tight confines under artificial or filtered light, ratio of edible to inedible mass, speed of growth, and ease of maintenance. For many species, there are many varieties and hybrids, and tailoring the genetics of these plants via selective breeding or direct alteration may lead to improved production or other tuned traits based on where grown. Dwarf species might be a better solution than the full-sized versions in open fields, since some may produce more food versus inedible parts. It is likely we will manipulate species for each location in space, directly in genes or by selective breeding. Some types of tomatoes, for example, might do better in lunar regolith soil versus Martian soil, or in free space settlement hydroponics versus aeroponics. Some types might do better when hand pollinated. Light cycles might also dictate plant variety selection, tailoring some plants to a 45-min day and night cycle as in Earth orbit, or even a 14-day cycle as on the Moon. Many plants might be tailored to produce different nutrients as well, e.g. more protein, less fiber, more vitamins, etc.

In general, a diversity of crop plants will be needed to meet nutrient minimums. Plants in general are great for making carbohydrates including fiber, less so for amino acids/proteins. While tomatoes are great to eat and hardy growers, like most vegetables, they are too low in calories to make a complete diet. A person would need at least 6kg of tomatoes to be close to the calories needed. Way too many wolf-peaches for me. Therefore, in addition to foods from other stages, calorie dense crops like potatoes (or similar) or grains could add the calories, while protein fixers like peanuts or legumes can add amino acids. This mix means either polyculture, greenhouse with multiple trays, or whole separate growth chambers optimized (via

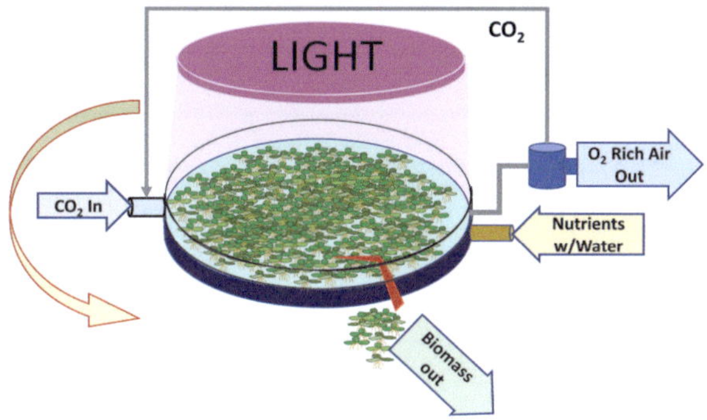

Fig. 7.31 Duckweed thin pond schematic

lighting, humidity, gas ratios, and root biomes) to maximize production for each species.

7.9 Duckweed Detour

Duckweed is a possibility for smaller space settlements since it can be grown with little space and might fit into tradeoffs for vegetable crops. Given the biology of this teeny highly reproductive plant, some interesting architectures emerge, that sit between hydroponics in this chapter and photobioreactors covered in Chap. 8. A grower could use a shallow disk set up, with stacked disks of growers to maximize space. My concept for trays is in Figs. 7.31 and 7.32. A very shallow dish of about 1 cm deep, with an air volume around 1 cm high circulates nutrient-rich water around the pond slowly, allowing the duckweed to reproduce. A simple skimmer after a period scrapes out biomass for use. The light could be a simple disk directly above the duckweed, closer to the surface, though with careful control to avoid burnout.

These trays could then be combined as in Fig. 7.32. Layers of grow areas where lights are on top of an aerial volume and tank with duckweed. Simple skimmer arms periodically pull in biomass for harvest. The duckweed floats on a slowly circulating solution tank with a thin volume of carbon dioxide-rich air below a light disk.

7.10 Summary

This chapter discussed technologies for growing vascular plants, such as traditional crops such as tomatoes or grains, assembled into growth chambers and then to stages. Soilless technologies such as hydroponics with or without substrate where roots bathe in an oxygenated solution of water and nutrients are most likely the first methods for early space farms. Aeroponics, where the hydroponic solution is replaced with a mist, can also be used for some crops to minimize mass. Some growth chambers

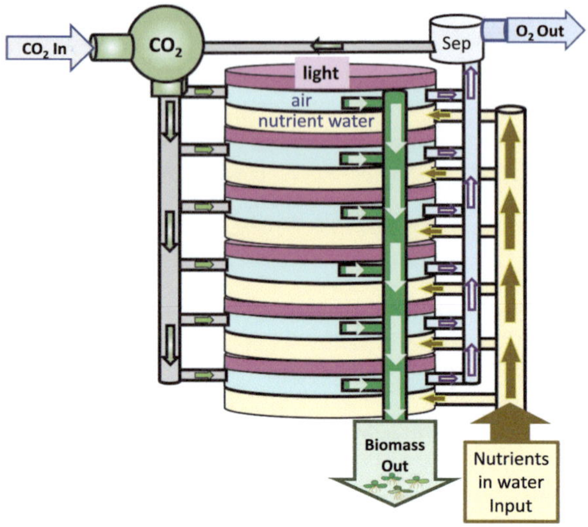

Fig. 7.32 Stacked disk duckweed photobioreactor schematic

may even use soil. Since each growth chamber in a stage should be optimized for the life stage of the crop inside it, all these plant root bed types might be used in the same stage. Lighting may also vary between growth chambers, some directly overhead, with a mix of white and pinkish lights, to bluish lights, to straight white light, overhead or all around, depending on life stage of the plants in each chamber. Each chamber will also have sensor arrays, and likely robots, to perform regular tasks, to check the health of the crop plants, and to respond to issues using AI with rules before the humans have to intervene. Plants will also grow outside of stages, in human habitats, using corner space, columns, walls, and hallways. How plants are selected, and which plants are grown in what stages, depends on the human menu and goals for the farm overall.

7.11 Questions

1. In addition to places to grow rows or pots of plants, what other machines are required in a growth chamber?
2. Given 120-day barley, with 4 growth cohorts for each 30-day interval in continuous harvest, what is the minimum number of growth chambers for the barley stage assuming each cohort has unique needs for lighting and humidity?
3. Given growth chambers for plants, what other devices may be needed to assemble the whole stage?
4. What are the differences between aeroponics, deep water substrate-less hydroponics, shallow water substrate hydroponics, and soil?
5. What are the tradeoffs between these hydroponic types: multiple plants per tray versus single plant pots?

6. What added machinery is required for hydroponic flows?
7. What is the tradeoff between fine particle substrate, and large coarse particle substrate?
8. What is in living soil other than material particles?
9. What is a green wall, and how might it be used inside the human habitat?
10. What process could a space farmer use to build soil from lifeless regolith?
11. Other than overhead lights, what other plant lighting options could I use to optimize growth? What are dangers and tradeoffs for using too much light? What about timing of lights?
12. What could I do with carbon dioxide concentration to increase production?
13. What are tradeoff between monoculture and polyculture?
14. What functions should control software and systems perform for a plant stage?
15. What functions could robots perform in a plant stage, and what types of robots could be used?
16. How could we grow melons and cucumbers inside the human habitat? What corner plants could improve the air in the human habitat?
17. What are two methods for growing oranges or limes given little space, and the need to produce fruit in a year after farm start?
18. Describe how duckweed could be grown in small areas to supplement the human diet?

References

Archer, P. D., Jr., Ming, D. W., Sutter, B., Hogancamp, J. V., Morris, R. V., Clark, B. C., Mahaffy, P. H., Navarro-González, R., McKay, C. P., Gough, R. V., & Fairen, A. G. (2019). Perchlorate on Mars—Overview and implications. In *Ninth International Conference on Mars 2019* (LPI Contrib. No. 2089) found at: https://www.hou.usra.edu/meetings/ninthmars2019/pdf/6233.pdf

Buckley, M. (2013). *Changes in epigenome control tomato ripening*, January 29, 2013, Cornell Chronicle, Ithaca, NY, found at: https://news.cornell.edu/stories/2013/01/changes-epigenome-control-tomato-ripening

Buckley, M. (2013). *Changes in epigenome control tomato ripening*, January 29, 2013, Cornell Chronicle, Ithaca, NY, found at: https://news.cornell.edu/stories/2013/01/changes-epigenome-control-tomato-ripening

EPA 560-K-16-004. (2016). *U.S. Environmental Protection Agency (EPA), Aquaponics Business Plan User Guide*, EPA 560-K-16-004, August 2016.

Fackrell, L. E., & Schroeder, P. A. (2019). Growing plants on Mars-potential and limitations of Martian regolith for in-situ resource utilization. In *Poster, Ninth International Conference on Mars*, California Institute of Technology (Caltech), Pasadena, California, July 22–25, 2019.

Handy, D., Hummerick, M., Dixit, A., Ruby, A., Massa, G., & Palmer, A. (2021). Identification of plant growth promoting bacteria within space crop production systems. *Frontiers in Astronomy and Space Sciences., 8*. https://doi.org/10.3389/fspas.2021.735834

Irons, M. A., & Irons, L. G. (2021). Terraform sustainability assessment framework for bioregenerative life support systems. *Frontiers in Astronomy and Space Sciences, 8*, Article 789563. https://doi.org/10.3389/fspas.2021.789563

Johnson, Jr., H. (1998). *Soilless culture of greenhouse vegetables.* UC Davis, Vegetable Research and Information Center, 1998, found 2022 at: https://vric.ucdavis.edu/pdf/hydroponics_soillessscultureofgreenhouse%20vegetables.pdf

Kacira, M., Giacomelli, G. A., Patterson, R. L., Furfaro, R., Sadler, P. D., Boscheri, G., Lobascio, C., Lamantea, M., Wheeler, R. M., & Rossignoli, S. (2012). System dynamics and performance factors of a lunar greenhouse prototype bioregenerative life support system. *Acta Horticulturea, 952*, 575–582. https://doi.org/10.17660/ActaHortic.2012.952.73

Klee, H. J., & Giovannoni, J. J. (2011). Genetics and control of tomato fruit ripening and quality attributes. *Annual Review of Genetics 2011, 45*(1), 41–59 DA—2011/12/15. https://doi.org/10.1146/annurev-genet-110410-132507

Lehmann, J., & Irons, M. (2020). *Soil health in space: Determination of gravitational effects on soil stability for controlled environment agriculture.* National Aeronautics and Space Administration (NASA), ISS Expeditions 63 and 64, PI.

Li, Q., Li, X., Tang, B., & Gu, M. (2018). *Growth response and root characteristics of lettuce grown in aeroponics, hydroponics and substrate culture.* https://doi.org/10.20944/preprints201809.0108.v1

Liu, M., Pirrello, J., Chervin, C., Roustan, J. P., & Bouzayen, M. (2015). Ethylene control of fruit ripening: Revisiting the complex network of transcriptional regulation. *Plant Physiology, 169*(4), 2380–2390. https://doi.org/10.1104/pp.15.01361

Mortensen, L. V. (1987). Review: CO_2 enrichment in greenhouses. Crop responses. *Scientia Horticulturae, 33*, 1–25. Elsevier Science Publishers B.V., Amsterdam, NL, found at: https://ucanr.edu/blogs/NurseryFlower/blogfiles/55782.pdf

OSHA-183. (2022). *OSHA Occupational Chemical Database Carbon Dioxide*, 183, found 2022 at: https://www.osha.gov/chemicaldata/183

OSHA-1926.451. *Occupational Safety and Fealth Administration (OSHA) Standard 1926.451—General requirements*, Scaffolds. Found 2022 at: https://www.osha.gov/laws-regs/regulations/standardnumber/1926/1926.451

Patil, K., Mane, M. S., Mahale, D. M., & Dekale, J. S. (2020). Hydroponic nutrient solution: A review. *Journal of Pharmacognosy and Phytochemistry, 9*(5): 2095–2099, found at: https://www.phytojournal.com/archives/2020/vol9issue5/PartAC/9-5-347-731.pdf

Poudel, M., & Dunn, B. (2017). *Greenhouse carbon dioxide supplementation*, March 2017, HLA-6723-6, Oklahoma Cooperative Extension Service, © 2017 Oklahoma State University, found at: https://extension.okstate.edu/fact-sheets/print-publications/hla/greenhouse-carbon-dioxide-supplementation-hla-6723.pdf

Schwarze, D. (1993). Carbon dioxide enrichment in the greenhouse. *Minnesota Flower Growers Bulletin, 42*(1), found at: https://hortscans.ces.ncsu.edu/uploads/c/a/carbon_d_52570d1851c70.pdf

Shevtsov, J. (2021). *Making soil for space habitats by seeding asteroids with fungi*, Feb 25, 2021, Trans Astronautica Corporation, NIAC Project, National Aeronautics and Space Administration (NASA), found 2022 at: https://www.nasa.gov/directorates/spacetech/niac/2021_Phase_I/Making_Soil_for_Space_Habitats/

Valle, S. S., & Kienzie, J. (2020). *Agriculure 4.0.* ISSN 1020-4555 Integrated Crop Management Vol. 24 | 2020 Agricultural robotics and automated equipment for sustainable crop production, Food and Agriculture Organization (FAO) of the United Nations, Rome.

Williams, A. (2024). Earth worm transformation of regolith to soil. In *Proceedings of the National Space Society (NSS) International Space Development Conference (ISDC)*, 24–27 May 2024, Los Angeles, CA, USA.

Chapter 8
Bioreactor Stages

Abstract This chapter covers how microscopic to macroscopic organisms are cultured in machines, called bioreactors, which are composed of the organisms with media, circulation, and insertion of gases and liquids to foster maximal development and growth. Bioreactors have a variety of designs and are used extensively on Earth for biofuel, pharmaceutical, chemical, beverage, and food production. Two major types are covered: photobioreactors that use light and yeast-bacteria bioreactors that use feedstock or wastes. Various designs, support, and mass flow for each bioreactor design to assemble a stage are key topics.

Keywords Bioreactor · Aerobic · Anaerobic · Photobioreactor · Algae · Bacteria · Yeast · Cyanobacteria

8.1 Introduction

Bioreactors replicate natural processes in very efficient ways. Photobioreactors are just efficient machines that use light and flow to consume nutrients (as from wastes) and carbon dioxide to produce more algae (or cyanobacteria) and oxygen. Natural photobioreactors are algae layers on stream beds or pond bottoms, or the phytoplankton clouds in the ocean. Yeast-bacteria bioreactors for this book are like the soil biome in nature that consumes solid or liquid wastes and, in some modes oxygen, to release carbon dioxide, water, and plant-usable nutrients. These machines can be as simple as an array of pumps and vats, or as complex as membrane mediated designs with custom cultures suited for each purpose in the farm. Each vat or machine is a growth chamber, and these chambers are linked together to form a bioreactor stage.

© The Author(s), under exclusive license to Springer Nature Switzerland AG 2026

B. L. Meyer, *Designing Farms in Space*, https://doi.org/10.1007/978-3-032-04395-5_8

8.2 Photobioreactor Stages

Photobioreactors are bioreactors that use light for algae and cyanobacteria, in photosynthetic metabolic reactions. Photobioreactor stages can have one or more photobioreactors, with all the support pumps, tanks, gas separators, thermal control, and light controls needed to move mass into and out of the stage.

8.2.1 What Is a Photobioreactor?

A photobioreactor simply a container with lights around it, or in it, to grow algae, cyanobacteria, or water plants. "Photo" means light, and "bioreactor" means a device to make living things grow. Ideally, a photobioreactor maximizes the ability of whatever algae, bacteria, or plant is in it to grow fast and convert carbon dioxide, water, and nutrients (usually wastes) using light into oxygen, clean water, and more living mass. For simplicity for now, I will refer to all the things you grow in a photobioreactor as 'algae'.

8.2.2 Replicating Nature: Muck, Slimes, and Seaweeds

In nature, algae take a variety of forms. Most common are the green or blue-green free-floating mucks, or surface sticking slimes. These mucks and slimes are made up of countless single cells, each with the ability to take light in certain wavelengths, pull in carbon dioxide dissolved in the water, water, and other things like nitrogen compounds, salts, and minerals. The core reaction is the same one discussed in the biochemistry sections above: carbon dioxide and water with light to oxygen and sugars. The sugars expand the cells, and they divide into more cells, which take in more carbon dioxide and water and light to make yet more cells. The nitrogen containing molecules, salts, and minerals are needed to finish cell building and balance the chemistry of the cells as they expand and divide. Given the right conditions, the speed at which algae can incorporate carbon and water is astonishing, as any fish tank owner can attest. Algae are classified by colors, though 'blue-green algae' are a simpler cell called cyanobacteria. Algae can also take a variety of complex forms, from filaments and columns, sheets, to branching forms in kelp and Sargassum. Cyanobacteria and algae are responsible for much of the oxygen humans on Earth breathe every second of every day, see (NOAA-NOS, 2021). Simple algae cells also partner with other organisms, including corals. Coral provides a welcoming habitat for algae cells, and the algae provide sugars to the coral. In a sense the smallest photobioreactor is a coral polyp.

Algae of course are not the only things in water that use light. In addition to diatoms at the microscopic level, algae can be clumps and strands, or larger complex

Fig. 8.1 Stream as Photobioreactor. In this case, Meramec Springs in Missouri, USA, Earth, which has green algae and other water plants as cleaners. Photo by author

structures like kelp. Photobioreactors can be built to grow kelp making seaweed salad a possibility; however, Elodia, leaf plants, or similar water plants are also possible. Such plants are efficient natural mechanical filters, and their branching shapes expose a high surface area for carbon dioxide absorption from water. They also are easier to make stay put in a tube or shape, and easy to mechanically harvest. At a minimum, a simple tank with lights and bubblers can be used to clean wastewater, as a simple photobioreactor. Many high nutrient content streams are cleaned and oxygenated by the natural version of this tank photobioreactor as in Fig. 8.1.

Due to the bounty that evolution has formed in algae, it is not too surprising that space settlers might want to make use of the biological machinery of algae. The photobioreactor can at a minimum recycle air and water in a space efficient way, replacing or minimizing the need for zeolite or other chemical filters. My data in one scenario showed that around 10 liters of live algae per person (paired with a yeast bioreactor to close the mass loop) was sufficient for full oxygen supply and the bulk of food for each settler. However, this glop menu would be unappetizing (even with some yeast). Algae are high in protein for volume, and grow very fast, doubling in the space of hours. Having eaten algae, I can vouch for the less than wonderful nature of a slime-based diet. If it was just a problem of oxygen and food, given enough volume, hydroponics might be a little bit less monotonous. However, in nature there are organisms that love algae, and prefer it even to complex plants. They are fishes, shrimps, and the like. More about the way this works later but adding fish or shrimp to a farm makes an algae stage very attractive due to mass flow and dietary variety. Further, until fish and hydroponics grow enough to be crops, possibly six months to a year, the photobioreactor will be the primary air, water, and back-up food source for an early settlement, so making algae grow fast in a small volume is very important, making photobioreactor design key.

Every photobioreactor concept has these parts (see Fig. 8.2):

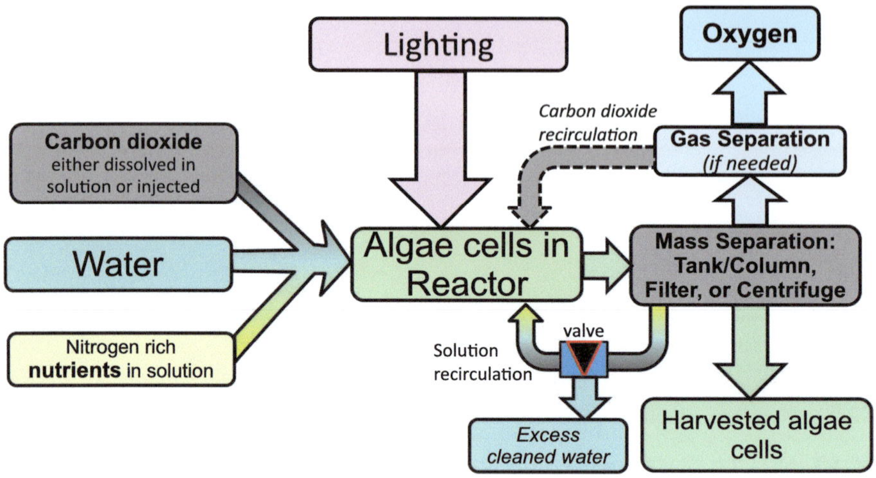

Fig. 8.2 Simplified photobioreactor function schematic

- input flow with either infused carbon dioxide or aerators that add carbon dioxide to liquid in the bioreactor,
- algae media with cells either free-floating or attached,
- mass collectors to harvest algae cells and send out cleaned water,
- lighting inside the liquid or outside a clear vat,
- pumps or oxygenators for recirculating fluids,
- and oxygen gas collection.

8.2.3 Growing Free-Floating Cells

On Earth, by the 1990s there were many commercially available photobioreactors that used cyanobacteria (Spirulina) or Chlorophyta to grow human or fish food, or to clean fish tank water. Several commercial models are available, and rough schematics for general types are below. There is no reason some of these models couldn't upgraded for space travel and be sent to a settlement (that has gravity of course). This adaptation has been suggested by your author and others (Meyer, 2015, 2017), (Fahrion et al., 2021). As noted, the simplest form is a tank that has either natural or artificial light. Biosphere 2 had such a simple photobioreactor via their ocean system as in Nelson and Dempster (1995). The design choices really depend on the selection of algae/plant species, and any animals that might consume the algae. Some algae become inedible once attached to a substrate, others spend all the time free-floating, others almost completely in shapes or structures. The quicker a species doubles in mass, i.e. doubling rate, the smaller a reactor can be, and the faster it can process carbon dioxide and water. Free-floating cells make designs somewhat simpler to maintain in that harvest is a simple as redirecting flow and filtering cells.

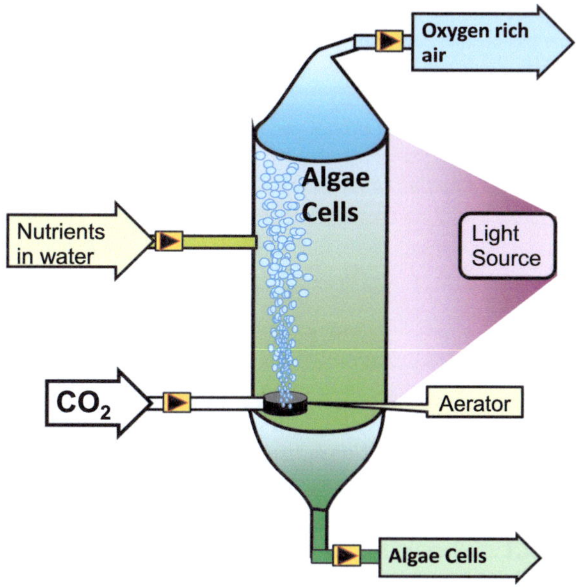

Fig. 8.3 Simple open tank (column) photobioreactor

But free-floating designs must make sure cells have the right exposure to dissolved gases, minerals, and light. Free-floating cells can also be smashed and sheared by certain pump types, requiring more inefficient pumps. The simplest photobioreactor is a tank with free-floating cells and a funnel bottom to collect cells, and an aeration block to both inject carbon dioxide and stir the tank, and a valve to extract oxygenated air as in Fig. 8.3.

8.2.4 Design Considerations: Light and Photosynthetic Machines

Given the drive on Earth for carbon capture, biodiesel, and waste treatment, many types of photobioreactors have emerged. In most cases on Earth, the reactor is simply a very broad and long shallow tank exposed to sunlight and aerated with carbon dioxide-rich outputs from powerplants or fermentation. In Doucha et al. (2005) they use coal plant flue gas over a shallow tank to capture carbon dioxide as in Fig. 8.4.

In space, these simple structures could work given a broad area, and the right clear covering to control sunlight and temperature. However, it is more likely for space that more compact artificially lit (or light channeled) structures that maximize algae production (and as a result, oxygen production), and minimize volume, mass, and energy. Initial set-up mass, i.e. lift mass, may be a driver for initial design, requiring compromises. It is well to note that algae can be very sensitive to temperature, lighting, and chemistry. A small perturbance can kill a crop, so it is wise to keep a

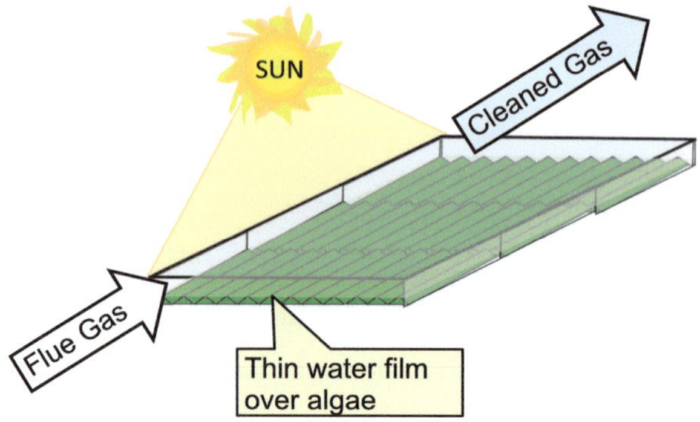

Fig. 8.4 Simple thin tank flue gas cleaner schematic. This is a guess at a design suggested in Doucha et al. (2005)

rescue culture ready to go. A wise space farmer can benefit from algae's high-speed growth, often doubling in hours, under the right conditions, to restock a lost culture.

8.2.4.1 Sheets, Sleeves, and Simple Tubes

A very mass efficient design uses thin-walled collapsible clear tubes filled with single-celled algae (including cyanobacteria) arranged in long horizontal rows or curtains, exposed to lines of lights tuned for the species used. At the beginning of the tube flow, carbon dioxide is added using gas injectors. The injectors can be a column with spirals and high-pressure aerators to fully dissolve the gas into the liquid nitrogen-rich nutrients that are added to the solution as required. The solution which contains algae cells is slowly pushed along the tube where light caused the cells to take in the dissolved gases and nutrients to make more algae cells. At the end of the tube, a settling tank or centrifuge or filter array extracts the bulk of the algae mass, leaving just enough cells to be recirculated to the beginning. The tubes or sleeves have a slight rise, so that oxygen collects at the top of the tube and gets extracted periodically and at the end of the tube to keep the algae healthy and prevent vapor lock in the tube. In lower gravity conditions, managing the balance of dissolved gases and bubbles will be very important, since bubbles will be harder to control as gravity drops. Temperature control is also key, due to the length and fragility of tubes, which expose the algae to whatever temperature surrounds the tubes. The long strands of LEDs or similar lights will add heat that must be controlled. Periodic surging of pumps will be required to loosen clumps that will form along the tube. Also note that the longer the tube, the less dissolved carbon dioxide will be available near the end of the tube. This design can be applied using reverse osmosis and sheet lighting in microgravity, like the design in Heinrich et al., (2019) on the ISS.

Since bubbles and solids are a concern, another concept uses long, clear tubes arranged vertically around lights, with bubblers on the bottom, water and nutrient injectors at various lengths or at the bottom, and a funnel mass collector at the bottom. Figure 8.5 is a simple design that can be married to pumps and separators and allows

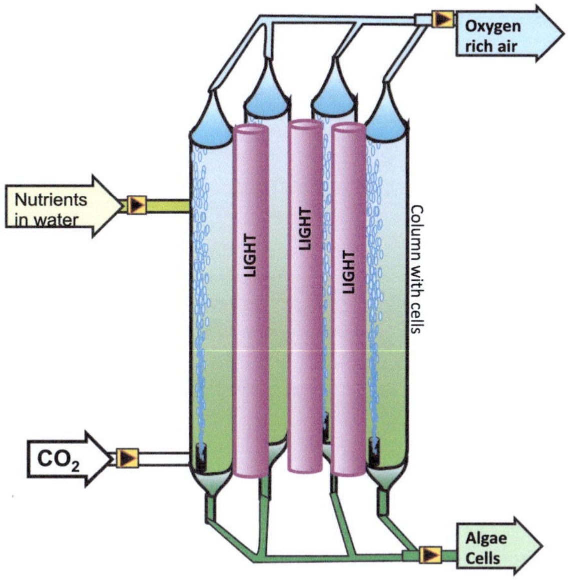

Fig. 8.5 Columns (tubes) and lights in a photobioreactor concept

resiliency by separating cultures. This could be enclosed in a reflective container to improve light penetration and insulate the reactor.

Gas bubbled from the bottom (or injected as dissolved) is collected at the top. The gas at the top will be oxygen enriched. Algae collected at the bottom is centrifuged to concentrate the cells, and water returned to the tubes as needed. Lights either surround the tube or the tubes surround light sources. The tubes can be straight, or spring-shaped coils around lights. Ideally, tube length is designed to maximize carbon capture by the top, without adding excess water volume. Again, lights may add heat, though thicker walls in tubes may add some insulation. The longer the tube the greater the chance of clogging as well. This concept is simple to implement, as it is an efficient extension of the classic glass bottle photobioreactor.

In a simple tank or glass bottle photobioreactor often the bubbler alone provides enough current to circulate the algae cells. Human intervention or a funnel structure is used to harvest algae clumps. The cost is a heavier design that will require frequent intervention. The simple open tank photobioreactor is available for home use now and is common for growing algae to feed animals, including oysters, or for human consumption. In this design, as in Fig. 8.6 a simple clear jar or bottle holds algae cells in a nutrient solution, and an aeration pump sends ambient air into the cellular solution. The oxygenated bubbles from photosynthesis simply bubbles back into the air. This is an easy way to reduce carbon dioxide content in the air near the bottle as well.

An extension of the open tank reactor is one where lighting is immersed in tubes into the tank, in lieu of placing lights around the tank, or putting the tank in a sunlit area. Immersed lights in this case will have to be frequently cleaned. Note also that

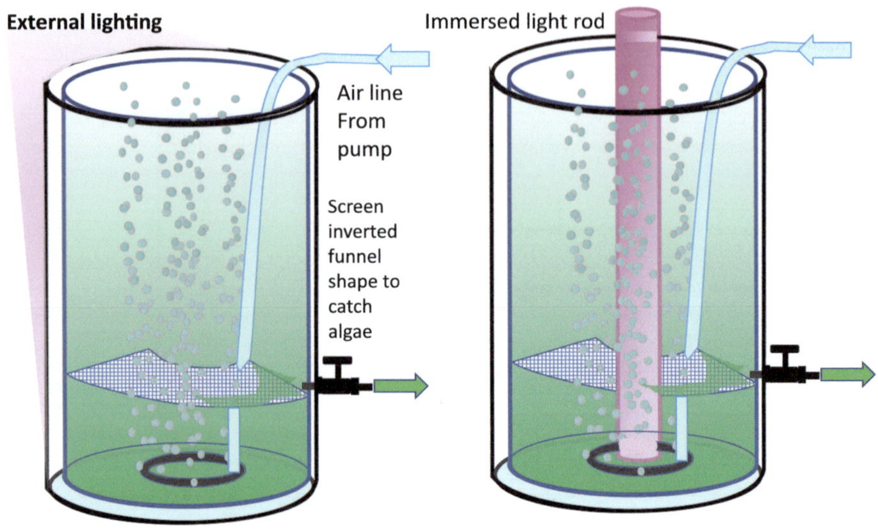

Fig. 8.6 Simple 'bottle' photobioreactors

algae will often attach to the wall of a tank near a light source, again requiring cleaning.

Circulating pumps have become interesting for free-floating designs. Most common pumps use a spinning arrangement of blades, an impeller, to push fluids. Assuming an aeration column is not sufficient, or a stirrer is not used, fluid non-impeller pumps are required. The pump must not chop up the cells.

Capillary pumps are one type of pump that does not crush cells, as in the design in Fig. 8.7. These types of pumps are common in the medical industry for blood movement. These pumps are more expensive than impeller-based pumps and tend to be more clog prone. If impeller pumps must be used, they are usually used in a one-way flow, where algae cells enter at one end of a tube, tank, or sleeve, and are in the sleeve long enough to double. The impeller pump then pulls the cell-rich fluid out for harvest, and new mother cells injected at the other end. Open stirrers are another option, where slow movements of fins stir the fluid in a tank, and the impeller pump only used at harvest.

8.2.4.2 Thin Channel Designs

In some designs, an algae 'chip', a network of thin ribbon-like channels hold cells that flow through the flat plate holding the channels. In some designs, semi-permeable membranes allow gas diffusion into the channels, eliminating the need for aerators (though replaced with pump energy). These plates can be stacked to take advantage of volume and light sources and have seen use on the International Space Station. Tradeoffs include the high chance of clogging in these narrow flows, and the inability

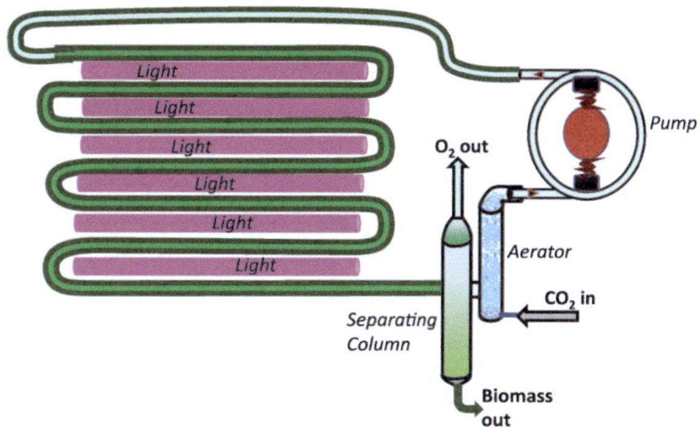

Fig. 8.7 Capillary pump and tube photobioreactor schematic. This design uses a capillary pump to preserve any cells in the tubes after aeration

to easily harvest the algae unless the plate can be pulled apart, adding a manual process. There would have to be very careful optimization of diffusion and cell size to limit clumping, as algae cells love to stick and clump if allowed. The chip plate method needs research with various varieties of cells and might be used to maintain a culture for backup. With further development, the chip-based designs can hold promise as a space efficient way to cycle air as a biofilter . A recent membrane photobioreactor concept is the chip bioreactor design funded by the European Space Agency expands on the above to make a chip concept that uses semi-permeable membranes. Membrane attached algae or cyanobacteria cells in channels, and light panels, process carbon dioxide into oxygen and produce cells in a layering concept. See Fig. 8.8 for an extrapolated design from this concept.

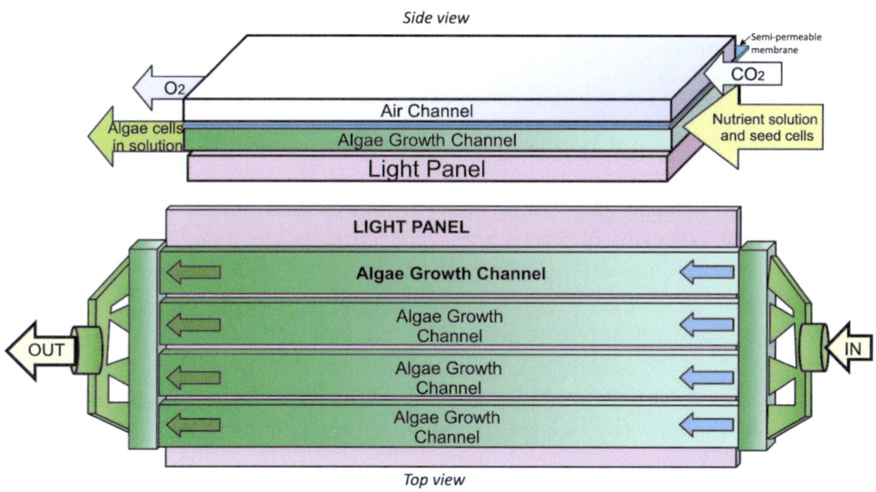

Fig. 8.8 Notional chip-channel photobioreactor schematic. An extrapolation of the algae chip photobioreactor design

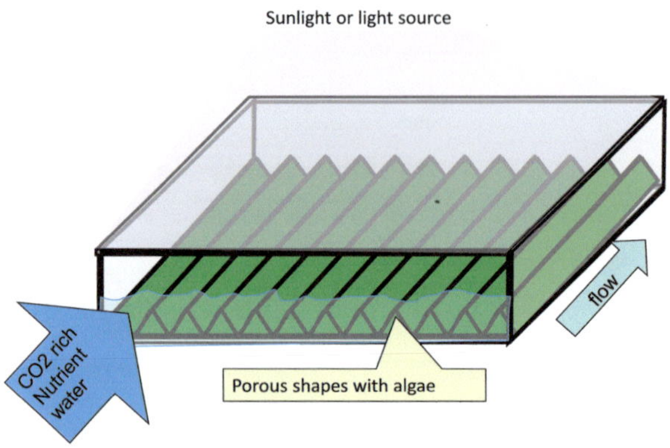

Fig. 8.9 Ridge-based high surface area photobioreactor tray schematic

A more common, less exotic, design is already in use for carbon capture on Earth. A flow channel of a few centimeters deep sits either in open air or under a transparent pane, and algae cells grab the bottom. I would humbly suggest using a stronger ridged design of porous material as in Fig. 8.9, and that once stringy enough, the algae clumps are pulsed out by increased flow to collection.

Other harvest options for photobioreactors like this include simple scrapers or squeegees that collect algae mass, or addition of fish, mollusks, or crustaceans to eat the algae, then harvest the animals. Some designs on Earth are strictly for carbon neutralization and water cleaning, so the algae may just be pitched. With or without animals, this model is like the slime layer that naturally forms in puddles or backwaters on Earth. Like the chip, these thin film reactors can be stacked, but will require either open air, an air gap with forced air, or aerated flow.

8.2.4.3 Membrane and Immersed Lighting Concepts

The architectures above are commercially available now and in use at industrial scale on Earth; however, if compactness and purpose are more important, other concepts can apply. One notional concept by your author attaches algae mother cells to membrane, which is then wrapped around a light source as in Fig. 8.10.

Clumps are pumped in pulses away from the mother cells as they double, and the surface area of the membrane optimizes growth, like the membrane mediated reactors for the yeast-bacteria bioreactor. The trick is to shape the membrane to optimize light and flow and pick the right material and support parts to minimize the size of the photobioreactor for space use.

Ease of harvest and the ability to use impeller-based pumps make membrane structures attractive. Since cells are largely attached to membranes, not floating about to the smashed by pump blades, the fluid can use impeller pumps. The shapes of the

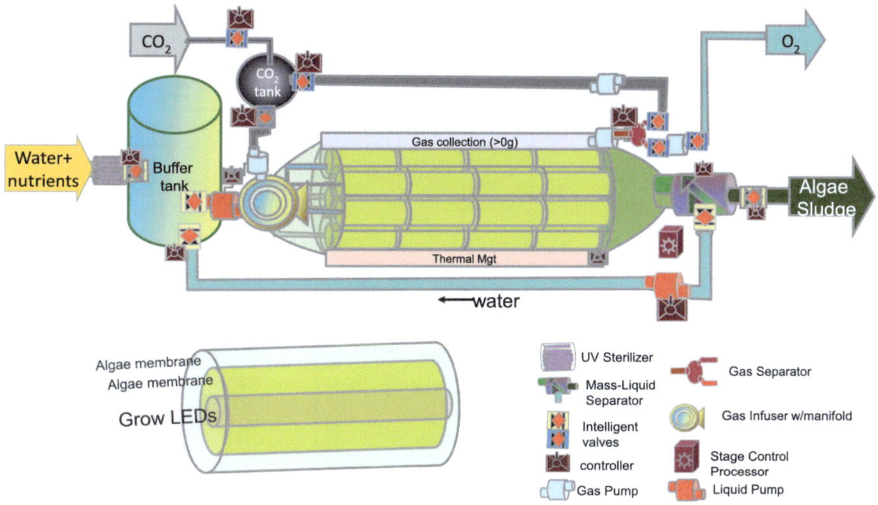

Fig. 8.10 Notional membrane-based low gravity design photobioreactor

membranes or meshes are made to optimize flow and minimize volume, while maximizing growth. As a result, meshes and membranes can be very complex. Attaching mother cells to the substrate is a challenge itself, but common to all membrane mediated bioreactors.

8.2.4.4 Light Management

Depending on the shape of the algae holding tubes, sleeves, tanks, or meshes, light intensity, heating, spectrum, and periodicity can be challenges. Two of the biggest challenges are heating and over-lighting (which kills cells). LEDs and other low power technologies thankfully have a variety of shapes and types for the bioreactor engineer to choose, to optimize light to algae and minimize volume and heat. In some designs, such as Fig. 8.11, a large rectangular light plate illuminates a film or tube array.

Fig. 8.11 Panel and tube photobioreactor schematic

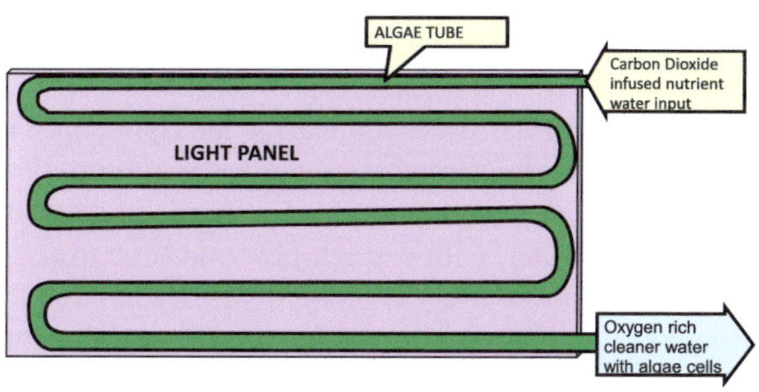

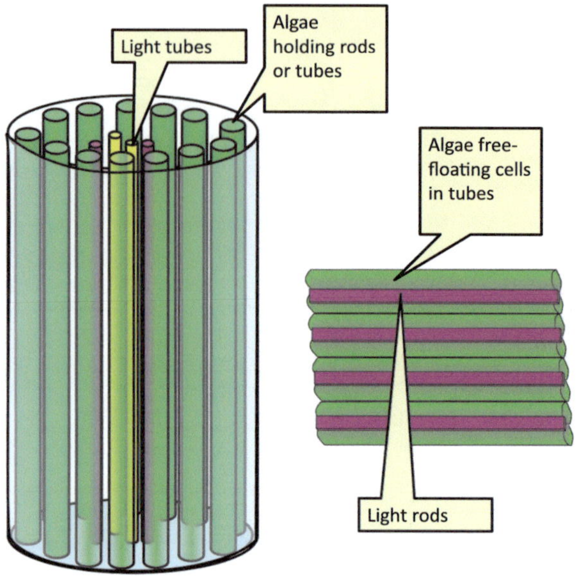

Fig. 8.12 Rod and tube photobioreactor schematic

In others, sleeves each have a light tube or light rectangle along their length. In some designs, tubes surround a light tube, or immersed light tubes are in a sequence of cylindrical membranes as in Fig. 8.12.

Regardless of architecture, the right amount of light needs to hit each cell to maximize growth, while not killing the cells. Lights also give off radiative heat, so heating of the fluid may require cooling of tubes or tanks. Cells that do not get enough of the right wavelength of light will have limited growth. If cells can, they will tend to attach to the lit side of a tank, eventually obstructing the rest of the tank, so special care may be required if the tank walls do not prohibit attachment. Note that some species of cells require a dark cycle, times with less light, though ideally species selection has led you to a farm crop that can grow 24 × 7.

8.2.4.5 Gas Injection and Extraction with Nutrients: Getting the Goodies to Every Cell

Just like getting light to all cells at the right level is critical, so is getting the right mix of dissolved gases and chemicals to each cell. Algae cells need carbon dioxide and a solution of nitrate-rich chemicals at the right concentrations. Various gas injection structures (see chapters above) can force carbon dioxide into the solution, though the aerator-tube design does this by shape. In principle, air pushed via small holes into flowing water, with enough flow and time to keep the bubble from making big bubbles will dissolve a gas into a fluid. If you have seen a lobster tank at a seafood restaurant, you saw such an injector. The cooler the water, the better gases stay

Fig. 8.13 Bubbles of oxygen formed around algae. Photo by author

dissolved. In nature, algae clumps pull the carbon dioxide dissolved in the water, then release dissolved oxygen that often forms bubbles as in Fig. 8.13.

In our systems, we would like to prevent bubbles as much as possible, but more critically, to keep the oxygen concentration low enough to keep from killing our algae cells. So, move the carbon dioxide to the algae cells at the right amount, then pull off the oxygen at the right amount.

8.2.4.6 Harvesting

As previously mentioned, centrifuges, funnels, settling tanks all provide methods for harvest. Really, it comes down to the two types of algae bioreactors: substrate or membrane and free-floating. If the algae cells are attached to a membrane, then pulsing pumps would release clumps that are large enough to be filtered out using an angled screen or settled out in a column.

With free-floating cells, harvest is a bit more complex. In gravity, cells settle to the bottom of the column unless stirred. This means the top part gets more see-through, the bottom opaquer. Every so often, the bottom can be drained, the cells separated from the water surrounding the cells, and the cells sent for processing. One way, that may have a higher energy cost and risk clogging, is to use a reverse osmosis membrane, where the pure water is cycled back to the bioreactor, and the cells concentrated. Another method is to feed the cells to a centrifuge which spins and concentrates the cells, water again sent back to the bioreactor. Concentrated cells can then be pressed or dried out to wring the water from the cells, leaving a film or jelly-like sludge that can be used as foodstuff.

For attached or unattached cells, the concentrated cells should not be dehydrated if they are to be fed to fish or shrimp. Since these animals evolved to feed on algae, they need the cells ideally in live states, so the concentrated but still living sludge is pumped to the aquatic stage and then pulsed into the current to feed the animals.

For people, films and dehydrated algae are easier to store than living cells and can be used as supplements for other foods, fermented to make digestive drinks or alcohols, or used as jellies for printed foods or creamy foods like puddings.

8.2.4.7 Water Plants Versus Algae Cells

Water plants and some algae like kelp or Sargassum are organized into recognizable structures that can result in different bioreactor designs. Kelps and structured water plants are options to clean water, cycle oxygen, and produce biomass for animals or humans. The duckweed bioreactor in a prior chapter is one version. The toughest plants are the ones on Earth with a history of invasion. The best choices are the ones that are invasive, but also edible to either human or animal. While some slow growing types are food favorites, such as kelps or other algae, other faster-growing plants like duckweeds and milfoil, can be fed to animals. In any case, structure of the photobioreactor can become simplified to take advantage of the water plant shapes. Due to the simplified structures, however, many plants require harvesting effort by manual work, or by robot.

In Fig. 8.14, a photobioreactor could grow water plants or large algae like kelp. Given this structure, a robot (arm) will be needed for trimming and maintenance, though this type of photobioreactor would also be an opportunity for polyculture with fish, crustaceans, or mollusks which could aid in cleaning if the aeration mix had a low carbon dioxide concentration and sufficient oxygen.

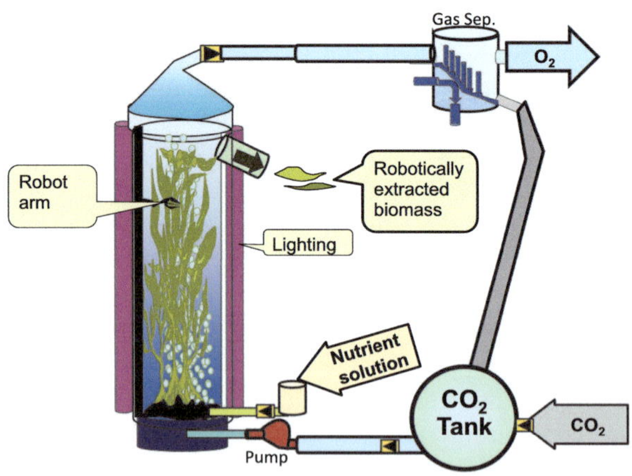

Fig. 8.14 Simplified kelp photobioreactor schematic. In this case, the tank is circulated by aerator, and a robot arm is used to trim and extract biomass

8.2.4.8 Combined Cycle Algae

Work at the US Department of Energy (DOE) and others as in DOE (2016), Joseph Rollin et al. (2018), and Tyler et al. (2019) found some freshwater algae species that can also eat cellulose. In essence, the photobioreactor with these species could take in plant wastes from the hydroponic or soil grows, and once finely ground and liquefied, input them to the photobioreactor as nutrient and carbon source. Algae in this sense reduces the need for the yeast-bacteria bioreactor to make animal feeds or human food. These algae don't eat lignin, so you would still need a fungus-based stage or incinerator, but cellulose eating algae would help. It would be a good research area to determine if this combined cycle photobioreactor is more energy or mass efficient for the farm than directly feeding waste to animals, or waste to aerobic bioreactor to algae to animals. It is also an area to figure out how this technology might apply to small settlements or affects footprint or oxygen production by the photobioreactor stage. A trade might be to combine the cellulose eating algae with other algae or cyanobacteria in the stage to improve efficiency.

8.2.5 Assembling Photobioreactor Stages

The photobioreactor stage is the combination of the growing volumes, lights, pumps, aerators, tanks, sensors, thermal controls, and separators, assembled to work in our settlement, as discussed above, and in schematic in Fig. 8.15. Buffer tanks of both gases and liquids are needed to make sure the flow rate into and out of the stage matches that of the farm as needed. These tanks might be initially centralized for many stages, then distributed as the farm grows to provide resiliency in the event of disasters. The diagram below is a vastly simplified schematic of what is no doubt a far more complex design. In the stage there are many parts beyond the photobioreactors. Tanks for carbon dioxide and oxygen hold at least a few days buffer, along with water input tanks, nutrient solution tanks, and tanks for algae cells. Numerous pumps and smart valves are coordinated by the farm. A culture lab holds backup cultures to restock the reactors. In front of the algae cell tank, there may be a centrifugal separator to get extra water from the extracted cells and return the water back to the nutrient tank or reactor tanks.

Pumps, valves, and sensors as part of the farms fluid and gas management system will move carbon dioxide, clean and dirty water, and oxygen, around the stages, and out to other stage types and the habitat. Algae can be very fickle to lighting, temperature, and chemistry, and the stage will need a culture lab to hold backup algae cultures and spores, especially given likely radiation exposure. Heat rejection from all the pumps and lights will require either a passive fin and pipe flow system or active controls with fans and maybe even heaters. It is critical to extract as much water as possible from harvested algae cells, even if provided live, to keep from dehydrating the photobioreactor stage, so it is likely a centrifuge separator (see machinery above) or at least another settling column is needed. As the photobioreactor might be a very

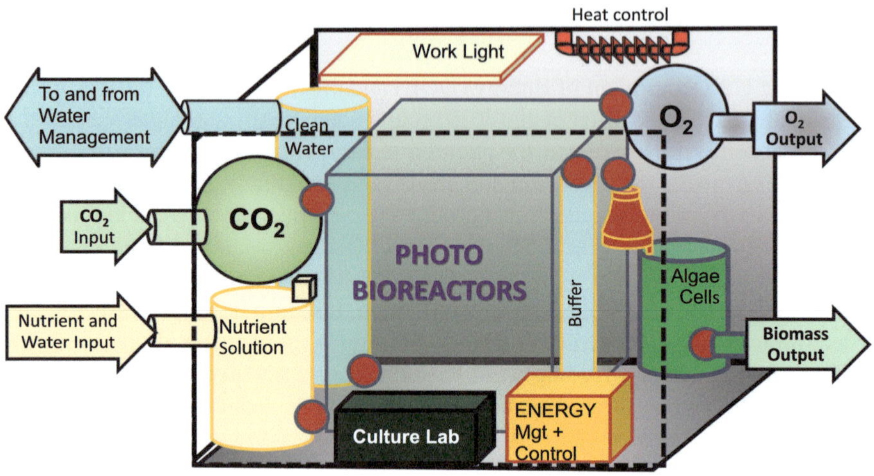

Fig. 8.15 Photobioreactor reactor stage schematic

small part of the overall farm footprint, it could be located centrally between other parts of the farm. As heat rejection is also a core requirement, pumps will need to move fluids through heat exchangers which may run to fins to cool tanks heated by lights. The photobioreactor tanks will also need to be sized for the full crop size after growth (with a bit of extra volume), to allow room for doubling cells to grow. It might also be advisable to use the photobioreactor as a flex point for food and feed production. Due to the high-speed growth of algae and cyanobacteria, it can scale up or down to make up for losses or overruns in hydroponic or animal stages or added humans in the habitat. A small version of the photobioreactor, with the initial culture, could be part of the life support system on the ship that takes settlers to the settlement, then this initial culture used to stock the larger photobioreactors in the settlement's farm. The growth tanks, which are lit, are fed, emptied by the buffer tanks, and circulated by aeration columns and pumps. Excess oxygen is extracted by shapes in the tanks, and recirculation by gas pumps sends unused carbon dioxide back to the cells. Reverse osmosis filters will extract excess water from growth fluids, and these may be pressurized by pumps as well. Sensor arrays will be in many parts of the overall reactor assembly. The overall assembly will need to be constructed to allow human access to replace parts, or to unclog pipes, with work lighting for the humans to do their work, and atmosphere for them to breathe. People working around the reactors and equipment need lighting and volume to work. In many space farm scenarios, there is a surplus of oxygen, but a deficit of carbon dioxide and water, so the photobioreactor may need other tanks and cultures to allow operation to be minimized when not needed. There should also be backup isolated cultures even beyond the culture lab for redundancy to restock the photobioreactor stage in the event of catastrophe.

Overall, the tanks may be constructed of in-situ materials such as silica glasses, pipes from polymers or metals or ceramics, but more complex parts will need to be imported until complex manufacturing can be done by the settlement. Luckily for photobioreactors, the complex items like pumps, reverse osmosis filters, sensors, valves, and lights are only a small fraction of the empty mass of the photobioreactor stage.

8.3 Yeast-Bacteria Bioreactor Stages

Yeast-bacteria bioreactors are machines that do not use photosynthesis, but instead use either anaerobic or aerobic metabolism.

8.3.1 Open Format Bioreactors: Septic Tanks and Fermenters

Current Earth technology contains many varieties of open format (i.e. vat) bioreactors that will find a home in space farms. The primary purpose of the aerobic bioreactor is to break down wastes to release carbon dioxide and water using aerobic digestion (see prior chapters on biochemistry and aerobic yeasts). The yeasts or bacteria will likely be tailored to optimize this process. See Fig. 8.16. Since aerobic respiration generates heat, cooling will be required to control the bioreactor, though the heat could be pumped where needed. Oxygen is pumped into the reactor with the wastes as food stock for the bioreactor. In the reactor, stirring or other circulation makes sure the cells have full exposure to the food and oxygen. Carbon dioxide produced (about a third more than the oxygen used) is sent to photobioreactors, or to gardens and can be isolated by a simple column from the water vapor produced. The water vapor or any other excess water to be returned to the clean water management system for the farm and habitat.

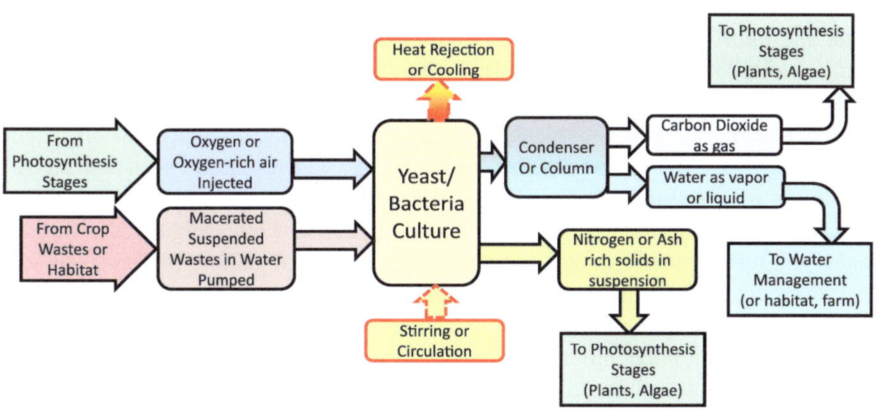

Fig. 8.16 Schematic for aerobic bioreactor function

Fig. 8.17 Super-simple aerobic bioreactor

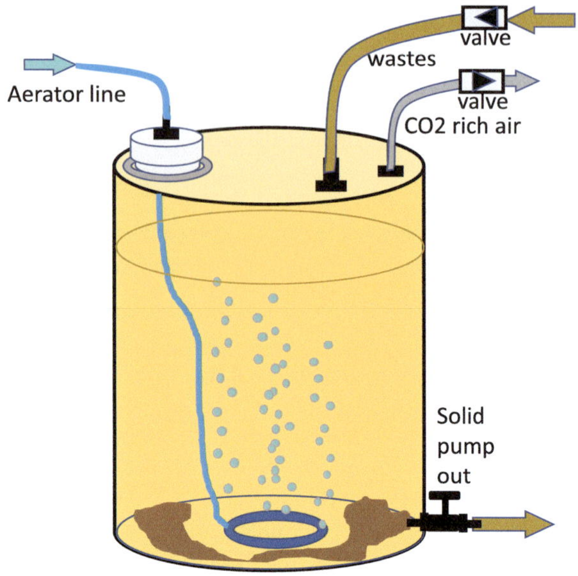

Septic tanks used as a first stage in sewer treatment are one such bioreactor. The simplest design is a tank with a bubbler and valves as in Fig. 8.17 and in it an oxygen aerator provides circulation and the culture of bacteria, maybe with enzymes. The bacteria in the tank consume cellulose and other wastes to release water and carbon dioxide.

A septic tank is simply a cylindrical tank and stirrer with a funnel bottom with pipes that input waste, inject oxygen, extract settled solids, extract processed liquids, and remove gases. In the diagram below is a simple schematic of what is around the bioreactor main tank. A variety of tanks, controls, pumps, and other machines are key to controlling the bioreactor. Stirrers and injectors circulate the living solution in the tank. Gases from the top are separated to remove the carbon dioxide and water and return any remaining gases to the reactor. Since aerobic reactions release heat, a cooling jacket (as in Fig. 8.18) may be required to control the reactions. This arrangement is like current chemical factory bioreactors or wine fermenters.

'Black water' (sometimes 'blackwater') is a term that means water and suspended waste, such as feces, food wastes, etc., and added wastes in solution such as urine and organic soaps. The purpose of the whole bioreactor stage is to complete the mass flow loop in a space farm by taking in wastes and oxygen and releasing carbon dioxide, water, and nitrogen-rich sludge or solids. It replicates the natural processes in soil, or in burning. Without this stage, or an incinerator, all the carbon in a farm would eventually be locked into inedible mass from uneaten parts of plants, and undigested material. Since animals cannot completely process fiber, cellulose included, the fiber must be cracked to release the carbon, hydrogen, and oxygen in it. Bacteria can accomplish this task using enzymes, with or without oxygen, though with oxygen the output is more carbon dioxide and water, and less alcohol and methane. Yeasts

Fig. 8.18 Vat aerobic bioreactor schematic

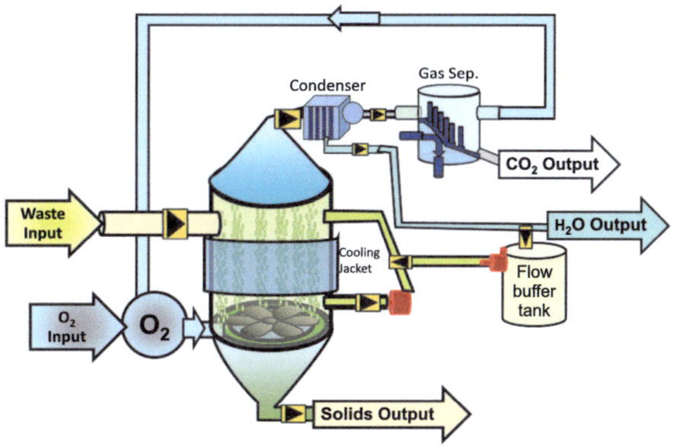

use a similar process, assisted by bacteria, and can produce yeast biomass suitable as high protein food for people or animals. Lignin is even more difficult to digest, though bacteria and yeasts could be engineered to efficiently break it down as well. Another type of open bioreactor is the same shape, but with yeasts in lieu of bacteria. Yeast reactors rely on a more acidic regime to limit bacterial competition but are less able to digest fiber.

Yeast instead prefers simple sugars, though adding enzymes from sprouted seeds (usually barley, called malt) or salivary amylase allows the yeast to crack starches into simple sugars. Figure 8.19 goes a little deeper into a stirred tank yeast-based anaerobic bioreactor. The primary outputs are liquids (with alcohol), solids in liquids, carbon dioxide, and water, from an input solution of nutrients (like sugars and minerals), and water. The design below T is like those used to make wines or beers and produce yeast solids.

Three common yeast-based bioreactors are mash production for distilled spirits (whiskey, vodka, etc.), malt and grain fermenters for beers, and fruit fermenters for wines. The fourth type is the reaction of yeasts in rising bread. A similar process is used for cheeses and to produce antibiotics like penicillin. Genetically modified

Fig. 8.19 Simple Anaerobic Yeast Bioreactor Schematic. * = Mostly carbon dioxide, though may have methane. ** = Mostly water but may also have alcohol. The result solution, which is wine or beer in some scenarios, may go to a distillation column to extract alcohol

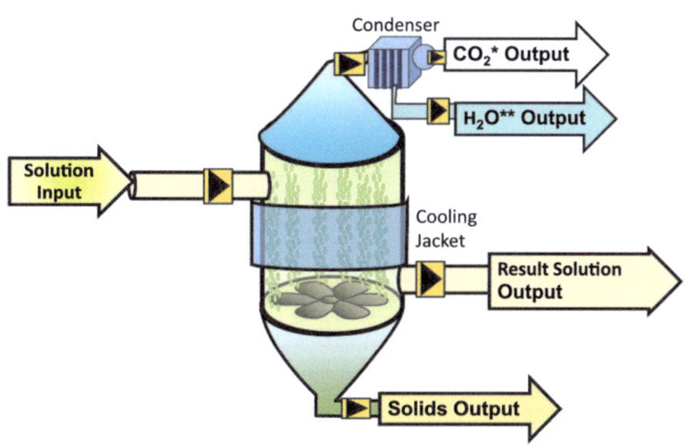

yeasts can make many organic compounds, including pesticides or custom pharmaceuticals. These are far from the only type, especially, and notably since yeast extracts are a common food additive.

This last function, yeast as food, is very significant for early-stage space settlements. Until farms can produce crops of plants or animals, the food web may be algae reactor to waste treatment to yeast vat to dinner to human wastes. The diet of green algae glop and yellow yeast glop might not sound fun but would close the mass loop with the human habitat and keep the people alive and (physically, mostly) healthy. In the event yeast solids are a primary output, the usual process involves an anaerobic digestion of sugars that have already been hydrolyzed from fiber by a previous aerobic bacterium or enzymatic stage or result from food wastes such as rotten fruit or excess grain. The result is alcohol and lots of yeast cells, exactly as for beer and wines. The glop could be used in inventive recipes, used in 3D printers, or fed to additional cultures to make better foods.

In most scenarios with at least a partially functioning vascular plant stage, and maybe an animal stage, the yeasts are combined with the bacteria to aerobically digest wastes from multiple sources to produce carbon dioxide and water and concentrate the remaining part of the waste stream into products for plant and algae fertilizers and animal feeds. Lignin is again a concern that may require tailored bacteria or yeasts given some crop choices, or even a follow-on incinerator. Clogging is a major concern for any flowing system, and doubly so for a bioreactor, so human intervention (and clog cleanout taps) is critical.

8.3.2 Membrane Mediated Bioreactors

While open format tanks are useful, advances have led to very high throughput reactors that use tailored bacteria and yeasts that are attached to membranes. These are membrane mediated bioreactors (or MMRs). The membranes are then wrinkled or rolled to maximize exposed surface area (and the organisms attached) to the flow of inputs to maximize efficient digestion to get desired outputs as in Fig. 8.20. Managing what sets of organisms get what inputs in what conditions, to get chosen outputs is the function of control systems and software. From beer to pesticide, membrane mediated bioreactors are common in the chemical industry, and a mature technology that can be adapted for space settlements quickly. MMRs are importantly also volume efficient and fast, keys for in space use. Since they can be precisely controlled, they become the multitool chemical balancer of the farm.

As with catalytic converters, the membranes could be combined into tubes or honeycomb shapes as well, as in Fig. 8.21. Tubes, as noted before, must be sized and flow controlled to avoid clogging, and the clogging issue is a concern for all membrane mediated designs.

Combining open format and membrane mediated tanks using valves, sensor arrays, temperature and chemical controls, and tanks to buffer flow rates makes what this book calls the yeast-bacteria bioreactor stage of the farm.

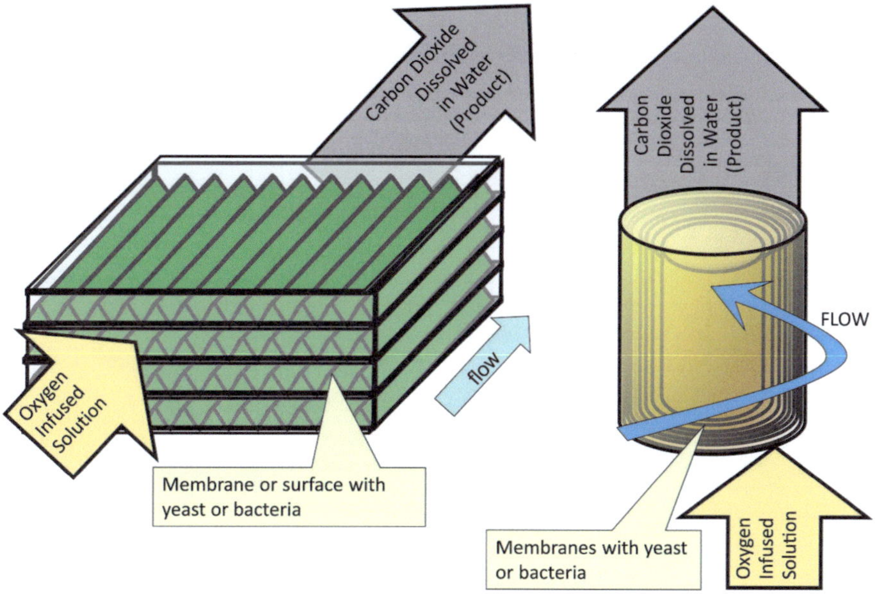

Fig. 8.20 Two membrane mediate bioreactor schematics. Crinkled membrane layer (left) and concentric (right) snail drum membrane bioreactors

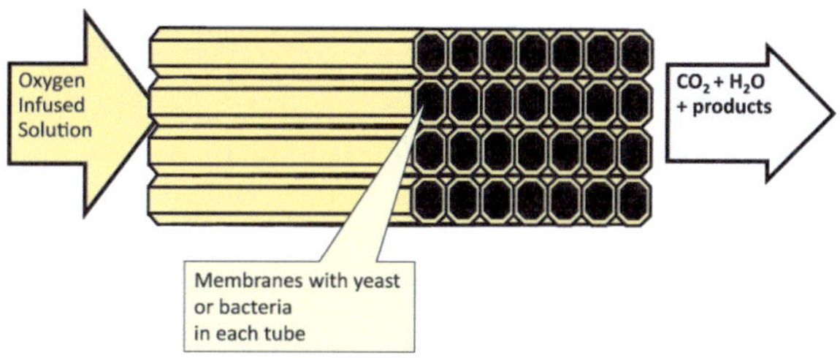

Fig. 8.21 Honeycomb or tube matrix membrane bioreactor

8.3.3 *Assembling the Yeast-Bacteria Bioreactor Stage*

Overall, the yeast-bacteria stage resembles a small biochemical factory, with the same concerns and structures. Tanks, valves, heaters, coolers, sensors, pipes, and automation all have purposes in the assembled stage. More complex components will be shipped pre-built from factories away from the settlement, then married to locally assembled components. Figure 8.22 is a much-simplified schematic for this stage. Each tank here represents several tanks for each type of product, output, and input for the array of bioreactors. A complex farm may have a complex set of bioreactors in this stage to balance the biochemistry of the farm. An incinerator may follow this stage.

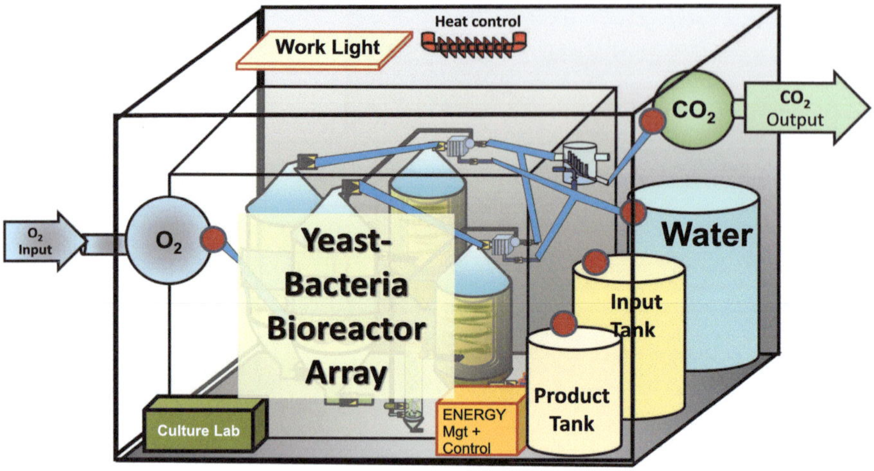

Fig. 8.22 A possible schematic for the yeast-bacteria bioreactor stage

Figure 8.22 is one possible design assuming some gravity. From the entry point, an open format tank array accepts black water and ground wastes as a sludge and is sized to accommodate normal flow with a significant safety margin (likely double a day's flow). Water is also in tanks to add as needed to facilitate flow, and store results from the stage. Oxygen tanks also buffer inputs from the farm's gas management system to hold enough oxygen (or oxygenated air from photosynthetic stages) to match the flow through the stage plus a safety margin. Stirrers and aerators in the tank stir the sludge, where bacteria begin aerobic digestion of the sludge. Gases are captured at the top of the first tanks and, if rich in methane or similar, are looped through a fuel cell or burner to get carbon dioxide and water and add electricity to the farm. Heavy solids settle at the bottom of the funnel and are either macerated further and recirculated or extracted and then dehydrated using heat to recover water which goes back to the water tank. The solids, sterilized, will be rich in minerals and nitrogen and can be used as fertilizer, or fed to animals or other parts of the bioreactor. The rest of the material in the tank is a partially digested soup, which is sterilized and pumped into other parts of the bioreactor. In most of the bioreactors, the soup is oxygenated and flows across MMRs to get carbon dioxide and water; any remainder is treated as the solids from the first tank. Carbon dioxide goes into tanks for distribution to other parts of the farm. Excess water is also pumped from the water tanks to other parts of the farm. In some cases, yeast biomass is needed as a food stock, so other tanks have fermenters to produce yeast sludges or other products. Pumps and pipes take the various products where they need to go. The stage may include an extra section as well, one that takes excess carbohydrates, and makes alcohol for consumption or other uses (i.e. the beer or wine factory), though food elements are usually separated from human waste treatment.

A core concern for all bioreactors is contamination control. Each tank a controlled environment for only what we want to grow, what we don't. Given the fact the mass flow in the farm is critical to human survival, inputs to tanks and flows between tank need sterilizers to keep out unwanted invaders. In the event of contamination, cultures are kept in isolated areas, including frozen storage of bacterial spores and yeast masses, to allow restart of tanks. Radiation is also a risk, necessitating mother cultures stored in a secure area.

8.4 Summary

Bioreactors are combinations of machinery and organisms to accomplish a purpose, such as waste processing, gas management, or making food for people or animals. Photobioreactors are bioreactors that use light and photosynthetic cells or organisms and take in carbon dioxide, water, and nutrients, while yeast-bacteria bioreactors use aerobic or anaerobic cells or organisms and take in wastes or other feeds, to produce water, carbon dioxide, and other products. Both the yeast-bacteria bioreactor stage and photobioreactor stages require many parts, where each bioreactor constitutes a growth chamber. Both will require careful machinery and assembly, possibly sourced from Earth or other more sophisticated settlements, though simple tanks that work well for many purposes can be built by even the smallest and simplest settlements. Both bioreactor stages are important for any settlement and space farm, to close the mass flow cycle.

8.5 Questions

1. What is a bioreactor?
2. What is the metabolic difference between a photobioreactor and yeast-bacteria bioreactor? What are the differences inputs between each type?
3. Which types of cells or organisms are grown in a photobioreactor?
4. What are natural photobioreactors?
5. What is a natural (fungi and) bacteria bioreactor?
6. What are the core parts of any photobioreactor?
7. How does doubling rate affect photobioreactor size?
8. What are the differences between a tank and membrane mediated bioreactor?
9. Is there any reason why different types of photobioreactors couldn't be combined in a photobioreactor stage?
10. Given the shape of the bioreactor in Fig. 8.10 what are concerns that could affect reliability?
11. What are concerns with light amounts in photobioreactors?
12. What other machinery is needed to make a photobioreactor stage beyond the photobioreactors?
13. What would be a function for a settlement filled by the yeast-bacteria bioreactor stage that is similar to a skeptic tank on Earth?

14. Why do aerobic and anaerobic bioreactors have cooling jackets? What is the difference in inputs for anaerobic versus aerobic bioreactors?
15. What is black water?
16. Why do beer fermentation vats have malt?
17. What are concerns with membrane mediated bioreactors?
18. What does a yeast-bacteria bioreactor stage need heat rejection?
19. Why do bioreactor stages need storage tanks? How might one determine what types of storage tanks for gases and liquids are required for a bioreactor stage?
20. What input tanks are different between a photobioreactor stage and a yeast-bacteria bioreactor stage?

References

DOE. (2016). *National algal biofuels technology review*. U.S. Department of Energy, Office of Energy Efficiency and Renewable Energy, Bioenergy Technologies Office. Found at (2022): bioenergy.energy.gov.

Doucha, J., Straka, F., & Livansky, K. (2005). Utilization of flue gas for cultivation of microalgae (Chlorella sp.) in an outdoor open thin-layer photobioreactor. *Journal of Applied Phycology*, 403–412.

Fahrion, J., Mastroleo, F., Dussap, C., G., & Leys, N. (2021). Use of photobioreactors in regenerative life support systems for human space exploration. *Frontiers in Microbiology, 12*, 699525. https://doi.org/10.3389/fmicb.2021.699525. PMID: 34276632; PMCID: PMC8281973.

Heinrich, R., Close, J., Boltz, G., Lechon, F., & Pikelj, M. (2019). *Photobioreactor: Oxygen and a source of nutrition for astronauts*. Airbus Defence and Space Press Release, Friedrichshafen, DE, 25 April 2019, found at: https://www.airbus.com/en/newsroom/press-releases/2019-04-photobioreactor-oxygen-and-a-source-of-nutrition-for-astronauts

Joseph Rollin, J., Vogler, B., Starkenburg, S., Sudasinghe, N., Schambach, J., Pattathil, S., & Barry, A. (2018) Characterization of plant carbon substrate utilization by Auxenochlorella protothecoides. *Algal Research*. https://doi.org/10.1016/j.algal.2018.07.001. https://research-hub.nrel.gov/en/publications/characterization-of-plant-carbon-substrate-utilization-by-auxenoc

Meyer, B. L. (2015). Of fruits and fishes: A space farm and recycling concept. In *AIAA 2015-4607, Proceedings of the AIAA SPACE 2015 conference and exposition*, 31 Aug–2 Sep 2015, Pasadena, California, found 2022 at: https://doi.org/10.2514/6.2015-4607

Meyer, B. L. (2017). Mass and volumes for a spectrum of multistage evolving space farms. *NSS Space Settlement Journal* (3).

Nelson, M., & Dempster, W. F. (1995). Living in space: Results from biosphere 2's initial closure, an early testbed for closed ecological systems on Mars. *Life Support & Biosphere Science: International Journal of Earth Space, 2*(2), 81–102. Bibcode:1996smgh.conf..363N. PMID 11538313.

NOAA. (2021). *How much oxygen comes from the ocean*? National Ocean Service website. https://oceanservice.noaa.gov/facts/ocean-oxygen.html. 02/26/21.

Tyler, C. R. S., Sanders, C. K., Erickson, R. S., Dale, T., Twary, S. N., & Marrone, B. L. (2019). Functional and phenotypic flow cytometry characterization of Picochlorum soloecismus *Algal Research, 43*, 101614. ISSN 2211-9264. https://doi.org/10.1016/j.algal.2019.101614. https://www.sciencedirect.com/science/article/pii/S2211926419300517

Chapter 9
Human and Animal Stages

Abstract This chapter covers how the human habitat interacts as a stage with the farm and how stages that contain macroscopic aerobic organisms, like animals and mushrooms, are assembled and interact with the rest of the farm. Human models, and how human water is managed and cycled inside the habitat, are key concerns for the space farm connected to the habitat. Various designs in the chapter foster growth in crop species, from aquatic species such as fish and shrimp, to terrestrial species such as rabbits and chickens, to mushrooms. Several growth chamber ideas are discussed and how these growth areas are combined into stages.

Keywords Life support · Aquaponics · Insects · Fish · Shrimp · Tanks · Mushrooms

9.1 Introduction

Aerobic organisms take in oxygen, water, and food, and release carbon dioxide, wastes in water or in solid form, and water in breath, or evaporated from the organism's surface. Cycling water inside the stage or habitat is key to minimizing the load transferred to other stages in the farm. In the case of humans, managing the water cycle for water vapor is key to comfort and health. For crop species, managing water, heat, the carbon dioxide/oxygen mix, and wastes are all key to maximize production. While growth chambers (areas) are discussed in detail in this chapter, the chapter also works beyond to discuss all the other parts required to build stages for aerobic organisms.

© The Author(s), under exclusive license to Springer Nature Switzerland AG 2026
B. L. Meyer, *Designing Farms in Space*, https://doi.org/10.1007/978-3-032-04395-5_9

9.2 Human Habitat Principles: The Human Stage

The human habitat is part of the overall ecosystem with the farm and is essentially a stage that provides wastes and carbon dioxide to the rest of the farm, and receives foods, clean water, and oxygen from the farm. Here, we consider how the habitat is viewed as a stage by the farm.

9.2.1 Air

On Earth, air at sea level is around 78% nitrogen gas, around 21% oxygen gas, and 0.04% carbon dioxide (or 400 ppm), the rest mostly inert gases, as per ISO Standard Atmosphere (see (ISO 2533, 1975) though adapted for modern carbon dioxide levels). See Fig. 9.1. Air pressure at sea level is roughly 100 kPa, though people can live at lower pressures to about the equivalent of 3000 m or around 80 kPa, as the Incans of South America can attest. To a person, the important parts are oxygen and carbon dioxide, if none of the more poisonous gases are around.

Humans also need some level of humidity, usually between 15 and 60% relative humidity, or we get nosebleeds and dry skin. Managing humidity is a core need in the habitat, and a source of water recycling.

In Fig. 9.2, a loose cycle with the space farm, habitat air has inert gas, carbon dioxide, oxygen, water vapor, and other volatiles. Water vapor from human breath, sweat, and other activities is captured by a dehumidifier which cycles the liquid water

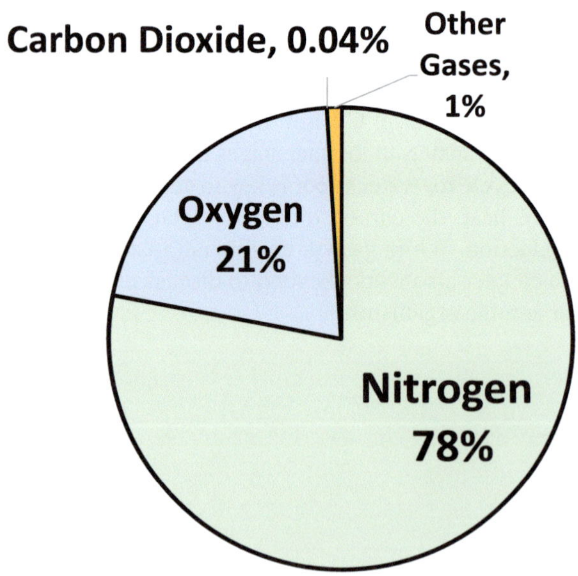

Fig. 9.1 Basic gas composition of sea level on Earth

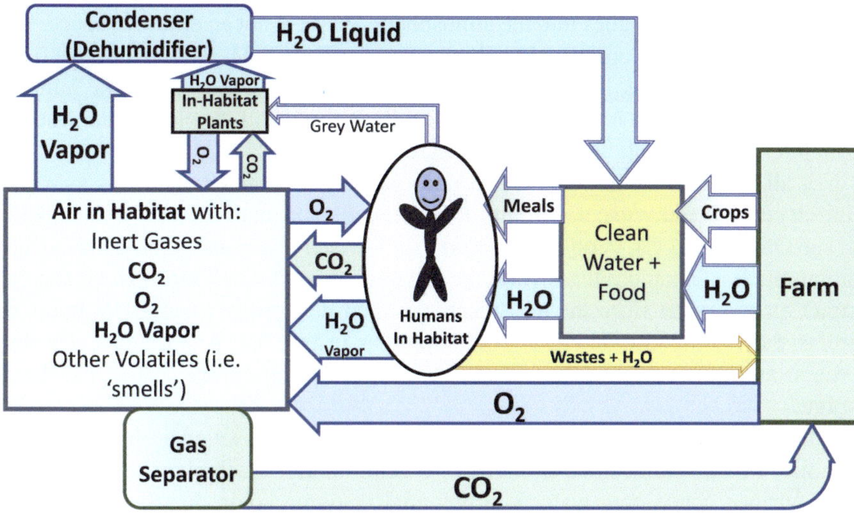

Fig. 9.2 Gas flow habitat and farm

back for use. Oxygen is used by humans, returned to the air as carbon dioxide, where a gas separator pulls out excess carbon dioxide and sends it to the farm. The farm processed wastes, carbon dioxide, and wastewater to produce clean water and food and oxygen.

9.2.2 *Water Types*

A human household uses clean (potable) water for most purposes, including drinking, bathing, food processing and preparation, waste disposal, and washing clothes and cookware. The used water that doesn't touch urine or feces, or large amounts of uneaten food wastes, is called grey water. The water in urine, feces, or from uneaten food, is called black water. One of the functions of the farm is to use grey and black water to get fresh food and clear water. In earlier space stations like the ISS, the process involved separating and compressing out the water, then collecting and pitching the solids. We cannot be quite that wasteful in a distant settlement. Grey water is water that doesn't contain lots of solids and isn't as stinky. This includes sink drains, drains in showers and bathtubs, and kitchen drains that don't have significant food waste. In practice in most residences, grey water is treated just like black water, with smaller pipes, and both are sent to either the municipal sewer system, or in rural areas to a septic system and leach field. In farms on Earth, all waste is treated roughly the same. Anything that flows like animal wastes is settled out, solids used as fertilizer, remaining water reused or released into fields or waterways. Solids that don't flow on Earth either go to recycling and landfill or are burned or composted.

What water evaporates goes into the atmosphere. Waste is not an option in space, and the farm can be a reuse engine, but habitats can recycle too. The typical outputs from humans by default look like Fig. 9.3, assuming very low water use. If the humans actually want to take a longer shower, wash their food well, or clean their clothes, the septic system would consume all of that water.

Not all water needs to go to the farm to get reused. Habitat air needs a certain humidity to be healthy, usually around 40% depending on many factors (OSHA 3430-04, see OSHA (2011)). People (and pets) exhale water vapor and release sweat, and animals such as people release water in metabolism. Dehumidifiers will be used to extract excess water from the air, which can then be sterilized using UV lights or similar and returned to the clear water supply as in Fig. 9.4. A smart use of walls, corners, and poles might also include growing vines and spice plants inside the habitat.

If these plants use grey water, they will also add to the water in the air as they transpire, which again can be captured by a dehumidifier. Each plant acts as a mini-water purifier. Since grey water is a potential resource going directly back to the habitat, it will be key to make sure biodegradable soaps and products are used by visitor and settlers alike. Any drain that captures grey water will also have to be outfitted with filters or settling tanks to collect solids (such as skin cells, hair, dropped contacts, etc.) which get sent to the black water cycle, and UV sterilizers to prevent contamination. Grey water can also be heated to evaporate clean water, or using reverse osmosis to separate water, and the remaining sludge sent as black water to the farm. In any case, careful control of the habitat-farm water balance may require

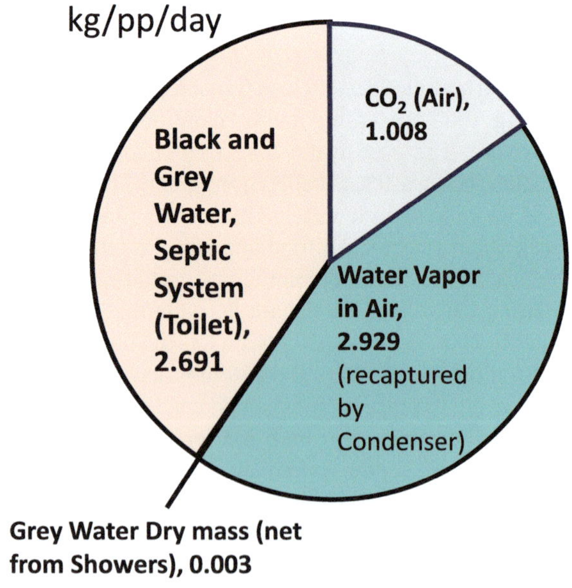

Fig. 9.3 Habitat water and waste handling per person per day. Using Meyer (2020)

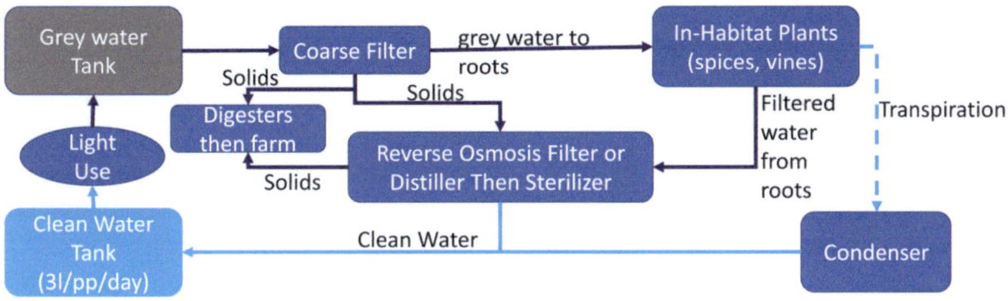

Fig. 9.4 Role of in-habitat plants for water

grey water to be fed to the farm, using complex control logic and control systems, like those in chemical factories now. Further, inside various stages, dehumidifiers can directly cycle condensed water (clean water). As plants transpire hundreds of times the water they absorb, it is key to making sure this dehumidifier to root cycle works well.

Black water, added to food wastes, and wastes from the farm, becomes inbound mass to the farm as a resource. From the habitat, it is likely that macerating pumps (i.e. pumps that grind up solids in liquids, such as a garbage disposer) will send waste sludge of black water and wastes to settling tanks that buffer mass flow into and out of the farm. These tanks will feed the first tanks of the Yeast-Bacteria stage. The possible pathway in Fig. 9.5 assumes all condensed water flows through the farm then back to the habitat, though condensed water will likely be directly reused in the habitat. All the numbers in Fig. 9.5 are from data in Meyer (2020). For Fig. 9.5, most of the water cycles through the condenser, which is evaporated sweat and water in breath, and ignores shower water, cooking water, or other grey water.

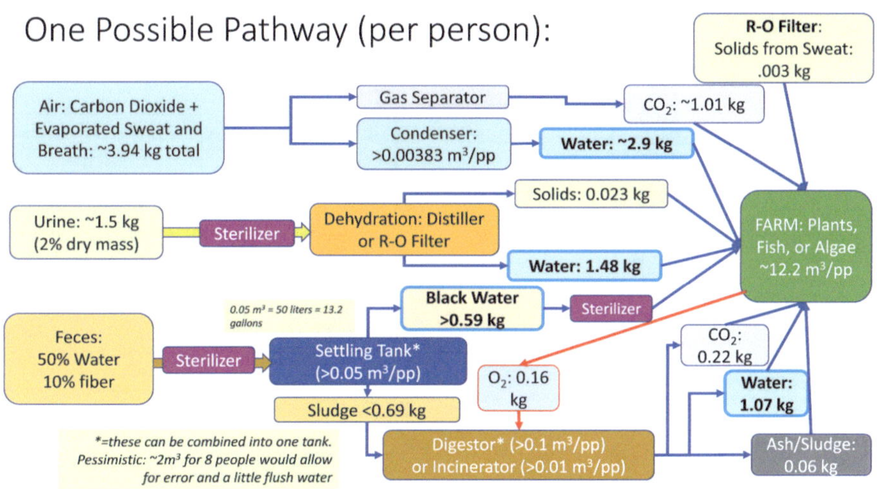

Fig. 9.5 One scenario for water treatment using the space farm

It will be critical that some combination of radiation or heat is used to break down pharmaceuticals in black water before they enter the bioreactors or other stages. Pharmaceuticals and some diseases are already a concern for many water systems on Earth and given the life and death criticality of the mass cycle in the farm and risks to organisms in the farm, black and grey water sterilization are a must. A wide variety of drugs, steroids, and hormones can wreak havoc on some organisms but can be broken up using radiation (including intense UV), heat, or other means.

Grey water sources are similar to those in Fig. 9.6. If grey water is pre-processed by filtering a bit to get out any clogging elements, and sterilized, it can be used for both watering in-habitat plants, and for uses like flushing toilets. Even so, some of the settled things in the grey water must get sent to black water processing as in Fig. 9.7.

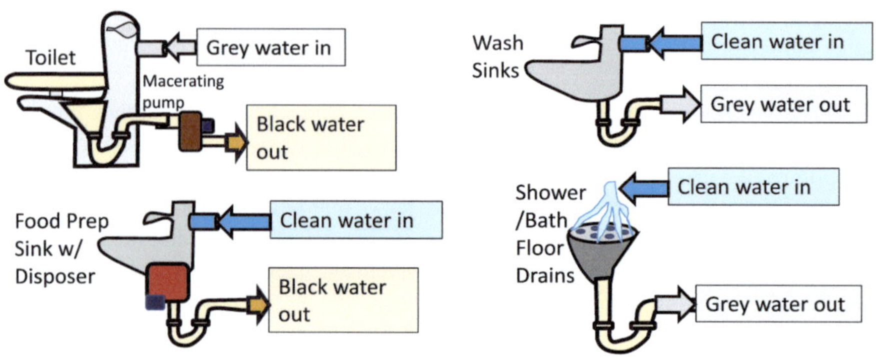

Fig. 9.6 Grey and black water sources

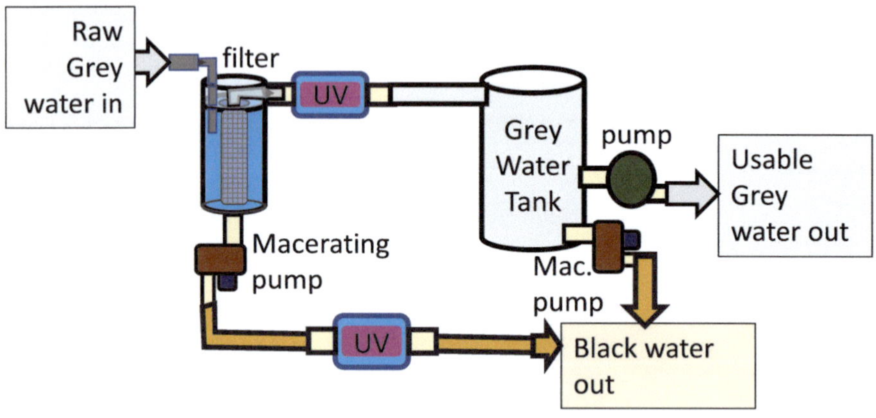

Fig. 9.7 Wastewater processing for grey and black water

An alternative given an excess of energy is to fully process the grey water into clean water by adding a reverse osmosis filter to clean out the useable grey water as in Fig. 9.8.

An incinerator is a valid option for waste breakdown to carbon dioxide, ash, and water for settlements with limited space or very low gravity. In Fig. 9.9, excess water could be extracted from the settling tank and then distilled, with the solids of that distillation sent to the incinerator in another loop. The condensed purified water is sent back to the habitat or to the farm based on mass flow needed.

Space does provide radiation in many forms for free, and in some settlements, a simple set of clear glass pipes exposed to hard sunlight might do the trick, followed

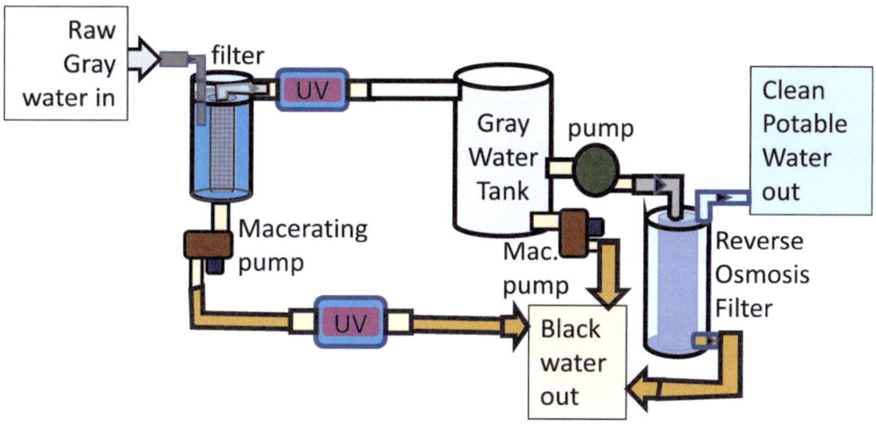

Fig. 9.8 Example of wastewater processing via direct reuse of grey water

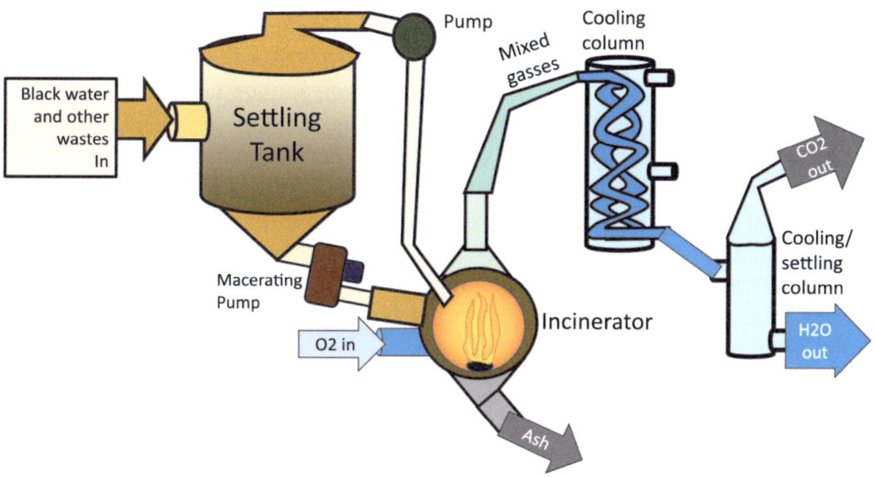

Fig. 9.9 Incinerator processing schematic for black water (solid and gaseous wastes)

by heat. In a bioreactor (e.g. digestor) option, the incinerator is replaced or supplemented by a sequence of bioreactors, which may include substrate (i.e. membrane mediated) designs. In the first tanks, complex control systems, agitators, and aerators will aid bacteria to begin the breakdown of solids into liquids and gases, and in many scenarios, eventually membrane bioreactors will use tailored bacteria or fungi to turn wastes into carbon dioxide and water. It is also possible that sludge from the initial tanks is pulled and sterilized to either make soil or to feed waste-consuming fungi and animals. Minerals such as calcium will accumulate in sludge and processing, especially in lower gravity environments.

The initial sterilized liquids from fermentation in the first tanks can also produce nutrient-rich water for plants and algae in the farm. Sludge from tanks can also be dried some then used to produce mushrooms or similar. More on what happens to water is covered for each type of stage and in the waste management section below.

9.2.3 Settler Dynamics and Habitat Mass Flow

Calculating the ins and outs from the habitat to and from the farm will depend on who is settling and who is just visiting. While much of my work has assumed a static population of equally sized average adult humans, settlement mass flow will vary based on the presence of kids and pups, presence of food vacuuming teens, and how much mass leaves on the waistlines of tourists. Luckily, life is resilient. A wise space farm designer will plan for the worst case, then assume food storage and export. In many cases, food can be frozen for a year without major loss of nutrients. If the tourists get the freshest chow, the settlers can occasionally tolerate the frozen or dried food that will act as a buffer. Additional dynamics such as gravity level or activity level affect mass flow to and from the settlement. Nutritional data assumes Earth gravity, and a degree of laziness. Assuming a true 'greenfield' settlement, it is possible that there will be a burst of reproduction, increasing mass flows, requiring in situ or shipped mass. Even without reproduction, settlers will have to be busy just setting up the farm and habitat and may need more calories than at steady state, especially if a transient construction crew is used. Seasonality may play in tourist-heavy settlements, as holiday times bring visitors. All this can further affect menus. It is possible one culture sets up the settlement, another emerges as the settlement evolves, and a third culture visits, all with different dietary requirements and ratios of adults of various ages to children. Tailoring the output of the farm to meet these competing demands will be a challenge and require expertise in various religious, social, or need/structure dietary requirements.

Table 9.1 Revised (notional) human model assuming a condenser loop in habitat

Component	Mass (kg)
Inputs	
Food (dry mass)	0.59
Oxygen	0.84
Water in food	1.45
Water to replace black water	1.14
Total inputs	4.02
Outputs	
Carbon dioxide	1.01
Wastes dry mass	0.32
Water in wastes	2.69
Total outputs	4.02

9.2.4 Revised Human Model

Given the ability to recycle inside the habitat to some degree, at least for condensing water from breath and sweat, and assuming urine and feces become black water sent to other stages, we can get a revised human model. The revised model in Table 9.1 assumes the in-habitat processing for grey water and black water in this chapter and now shows how the farm views the habitat.

This model of course is just a rough model and would need to be adapted and corrected for various conditions, menus, and settlers.

9.3 Aquatic Animal Stages

Aquatic animals include fish, shrimp, and mollusks. Stages for these animal crop therefore feature grow-out tanks which are growth chambers, with extensive management of liquids using buffer tanks/columns, pumps, and filters. These devices are arranged and connected into the stage for each crop.

9.3.1 Tank Types and Dynamics

As noted in previous chapters, there are two primary aquatic animal holding tank types, round and raceway. While this section is loosely based on my own experience (see Meyer (2019)), a great reference is (Bronzi & Parrini, 1986).

9.3.1.1 Raceway Tanks

Raceway tanks replicate a stream and have flow from one side through a trough to the end. In implementation, the current comes in via a spray bar, so water under pressure is pushed through a pipe with holes or spray nozzles into the water from above the surface, pushing air from above the tank into the water. This of course means the air above the water must be maintained at atmospheric oxygen levels. The spray bar might be supplemented by, or replaced by, with oxygen injector aeration bars at various points in the raceway all the way to the end. At the end, a screen prevents animals from exiting the raceway and then pipe takes in the water flow to recirculate it. The filters may be replaced with a loop through a hydroponic root bed, or water plants, making an aquaponic system as in Fig. 9.10.

Historically, raceways on Earth were add-ons to a stream, where water was channeled from upstream to the top of the raceway, then water returned downstream to the stream, and gravity was the force to move the water along the raceway. The fountains of L'Este in Tivoli, Italy, are powered like this, as is the Shepherd of the Hills Fish Hatchery in Branson, MO, USA. These arrangements are open format and won't quite work in space.

Raceway length is a tradeoff. Animals require the right mix of partial pressures, chemical concentrations, and volume for swimming space, to grow. While raceways can provide access to enough oxygen, raceways are efficient because they encourage the fish to orient the same direction, and flow pulls away carbon dioxide— and nitrogen-rich wastes. If the right species are selected from fish that like to swim in tight schools, raceways are a way to reduce volume, at the cost of extra energy to push and cycle the water. Even so, a few dynamics for raceways emerge. Fish and shrimp need energy to maintain position in the raceway, so careful current management is required to not overwork the animals to the point that they have no energy left to grow, or that slower animals end up squished against the screens at the end.

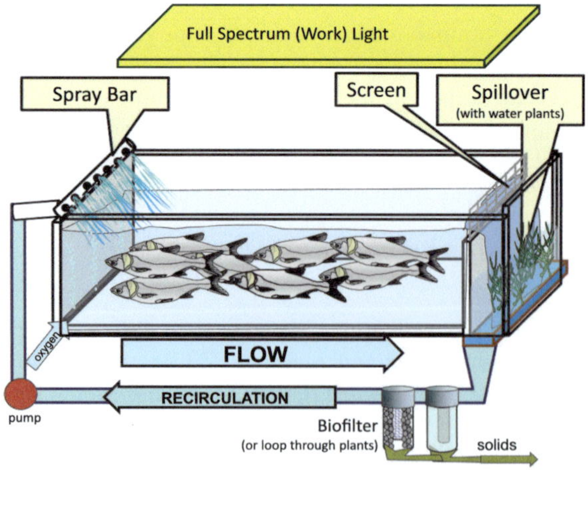

Fig. 9.10 Simplified raceway schematic

A too long tank will have slow low oxygen water at the end, and fast water at the front. Additionally, some fish can miss out on food if they are overcrowded, and due to tight packing get shoved away from food and light, so raceway depth is a tradeoff as well. Designs need to make sure flow is consistent throughout the cross section of the tank. Feeding in the front can result in jockeying for position by fish downstream, which is not a bad thing if not excessive, but can result in stunted fish downstream in extremes.

Raceways do not have to be smooth troughs. The sides of the raceway should have a surface to allow bacteria, algae, and ideally snails, and maybe isopods, to attach as in Fig. 9.11. Bacteria will use some of the oxygen to balance the biochemistry of the tank, as will some algae growth, which will consume some of the nitrates and add oxygen. Snails will keep the algae in check. The cost of these wall hangers is that they can overgrow and clog screens and filters. Deep bacterial mats can lead to animal infections. Isopods can be used to partially mitigate overgrowth of bacteria. Overfeeding of the animals can lead to overgrowth of bacteria. It might also be possible to add a secondary crop in polyculture to the bottom of a raceway in gravity. Mussels can help clean water, and some crustaceans, e.g. shrimp or crayfish, can provide a second crop and clean the tank if they are not a threat to the primary crop animals. Note also that a boundary layer of calmer water forms along surfaces of the raceway, so slower flow allows these additional organisms to survive, directly proportional to the speed of current in the raceway.

Geometrically the volume of the raceway is a classic rectangular box, with volume of length x width x height. Useful volume is a subset of this volume, usually around 80–90% depending on species and flow. Above the raceway will be a volume of air, roughly at least the same size as the tank itself. It is common for a 10:1 or more length to depth ratio. The walls of the tank are dependent on material and gravity field. If you care to look at the numbers, walls along the side have areas of length x

Fig. 9.11 Algae and bacteria growth in a raceway in Branson, MO (see shape in front of fish). Photo by author

height (two of these) plus a bottom section of length x width, plus two ends of areas width x height. Total tank mass is found by multiplying each area by the width of the wall, and the mass of the walls are the volumes of the walls times the density (d) of building material i.e. (2*l*h + l*w + 2*w*h)*d. Assuming polycarbonate plexiglass for example, which is around 1.2 g/cm^3 (or 1200 kg/m^3), assuming tilapia packed in (very optimistically) at 50 kg/m^3 and assuming 500 kg of living tilapia (around 1000 adult fish) we can do some simple calculations. Similar numbers and calculations can be found in these sources if you want a second opinion: Kungvankij and Chua (1986), Sinha et al. (2012), Altan (2012), and Green (2013). Let's assume the width and height are the same, around a meter. At 10 m^3 required (i.e. 500 kg/50 kg/m^3), the raceway will be 10 m long. Assume the walls are 3 cm thick, then the walls will have a mass of around 1224 kg. Fill it up with water, and the volume of 10 m^3 will need at least around 11,000 kg of water and minerals, plus around 50% more for buffer tanks and side parts. The raceway will take up at least length*width * 1.15 in area or around 11.5 m^2 (i.e. a living room).

The end of a raceway post screen might have a section with water plants to partially clean water before the mechanical pumps pull the water into the pipes. Plants should be of types that do not fragment into the filters.

9.3.1.2 Round Tanks

Round tanks replicate a pond on Earth and can be circular or ellipsoid in cross section from above and are usually cylindrical. A common type of recirculated tank is a home aquarium, or in a larger scale, the above ground swimming pool. For fish and shrimp, the round tank as in Fig. 9.12 creates a whirlpool flow with a spray bar on the side and an intake pipe with screen in the middle (i.e. the center of the vortex).

Like in raceways, current control is critical. For species that must swim long distances to be healthy, the circle creates an infinite treadmill; however, the densities in round tanks are lower than raceways in most cases, since the animals gravitate to

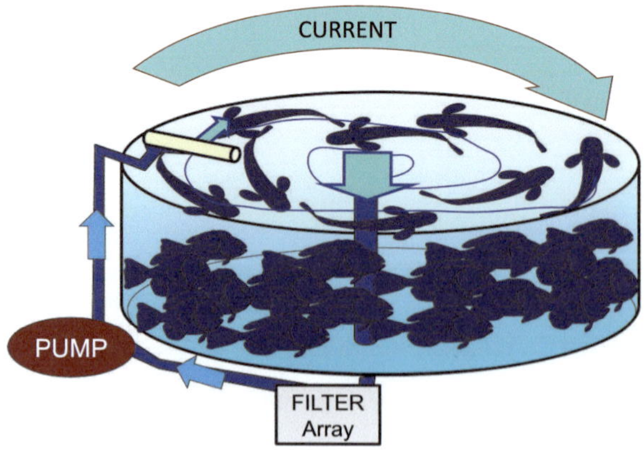

Fig. 9.12 Round tank flow schematic. Fish face the circular vortex current made by the spray bar. Water drains out the skimmer pipe in the middle

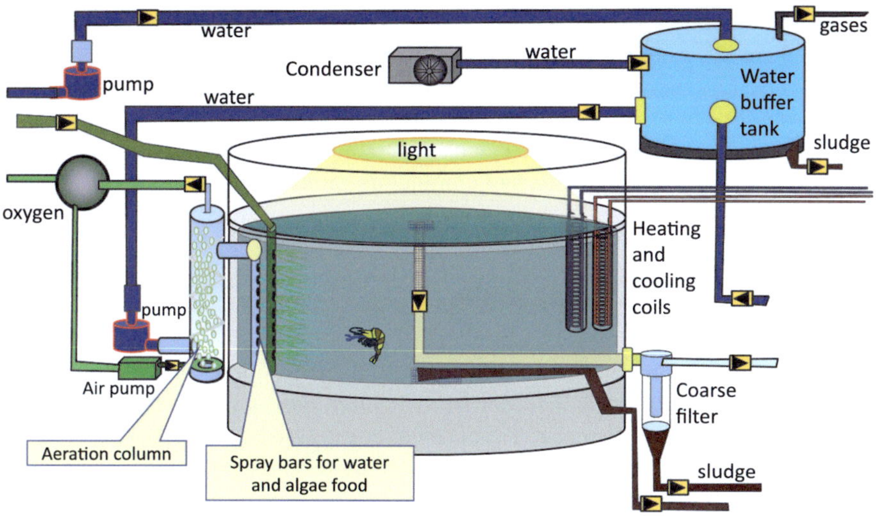

Fig. 9.13 Round tank schematic with simplified support equipment

a single depth and form a single school. Feeding is simpler in that food injected on the sides and top has multiple chances for the animals to eat it before it settles to the bottom. Cleaning the bottom will be more critical in a round tank versus raceways, so that debris does not accumulate too deep in the center or around the edges.

Geometrically, round tanks have volumes of pi*radius*radius*depth. Walls will have an area of 2*pi*radius*(depth + 1), multiply by thickness and density to get mass of the walls. In the example above, assume tilapia can be packed at 30 kg/ cm^3 in a round tank (same polycarbonate material and thickness) 500 kg of living fish. Assume a 1.5-m depth since it is common for round tanks to be deeper than raceways. I need a volume of around 16.7 m^3 (= 500 kg fish/30 kg/m^3). Radius of the tank is then SQRT [volume/depth/pi] which is around 1.88 m. I will need around 18,370 kg of water and minerals in the tank. The walls will mass 2*pi*1.88*(1 + 1) * 0.03 * 1200 = around 850 kg. The area covered by the tank is 2*pi*radius^2 *1.15 or 25.5 m^2 (i.e. a big living room). If the fish however could be packed at the same density, the volume of the tank is the same as the raceway, but the round tank takes up a smaller footprint. The support tanks and other devices for a round tank are shown in Fig. 9.13, which add to the footprint.

9.3.1.3 Tradeoff: Round Versus Raceway

A few tradeoffs are required to choose a tank shape. If area (footprint) is the primary concern, round tanks are more efficient. If the species are oxygen sensitive, raceways work better. Some fish need to do laps in a round tank, while some do better if they are in a riverine environment as in a raceway. It is likely round tanks will be the first choice for early settlement just due to the lower wall mass versus water volume contained (in the example the round tank's walls were 70% less massive). Round tanks are also less energy intensive for heating and cooling and require shorter piping. It is also possible that packing in a raceway is more efficient for mass, so the 50 kg/m^3 rule might be 20–30 kg/m^3 in a round tank depending on flow. This trade is amplified when looking at age composition and continuous harvest.

9.3.2 *Breeders and Tank Staging*

Even without trying to always have fresh food, there are more than just a tank set for harvestable crop animals. Breeder animals often need different conditions than simply growing animals for mass. Tilapia breeder fish (see Fig. 9.14) for example are older and larger than harvest sized fish and need more space for digging nest craters in sand or gravel than provided for crop grow-out.

Shrimp are similar, in that breeding shrimp need extra room and longer time to reach breeding size. Once fish are hatched, they need to be separated by size cohort to prevent predation or crowding by larger fish. As a result, animals will need a sequence of tanks, culminating in a final grow out tank. If we want always fresh fish, then a continuous harvest approach requires even more space. A last consideration is keeping a refuge population of animals in case disaster or disease strikes the main crop, though this can be a frozen set of gametes at minimum. Ideally, multiple tanks with sterilizers to prevent cross flow allow isolation if a population gets sick or is lost. This concept is not unlike the continuous harvest concept.

Fig. 9.14 Tilapia fish on nest (Titusville, FL, USA). Photo by author

While many tanks beyond the final grow out are required, the extra tanks are usually a fraction of the volume of the grow-out tanks. Few breeders are needed to produce many crop animals in the case of fish or shrimp. Each breeding couple can produce hundreds to thousands of juveniles for grow out. Even with a mortality rate of 10% to maturity, the production of juveniles given even a small fraction of breeders will be sufficient. Trickery may be required to make the breeders produce at need, such as light timing and spectrum, or use of hormones, but these factors are well known and in use now for most crop species. Also given the fast growth of crop species, the volume of extra tanks is manageable, usually a small fraction of grow out tank sizes.

9.3.3 Age Composition in Tanks

Fish and shrimp can't typically be in mixed age tanks, with exceptions for nesting. Plant and algae eating fish can handle living with younger and older fish, but omnivorous animals, especially species like catfish, trout, or prawns, cannot be in tanks of mixed aged animals, or the smaller animals will 'disappear'. I knew fish farm newbies that had a single raceway of rainbow trout. As they pulled out 30 cm fish, they replaced them with 6 cm fish, then wondered why they were missing fish in their count. Turns out that for many species, nothing is as tasty as a small member of the same species, rainbow trout notoriously so. The small rainbows became a favorite food of the larger rainbows. Prawns are even more cannibalistic. A slightly weaker or smaller prawn will be dismembered and gobbled up by their neighbors if the tank is crowded, even if food is plentiful. Any prawn that dies is recycled into the bellies of their neighbors as well. This mortality must be added to any other mortality source in your population calculations and increases based on the nature of the animals, very low for tilapia or silver carp, very high for prawns or piranhas.

Affects due to age mixing are not just as obvious as cannibalism. Even in more passive species, smaller animals can get pushed away from food, or intimidated and stressed by larger animals. This will result in slower overall mass accumulation for the crop, but can be minimized given a good food dispensing strategy (i.e. making sure all animals get enough) and adding volume to the tank.

As a result, to implement continuous harvest, a space farm will need to have a range of age segregated tanks or screened off areas, usually two or three for tilapia or carp, four or more for more aggressive species.

A scenario for tilapia or silver carp might in one-month to three-month cohorts, with a very small tank for hatching and fry, a backup tank of alternate breeders, and a further tank for breeding fish. Using the 30 kg/m^3 rule, assume it takes 180 days for a tilapia crop, and the settlement needs 100 kg of fish at 180 days, or roughly 0.56 kg/day (about a fish a day). Assume 5 kg breeders. This is also the minimum mass to even have a tilapia fish crop. This assumes no mortality of course, which would add about 3% to all totals (to account for 3% loss). For a big bang harvest,

Table 9.2 Round tank cohort scenario table

Tank	Volume (m^3)	Volume (l)	kg fish in tank
0–14 days	0.0042	4	0.13
14–30 days	0.0197	20	0.59
30–60 days	0.0672	67	2.02
60–90 days	0.1313	131	3.94
90–120 days	0.2164	216	6.49
120–180 days	0.8010	801	24.03
Breeders	0.1667	167	5.00
Total	*1.4064*	*1406*	*42.2*

this would mean a big tank of 3.33 m^3, or 3333 l plus a baby fish tank of about 10 l, and a 167-l breeder tank.

To explore how the continuous harvest model applies, the same scenario works out a bit differently. We might likely still use just a few tanks for this small case, but let's pretend we had to optimize. Given cohorts of 0–14 days, 14–30 days, 30–60 days, 60–90 days, 90–120 days, 120–180 days, and a breeder tank with 5 kgs of fish. The table looks as in Table 9.2, with a total of less than half the big bang harvest model.

To match the tanks to the table, relative tank size looks somewhat like Fig. 9.15.

Raceways would be a bit different. Packing can be tighter, to around 40–50 kg/m^3, though it could be more like 10 kg/m^3. Assuming 45 kg/m^3, the raceway table is Table 9.3.

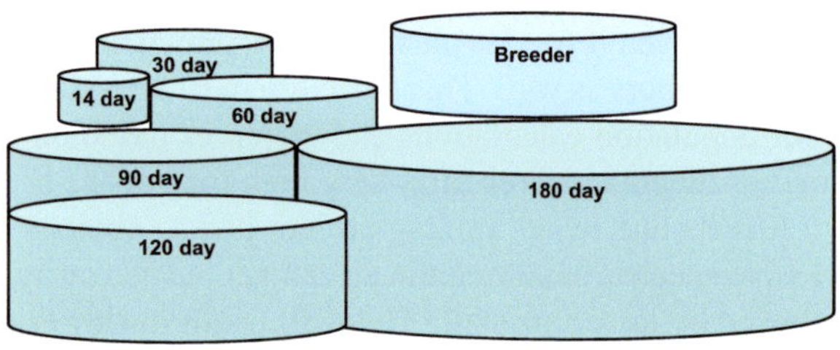

Fig. 9.15 Relative tank sizes for round tanks

Table 9.3 100 kg per 180-day raceway scenario

Tank	Volume (m^3)	Volume (l)	kg fish in tank
0–14 days	0.0028	3	0.13
14–30 days	0.0131	13	0.59
30–60 days	0.0448	45	2.02
60–90 days	0.0875	88	3.94
90–120 days	0.1442	144	6.49
120–180 days	0.5340	534	24.03
Breeders	0.1111	111	5.00
Total	*0.9376*	*938*	*42.2*

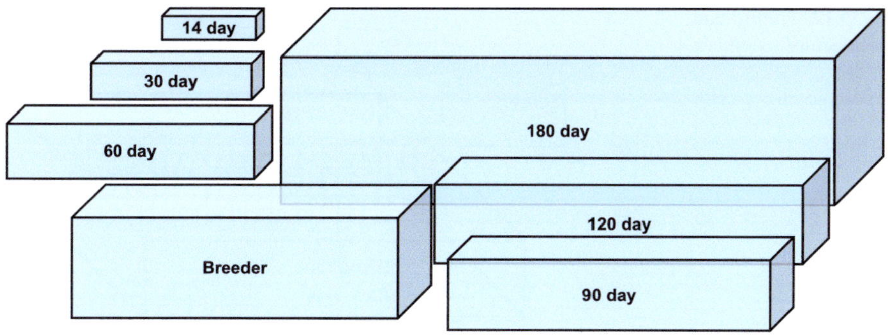

Fig. 9.16 Relative tank sizes for raceways

Relative tank sizes look like Fig. 9.16 for raceways. The relative volumes for each class are similar to the ratios for round tanks. The (120-)180-day tank is bigger in volume than the combined size of the 14, 30, 60, 90, and (90-)120-day tanks.

There is a problem with this diagram. Breeders for tilapia likely need a round tank, not a raceway, so even if crop fish are in raceways, you might still need a round tank for the breeders. The raceways are smaller in volume than round tanks but will require more energy due to flow. It may be that smaller fish are more likely to be deprived of food by slightly bigger fish in a raceway versus a round tank for tilapia. For silver carp, it is possible the raceway increases growth since they are filter feeders, and fewer tank cohorts may be needed. The space farm engineer will need to consider factors like these when doing cohort layouts by age, and trading off round tanks versus raceways.

As stated before, a 100 kg crop in 180 days grown for very passive fish like blue tilapia could use just 1 grow out tank, with a breeder tank and baby fish tank. That said, a larger harvest of 1000 kg/180 days or more might use something like the scenario with many tanks, a set for each cohort. If we swapped tilapia for a more aggressive fish, the many tank scenario applies, as noted previously, bigger fish (or shrimp) happily eat smaller animals of the same species.

Given the tilapia case, there is a compromise architecture using partitions as in Fig. 9.17. Fish can get stressed when moving them, to minimizing tank count can be important, especially for small total crops. Simple screen partitions add surface area for bacteria and algae and can let smaller fish get to food easier. In this case the design might be as in the figure below, at roughly 2 m^3 total tank volume. This design minimizes complexity and fish moves yet allows growth.

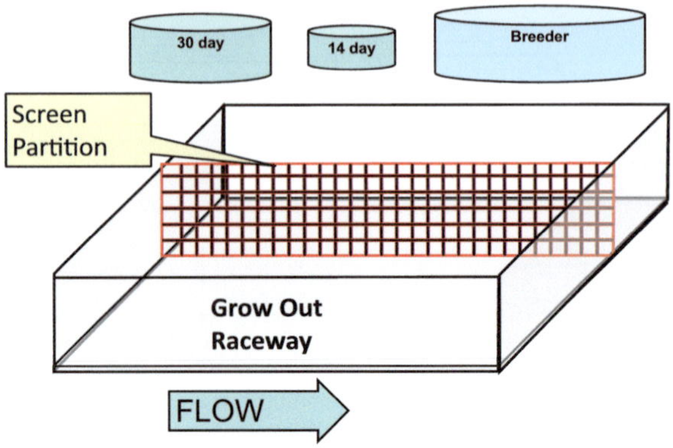

Fig. 9.17 Hybrid tank architecture schematic

9.3.4 Buffer Tanks, Filters, and Pumps: Systems in the Stage

Ordinarily a fish or shrimp tank can operate with just a filter set, spray bar, and temperature control. In a space farm, the same tank might have other parts due to its role as nutrient source for photosynthesis, and as a recipient of wastes fed to fish and shrimp.

The most complex form may include a variety of machinery around, and connected to, the tank. However, the form could be as simple and small scale as the set-up I used to make my shrimp grow-out, as in Fig. 9.18. This simple arrangement features an electric heater, overhead light, spray bar, aerator, and recirculation pump loop (via center intake) through a biofilter and fine filter. The air over the tank needs to be kept at the right oxygen balance so that oxygen permeates the tank, and the air needs to be monitored for ammonia.

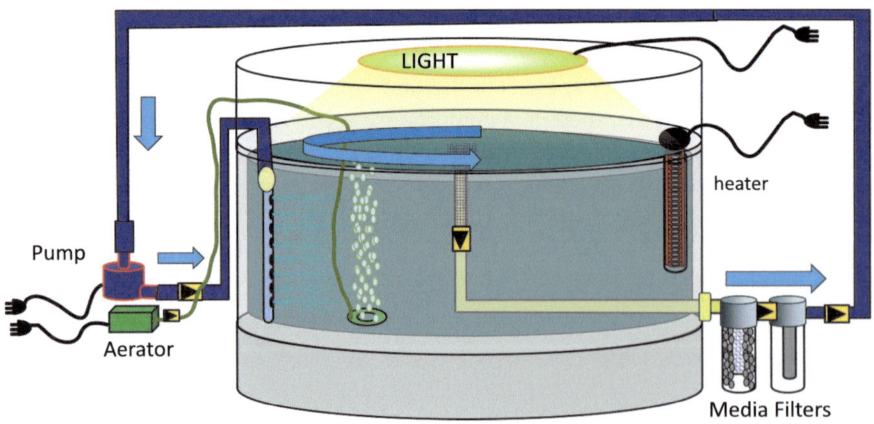

Fig. 9.18 Minimal small round tank

Typically, at a larger scale, it is advisable to have a buffer tank in front of flow and behind flow to account for variations in flow from other stages and systems. Intake buffer tanks take recirculated water, and water added from other stages in the farm to keep an even flow for the water animals. The buffer tanks (columns) altogether hold roughly at least half the volume of the main tank/raceway, to maybe a day's input. A primary and secondary pump for each tank or set of tanks is needed to push pressure into the tanks and is sized to the flow rate for the raceway or round tank, typically oversized to reduce workload on the pump. Aerator pumps are used to add oxygen to the water pushed into the animal holding tanks, using a column or in one of the buffer tanks, or even in the round tank or raceway itself. Aerated water needs to have fully dissolved bubbles, if possible, because water with too many bubbles can clog gills with bubbles and suffocate fish! Aerators and pumps need to be controllable to make sure the right amounts of oxygen and flow end up in the tanks with animals.

Water exiting the tank will need at least a coarse filter to catch clumps of wastes and scales or shells, and humans or machinery will need to flush this coarse filter, usually screens or gravel, to keep it from clogging later filters and pumps. A tank accepts the result. Settled wastes are pulled out by a macerating pump and sent to photosynthetic or bioreactor stages as needed, while the water is cycled back into the animal tanks (recirculated) via the input buffer tank.

Controls might also send wastewater to the roots of hydroponic plants or the algae bioreactors if it is too rich in ammonia or other wastes for recirculation. In the simplest form, an engineer may replace several mechanisms leaving the animal tanks by a simple pump and loop through the root bed of a hydroponic garden, creating an aquaponic system as in Fig. 9.19.

In most cases however, gases that emerge in the tank are separated, carbon dioxide sent to photosynthetic stages, water condensed and sent to input tank or if in excess sent to other stages, though typically water will be needed to grow fish and shrimp and lost in wastes, is above any generated in aerobic respiration.

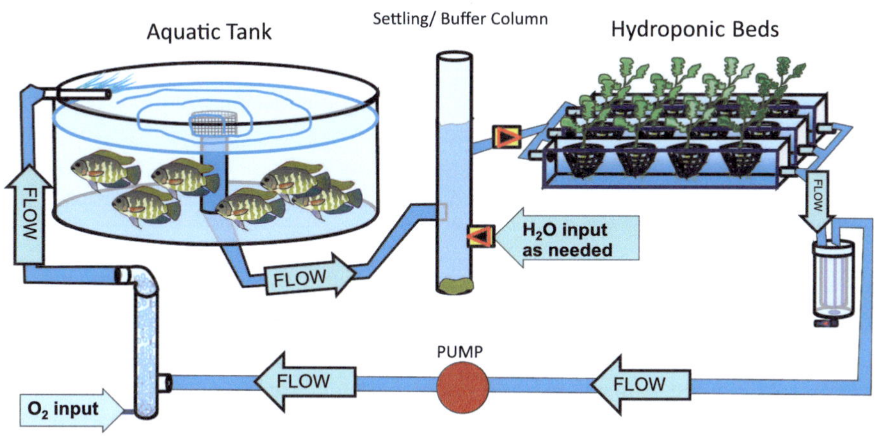

Fig. 9.19 Simplified aquaponic flow schematic

It should be noted that the air above the tanks will need to be dehumidified at some level to prevent mold growth and fog and be cycled and circulated to allow humans to work the tanks and need atmospheric oxygen pressures for human workers.

9.3.4.1 Use of Temperature Control and Lighting

Temperatures are critical to animals and plants. Metabolism in animals in tanks is controlled by water temperature, and to optimize growth temperatures need to be maintained in ranges for each species. For example, tilapia are tropical species, and grow best in temperatures in the 20 °C range (see (Henson et al., 2018), and (FAO, 2022). Invertebrates can be very temperature sensitive and stunt in too cold water, which I witnessed for freshwater prawns. Breeding temperatures can vary from growth temperatures, and juvenile animals might need a different regime than adults. Given all this, it is likely a tank will be exothermic due to pumps, lights, and aerobic respiration, and need to be cooled using loops and exchangers. If somehow heat is needed and can't be sent from other stages by fluid-filled pipes and pumps (then use loops or fins at circulation points to warm the water), electrical heaters similar to those for commercial aquariums can add degrees, at the cost of energy to the stage. If tanks are set on ground in planetary farms, the ground will act as a heat sink, and in many space scenarios insulation will be required to keep the tank water from losing all its heat. If the tanks are in sunlight from windows, the tanks may require additional cooling during daylight cycles, in addition to algae growth handling.

Like temperature, lighting is important. Light is how most water animals orient, usually with the backs/tops facing light, so for the fish, up is where the light comes from. However, swimming directions aren't the only need for lighting. Animals locate food using their eyes in addition to scent, and light tells animals when to feed. Just like decreasing light to blue and purple wavelengths trigger yawns in humans, light trends, levels, and colors trigger behaviors in fish and shrimp. Most fish and shrimp in nature feed best in the hours before and after sunrise, and the hours before and after sunset, so setting light levels and spectra to the key feeding levels for animals can increase feeding and mass gain. Just like some plants, some animals need dark intervals to trigger hormonal changes and inactivity.

Since the idea in a space farm is to optimize growth of crops given mass and energy, it is a must to know the light and thermal needs of your animal crops at the various life stages of the animals.

9.3.4.2 Chemical Control in Tanks

As previously noted, salinity, pH, and mineral content are critical to aquatic animal health and growth. Freshwater environments are not pure water, but instead contain a series of salts and minerals in solution. Tank water therefore contains a solution of various dissolved gases and ions, including salts and minerals, which add to the mass of the tank water, and need to be maintained as animals are harvested, wastes removed, or food is added. Typically, when a tank is set up, a mix of salts and minerals is added, then refreshed from dispensers or by manual work as mass is removed from the tank. Freshwater prawns for example, need calcium carbonate to grow shells (see (Hicks & Pierce, 2022)). Calcium and chlorides are the most important, along with pH, which can't be too high or too low. Evaporation can concentrate hardness and salts in tank water, though this should be directly replaced in the chamber by condensers. The biggest loss of water and minerals with a condenser loop is therefore harvest and removal of solid wastes. A carp for example incorporates at least 9 g of minerals like calcium, potassium, and others per kilogram (from previous chapters), lost when the fish is taken from the stage. Usually, minerals are in salts, biomolecules, or carbonates when the animal takes them in. Any food not incorporated into the animal or released in as carbon mass in carbon dioxide ends up as wastes, significantly in solids that are removed. These two losses will require a constant balancing act to restore the water solution balance. To find out how much and when, commercial pH, salinity, and hardness sensors should inform control system software that can either tell humans to add stuff or have automated dispensers add to the circulated tank water (though the dispensers will need to be refilled too). Adding lime, a sea salt mix, or potash are common balancers. In most situations, the added mix is biggest if seasonal (big bang) harvest versus continuous harvest is used, or as animals are moved between grow out tanks. Adding food also alters the chemical balance, for example green algae send in roughly 1% of its dry mass as ash, some amount incorporated into animals, the rest as wastes or dissolved into the water. Some minerals such as limestone (calcium carbonate or $CaCO_3$) can buffer water chemistry to a degree, but limestone might be scarce at a place where there was never a living ocean. Note also that chemical balances are affected by, and affect, the levels of dissolved oxygen and carbon dioxide in the water, further complicating the balance problem, as salts and minerals chemically react in the tank.

It is not always an additive situation, however. Ammonia is a key factor for health and must be kept below thresholds. Ammonia will affect pH. As water evaporates from a tank it concentrates the minerals in the tank, including ammonia, though some ammonia may end up in the air above the tank. From personal experience, adding hard water to replace evaporated water over time can kill crops, especially ammonia sensitive animals such as crustaceans. Boney fish are somewhat less sensitive due to having kidneys.

What to do? Fortunately, nature has a way to smooth tank chemistry. Bacterial and algal colonies in the tank aid in keeping the balance, dying and growing based on chemistry, and are a core reason to manage the microbial culture in the tank. Cycling water through plant roots in hydroponics can help too but add additional considerations as well. Culture or roots do not eliminate the need for sensors and control systems, but they do aid in stability.

9.3.4.3 Feeding

When to feed and how to feed are criteria to grow animals and vary by species and life stage of each animal. Too little food will reduce growth and increase aggression, while too much adds to wastes which can decompose in the tank to affect methane, ammonia, and oxygen. Most of the time, a slight overfeed is used, and the farmer counts on the course filter to remove the excess. If live algae is a feed (part of a virtuous energy-mass efficient flow), it must be concentrated from the photobioreactor, then pushed by pump (a non-impeller pump in most cases) to near the spray bar area of the tank, and ideally pushed in a mesh through the water column, not just on the surface. Feeding is not usually a 24 × 7 affair, so for live algae feeding, a buffer tank will need to hold algae until the animals are ready to eat it, usually during 'daylight' hours.

For more solid foods, the process is a bit simpler. Waste biomass, or dried algae, or dried yeasts are mixed by human or robot in a chopping mixer, then put into a holding hopper/tank. A set of pipes and corkscrews or pumps move the food to the tank near the surface. The feeding spot can be just down from the spray bar during set times, which may be hourly, a few times a day, or somewhat continuous as per crop species. Tilapia may feed continuously during daylight hours, while common carp, catfish, shrimp, and trout might be fed two or three times during the day, based on levels programmed into the lighting system. Ideally a timer clicks on the feed dispenser system, and sensors (including cameras) determine if the animals are satiated, though a simple feed table might be used instead. The table usually shows how much food drops in based on mass and age of fish or shrimp, and water temperature.

9.3.4.4 Aerial Section Control Including Fans and Extractors

A volume of air above each tank, and around each tank is a must have. Access areas are needed to service the various machines around each tank, to get to the animals in the tank, to remove dead or sick animals, to clean the tank, and for moving animals to grow out or harvest. The area above the tank will have lights (essentially lights for the animals and work lights for people) to orient and regulate the animals and will need to be humidity controlled to prevent mold growth and return evaporated water to the tank. While aerators and spray bars for the tank will likely make this area breathable in most cases, it is possible it could also harbor high levels of methane or ammonia. From my own experience in losing a crop of prawns in a summer heated garage, the air above the tank can become unbreathable, or at least unpleasant, quickly. Therefore, the air will need sensors and be connected to pumps and fans to circulate the air out and in as required. A cheat may be to pump air from this section to aerate algae or plants to clean it. A few vine plants could be used to increase the suitability of this air, or the air above the tanks could be in a shared volume with a hydroponic garden.

9.3.5 *Overall Aquatic Stage Structure*

The overall aquatic animal stage structure will have a series of tanks, sized as appropriate for the species, and for the cohort (or life stage), with columns, thermal controls like heaters or heat rejection fins, and aerators as in Fig. 9.20. Above the tanks are lights for the animals, and the tank may have automatic feeders. A series of sensors are in the water and at various parts of the support tanks and pipes to ensure flow is maintained as needed. Beyond the tanks, the stage has tanks for clean water, tanks to hold oxygen or air and carbon dioxide extracted from the water. Waste tanks hold wastewater for use in hydroponics, or further settling and processing. Storage holds spare parts, and a fan and condenser manage humidity in the stage to prevent mold or corrosion. Above human head height are work lights that allow people to work inside the stage, and thermal control for the air in the stage makes the temperature acceptable for the humans and equipment.

9.4 Terrestrial Livestock Stages: Micro to Macro

Designing stages for animal crops that live in open air is simpler in some ways than for aquatic livestock, and usually the growth chambers have an atmosphere that is fully breathable by human workers, though may vary in levels of comfort depending on species. However, the air around animals will also be rich in dust from skin flakes, shell fragments, or feathers, and present a problem in air filtering and a flame risk. These and other issues are considered in this section.

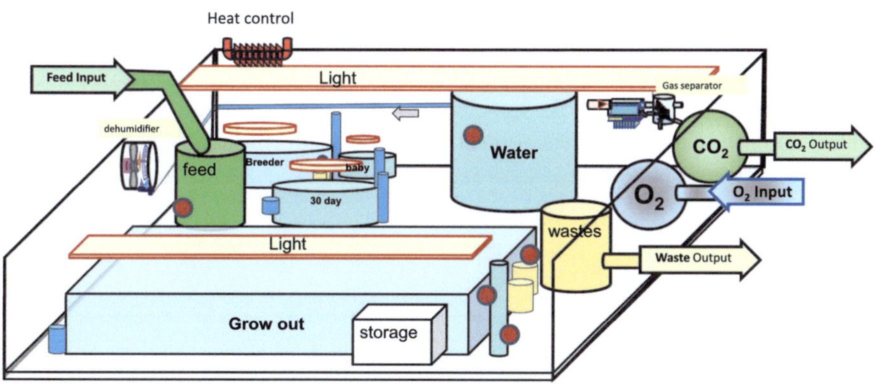

Fig. 9.20 Sample simplified aquatic stage layout

9.4.1 Insect Stages

While already noted in prior chapters, insects may not be a prime pick food for Europeans and North Americans, they are already on the menu for much of Earth. Given the right insect, an insect stage can be constructed in a small volume. The three model species are crickets, mealworms, and cockroaches, though there are many others discussed in prior chapters. In short, most species can be grown using an entire volume of honeycomb, egg crate, or mesh shapes as in Fig. 9.21. The shapes, made from cellulose or similar material, should have water drip taps, feeding tubes, and circulating pumps and fans. The volume will also need lights and thermal controls. Humidity and oxygen levels are very important to insects, and they will produce prodigious amounts of methane and ammonia that will have to be circulated out with carbon dioxide. The grow box is like the aquatic tank concept. Fans will recirculate air, adding oxygen as needed and extracting other gases to a plant or algae stage. Like aquaponics, the insect tank would ideally be paired with a sized photosynthetic stage. Vegetable waste or dried algae is dropped in between each wall to allow the insects to feed, and drip lines between walls will provide water drips via lines with pinholes. The interior walls are pitted like egg crates to allow larvae to grow, and a fan would circulate air to control humidity and ammonia and methane. Bee growers will recognize the arrangement. Wastes will need to be occasionally pulled out, usually during harvests.

Another type of farm for insects, especially mealworms, uses stacks of shallow bins filled with dried vegetable wastes, and a sifter is used during harvest to pull out worms and wastes as in Fig. 9.22. As in the diagram below, fans and sensors circulate air and ducts will need to bring in oxygen and remove carbon dioxide and other gases. The bins are dried and semi-dried vegetable wastes and semi-dried algae, and sponges for water fed by lines only as needed. Harvest and cleaning are by manual sifting, as is feeding.

Overall, the stage will need to be more than the box or bin growth chambers but will consist of machines like fans, condensers, tanks for water and wastes, gas control and separation, and heat control devices that are similar to larger livestock as in Fig. 9.23 just smaller.

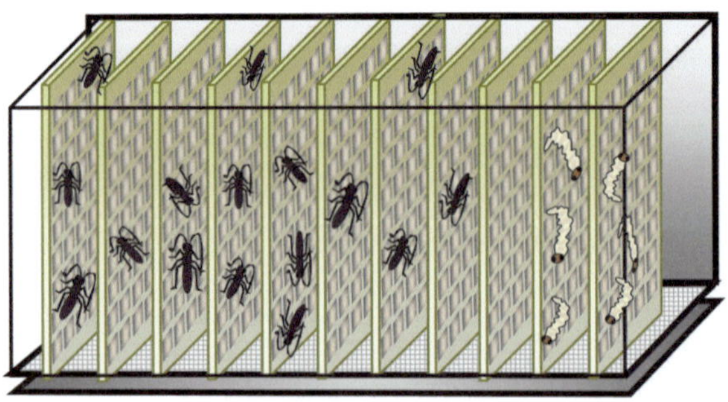

Fig. 9.21 Insect grow box

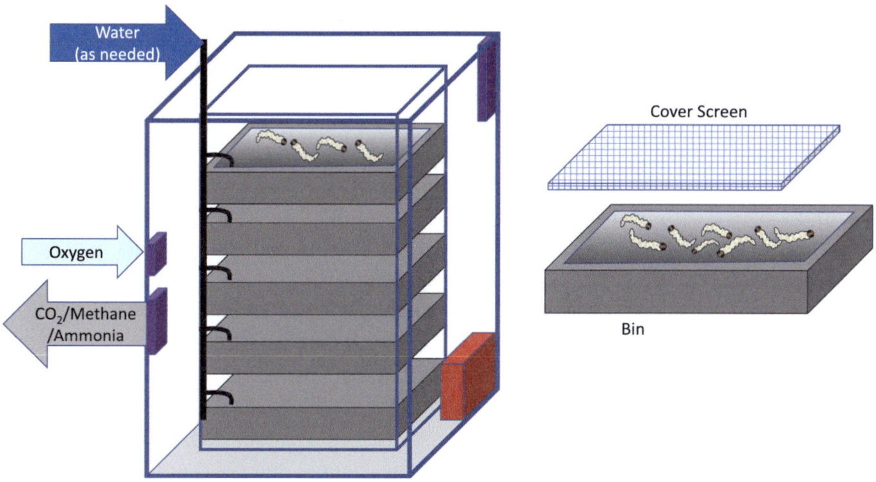

Fig. 9.22 Bin-based mealworm or cricket rack

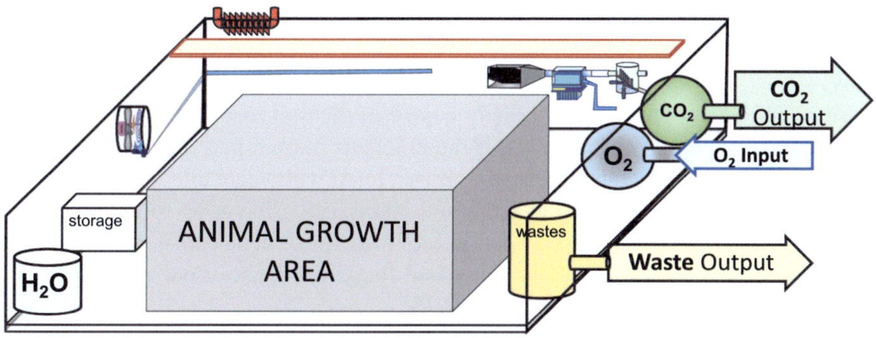

Fig. 9.23 High-level terrestrial animal stage schematic. Note heat rejection, condensers, and gas separation elements

9.4.2 Bird and Mammal Stages

As mentioned previously, it may take a very developed settlement before farm animals become important. The stage for terrestrial animals is like other aerobic stages, in that there will need to be tanks for oxygen, carbon dioxide, water, and feed, and waste collection bins and tank (since animals are not trained to use a toilet cleaning is not fun). As in the simplified schematic in Fig. 9.23, inputs are feed, water, and oxygen. It is likely that feeding and waste removal are manual tasks. Further, it would be a good idea to add plants to this stage to clean the air.

Smaller animals such as rodents or chickens will be available before larger livestock such as cattle, emus, and pigs. That space steak may be an import or printed from a cell culture for a long while, but an omelet with shrimp or grilled guinea pig may be sooner. One way to do terrestrial animals like chickens or goats is to let them roam in pastures for other crops grown in soil, where the wastes then can be directly added to the soil complex, but adding to human labor to control the animals and mix in the wastes. This is nearly the classic farm structure, where soil is planted with forage, and the animals feed upon that, requiring many square hectares. Manure is either collected or left to be recycled in the soil by a biome with insects and worms. The crops in the pasture would need to be tolerant of animals, or benefit by having them, such as grass, maybe with some trees or bushes. This farm model favors ground birds and goats, and especially egg laying hens. Another option is higher density pens, where food is provided for the animals from other stages, and the animals have just enough room to remain healthy given constant care. Super high-density pens, like laying cages and the like might seem sensible, but have drawbacks including disease epidemics, reduced mass yield, and reduced reproduction, leading to less happy animals. Micro-livestock, such as rodents and chickens are the easiest to grow in crowded conditions, in a mass efficient manner. Large animals might also be grown in a polyculture pasture, with cattle, chickens, and smaller mammals all sharing a large expanse over a soil biome as in my great grandmother's farm of chickens, guinea fowl, and cattle over a rotated area of grass, surrounded by fruit trees and berry bushes, though this classic farm in space is a very, very, long way away.

Since egg laying chickens will likely be one of the first medium-sized animals in a space farm, a coop is a must even if the chickens roam a park or pasture. Coops provide security of structure and nesting boxes lined with vegetable waste, allowing hens to roam outside the coop to feed, and come inside the coop to sleep and lay eggs. Each nesting box is mostly open in the front so the hens can step down to the floor, where there is supplemental water and food. The back of each box opens to allow humans to collect eggs, as in Fig. 9.24.

The coops are then combined with a small yard to allow the birds to roam, and likely a few plants will be in the same area to moderate air in the stage as in Fig. 9.25. These are combined with the rest of the stage machinery and tanks. Heat and humidity

Fig. 9.24 Simplified chicken coop for egg laying hens

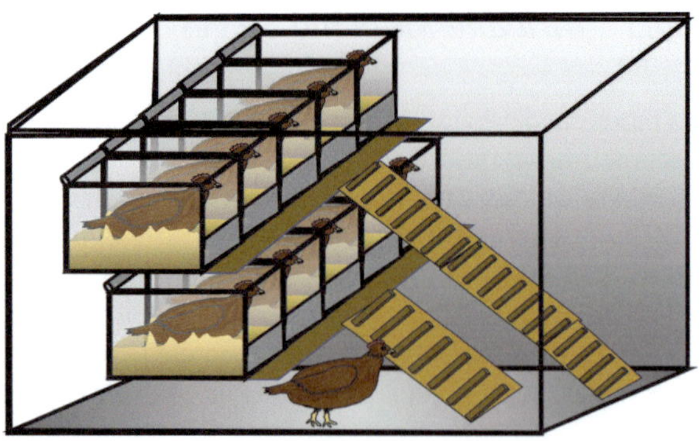

are controlled, sensors monitor temperature, gases, light, and each animal's health. Carbon dioxide is extracted, oxygen and water added as needed. Wastes are washed into a tank, as are dead animals that are not harvested for food, leftover feathers, and other solid waste, which are macerated and sent to the waste system.

Rodents can use a similar coop structure, though a simple cage works well for many smaller rodents such as guinea pigs, rats, and rabbits. Small rodents are a prey species and grow best in an area just big enough to stretch and hide, with lots of nooks and connecting tunnels. Rabbit cages linked together are called hutches, and a diagram is in Fig. 9.26.

As the atmosphere provided to mammals is in the same chemical balance as for humans, smells notwithstanding, the enclosed volume could be shared with the habitat, though likely not given smells (from urea and feces). It would be a good idea to put some of the herbs or vine crops, or even plants like peppers or bushes,

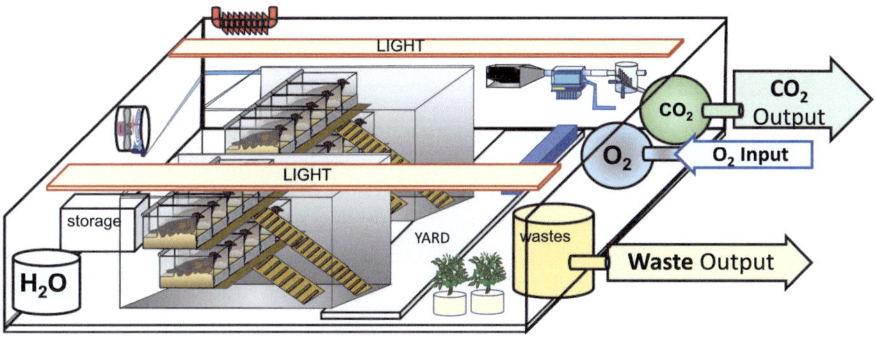

Fig. 9.25 Chicken stage schematic

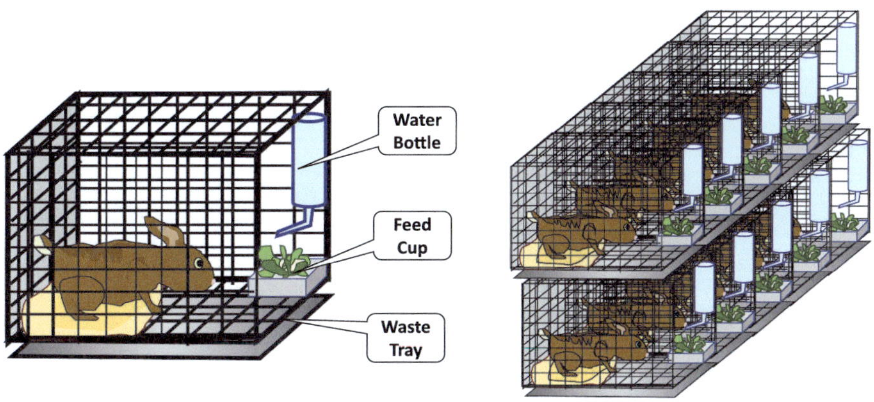

Fig. 9.26 Classic rabbit hutch cage layout

in the stage to clean the air. The enclosed area will need to be heavily checked for gas levels, humidity, and temperature, with air being likely pumped into plant growing areas to use the carbon dioxide from the animals. As aerobic metabolism also produces water, humidity control, using dehumidifiers will be needed to keep the animals healthy. Fans are needed to ventilate the animals, and to control temperature, and heat control fins and heat exchangers to circulate out excess heat, and infrared heaters may be needed to add heat. Natal and rearing areas may also be needed to increase reproduction, like egg incubators for poultry, or nursing pens for mammals. Often the natal areas have enclosures for offspring and mothers, with extra heat and air control to minimize mortality, and prevent stress. Your author has experience with sow piglet operations and can attest to the general meanness of sows when it comes time to vaccinate piglets! Other areas in such a stage will include veterinary areas for isolating sick animals, slaughter areas for processing (unless harvest is just milk or eggs), and areas to store feed and materials.

9.5 Mushroom Stages

Mushrooms are much like insects that don't move when considering stage design, though can be arranged like plants. On Earth, mushrooms and other edibles are either harvested in the wild or grown on a soil-like mix of worm casings, sawdust, or paper nuggets. Mushrooms and similar fungi have rootlike structures that invest whatever they sit on, slowly dissolving the material using colonies of cooperative bacteria and, as a result, may be grown on many shapes and media. Farmers can use sponges or substrates of various types with taps to add nutrient-rich water, or simple porous bags of dried plant wastes as in Fig. 9.27.

The sponges or substrates could be in rows like hydroponic trays on top of slowly circulated nutrient water, or simply fed by tap lines. The trays could then be arranged in enclosed tubes or boxes that have oxygenated air pumped in, and carbon dioxide-rich air pumped out. The fungi garden could be connected to a hydroponic stage

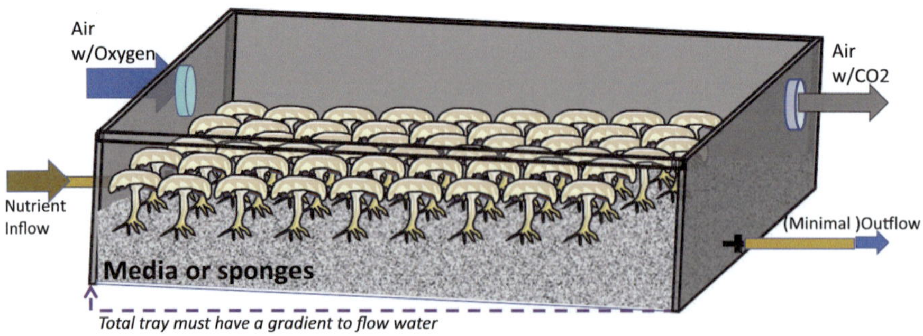

Fig. 9.27 Simplified mushroom growth chamber schematic

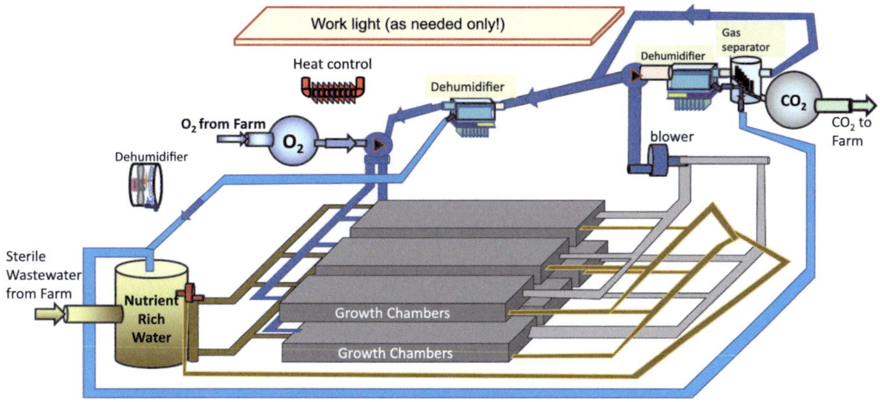

Fig. 9.28 Mushroom stage schematic

to cycle and reoxygenate the air, though mushrooms do not need, and in fact can be harmed by, light. This aversion to light is why mushroom farms are often in old mines. Temperature and humidity control are very important to fungi as it is for most life, so sensor arrays to control both, linked to humidifiers, dehumidifiers, heat exchangers, and fans, are must haves as in Fig. 9.28.

If soils are developed, deeper soil growing fungi such as truffles, can be developed in the dirt and dug up for harvest as part of the growing cycle of the overall farm. Some may require a tree's root structure, though others may not. Given a simple flag, a trained sow's nose would not be required to find the tasty morsel. Tourists might even seek Martian truffle soufflé for a splurge!

9.6 Summary

Overall, aerobic metabolism is the driver for both the human habitat and animal crop stages. As a result, control of the atmosphere in these areas, especially oxygen, carbon dioxide, and humidity, is key. Human and terrestrial animal areas have a cycle where condensers return water from breath and sweat, fungi from surface evaporation, and aquatic animal areas from evaporation from tanks. Inputs in all animal stages include oxygen, water, and feed, with outputs of carbon dioxide, harvested crops, and wastes. Buffer tanks for all gases and liquids are necessary because of changes in mass flow between stages, including the human habitat.

9.7 Questions

1. Why do animal and human stages have a loop of condenser (dehumidifier) to clean water back to human/animal?
2. How could grey water be treated differently than black water?
3. When and why would the Habitat-Farm system need an incinerator?
4. Why might reducing the average age of the settler from 50 years old to 25 years old affect the space farm?
5. What are the tradeoffs for raceway tanks versus round tanks? When might you have a stage with both types?
6. Why should fish and shrimp tanks have a wall biome of algae and bacteria?
7. How is light used in all animal stages? How can light improve production?
8. Why does a space farmer not mix 1 cm shrimp with 5 cm shrimp?
9. What is aquaponics, and how does it benefit fish or shrimp and plant production?
10. Why is it critical to replace evaporated water from aquatic animal tanks?
11. What are two ways to make growth chambers for insects?
12. Why is air filtering important for chickens or guinea pigs?
13. What are similarities and differences between growing mushrooms and growing crickets?
14. What are the similarities and differences between growing mushrooms and growing vascular plants?

References

Altan, O. (2012). Diversification of aquaculture activities and production methods. FAO—sub regional Office for Central Asia, Food and Agriculture Office of the UN, 2012.

Bronzi, P., & Parrini, F. (1986). Technological and energetic aspects of water management for intensive aquaculture. Technics used for intensive rearing and alimentation of fish and shellfish, Mediterranean regional aquaculture project, FAO, March 1986, found (2024) at https://www.fao.org/4/af011e/AF011E00.HTM

FAO Fisheries and Aquaculture. (2022). Aquaculture feed and fertilizer resources information system, species profile, Nile tilapia. Food and Agriculture Organization of the United Nations, captured 2022, found at https://www.fao.org/fishery/affris/species-profiles/nile-tilapia/nile-tilapia-home/en/

Green, B. W. (2013). Growing stocker channel catfish in a biofloc technology production system. *Aquaculture North America, 4*, 24.

Henson, M. N., Aday, D. D., & Rice, J. A. (2018). Thermal tolerance and survival of Nile tilapia and blue tilapia under rapid and natural temperature declination rates. *Transactions of the American Fisheries Society, 147*, 278–286. https://doi.org/10.1002/tafs.10023

Hicks, C. H., & Pierce, R. A. (2022). Freshwater prawn production in Missouri. University of Missouri Extension, April 2022, found 2024, at https://extension.missouri.edu/publications/g9471

ISO 2533:1975 Standard Atmosphere. International Standards Organization, Publication date: 1975-05, found 2022 at https://www.iso.org/standard/7472.html and on Wikipedia: https://en.wikipedia.org/wiki/International_Standard_Atmosphere

Kungvankij, P., & Chua, T. E. (1986, June). Shrimp culture: pond design, operation and management. NACA Training Manual Series No. 2. Food and Agriculture Organization of the United Nations and Network of Aquaculture Centres in Asia (NACA), found at http://www.fao.org/docrep/field/003/ac210e/AC210E00.htm
Meyer, B. L. (2019). Bio-sustainability: Role of aquaculture in space. National Space Society (NSS) International Space Development Conference (ISDC) 2019, presented 7 June 2019, Arlington VA.
Meyer, B. L. (2020). The messy details of going from excrement to entrée. AIAA ASCEND 2020, 16–18 November 2020 (Virtual), TECHPRES-23, Control ID 3418246.
OSHA. (2011). Indoor air quality in commercial and institutional buildings. Occupational Safety and Health Administration, U.S. Department of Labor, OSHA 3430-04, 2011.
Sinha, A., Datta, S., & Mahapatra, B. (2012). Diversification of aquaculture.

Chapter 10
Assembling the Farm from Stages

Abstract This chapter covers design concerns for connecting stages in the farm together to make a complete mass processing capability that can fully recycle mass. This section also shows how to mathematically link the farm together to find footprint, energy requirements, and enclosed volume using three examples, with a discussion on element balances and using machine learning to solve the balance problem.

Keywords Balance · Mass flow · Troubleshooting · Systems · AI

10.1 Introduction

Every space farm has a hierarchy, where the farm has stages linked together by systems, and inside each stage are growth chambers and the other equipment to manage the stage, and in the growth chambers are organisms in an environment that maximizes production. In this chapter, we assemble the stages together to make a comprehensive mass flow that with the habitat form a synthetic ecosystem that balances mass. Systems that manage how mass flows, and how information and energy move around the farm, are combinations of devices and software, that link stages and the habitat. Working out how the mass flows is not always easy, but back of the envelope level guesses can be explored in examples. Making the elements balance is harder yet, and again this is explored in this chapter, along with a discussion on where to get the starting materials for the farm.

10.2 Making Stages into Managed Engines

The space farm stages and human habitat will require designs to link the mass flow, control system, electrical systems, and processes to form the settlement. Stages in this tense become mass flow engines, much like elements in a factory.

© The Author(s), under exclusive license to Springer Nature Switzerland AG 2026

B. L. Meyer, *Designing Farms in Space*, https://doi.org/10.1007/978-3-032-04395-5_10

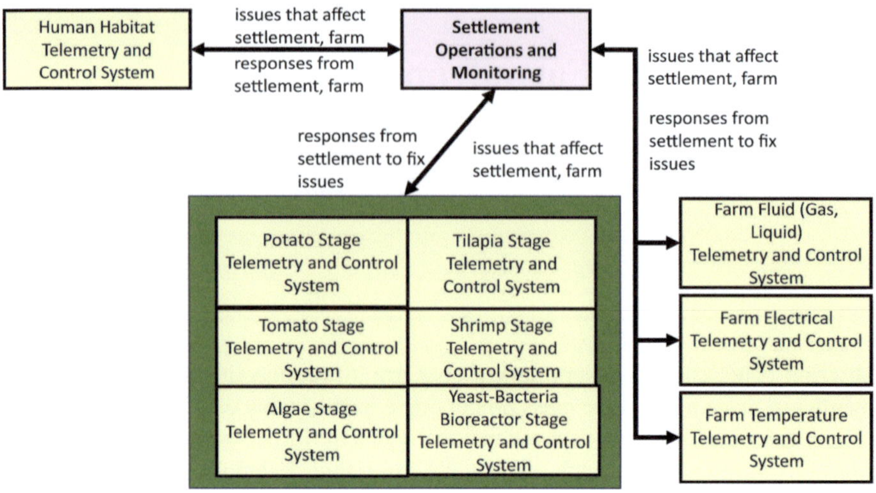

Fig. 10.1 Troubleshooting from stages to overall farm and habitat

10.2.1 Flow Systems and Troubleshooting

Troubleshooting issues in farm machinery and biochemical mass balance are life and death and will make or break the independence of settlements. Luckily many templates exist for the software and processes needed to efficiently find issues and diagnose causes before they become lethal. These models can emerge from chemical factories, backbone networks, and current space exploration methods. A hierarchy of troubleshooting systems and networks will be designed where issues are handled as close to the issue as possible, and only issues that affect other levels are treated as urgent messages. This is doubly so for mass flow or atmospheric issues that won't wait. Artificial intelligence will very likely handle most issues inside each stage and use rules and context to decide if notifications are urgent to higher levels, as in Fig. 10.1.

Essentially the farm needs at least control systems to manage mass flow for gases and liquids between stages and the habitat, with the networks, sensors, tanks, and valves to allow the software to keep the farm working. These systems only work if processes and rulesets are loaded into them, and humans follow them, even with artificial intelligence.

10.2.2 Getting to Whole Stage Mass Flows

In many cases, the mass flow for whole stage is a multiple living kilograms in the stage. This assumes internal water flows from condensed breath or transpiration, recirculation, and other internal movements are inside the stage. Only net mass movements to and from the stage are counted when looking at the farm level. The exceptions are the more mechanical stages like yeast-bacteria bioreactor stage and incinerator stage which are sized based on worst-case mass flow need.

To simplify how we look at mass flow and biochemical connections between stages, we can assume a few kinds of stages:

- Habitat Stage: The whole reason for the settlement. Net inputs are edible crops, oxygen, and clean water, while net outputs are carbon dioxide, and wastewater with plant/algae nutrients in solution with suspended solids. The solids can be settled out and partially dehydrated to recover some water, leaving a net of semi-solid waste. The nutrients in solution, minus suspended solids, become a critical wastewater source for photosynthesis. See Fig. 10.2.
- Incinerator Stage: Stage that uses heat to combust solids or liquids. Inputs are oxygen, and a combination of wastewater solids (or mostly solids) and inedible crop parts. Outputs are primarily clean water and carbon dioxide, and ash which can be used for soil base, or placed into solution and possibly further treated to feed photosynthesis. Quick and compact, though the ash is not perfectly useful for plants or algae. Also, critical if lignin is a significant output from crops, or if material is diseased or otherwise contaminated. This stage can produce energy if combined with a fuel cell or turbine. See Fig. 10.3.
- Yeast-Bacteria Bioreactor Stage: Usually an aerobic bioreactor, but may also have anaerobic elements. Composed of bacteria aerobic and anaerobic elements, tanks or membrane bioreactors with custom yeasts or bacteria, and reactors to make yeast solids or digest materials to make nutrients useful to algae or plants. Inputs are usually inedible crop parts, semi-solid wastes from black water, and oxygen. Most of the cells in the stages are already in use on Earth, and in nature. This stage can have a variety of outputs including carbon dioxide, water, and biomass types from nutrients for photosynthesis, including ash- and nitrogen-rich compounds, to edible biomass like yeast solids, to alcohols, to feedstock for food printers, or other products. Can produce energy if combined with a fuel cell or turbine. Each

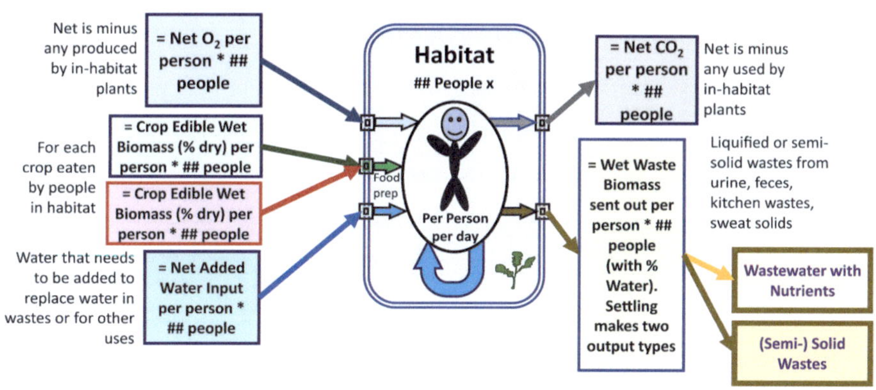

Fig. 10.2 Habitat stage mass flow schematic

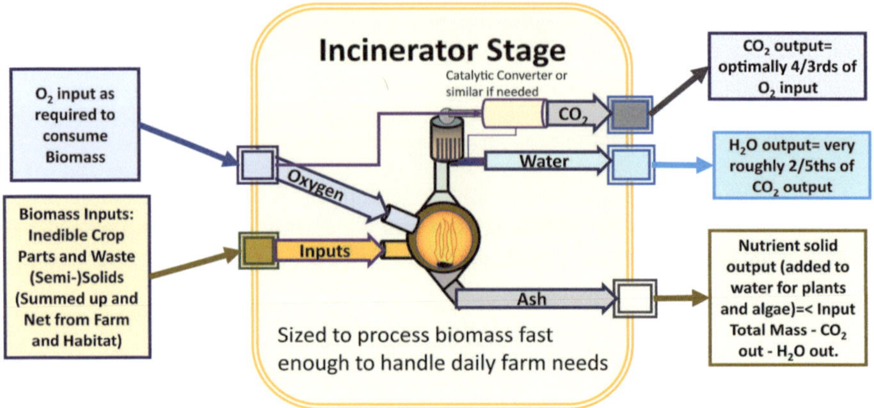

Fig. 10.3 Incinerator stage mass flow schematic

function is regulated using complex rules and machines to ultimately balance the biochemistry of the habitat-farm combination. This could be as simple as a tank and pump or far more complex for larger farms. In most scenarios to date, the yeast-bacteria bioreactor stage is an aerobic processor to convert biomass and oxygen into carbon dioxide, water, and ash in sludge. See Fig. 10.4.

- Yeast-Bacteria-Incinerator Stage: A combination of yeast-bacteria bioreactors and a follow-on incinerator, used especially in very large farms, to completely process inputs into selected outputs. See Fig. 10.5.

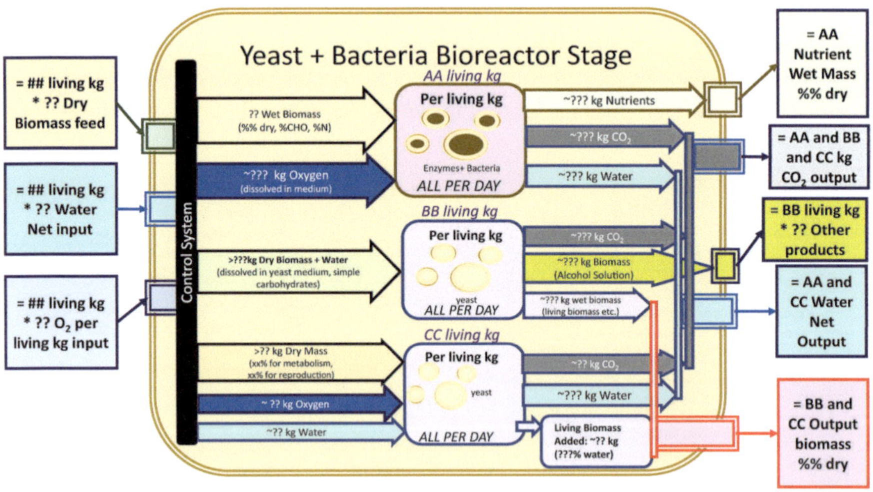

Fig. 10.4 Yeast-bacteria bioreactor mass flow schematic. The yeast-bacteria bioreactor is a combination of a variety of yeasts and bacteria in various tanks and membrane arrangements, for anaerobic and aerobic processing

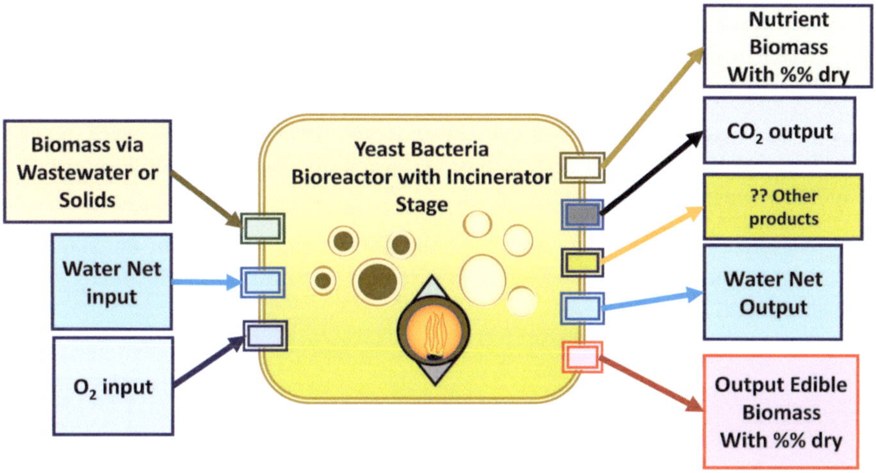

Fig. 10.5 Yeast-bacteria-incinerator stage mass flow schematic

- Photosynthetic Stages: These stages use light to convert inputs of carbon dioxide, water, and nutrients in soil or in wastewater to make carbon dioxide and living biomass. Plants with roots can also cycle and filter wastewater into cleaner water, minus any water used to make more plant.
  - Photobioreactor Stages: Use a photobioreactor and microorganisms or macroalgae to convert carbon dioxide, water, and nutrients into oxygen and biomass. Note that wastewater inputs can cover both water and nutrient inputs. While the outputs are almost always edible, they may be sent to an aquatic stage or other animal stage as feedstock. See Fig. 10.6.
  - Plant Stages: (Vascular) Plant stages take inputs and release outputs based on the proportion of living biomass in the stage, taking nutrients, water, and oxygen to output inedible and/or edible biomass and oxygen, using light. Note

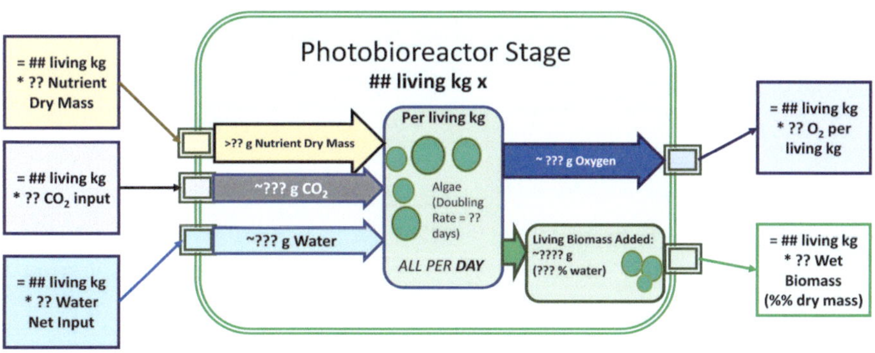

Fig. 10.6 Photobioreactor stage mass flow schematic

that wastewater inputs can cover both water and nutrient inputs. Any excess water in the stage, usually in an aquaponic set-up or grey water process, can be output in some scenarios. See Fig. 10.7.

- Aerobic Stages (Animal or Fungi): These stages rely on the conversion of oxygen, water, and biomass into carbon dioxide and edible and inedible biomass that includes wastes in liquids and solids.
  - Animal Stages: Outputs are proportional to living biomass in the stage and may be animals harvested, or other products like milk, manure, and eggs. See Fig. 10.8.

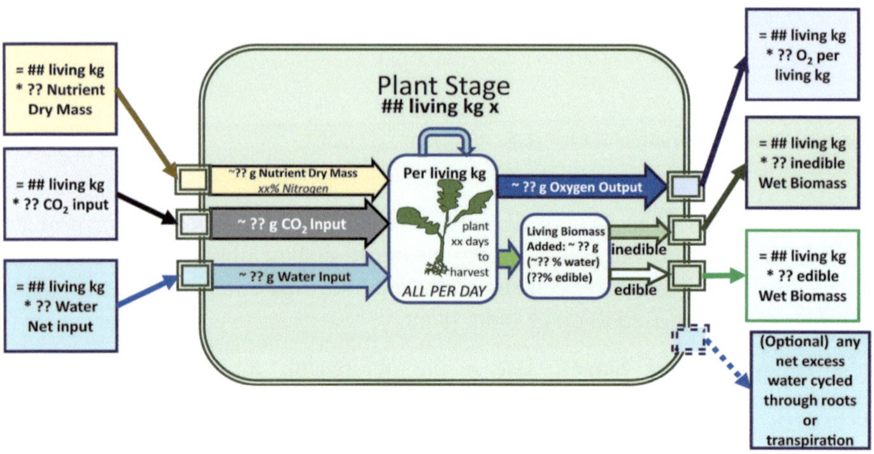

Fig. 10.7 Plant stage mass flow schematic

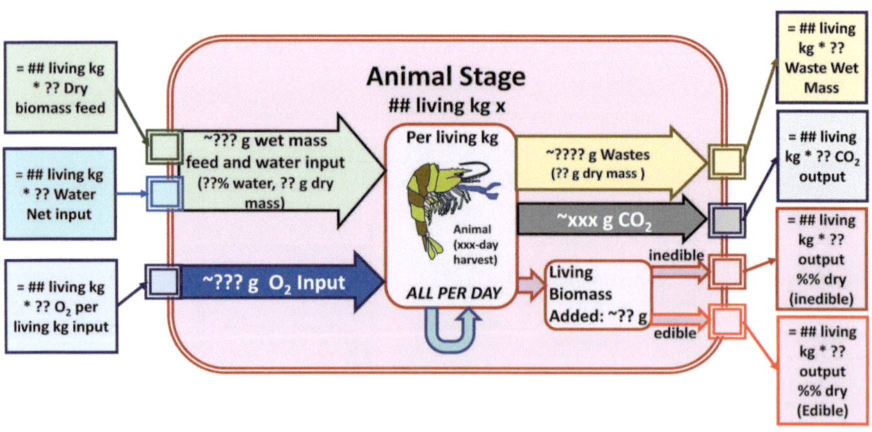

Fig. 10.8 Animal stage mass flow schematic

- Fungi (Mushroom) Stages: In this case, large multicellular fungi such as mushrooms, morels, or truffles as a crop. This stage could use mushrooms to process lignin. Input biomass is usually solid waste or more likely, macerated crop inedible wet biomass, with water and oxygen added as needed. Outputs are proportional to living biomass in the stage. See Fig. 10.9.

- Direct Assembly and Bioengineered Products Stage: Advancements in machinery, biology, and genetic manipulation could allow a variety of products with any number of inputs. Examples to date include direct assembly of carbohydrates from carbon dioxide and water, or cultures of bacterial or yeast cells to make meat, egg, or dairy products. The tradeoff is complexity and energy use for each product. Complex machines or isolated specialized cells may be very difficult to maintain in higher radiation or other space environment hazards. See Fig. 10.10.

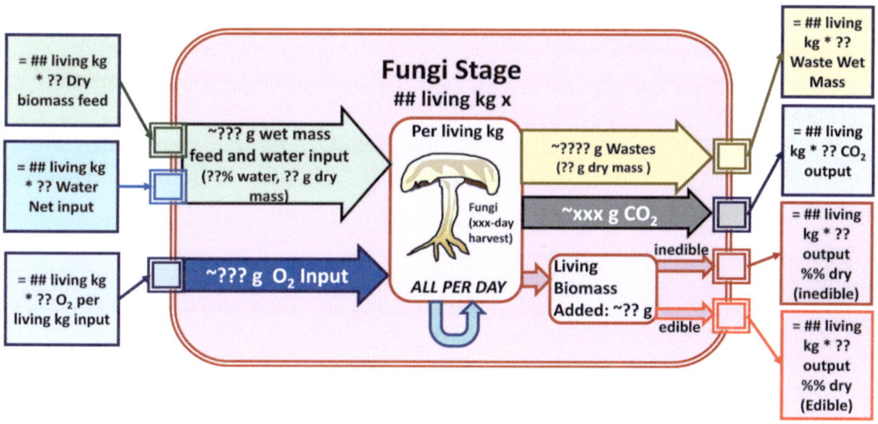

Fig. 10.9 Fungi stage mass flow schematic

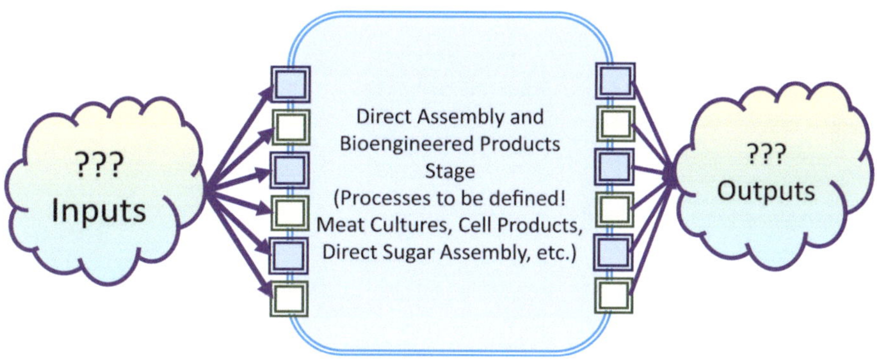

Fig. 10.10 Direct assembly stage mass flow schematic. Ins and outs will depend on exactly what happens inside the stage

10.2.3 Mass Distribution Systems

As mentioned previously, the aerobic (yeast-bacteria) bioreactor stage is the core of waste management, but not the only part. There are multiple systems needed in the farm to make sure mass of various types get to the stages where it can be used, and to the habitat where it is needed as in Fig. 10.11. These systems may be combinations of pipes, tanks, sterilizers, pumps, filters, and columns. The mass distribution system will also use other machines, and humans or robots manually moving crop products or wastes. Overall, the mass movement systems can be grouped like so:

- Liquids and Loose Solids:
  - Clean Water System: This system moves clean water, such as potable water, to various locations, including storage for the overall settlement.
  - Wastewater (Liquid Wastes) Distribution System: This system moves wastes like urea, ammonia, and other photosynthesis ready nutrients around the farm, from animals, humans, and yeast-bacteria bioreactors or incinerators to algae and plants.
  - Solid Waste Distribution System: Moves sludge and semi-dehydrated wastes from feces or other black water sources, to either soil development or incinerators or yeast-bacteria bioreactors.
- Biomass Distribution:
  - Edible Crop Distribution: Methods to move and store harvested food parts of crops, including whole algae, grains, fruits, tubers, mushrooms, and animal parts. This is likely a human or robot manual system. Excess food items can flow to the inedible crop system, especially if algae are to feed shrimp, fish, and mollusks. Excess edible crops could also flow to aerobic bioreactors or

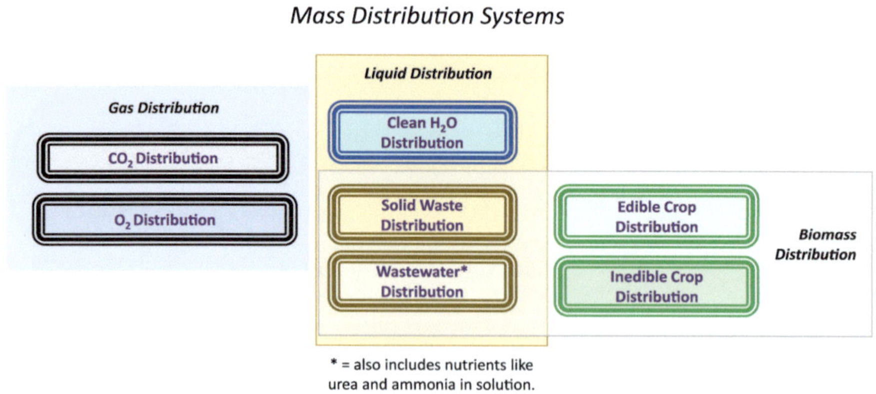

Fig. 10.11 Mass distribution systems for the farm

incinerators if more carbon dioxide or water is needed, or to ferment into products like wines and beers.

- Inedible Crop Distribution: Methods to move and store parts humans can't eat from harvests. These can be sterilized and flowed to animal or fungi, ground and composted into soil, or likely sent to incinerators or aerobic bioreactors to recover carbon dioxide, water, and ash.
- This also intersects the waste systems above.

10.2.4 Wastewater as Mass Flows

Like the flows in the habitat, farm water has three types: clean, grey, and black. Waste flow comes in two types in the parlance of poop: black water and grey water, both of which are wastewater. Potable water is the stuff you can drink, use for bathing, or use for food prep, fish tanks, bioreactors, etc. What gets flushed down the toilet is black water, a combination of solids suspended in water, chemicals in solution, and water to make it move. These types in turn get pre-processed by settling or similar process into two large categories: solid wastes and wastewater as in Fig. 10.12.

In space, we can take advantage of the different natures of grey and black water but must also be more careful since we will be drinking and eating what once was wastes, and resupply is costly.

First let's cover the movement of potable (drinkable) water. Potable water is moved in the farm and to the habitat by the Clean Water (Clear Water) System. In all stages of the farm, water that evaporates into the air is condensed out, minus any humidity needed in the air, then sterilized and reused right in the stage as potable water, suitable for any use. This loop is very critical in the plant stages where transpiration occurs, since the right humidity is required for each variety to grow well. Note that transpiration is a good way to purify water, and in small settlements, is a nice compact

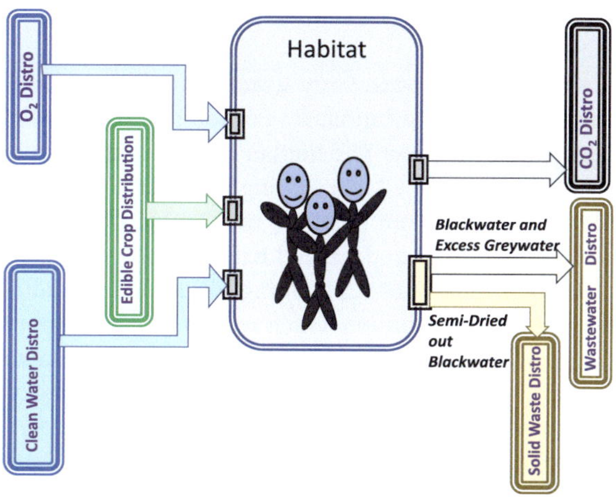

Fig. 10.12 Mass distribution systems and human habitat

way to turn grey water into potable water. In the human habitat or terrestrial animal stages, humidity is maintained, and the rest of the excess water is moved to tanks. Note that aerobic respiration in animals produces some water. In aquatic stages or in bioreactor stages, the humidity in the air above and around the tanks is maintained high enough to allow human work, but maintained low enough to discourage mold growth. The condensers will have a sterilizer, and then the condensed water is sent to a storage tank. Eventually the clean water is sent to the input water buffer tank for the stage to be used in the stage. Another source of water will be food processing, where plant, fungus, or animal parts are dehydrated, then eaten by humans or animals. Again, the evaporated water is condensed and reused. All those methods do not prevent either a net need or surplus, which is why the Clean Water System is used everywhere. Excess portable water, that is excess beyond what is needed by each stage, is sent to the potable water buffer tank array for the farm itself, to be pumped where needed in the farm or habitat.

A major source of potable water, however, will be the aerobic bioreactor stage with or in addition to an incinerator stage in most scenarios. Wastes sent to the bioreactor are broken down by the microorganisms using oxygen into carbon dioxide and water. Alternatively, enzymes or incinerators do the same. This returns the carbon dioxide and water consumed by plants or algae back to the water and gas systems. Aerobic bioreactors and incinerator stages become the biggest source of clean water in the farm. Don't worry about the oxygen to feed the breakdown. The plants and algae produce more than enough for you, the animals, and the bioreactor or incinerator, trust me. All proteins, fats, and carbohydrates consist mostly of carbon, hydrogen, and oxygen. Once digested by the bioreactor, the complex molecules become mostly carbon dioxide and water. The plant and algae consumption leads to food for animals and humans, which leads to wastes from both, which goes to the bioreactor, releasing carbon dioxide and water that started as material for plants and algae. Without releasing the carbon dioxide and water from the wastes, the farm dries up, the plants and algae die, and so do you. That said, the potable water produced by the bioreactor or incinerator is again added to the potable water tanks for the farm and habitat. The whole habitat and farm water cycle therefore looks like the diagram in Fig. 10.13.

Grey water is next. Grey water has a few solids suspended in it, and some salts and chemicals in solution. It is called grey water because that is how it looks, clear but milky or murky. The human habitat is the primary source of grey water. Grey water can be pumped to a settling tank which pulls out most of the solids. The solids next get pumped out as black water. One possibility after settling is that the grey water is sterilized and filtered using plant roots and sand/gravel filters. The habitat plant roots remove any remaining solids, and any nutrients. Next the water is used by the plant in transpiration which results in water vapor that can be condensed as clean water. This loop assumes nothing nasty has been put into the grey water and means habitats must use biodegradable soaps. Excess settled grey water can be pumped to farm plant stages for root water, after some electrolysis to remove excess salts. Salt control in all water types will be a challenge in all places in the farm. Alternatively, the grey water can be evaporated, or reverse osmosis filtered to potable water, and

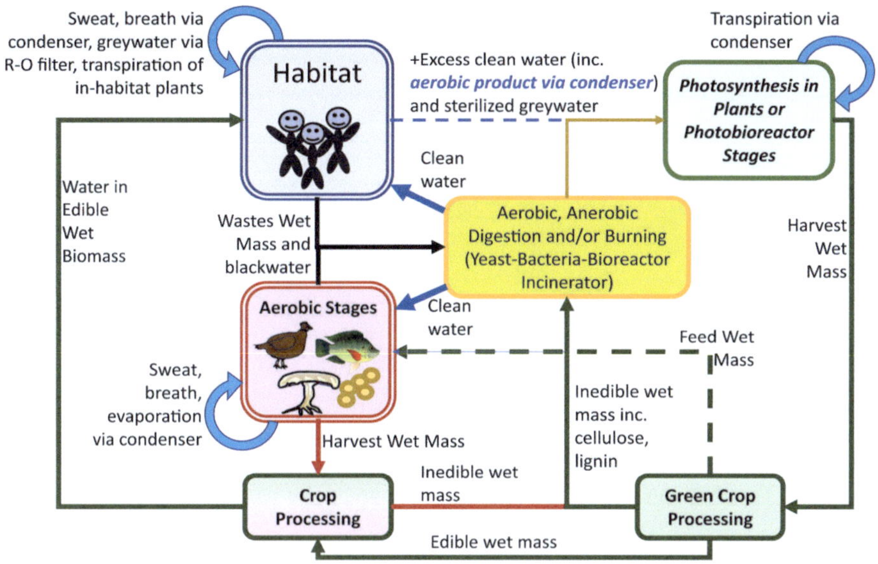

Fig. 10.13 Farm water cycle

any remaining leftover mass from salts, solids, etc., is sent out as black water or used to balance the farm's chemistry. There are many options for the space farm designer!

Now we get to the stinky stuff, the black water. Black water loves to clog pipes, make gases that can explode, and generally make a mess of things, but it is also a critical component of mass flow in the overall farm. In many cases, the thick sludge in settling tanks goes to bioreactor or incinerator, but really thin stuff at the top can go to photobioreactors or hydroponic gardens. Treating the black water is the role of either a yeast-bacteria bioreactor stage, the same with an added incinerator (yeast-bacteria bioreactor with incinerator), or in some small outposts, directly to an incinerator stage, as in Fig. 10.14. Processed black water mass gets shunted into CO_2 distribution, (via aerobic respiration, anaerobic respiration, or burning), clean water distribution (same reactions), or wastewater distribution once sterilized (via initial settling, plus any added ash, or fully digested sludge now nutrients in solution).

Once a settlement is big enough, soil comes into play. That solid waste material, once sterilized and dried is great starter for soil, or fertilizer for soil plants. If it is used as a starter, it is mixed with a seeding of bacteria and fungi to establish a soil biome, maybe with worms and insects as well. The seeded starter is then added to more inert materials from in situ such as sand, dust, and fine gravel, or silica and cellulose wastes from packaging, and any crop wastes. All laid together in a planting bed of usually at least 10 cm, then planted with an initial plant crop. The soil over time does the same work as a slow yeast-bacteria bioreactor, releasing nutrients, carbon dioxide, and water to the plants growing in the new soil, exactly as in the prairie example in the introduction. Soil in the plant stage both adds and subtracts to

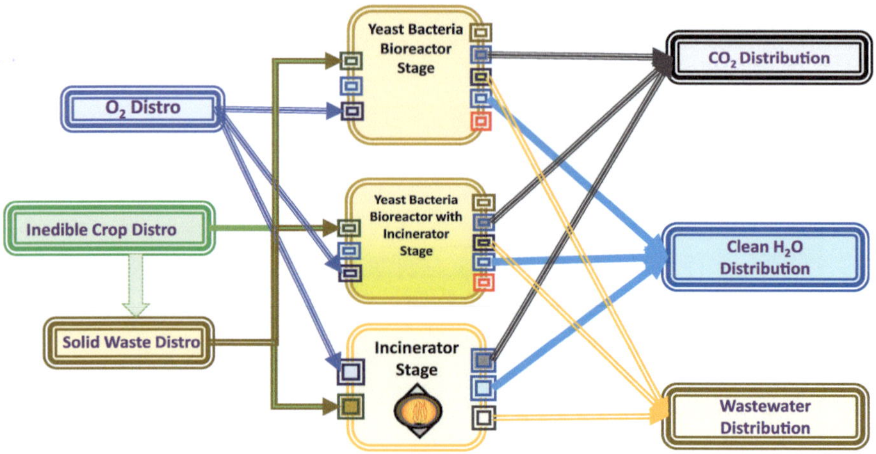

Fig. 10.14 Yeast-bacteria and incinerator stages mass flow to distribution systems

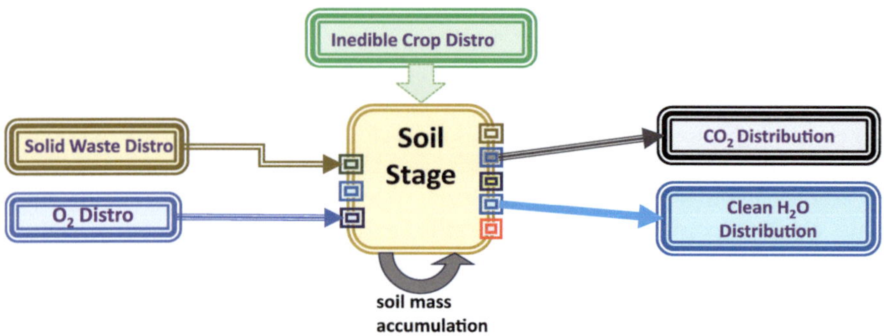

Fig. 10.15 Soil as a stage in the mass flow systems

the net mass flow for those stages, and acts as a yeast-bacteria bioreactor for mass flow, except for soil accumulation. See Fig. 10.15.

10.2.4.1 Aquatic Stage Wastewater

Most of the water in the fish/shrimp from the middle or top of animal tanks is sent to a filter set in small holding tanks, or a settling tank in larger aquaculture stages. Most fish and shrimp waste that is black water will settle near the bottom, either in the center in round tanks or at the end in raceways. Macerating pumps chop and pump the black water from its source at the bottom of the fish or shrimp tank to a settling tank and coarse filters in aquatic stages. The settling tank or filter pulls out any solids that might clog pipes, then sends the water through oxygenators and sterilizers back to the tank, or if too loaded with urea or ammonia, to a hydroponic root tray then

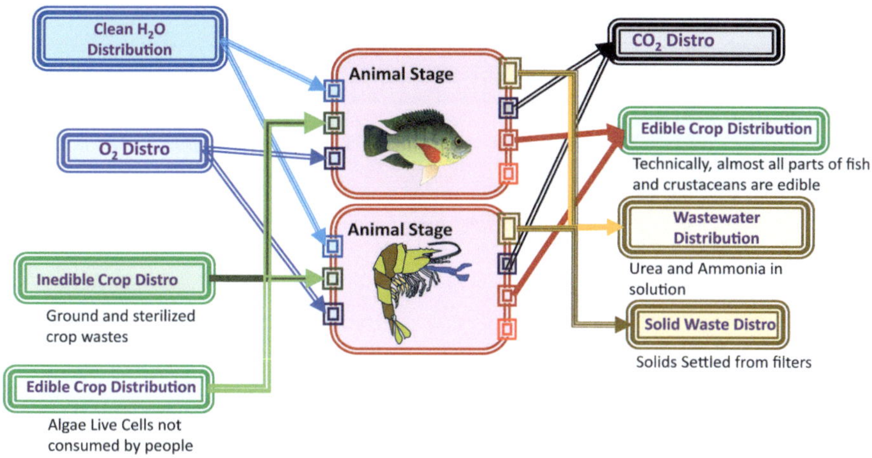

Fig. 10.16 Animal stage and the farm mass distribution systems

back to the tanks in an aquaponic process. The stuff that is filtered out or settled out is black water sent to bioreactors or incinerators or other uses. The thicker stuff at the bottom of the tank gets sent to the yeast-bacteria bioreactor's first open tank. The thinner stuff goes to a second set of filters that get out the rest of the clogging mass, which is also sent out to the bioreactor array. A percentage of the water is simply aerated and circulated in the tank, while some is sent out for filtering, based on the chemical needs of the fish or shrimp. See Fig. 10.16.

Bioreactors and incinerators turn nasty black water sludge into carbon dioxide, clean water, and ash. If you are making soil however, the sludge gets added to your mother soil and other substrate to make good dirt as discussed in the prior section.

Essentially, how you deal with black water depends on scale. In a very small farm, a direct aquaponic system is easier to manage, though this reduces the nitrogen available to algae. In this system, the water once coarse filtered in a settling tank is routed to trays elevated above the tanks where gravity and pumps send it through the roots of plants in the garden to be filtered and have nitrogen removed. The water is then aerated before being returned to the fish tank.

Another system uses three stages, with a percentage again being directly recirculated after aeration and sterilization, and a percentage then coarse filtered. Then pressure is added against a reverse osmosis array that pulls pure water out to resend back to the tanks as needed. Next, the filtered wastes and sent to a settling tank. The thicker material will go to the yeast-bacteria bioreactor, and the thinnest material to photobioreactors or hydroponic roots as needed by the farm. The system, as in Fig. 10.17, is more intensive but allows more fine-grained control of the biochemistry of the farm.

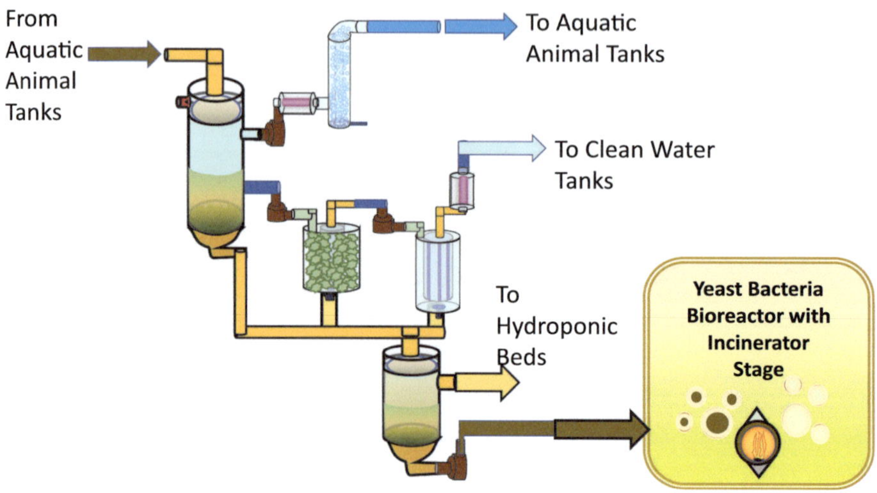

Fig. 10.17 Three-stage aquatic tank wastewater treatment

10.2.4.2 Black Water from Terrestrial Animals

Stalls and cages will have to be cleaned to remove animal wastes, from small livestock like insects and rodents and chickens, to larger animals like cattle and goats. Their waste is treated like human wastes. Solid in suspension, liquids, and inedible parts from processing, are macerated and sent to a settling tank, where the thicker material is either sent to the aerobic bioreactor stage or incinerated to recover carbon dioxide and water. The thinner material is sterilized and either sent to photobioreactors or hydroponic gardens. One option is to take animal effluent, heat, and dehydrate it to recover sterilized water and take the resultant solids to make soil. The soil option of course requires mass inputs at some point via in situ or resupply sources. See Fig. 10.18.

10.2.5 Tracking Water

All the water in the farm-habitat system is the sum of wastewater from animals and people, water in wet biomass from crops, water produced from bioreactors, minus water consumed by crops to make wet mass, minus water used and consumed by people, which eventually becomes wastewater again. The water is in various forms, and only clean water or water in food is used by people. Wastewater (grey water) is usable by photobioreactors and plants, and only once distilled or run through reverse osmosis filters is it usable by people. Water locked into wet biomass that is not people food, including suspended solids in black water, is usable by fungi and bioreactors, and could be dehydrated out to make clean water and dry biomass.

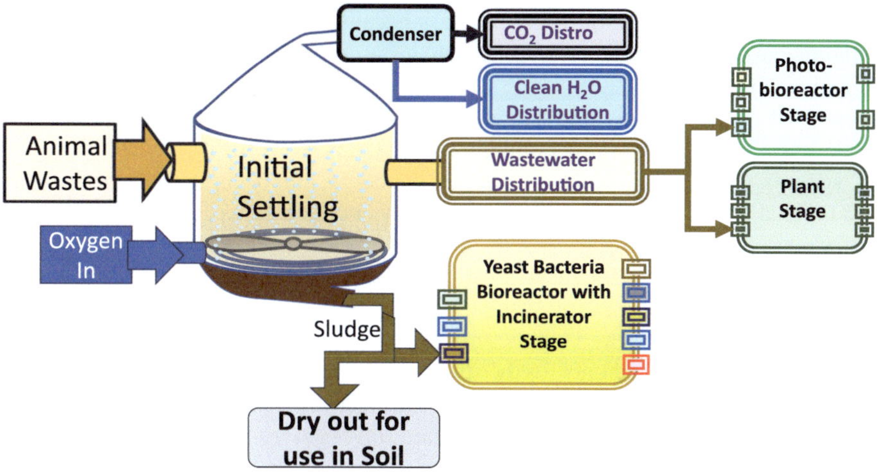

Fig. 10.18 Mass flow schematic for animal wastes. Animal wastes (black water and solids) are first put into a settling and storage tank then fed to various other parts of the farm

Bioreactors can crack dry biomass in animal or human wastes or excess crop dry mass into water, using oxygen in the process. Just like a balanced budget in accounting, the water sources and uses need to balance, and usual process to close the loop is via bioreactor and incinerator. Overall, one of the key balances for the whole farm is tracking water as it is consumed, processed, used, and converted into biomass.

10.3 Farm Design: Linking Stages with Systems

This chapter covers how to balance the overall farm with three examples of how picking crops and adding complexity affect the farm size, mass flow, initial supply, systems, and energy needs.

10.3.1 Stage Basics Again and Mass Balancing Principles

As a reminder, the concept of a stage in this book, is defined as the combination of living crop biomass, with the artificial environment and machinery required to optimize crop growth. The stage can have a physical representation as a closed off volume of just one kind of crop, or a virtual representation where a crop is distributed with other crops in the same enclosed volume. The idea is to have some set of living biomass + environment + machines to allow calculation of how much goes in, how much goes out, how much room is needed, how much energy is required, and how many calories are produced.

The analysis is by unit of living biomass because it gets around having to count individual organisms. In the end, we want to grow crops to feed people and to recycle air and water.

In words, a stage at any moment in time has a mass of living organisms, takes in mass, lets out mass, grows (accumulates mass), has mass for its environment (i.e. air, water in tanks or root beds, etc.), requires energy, and occupies volume. A stage also is sized to include nurseries, spare capacity, working space, and tank volume to buffer mass flow.

Engineers may design the result a variety of ways, but mathematically, the stage is a container for data about a crop species in the farm. Each unit of living mass sums to the whole stage, and acts as a biochemical factory. The ins and outs for the mass are summed to get the ins and outs for the stage. One of the outputs of the stage at steady state is living mass in the form of the crop. The stage occupies volume and footprint area. The stage consumes energy. When the farm is set up, the stage has very little living mass and takes in little mass and then exponentially slopes up to steady state rate, where mass in is balanced by mass out. At steady state the continuous harvest model is used to simplify sizing the stage and allows for a continuous output. You could do the analysis as a series of sawtooth outputs with respect to time, but that makes it very hard to make the farm mathematically converge. We have to perform these mental gymnastics so that we can plan for the farm, assure it can recycle for the human habitat, provide diverse menus, and enough nutrition in calories and by nutrient type. Using the stage method, we can optimize how many of what crops need to be grown, how big tanks are, how much material and gases are needed to get going, and what dinner I get in the end.

10.3.2 Habitat Source Balancing

The whole point of the farm is to keep the people alive, so we end up with the following flowchart cycle of processes to find the mass balances. The habitat is the only really fixed amount, given the number of people. The other given is the types of crops to get the average daily meal content for the Habitat. Many possible farms can accomplish these must have rules:

(1) Must meet clean water and oxygen needs for the habitat.
(2) Must provide nutritional minimums in terms of calories, fiber, fat, protein, and carbohydrates.
(3) Must use as input to the farm, the outputs from the stage, plus any assumed resupply mass.
(4) Must produce at least some of each crop species.

Optimizing factors for these are usually:

(1) Must mass balance with a minimum resupply.
(2) Minimum total mass.
(3) Minimum total footprint.
(4) Minimum total energy need.

There are lots of mathematical techniques to get to the combinations of living biomass in each stage (i.e. per crop). You might need these assumptions:

(1) Assume the yeast-bacteria bioreactor array (or incinerator if chosen) and tanks can convert any combination of waste materials, given oxygen, into any nutrient input, carbon dioxide, and water, at the cost of size and energy.
(2) Assume water from transpiration and exhalation and evaporation is recycled inside the stage, and only excesses leave the stage. Cost is in energy to condense, filter, and pump.
(3) Assume mass is moved perfectly as needed per day between stages using pipes, pumps, tanks, and flows. This means if rice stalks and leaves and roots get plucked from the rice stage, they get moved to the bioreactors fast enough that that stage can convert them into carbon dioxide and water exactly as needed.

Figure 10.19 provides the top-level process to get a balance starting with the needs of humans in the habitat.

The easiest way to visualize what the whole farm flow looks like if you link all those mass flow blocks together into a flow diagram, then run some accounting. With that done, then real engineering begins when you build designs that can make the flows work.

The real payoff for doing all the math is to find the size and mass of a space farm to support a settlement of a known size and with a known set of crops. In most cases, the

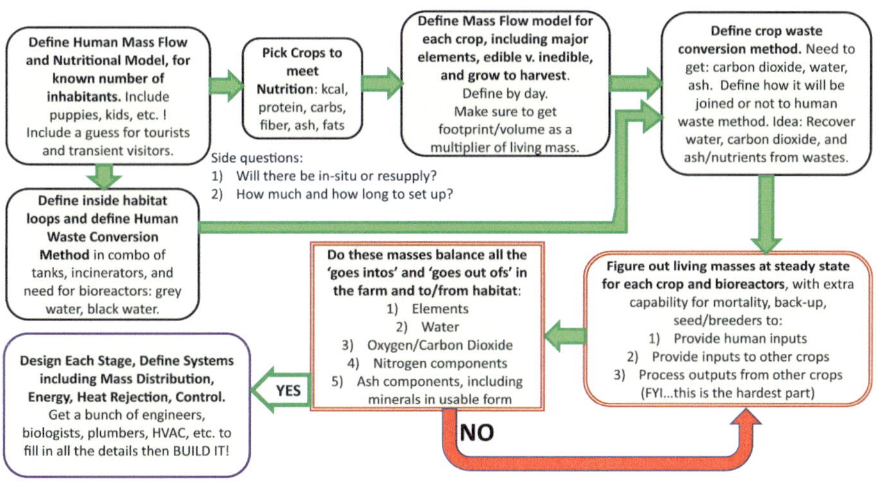

Fig. 10.19 Balancing the farm as a process

goal is to find the possible farm crop and mass combinations that are minimal in size yet come as close to possible to needing no resupply once at steady state. Overall, the balancing of the farm is an accounting exercise, tracking each type of output and input in total to meet habitat needs. At the core, there are key transformations and exchanges that occur using living organisms and machines, boiling down to biomass and oxygen to make carbon dioxide and water, and the reverse, which is carbon dioxide and water to make biomass and oxygen. At the fine-grained level, tracking elements makes sure that they balance, and using human nutrition inputs based on US RDA or similar determines the amount of living biomass in each stage. Each into and goes out of for each stage is balanced, and the sum of stages and habitat inputs and outputs balance ideally. This means the mass flows are tracked just like a bank account, with each stage a pair of transactions, with additions for outputs, and subtractions for inputs, in the account. There are multiple 'ledgers' here, since the matter types count.

10.3.3 Multistage Pathways and Flows: Three Examples

If you wanted to simulate how a space farm fits into a settlement, the SIMOC package (SIMOC, 2022) is a great start. That said, this section will also show you some simple ways to start looking at how big the farm needs to be. So how do we get at least a 'back of the envelope' guess at what a space farm's mass flow and size looks like? The least mathematical way to look at hooking together all the stages and systems is to draw a sample diagram, then pretend all the parts are part of the budget for the big items like CO_2, O_2, H_2O, nutrient biomass, waste biomass, and food, for a day, assuming everything is at a steady state. Going deeper the same approach down to the major element level (at least C, H, O, N, and a few others) gets an even tighter balance. For each stage, we will have to assume a bit of extra size for backups and mortality, and we will have to assume the mass systems in the farm can ideally move mass where it needs to go exactly when it needs to be there. Lastly, we will have to assume the magic chemical balancer is the yeast-bacteria-incinerator stage. The simple version of the table would look something like the table below, with inputs as minuses, and outputs as pluses (Table 10.1).

Given a table like this, we can look at a few examples.

10.3.3.1 Multistage Pathways and Flows: Example 1

Example 1 is a very small, simplified, example with 100 settlers using the human model, and with a horrible menu of cyanobacteria, yeast, and bell peppers (read this as 'glop, glop, and salad') as in Fig. 10.20.

Table 10.1 Simple stage balancing table

Stage	CO_2	O_2	H_2O clean	H_2O Waste	Waste dry mass	Nutrient dry mass	Food wet mass
Habitat	(per person * people)	(per person * people)	(per person * people)	(per person * people)	(per person * people)	(per person * people)	(per person * people)
(Stage)	(per live kg in stage * mass in stage)	(per live kg in stage * mass in stage)	(per live kg in stage * mass in stage)	(per live kg in stage * mass in stage)	(per live kg in stage * mass in stage)	(per live kg in stage * mass in stage)	(per live kg in stage * mass in stage)
Totals	Sum for all stages	Sum for all stages	Sum for all stages	Sum for all stages	Sum for all stages	Sum for all stages	Sum for all stages

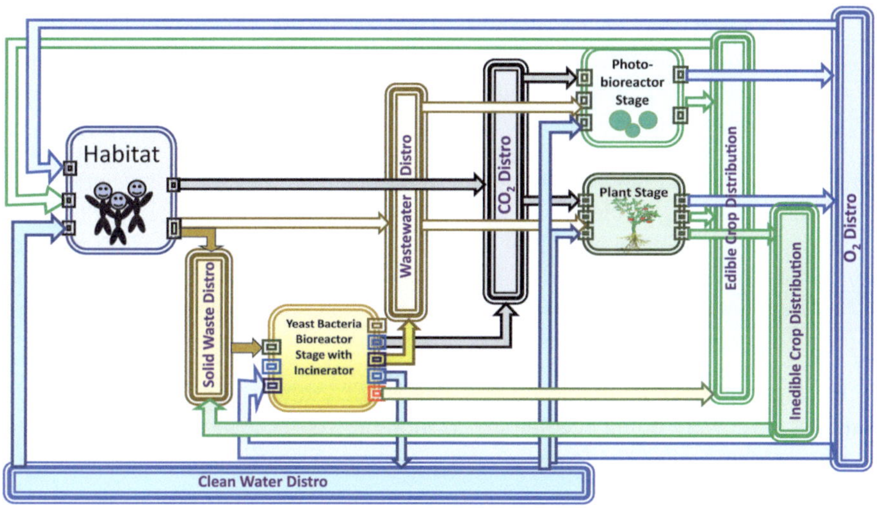

Fig. 10.20 Simple mass flow schematic for Example 1. Diagram shows mass flow for habitat, and farm with algae, red bell peppers, and yeast-bacteria-incinerator

For this simple case, assuming 100 people, and making a few guesses, the mass balance table per day might look like Table 10.2, once we add some data:

1. First, we need to pick an average menu and human model to calculate mass flow to and from the habitat. The menu is important because it tells me how much food we need from each stage per day. This menu and a human model together in turn provide the waste we send to the farm.
   a. For this case let's assume each person takes in 7 kg cyanobacteria, 0.5 kg peppers, and 0.1 kg yeast wet mass. This example is so high in cyanobacteria because otherwise there will not be enough oxygen for humans and assuming that there is not infinite space for a pepper stage. Roughly, it follows the

Table 10.2 Example 1 menu and dry mass per person per day

Crop	kg Food Wet Mass	Dry Mass %	Wet Mass %	Dry Mass (kg)	Water in Food (kg)
Cyanobacteria	7.00	0.09	0.91	0.66	6.34
Peppers	0.50	0.08	0.92	0.04	0.46
Yeast	0.10	0.24	0.76	0.02	0.08
Total	7.60			0.72	6.88

process in Fig. 10.19. This means that the slime from the algae reactor is likely processed extensively in the habitat's kitchens to distill out the water and may be fermented to make vinegar or other products. They will likely be other chemical processes to make the mass of slime edible. Given the peck of peppers, the peppers might need processing too. You could come up with other mass guesses too. This food input, using the mass models in Chap. 3 (and Figs. 3.51 and 3.52) and Chap. 4 (Figs. 4.10, 4.11, 4.39 and 4.40), make a table like Table 10.2.

b. Given the human model example in this chapter, we can guess the settlers will need a net 2.6 kg of clean water per person. We can then take a rough estimate at our average human model for example as in Table 10.3 which alters the food masses from the earlier human model (especially water in food), and therefore its compliment in water in wastes. We will also assume some tough settlers, and some very capable machines and cooks to make that glop edible.

We then multiply this table by the number of settlers, i.e. 100, and show the result as if we were doing accounting, as in Table 10.4. Clean water is assumed to be distilled from food and from in-habitat cycling of water vapor from breath and

Table 10.3 Example human model

Component	Mass (kg)
Inputs	
Food (Dry Mass)	0.72
Oxygen	0.84
Water from food*	6.88
Total Inputs	*8.44*
Outputs	
Carbon Dioxide	1.01
Wastes Dry Mass	0.32
Water in Wastes	7.11
Total Outputs	*8.44*

* = this will likely be distilled to get drinking water

evaporated sweat. If food requires rehydration, this will mean clean water in the H_2O clean column, and an equal amount added to the wastewater (H_2O waste) column.

2. Given the food inputs to the habitat, next we figure out the inputs to the living mass in the food-producing stages that make that much food per day ideally using the model for each organism. Initially, ignore the yeast part. The food outputs entail other outputs too. We can use a guess at living mass in the stage for the crop, then expand for the whole stage. This means also adding in a fudge factor for waste factors that end up as wastes. For cyanobacteria, we again use the model in Chap. 4, and for the bell peppers, we again use Chap. 4's bell pepper mass model. The minimum mass to get to the food is the food to the habitat and divided by per living kilogram food number.

   a. For cyanobacteria, we can start with the model from the figure and put it in a table, as shown in Table 10.5.

To get the needed 700 kg (7kg * 100 settlers) from the cyanobacteria row in the menu in Table 10.2, we can divide the 700 kg by the living biomass output per living kilogram in Table 10.5 (275.1 g/living kg) to get the rough cyanobacteria photobioreactor stage ideal living mass: 700 kg/(0.2751 kg/living kg) = 2544 living kg. This is of course not reality due to various efficiency problems, so assuming 1% crop death, we can roughly say the total living mass in the stage is at least 2570 kg.

Table 10.4 Example 1 human per day model for 100 settlers

	CO_2 (kg)	O_2 (kg)	H_2O clean* (kg)	H_2O waste (kg)	Waste dry mass (kg)	Nutrient dry mass (kg)	Food wet mass (kg)
Habitat	101	−84	–	711.3	31.8	–	−760

* = Water will likely be distilled from raw food wet mass for this scenario

Table 10.5 Cyanobacteria mass flow model per living kilogram

Cyanobacteria per living kg	g per day per living kg
Inputs	
Nutrient dry mass	4.2
Water (sterilized wastewater or clean)	265.4
CO_2	40.5
Outputs	
O_2	34.9
Living biomass (food)	**275.1**
% water in wet mass	90.60%
Dry biomass out	25.9
Water out in biomass	249.2

Table 10.6 Cyanobacteria photobioreactor stage mass flow

Cyanobacteria stage	kg for stage
Inputs	
Nutrient dry mass	**10.8**
Water (sterilized wastewater or clean)	**682.1**
CO_2	**104.1**
Total inputs	*796.9*
Outputs	
O_2	**89.7**
Living biomass (for food)	**700.0**
Living biomass (waste)	***7.0***
Total outputs	*796.7*
% water in wet mass	*90.6%*
Dry biomass out in food	*65.8*
Water out in food	*634.2*
dry biomass lost	*0.7*
water in lost biomass	*6.3*

With this figure, we can multiply by the per living kg table to get a table (Table 10.6) for the whole cyanobacteria photobioreactor stage.

b. We can use a similar process for the bell peppers. Starting with the mass flow model in Chap. 4. We can make a table per living kilogram as in Table 10.7.

Using a similar process as for the cyanobacteria above, using the menu in Table 10.2 we need 50 kg of bell pepper fruit (i.e. 0.5 kg * 100 settlers), and each living kilogram of pepper plants produces 2.54 g of fruit, along with 3.04 g of inedible biomass (such as trimmings). The ideal living kilograms in the stage is then 50 kg/(0.00254 kg/living kg) which is 19,686.5 kg living mass. For perspective, 20 metric tons of plants need roughly a hectare in soil, assuming 3 m tall plants. Assuming 1% of added waste, the total size in the stage is at least 19,883.4 kg living mass. Just like above, we can then multiply the per living kg table above by the living mass in the stage to get Table 10.8. Note the biggest input is water for photosynthetic crops, since the water goes to make dry mass, and inflate the dry mass into wet mass. Water to make both the crunch of the pepper and the juice of the pepper so to speak.

c. Given the results for the two photosynthetic stages, we can add that to the habitat data to get the intermediate table, Table 10.9. Note that the cyanobacteria, sometimes called blue-green algae, uses a higher percentage of CO_2 to water than the bell peppers.

Looking at plants versus algae shows some interesting tradeoffs in chemical efficiency. See Tables 10.10 and 10.11. The higher carbon dioxide and nutrient (mostly

Table 10.7 Bell pepper mass flow model per living kilogram

Bell Peppers per living kg	g per day per living kg
Inputs	
Nutrient dry mass	0.022
Water (sterilized wastewater or clean)	5.38
CO_2	0.65
Total inputs	*6.05*
Outputs	
O_2	0.47
Living biomass (total)	5.58
% water in biomass	***92.20%***
Dry mass out	0.00958
Water out in biomass	5.57
Total outputs	*6.05*
% edible mass produced	45.50%
Living biomass (food)	**2.54**
Dry mass out in food	0.2
Water out in food	2.3
Inedible biomass	**3.04**
Dry mass out in inedible	0.2
Water out in inedible	2.8

nitrogen-based) inputs for cyanobacteria reflect the higher protein in the cyanobacteria produced versus in pepper plants including fruit. Bell peppers, like most land plants, need more water for each unit of carbon dioxide, but are better at producing dry mass for that carbon dioxide by a small fraction. Even so, the cyanobacteria are more efficient at converting inputs to dry mass, and over twice as efficient at making food dry mass than peppers since only 45% of the dry mass made by peppers is edible. There is a cost for getting crunch in lieu of glop!

d. Next, we need to make a few more assumptions about how we get the yeast food mass. Ideally, part of the yeast-bacteria-incinerator stage has an aerobic membrane mediated (bio)reactor (MMR) to produce a continuous flow of yeast cells. Given this assumption, and data from the yeast section above, we can use this per living kg (per day) table, Table 10.12.

As before, we divide the needed yeast biomass from Table 10.2 (0.10 kg *100 settlers =) 10 kg, by the living biomass produced in Table 10.12 to get an ideal value of (10 kg/6 kg/living kg =) 1.67 kg. That sounds low, but yeasts are very efficient when arranged right, and under the right conditions, as in a very advanced MMR. Let's assume there is insignificant loss for this little reactor, so we can get the following table by multiplying the 1.67 kg figure to get the next table, Table 10.13.

Table 10.8 Bell pepper stage mass flow

Bell Pepper Stage	kg for stage
Inputs	
Nutrient dry mass	0.4
Water (sterilized wastewater or clean)	107.0
CO_2	12.9
Total inputs	*120.3*
Outputs	
O_2	9.3
Living biomass (edible + inedible)	111.0
Total outputs	*120.3*
% water in wet mass	92.20%
Dry biomass out in total biomass	8.7
Water out in total biomass	102.3
Food biomass	50.0
Dry mass out in food	3.90
Water out in food	46.10
Inedible Biomass (inedible from productivity +losses)	61.0
Dry mass out in inedible	*4.76*
Water out in inedible	*56.23*

Table 10.9 Intermediate summary table with habitat and photosynthetic stages

Stage	Living kg in stage	CO_2 (kg)	O_2 (kg)	H_2O Clean* (kg)	H_2O Waste (kg)	Waste dry mass (kg)	Nutrient dry mass (kg)	Food wet mass (kg)
Habitat		101	−84	0	711.3	31.8	–	−760
Cyanobacteria	*2570*	−104.1	89.7	−682.1	6.3	0.658	−10.8	700.0
Bell Peppers	*19,883*	−12.9	9.3	−107.0	56.23	4.76	−0.4	50.0

* = Clean water or for plants and algae, sterilized and filtered processed wastewater

Table 10.10 Input ratio comparisons between cyanobacteria (from a photobioreactor) versus bell peppers

Inputs	Cyanobacteria	Bell peppers
CO_2 (%)	13	11
H_2O, total (%)	86	89
Nutrients (%)	1.4	0.4

Table 10.11 Mass ratio comparison between cyanobacteria and bell peppers

Ratios	Cyanobacteria	Bell peppers
H_2O/CO_2	6.55	8.28
Total dry mass Out/CO_2 (%)	64	67
Total dry mass out/inputs (%)	8.34	7.19
Food dry mass/inputs	0.083	0.032

Table 10.12 Yeast MMR mass flows per living kilogram

Yeast as crop	kg/living kg
Inputs	
O_2	3.94
Feed dry mass	5.12
Clean water	2.36
Total inputs	*11.42*
Outputs	
CO_2	5.42
Living biomass	6.00
Total outputs	*11.42*
% water in biomass	76%
Dry mass out in food	1.44
Water out in food	4.56

Table 10.13 Yeast MMR stage (or sub-stage) productivity

Yeast MMR for Aerobic Production	
Inputs	
O_2	6.6
Feed dry mass	8.5
Clean water	3.9
Total inputs	*19.0*
Outputs	
CO_2	9.0
Living biomass	10.0
Total outputs	*19.0*
% water in biomass	76%
Dry mass out in food	2.4
Water out in total biomass	7.60

Adding this data to the total for the farm so far, we get this next intermediate accounting table in Table 10.14.

It is easiest to assume the yeast-bacteria-incinerator stage that is constructed in such a way to make the outputs we need, and process all the inputs as needed, including the conversion of waste inputs to nutrients. This makes a bit of sense with the numbers above. I need around 7 kg of carbon dioxide, I have 8.5 kg extra oxygen, have 773.9 kg of water in wastes of various types, and need 793 kg of sterilized and filtered water. Given the dry mass and need for nutrient mass, the array of bioreactors and incinerator in that stage should be able to make the balance using anaerobic and aerobic processes. Given this assumption, we can get a summary table for the whole farm, in Table 10.15.

Looking at Table 10.15, we can see the way mass flows overall. Since we chose to have bell peppers, the waste mass of the trimmings goes into yeast-bacteria-incinerator stage to make carbon dioxide and water. We could have just gone for glop and made extra algae to burn so to speak, resulting in a far smaller farm, but a less tasty menu. It is possible to eliminate the bell peppers as the source of waste mass in lieu of a slightly larger photobioreactor stage, eliminating even the tasty crunch of a salad fruit. Somewhat ignored by many science fiction novels, the glop diet's effect on crew psychology would be slowly devastating. That said, the overall flow for our example algae + yeast + bell pepper farm is shown in Fig. 10.21.

3. After we get a table such as Table 10.15, we can then take guesses at sizing the systems in the farm, then the farm itself.

    a. Stages, except the yeast-bacteria-incinerator stage and habitat are sized by using a multiplier of total living mass in the stage.

4. Photobioreactor stages have to have enough tank volume for all the inputs and outputs, plus enough for the base culture, living biomass right before the outputs. Membrane mediated designs are smaller than vat designs. That said, we figure out above that the total living biomass in the stage is 2570 kg, or roughly about

Table 10.14 Intermediate table of mass flows before the yeast-bacteria-incinerator (YBI) stage

Stage	Living kg in stage	CO_2 (kg)	O_2 (kg)	H_2O Clean* (kg)	H_2O waste (kg)	Waste dry mass (kg)	Nutrient dry mass (kg)	Food wet mass (kg)
Habitat		101	−84	0	711.3	31.8	–	−760
Cyanobacteria	*2570*	−104.1	89.7	−682.1	6.3	0.658	−10.8	700.0
Bell peppers	*19,883*	−12.9	9.3	−107.0	56.23	4.76	−0.4	50.0
Yeast Aerobic MMR	*1.7*	9.0	−6.6	−3.9	0.0	−8.5	0.0	10.0
Subtotal before YBI		*−7.0*	*8.5*	*−793.0*	*773.9*	*28.7*	*−11.2*	*0.0*

* = Clean water or for plants and algae, sterilized and filtered processed wastewater

Table 10.15 Example whole farm mass flow table using simplified chemistry

Stage	Living kg in stage	CO_2 (kg)	O_2 (kg)	H_2O Clean* (kg)	H_2O waste (kg)	Waste dry mass (kg)	Nutrient dry mass (kg)	Food wet mass (kg)
Habitat		101	−84	0	711.3	31.8	–	−760
Cyanobacteria	*2570*	−104.1	89.7	−682.1	6.3	0.658	−10.8	700.0
Bell peppers	*19,883*	−12.9	9.3	−107.0	56.23	4.76	−0.4	50.0
Yeast Aerobic MBR	*1.7*	9.0	−6.6	−3.9	0.0	−8.5	0.0	10.0
Yeast-Bacteria-Incinerator (Processing)		7.0	−8.5	793.0	−773.9	−28.7	11.2	0.0
Total		*0.0*	*0.0*	*0.0*	*0.0*	*0.0*	*0.0*	*0.0*

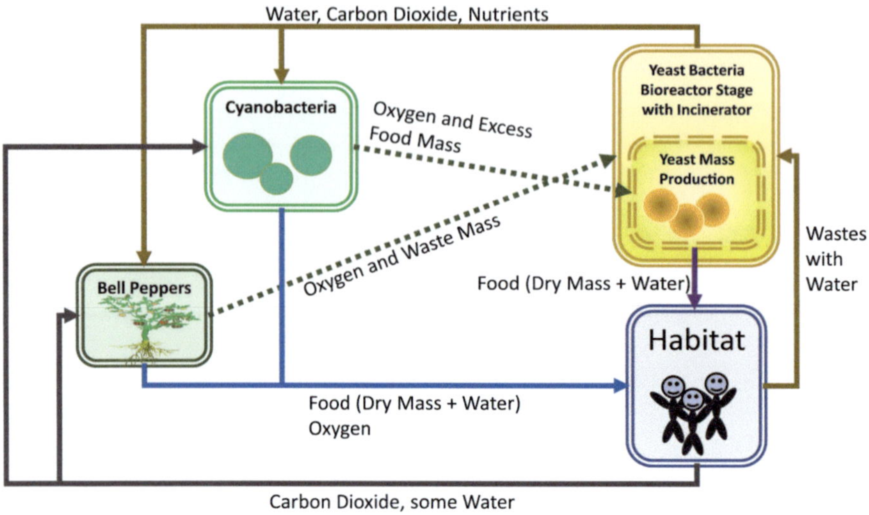

Fig. 10.21 Key mass flows for Example 1

2600 L of live cells, with a buffer of room around the cells, usually in most cases for vats about 50% to 100% more than raw cell volume. So, 3900 L to 5200 L, assuming thin-walled tanks. Adding in the inputs and outputs: Inputs are 104 + 682 + 10.8 or around 800 L input, though for CO_2 it may mean a twice as big gas tank in lieu of liquid tank so 880 L for input tanks, assuming some decompression. Roughly double that to add in outputs (mass in = mass out) so total tanks around 1800 L. Add in a factor of around 30% for the whole living tank mass plus input and output tanks, so (1800 + 5200) * 1.3 so around 9100 L, or 9 m^3, or the size of a bedroom in an average US house, at least.

5. The bell pepper stage is large, especially since these bell peppers are likely semi-perennial, and using the soil data in Table 10.2, bell peppers use around a square meter for 1.68 kg of plant though there are many factors that may add to that guess. Dividing roughly 19,883 live kg from the table above by the 1.68 yields a ground cover of around 12,000 m^2, or 1.2 ha, assuming 1-m-tall plants as in a commercial soil farm. One could stack the plants or stretch out the vines, shrinking the area to around 4000 m^3 at 3 m tall, which is still the size of a large lawn and house. The tanks for the ins and outs are trivial compared to this figure, but a few cubic meters would cover those. These settlers might want to reconsider how many peppers they want and beef up the photobioreactors!

   b. Assume the yeast-bacteria bioreactor must be sized based on an efficiency factor for total mass processed, plus tanks for ins and outs.
      i. For the food yeast, the total size is around 2 cubic meters at worst…you could double production and likely still fit into the 2 m^3 figure.
      ii. For the yeast-bacteria-incinerator, adding the total inputs is around 800 kg, for a guess of around 1000 L given gas inputs, times 2 or 2000 L.

Assuming a multiplication of 2 to 1 for all that machinery, 4000 L or 4 m^3.

iii. Adding the bioreactor sets and tanks, 6 m^3, which is the size of a bedroom, especially adding is a bit extra for reserve cultures.

c. Systems are sized based on how much total mass they handle per day. This means assuming all the pluses and minuses are all pluses.

i. Liquids: Adding all the clean and dirty water, this farm moves 1563 kg per day, or around 1.6 m^3 of tanks. Assume 2 m^3 to be safe, scattered over 2 m^2.

ii. Gases: this farm moves 31 kgs of gases, or at worst 24 m^3 at sea level pressure, though most tanks are pressurized between 10 and 50 atm. As a result, the tanks in the system are around 3 m^3, scattered over around 3 m^2 as a rough guess.

d. Adding all this up, as in Table 10.16, the total footprint area is around 0.4 ha (or around an acre in Missouri farmer-speak) for the whole farm, at around 3 m tall, would cover it all, in pessimistic terms. One-hundred people fed (sort of), watered, and waste treated in an acre is not too bad, not counting living space of course.

For each stage, the energy flows based on the types of machines or lights used. An example for a photobioreactor stage is in Fig. 10.22.

We can look at the same types of electrical energy use for the yeast-bacteria bioreactor stage, as in Fig. 10.23.

The last stage type to examine is the plant (bell pepper) stage, with a similar energy breakout shown in Fig. 10.24.

4. A very rough energy guess assumes all gases use high-pressure gas pumps to move into and out, and all liquids are pushed by macerating pumps. Fans are fans of course, and a multiple by the volume in the stage time how often it gets flipped (about 360 times per day for most places, so if air in the stage is around 1.2 kg/m^3, then the energy to flip it per day is volume in stage * 360 * 1.2* fan energy in Chap. 5 or (Meyer, 2023). We will need another table, where we add up each energy type and the masses of gases, liquids, and solids. Assume 50% extra water in tanks with yeast or algae for recirculation. (Meyer, 2023) tells us the recirculation rates for the photobioreactor are around 5760 times per day, root volume liquid in the hydroponic bed is 126 times per day, and the yeast-bacteria bioreactor at around 769 times per day. We can guess liquid root water to be about 16% of plant living mass for a good hydroponic system, so here a guess is *19,883 kg * 16%* = 3,181 kg of liquid in the roots getting flipped by a liquid pump 126 times a day for the peppers (a worst case, likely far less in reality). For photobioreactor liquids, pessimistically, 9m3 *(1 + 50%) * 5760 times per day * 1000 kg/m3 is kg/day liquid recirculated. For the yeast bioreactor, using data from the machinery section in Chap. 5, the electrical energy is computed in Table 10.17.

Table 10.16 Footprint assumptions and calculations for Example 1

Stage	Living kg crop in stage	Multiplier	Volume in m^3	Area in m^2	Notes
Cyanobacteria	*2570*	0.0015	9	3	Assume 50% extra for room, 1 kg/liter, then add inputs and outputs at 50 atm for gases, 1 kg/liter for liquids
Bell Peppers	*19,883*	0.595238	11,835	3945	1.68 kg/m^2 in soil, assume 3-m stack
Yeast Aerobic MBR	*1.7*	0.0015	1	1	Same rules as for Photobioreactor
	kg of reactant mass	Multiplier	Volume in m3	Area in m^2	Notes
Yeast-Bacteria-Incinerator (Processing)	*1608*	0.002	4	2	Rough multiplier of (total inputs and outputs) * 2 * 1 kg/liter/1000 L to m^3
Systems	Mass per day (kg)	Multiplier	Volume in m^3	Area in m^2	Notes
Liquid movement	*1563*	0.001	2	2	Multiplier is kg to liters to cubic meters, round up
Gas movement	*31*	0.076923	3	3	Multiplier assumes 1.3 kg to cubic meter at 1 atm, and 10 atm pressurization, round up
Total			*11,854*	*3956*	About 0.4 ha or 1 acre, at 3 m tall

What do all these tables and numbers tell me? This farm uses lots of space and energy just growing bell peppers! Even if we swapped bell peppers for tomatoes or another perennial crop like those, the result is similar if we really prioritize those fruits. We could just nip down the number of peppers and force the algae into the yeast vats and yeast-bacteria-incinerator array, eating more glop and get a far smaller farm, at the cost of a more slime-based diet. Let's stick with the farm above and forgo the slime-fest for now. That said, the glop comes at an energy cost due to recirculation. Bioreactors of any type, but especially photobioreactors need lots of recirculation (i.e. stirring) in the tanks. More glop costs more pump energy but takes up little room, and less photosynthetic light than vascular plant crops.

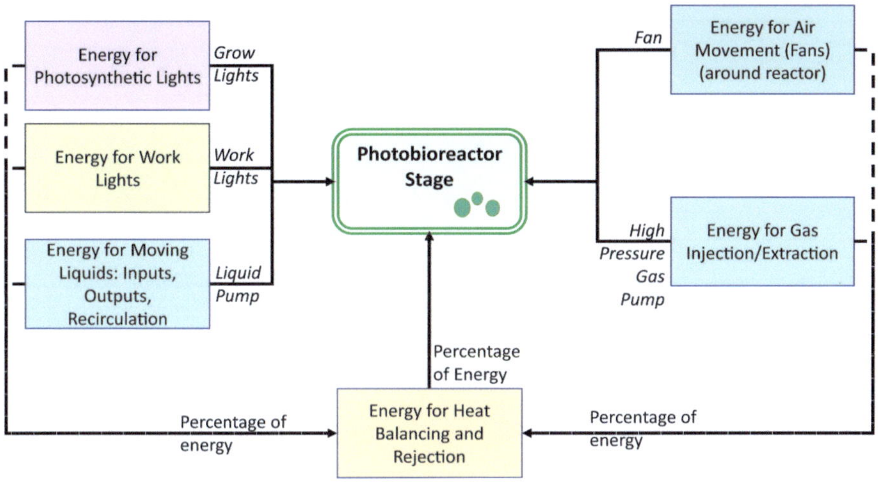

Fig. 10.22 Example 1 energy types for the photobioreactor (cyanobacteria) stage

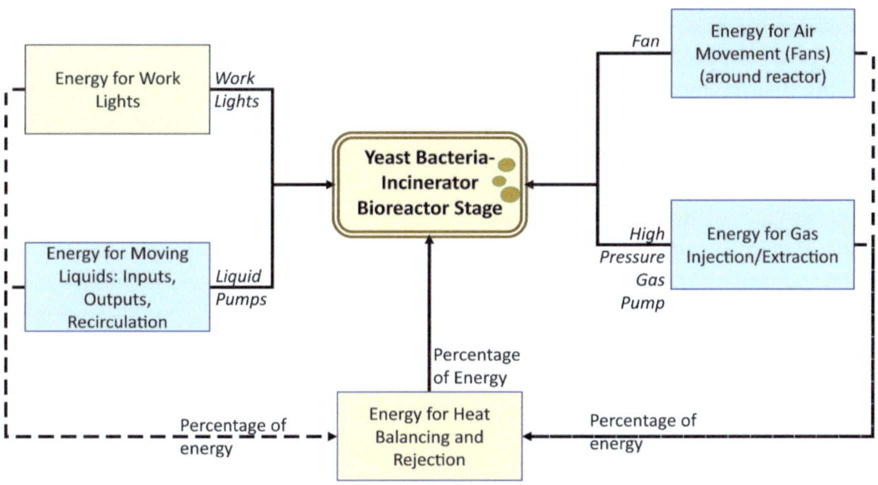

Fig. 10.23 Example 1 yeast-bacteria-incinerator bioreactor stage energy types

To get an initial supply to set up the farm as shown above, a super-pessimistic engineer would assume a year's worth of inputs, or around something like Table 10.18, which means you would need at worst ~ 910 metric tons of input stock (including food, oxygen, and mostly water, for 100 people).

In reality, the farm would need only a fraction, maybe less than half, the 908 metric tons figure since as the crops begin to grow, the humans begin to process, but I would like to make absolutely sure that my settlers survive, and crops get going. This starter amount is even lower if in situ water is available, and more mass efficient

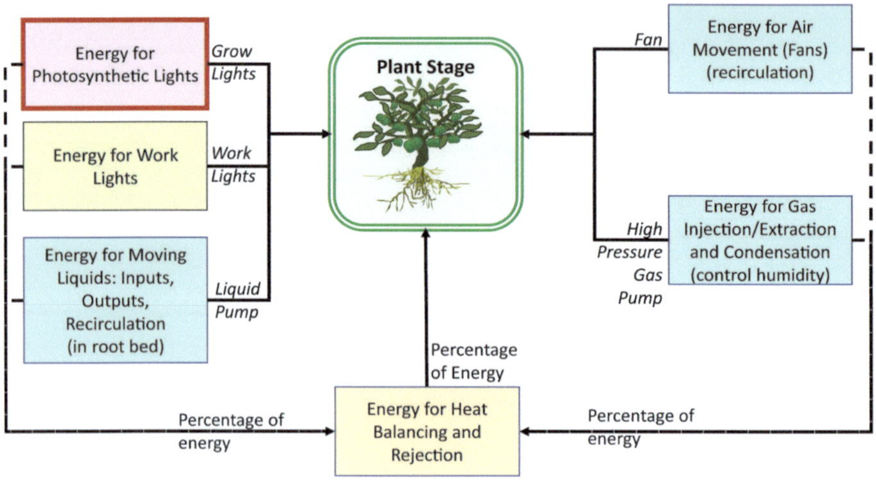

Fig. 10.24 Example 1 plant stage energy breakout

rations for the first year are available (having eaten meals ready to eat or MREs for short, given the glop, I would take the MREs).

This first example shows what is really a back of the napkin estimate for one scenario. It also shows the many tradeoffs that would need to occur in planning and designing a space farm. As in the optimization process far above, we would always start with human needs, including oxygen, water, and a menu. Next guess at photosynthetic crops that meet the oxygen need and all the plant- or algae-based foods. If there are aerobic crops, add in extra photosynthetic crops to feed the aerobic crops and provide oxygen. The rest rolls into how the bioreactor-incinerators handle the rest. In most cases I use a series of math models like Monte Carlo, regression, or genetic algorithms to guess the masses for each crop, then perform more tradeoffs.

To get a tighter model, we would need to look at elements, the actual nutrition of crops, and crops that use the continuous harvest model as annuals.

10.3.3.2 Example 2: More Crops, Growth Curves, and Continuous Harvest.

Using the same human model and assumptions as in Example 1 above, we can assume the settlement has gotten used to moving away from glop to 'real food' and decided to make more use of gardens in lieu of photobioreactors. Yes, this would be a bigger farm, but adding a protein crop like pinto beans adds a wide variety of menu options, since beans can be eaten whole boiled, dried, rendered for flour to make noodles or flatbreads, made into a sort of cheese, and pressed for oils, among many other uses. Legumes like pinto beans need lots of nitrogen-rich nutrients, and make a lot of silage, i.e. inedible stems, leaves, and roots. Legumes are also seasonal plants,

Table 10.17 Energy summary table for Example 1

	Watts/day						
Stage	High pressure gas	Fan	Liquid	Photosynthetic light	Work light	Heat balancing and Rejection	Total kW
Cyanobacteria	88	12	78,279	540	0.38	3946	83
Bell Peppers	10	15,849	606	710,100	493.14	36,353	763
Yeast Aerobic MBR	7	1	1161	0	0.13	58	1
Yeast-Bacteria-Incinerator (Processing)	7	0	1	0	0.25	0	~ 0
Total	88	15,863	80,047	710,640	494	40,358	848

Table 10.18 Very pessimistic initial supply, for Example Farm 1

	Inputs for a year (kg)				
Stage	CO_2 (kg)	O_2 (kg)	H_2O clean (kg)	Nutrient dry mass (kg)	Food wet mass (kg)
Habitat		30,660	259,625		277,400
Cyanobacteria	37,991		252,717	3,940	
Bell Peppers	4717		39,045	160	
Yeast Aerobic MBR		8.5	1438.2		
Total (kg)	42,708	30,668	552,825	4099	277,400
Total (T)	43	31	553	4	277

i.e. annuals, and will need to be grown for continuous harvest, unlike the cheat for possible perennials like modified bell peppers. The mass distribution schematic for Example 2 is in Fig. 10.25.

As before, the first concern is oxygen for humans (see Table 10.19) then calories and nutrition, then getting the balance in water and carbon dioxide. We will let the rest fill in as we get a rough average food input.

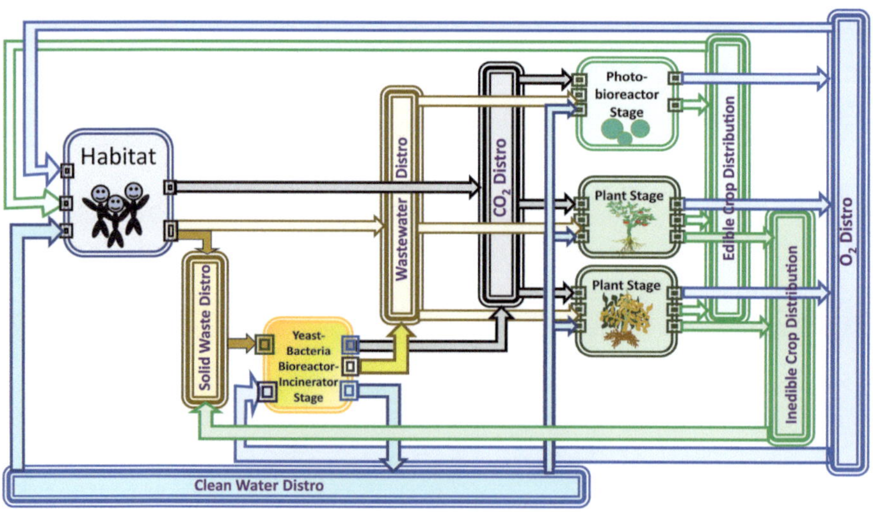

Fig. 10.25 Example 2 farm mass flow schematic with multiple plant stages

Table 10.19 Starting habitat data Example 2

Stage	People	CO_2 (kg)	O_2 (kg)
Habitat	100	101	−84

We already know the per living kilogram numbers from above for cyanobacteria photobioreactor and bell pepper stages, as in Table 10.20.

We can also use data as from Chap. 4 (Fig. 4.26), per living kilogram for pinto beans, in Table 10.21. This mass flow model assumes (very roughly and possibly wrongly) that there is the same ratio of water in the stems, roots, and leaves as in beans near harvest. This is even more risky in earlier growth when the plant is greener, but we need a rule of thumb to use. This illustrates the need to collect more experimental data.

It would be tempting to do a simple multiplication for beans like we did for the algae and peppers, but that would be a bit inefficient, especially since continuous harvest shrinks the farm down. More on that after we look at making sure the humans can breathe. The truth is that oxygen is easy in this scenario if we make enough food. The big problem is carbon dioxide. Given the mass of greenery in our farm, carbon becomes the driver, and cycling in water and carbon dioxide from wastes generated by the farm itself becomes critical, with the habitat becoming just a small aerobic stage compared to the need to produce carbon dioxide from all that silage. Doing a balance for even three crops gets complex, but we could make a few guesses. Let's start with some simple multipliers for the oxygen part. Let's say you just picked one crop to meet the oxygen needs for the humans. Dividing the oxygen input for the habitat by the per kilogram oxygen output yields 2.4 tons of cyanobacteria, about 8 tons for pinto beans, and many, many, many tons of bell peppers. This tells us that the bell peppers are going to be minimized right away to a side dish at best, but we can substitute the bean mass at around 4 to 1 for the cyanobacteria.

Next, we look at the food part. We handwaved over nutrition last time, since I gave you some numbers that met the minimums in Example 1. Let's get a bit more technical this time. First we can remember what happens when humans in the habitat take in food and oxygen from green stages like cyanobacteria, beans, and peppers. Food produced by the crops comes in as wet mass, which has dry mass and water parts. Humans, and all aerobic organisms, take in oxygen, in this case from the plant and algae stages, and use it to digest the food, combining carbon from the food's dry mass with the oxygen to release carbon dioxide (roughly). The rest of the food dry mass, water in the food wet mass, and any additional clean water provided to the habitat, become wastes. Like the food, the wastes are wet masses, with dry mass parts that come from the remaining food dry mass after digestion and minus the carbons

Table 10.20 Cyanobacteria and bell pepper stages per living kilogram for Example 2

Stage	Mass per living kg in stage per day						
	CO_2 (kg)	O_2 (kg)	H_2O Clean (kg)*	H_2O waste (kg)	Waste dry mass (kg)	Nutrient dry mass (kg)	Food wet mass (kg)
Cyanobacteria	−0.0405	0.0349	−0.2654	0.0025	0.0003	−0.0041	0.2724
Bell Peppers	−0.0007	0.0005	−0.0054	0.0028	0.0002	0.0000	0.0025

Table 10.21 Pinto bean mass flow in grams per living kilogram

Pinto Beans per living kg	g per day
Inputs	
Nutrient dry mass	0.48
Water (sterilized wastewater or clean)	7.2
CO_2	13.5
Total inputs	*21.18*
Outputs	
O_2	10.5
Living biomass (total)	10.68
% water in biomass	***18.00%***
Dry mass out	4.20365
Water out in biomass	6.48
Total outputs	*21.18*
% edible mass produced	38.00%
Living biomass (food)	**4.06**
% water in food	***18.00%***
Dry mass out in food	3.3
Water out in food	0.7
Inedible biomass	**6.62**
% water in inedible	***18.00%***
Dry mass out in inedible	5.4
Water out in inedible	1.2

for carbon dioxide. The water in the waste dry mass comes from digestion of plant dry mass and wet mass in the plants and the added clean water. See Fig. 10.26.

Since we know oxygen is likely easy here, let's look at calories. Let's say I want just enough algae as a food additive (and as a backup in case of crop failures for beans or peppers). Let's assume most of the calories come from the beans.

First, Table 10.22 has calories for wet mass of edible wet mass for each crop, best guess. We want at least 2500 kcal per person. Getting most of the calories from beans sounds pretty good given some mostly dry beans. Better than glop for certain! Beans are also very close to the being a nutritionally balanced food, though if we start chasing that tail, you will need much more complex math to guess a balance.

Next, we need to balance a few key tables and make a goal. We used pinto beans for the primary caloric source to around 90%, but we need to make sure that there is a good amount of dry mass for the habitat to account for the flows in Fig. 10.25. This means that the dry mass produced by all the crops needs to account for carbons out (mass CO_2—mass O_2) and aerobically produced water (rough guess is 41% of the CO_2 produced mass for a very worst case of super-efficient digestors). The habitat mass must balance just like every stage, so habitat waste dry mass will be the sum of

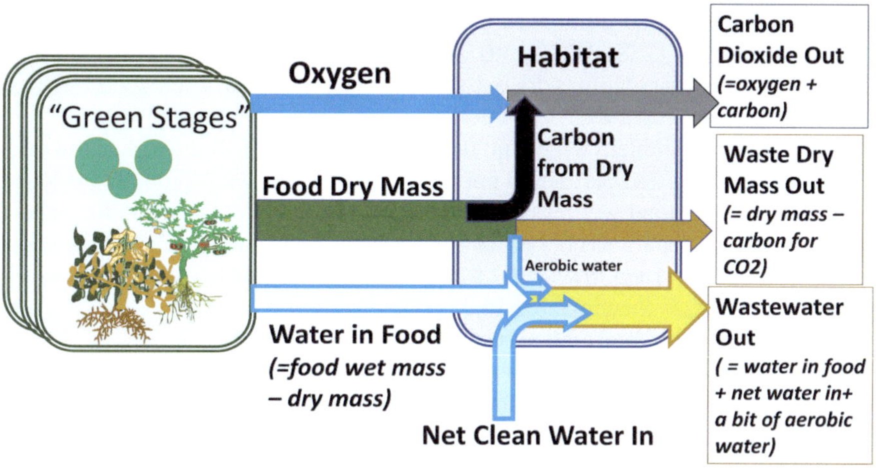

Fig. 10.26 Rough mass flow from green crops through the habitat

Table 10.22 Food calories per kilogram for wet mass

Crop	Kcal per kg food wet mass
Cyanobacteria	260
Bell Peppers	290
Pinto Beans	2,881

all the food dry mass produced per crop, minus (mass of CO_2–O_2) from Table 10.19, minus 41%*CO_2 from Table 10.19.

Your prime criteria for designing a farm start with habitat inputs and outputs, and some loose guidance on volumes and energy; rest might be up to you! An engineer must deal with the farm balance from there, and that is what I did in my papers. However, we can be a bit speculative here.

To get to the balance given all this data, you could start with Table 10.23, and then keep guessing the mass of the crops to get a balance of calories and habitat mass flow. You can do this in a spreadsheet as a rough guess.

You need to know how much of each crop's food is dry mass versus wet mass.

The combination of Tables 10.19, 10.20, 10.21, and 10.22 makes the next table. The first column is the wet mass of food produced per living kg per day (4.06 g =

Table 10.23 Food dry mass percentage of food wet mass

Crop	Food dry mass %
Cyanobacteria	9.4
Bell Peppers	7.8
Pinto Beans	82.0

.00406 kg in the case of pinto beans). The next column is the multiple of kcal per kg, times wet mass (2881 kcal/kg *0.00406 kg). The next column is copied from Table 10.23. The last column is dry mass % for each kg wet mass* the food wet mass column to get Table 10.24.

The pinto bean problem will haunt us later, for now our guesses will be the mass of total plants harvested per day to get the food mass we need. From Table 10.24, 38% of the plant's total wet mass is in the bean itself, 61% plant parts like stems, leaves, and roots I can't eat. If I guess I want 77.11 kg of food wet mass per day, I need to pull out 77.11 kg/38% = 202.92 kg of plants. We will use this in the harvest models to get a mass flow in Table 10.25.

Now you can guess total crop masses in the next table, adding in the human habitat data. I guessed for you to get the crop totals below. To make this diet work, we will need some salts and minerals too, about 4 g per person at least (or > 0.004 kg). This salt is why the balance always looks odd. For now, we leave off the salts. This guess is a big guess, likely incorrect. Since people's metabolism is FAR more complex than depicted, this back of the envelope estimate is for sizing the farm, not detailed nutritional guidance for settlers. You will need a nutritionist for that, and it would be

Table 10.24 Example 2 per living kilogram per crop dry mass table

Crop	Food Wet Mass (kg)	kcal Produced for food wet mass	Food dry mass %	Food dry mass produced
Cyanobacteria	0.27238	70.8	9.4	0.025603
Bell Peppers	0.00251	0.73	7.8	0.000196
Pinto Beans	0.00406	11.69	82.0	0.003328

Table 10.25 Dry mass balance and food calories between crops and habitat

Crop	Minimum living mass in stage (kg)	*Total food wet mass* (kg)	*Total food dry mass*	kcal Produced for Food Wet Mass
Cyanobacteria	250	*68.094*	*6.401*	*17,704*
Bell Peppers	14,000	*35.205*	*2.746*	*10,210*
	Mass of plants harvested per day (kg)	*Total food wet mass (kg)*	*Total food dry mass*	kcal produced for food wet mass
Pinto Beans	202.92	*77.110*	*63.230*	*222,121*
Sum of crops		180.409	72.377	250,035
For each person	*100 people*	*1.804*	*0.724*	2,500
Habitat balance	*Carbon mass from CO_2 out*	*H_2O mass from dry mass*	*Total food dry mass*	*Net habitat waste dry mass*
Habitat waste dry mass balance (>0?)	17	41.355	72.377	14.022

likely that many vitamin and mineral supplements will be in that initial supply. For the purposes of guessing, we will play along.

Now we need to guess clean water in and use that to calculate the wastewater, again a likely guess, and less than reality, but again, for the purposes of the example, shown in Table 10.26.

This results in a habitat stage mass balance table as in Example 1, shown in Table 10.27. If we wanted clean water net input, it would be added to the wastewater estimate too.

Now, we need to figure out the difference between the stage living mass for pinto beans, and the mass flow for the individual living kg.

Since we already assumed all the mortality and stage mass flow for the cyanobacteria and bell peppers, the pinto beans are all we need to estimate as a stage. Pinto beans are an annual vascular plant and therefore should use the continuous harvest methods in Chap. 6. The minimum living mass for pinto beans is the daily crop needed to produce the required food wet mass harvest. The actual amount of pinto beans in the stage total must be higher to get enough food for the total growth time of 94 days. The worst case is the big bang harvest model. From Chap. 4, it takes 94 days to get a crop. Harvesting 202.92 kg of bean plants per day, assuming I store the harvest perfectly, means I need 94 days *202.92 kg/day = 19,074 kg total plants in stage. That means I harvest 19 metric tons of plant every 94 days, yielding 7.2 metric tons of beans, that I store, and dole out in roughly 77 kg daily amounts. The 19 metric ton living mass assumes perfect growth, no losses, and lots of other interesting things, and takes a lot of room, at roughly 0.9 m^2/kg in soil as for the crop table in Chap. 1, this means 19,074 kg * 0.9 m^2/kg = 17,167 m^2, or 1.7 ha in an ideal field. The mass flow will vary from near nothing to days to harvest times inputs at maximum crop mass. That said, enclosed volume has a cost and for all the interesting reasons listed in the section on continuous harvest far above, we can do better. We will do some analysis here as in Chap. 6.

Table 10.26 Very rough wastewater estimate. Assuming 0 clean water net input

Total food wet mass (kg)	180.41
Total food dry mass (kg)	72.38
Water in food wet mass (kg)	108.03
Metabolic water (kg) est	41.35
Total habitat wastewater (kg)	149.39

Table 10.27 Habitat stage mass balance table for Example 2

Stage	CO_2 (kg)	O_2 (kg)	H_2O clean (kg)*	H_2O waste (kg)	Waste dry mass (kg)	Nutrient dry mass (kg)	Food wet mass (kg)
Habitat	101	−84	0	149.39	14.02	0	−180.41

First, we need a growth model for pinto beans, and the graph in Fig. 10.27 is a reasonable curve for 94-day beans, though notional. Note that there is an exponential growth phase to day 45, then a linear growth phase to dry out day 85, then an assumed zero mass flow during dry out right before harvest. We need this to get footprint and volume. For footprint and volume, we need the cumulative area under the curve in Fig. 10.27, and more math than you want here. The area under the curve is around 8279.48 kg (assuming 100% = 1 kg), and while unrealistic, results in a field of around 0.74 ha. That 0.74 ha, one layer of plants deep area, can be considered a lower bound of size for continuous harvest. For comparison, a field of pinto beans on Earth to meet our 77 kg per day need, assuming big bang harvest, would need around 2.2 ha using data in the crop table in Chap. 1.

To get mass flow, we also need a mass flow model for the pinto beans that captures the amount of material that flows in and out during the whole harvest time, per kilogram at harvest. Simplifying Table 10.21, we get Table 10.28.

So for every kilogram of food beans, over 94 days, it takes 3.326 kg carbon dioxide from the air, 1.774 kg of water via roots, not counting transpiration, and 0.118 kg of nutrients dissolved in the water via the roots. Over the same 94 days, the bean plant makes the 1 kg of food beans, but also 1.632 kg of inedible plant (silage), and releases 2.587 kg of oxygen to the air. We can assume that the silage is at least 18% water, so it will be 1.632*(1–18%) waste dry mass, and 1.632*18% wastewater.

Fig. 10.27 Pinto bean combined growth model with dry out

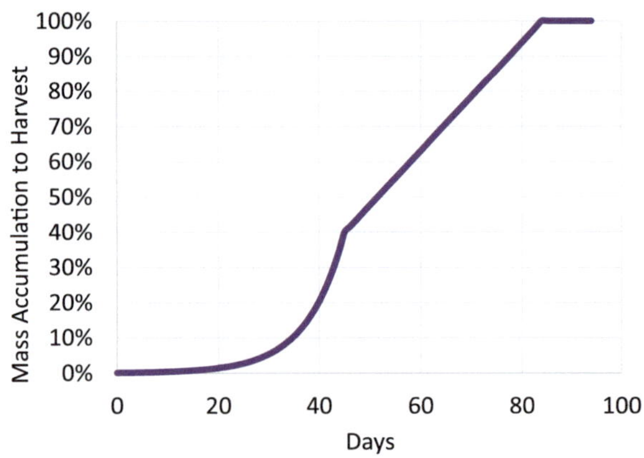

Table 10.28 Simplified pinto bean mass flow per kg food over 94 days

Pinto bean per harvested bean kg model			
Inputs		Outputs	
CO_2	3.326	Beans	1.000
H_2O	1.774	Silage	1.632
Nutrients	0.118	O_2	2.587
Input sum	*5.219*	*Output Sum*	*5.219*
94-day Harvest			

The daily beans needed is 77.110 kg, so multiplying the table immediately above, and assuming 1% loss in the stage, we can get Table 10.29.

Finally given these guesses, we get the big mass balance table in Table 10.30 just like in Example 1. As before each column totals 0, indicating closed cycle mass balance.

This mass flow means the yeast-bacteria bioreactor array with the incinerators need to aerobically process the excess oxygen and waste dry mass to get nutrients, carbon dioxide, and clean water. Total mass in and out of the yeast-bacteria-incinerator stage is around 935 kg per day.

To get footprint, the calculations for bell peppers and cyanobacteria are straightforward multipliers given the same reasoning as the first example, but the pinto bean continuous harvest creates a more interesting estimate. For the bell peppers, 14,000 kg, using a multiplier of 1/1.68 kg/m^2, and adding a 10% factor for tanks, walking area, etc., yields a volume of 9167 m^3 and a footprint assuming 3 m high areas, of 2778 m^2 (or 0.3 ha). For the cyanobacteria, the result is very much smaller than in example 1, but since we assume the photobioreactor used to process more material, we can guess the same dimensions as in example 1, i.e. 9 m^3 and 3 m^2. Now for the more complex pinto bean case. The first decision is how many cohorts, i.e. replanting or moves, should the pinto bean crop endure. The more cohorts, the lower the footprint and volume in the stage, but the more each plant is replanted or moved, increasing mortality before harvest, and increasing complexity in growth chamber design; however, the more cohorts also allow multiple growth chambers to optimize conditions for each stage in the pinto bean lifecycle. If plants were moved optimally, the closer we can get to the 7.5 ha optimum. There are at least three big breakpoints: 0 to 45 days for the exponential phase, 45–85 for the linear phase, and 85–94 for the dry out. Note that if we did this, the sizing for the 45–85 day and 85-day to 94-day cohort is essentially a full-sized plant size. So we could break up the linear phase at least at around day 65, making 4 cohorts: 0–45 days, 46–65 days, 66–85 days, and 86–94 days, as in Fig. 10.28.

Therefore, given these four cohorts, the footprint calculation looks like these steps, as in Table 10.31.

First, add up the 'squares' above: 40% * 45 day s + 71% * (65–46) days + 100%*((94–86) + (85–66)) = 58.4. Then we multiply the daily need for pinto bean plant, 202.92 kg = 12,858 kg. Assume 1.12 kg per m^2, which estimates that plants that are about 1 m tall. 12,858 kg/1.12 kg/m^2 * 1 m = a volume of around 11,500 m^3. This is above the optimum 0.74 ha ideal, but more realistic for our farm. Given

Table 10.29 Pinto bean stage mass flow for Example 2

Stage	CO_2 (kg)	O_2 (kg)	H_2O clean (kg)*	H_2O waste (kg)	Waste dry mass (kg)	Nutrient dry mass (kg)	Food wet mass (kg)
Pinto Beans	−259.07	201.50	−138.17	23.01	104.83	−9.21	77.11

Table 10.30 Example 2 farm and habitat mass flow table

Stage	Minimum living mass in stage (kg)	CO_2 (kg)	O_2 (kg)	H_2O clean (kg)*	H_2O waste (kg)	Waste dry mass (kg)	Nutrient dry mass (kg)	Food wet mass (kg)
Habitat	100 people	101	−84	0	149.39	14.02	0	−180.41
Cyanobacteria	250	−10.13	8.73	−66.33	0.62	0.06	−1.05	68.09
Bell Peppers	14,000	−9.10	6.58	−75.32	39.59	3.35	−0.31	35.21
Pinto Beans	8279	−259.07	201.50	−138.17	23.01	104.83	−9.21	77.11
Yeast-Bacteria-Incinerator	> 935 *kg processed per day*	177.29	−132.80	279.81	−212.61	−122.26	10.57	0.00
Total		0	0	0	0	0	0	0

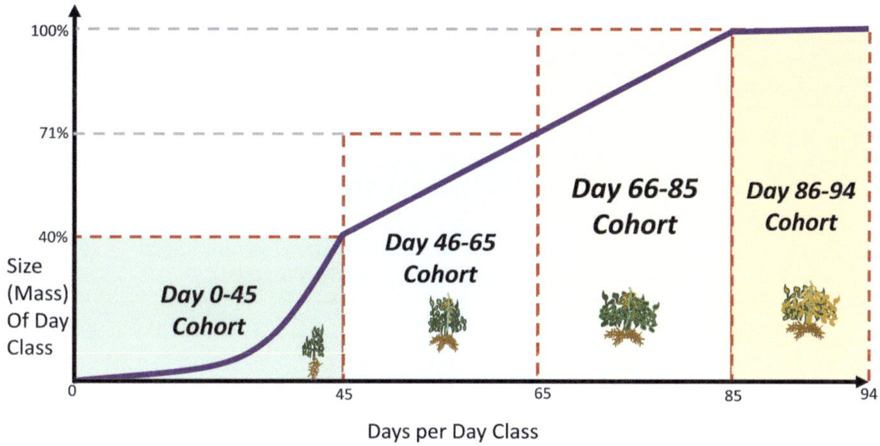

Fig. 10.28 Pinto bean cohort graph for Example 2

Table 10.31 Pinto bean cohort growth schedule and footprint for Example 2

Cohort	% Growth	Area
0–45 day	40%	18.00
46–65 days	71%	13.45
66–85 days	100%	19.00
86–94 days	100%	8.00
Sum		*58.4*
Mass Needed (kg/day)	202.92	12,858
	Kg per sq. m. plant	*sq. m. total*
Footprint Area	1.12	11,505 (1.15 ha)

a very rough 10% factor for tanks and areas around the growth chambers, and some fudging, this leads to a footprint of around 12,628 m^3, area depends on if we stack the plants or not, but unstacked, 1.3 ha roughly. All this analysis leads to Table 10.32.

For the yeast-bacteria bioreactor and incinerator stage, the sizing is based on mass processed, roughly a multiplier of (total inputs and outputs)*2 * 1 kg/liter/1000 L to m^3. This leads us to 935 kg * 2 * 1/1000 = 1.87 m^3, conservatively 2 m^3, though I would add in a cubic meter to be safe, being a pessimistic Missourian myself, to 3 m^3.

For the connecting systems, we move 857 kg per day by adding up all the liquid inputs and outputs from each crop stage, and 805 kg per day in gases, adding up the input and output gases for each of the crop stages, adding in the habitat inputs and outputs makes a total of roughly 1000 kg in liquids, and 990 kg gases, as shown in Table 10.33.

Given that the size of the vascular plant crop stages, the systems, tanks, and bioreactor stages are teeny, all the systems and bioreactor stages together are 77 + 2

Table 10.32 Footprint and sizing for Example 2 crops

Stage	Sizing method	Area (m^2)	Volume (m^3)
Cyanobacteria	*Reused from Example 1*	3	9
Bell Peppers	*Multiplier of living kg in stage (14,000 kg), 1/1.68 m^2/kg, assume 10% for tanks, walking area,* etc	2778	9167
Pinto Beans	*4 Cohort Continuous Harvest Model yields 11,505 m^2 at 1 m high, add 10% for tanks, walking area,* etc. *No Stacking*	12,628	12,628
Yeast-Bacteria-Bioreactor Incinerator	Total Mass Processed, (total inputs and outputs) * 2 * 1 kg/liter/1000 L	2	3

Table 10.33 Liquid and gas movement by systems in Example 2

Systems	Mass per day (kg)	Multiplier	Area in m^2	Volume in m^3	Notes
Liquid Movement	***1000***	0.001	**2**	**2**	Multiplier is kg to liters to cubic meters, round up
Gas Movement	***990***	0.0769	**26**	**77**	Multiplier assumes 1.3 kg to cubic meter at 1 atm, and 10 atm pressurization, round up, and 3 m tall

+ 9 = 88 m^3, versus 9167 + 12,656 = 21,795 m^3. Around 2.2 ha of total farm size would cover all bases and feed all 100 humans.

Something very critical to note, the flow of mass inside the farm is around four times greater than the flow to and from the habitat. See Fig. 10.29. This is expected once the menu starts looking like something beyond glop and has been seen in all my prior work. On Earth, the flow comes from water and the atmosphere largely, and some from the soil, but in space we have to account for all mass flow, and this illustrates the cost of a better dinner, anywhere.

For energy, we use the same method in Example 1 above, breaking out the type of mechanical action, and lighting, for each stage and the farm. Same multipliers as Example 1 too. See Table 10.34.

As in many cases, light drives the energy bill. Again, we are paying for what we get for free on Earth, i.e. sunlight and wind. Adding a protein vascular plant crop added to the energy requirement over Example 1, so not just footprint but energy are costs to expand the menu. This may drive a switch from solar to nuclear power for the settlement.

Starting mass is again computed as it is for Example 1 above, by figuring out a year's worth of inputs (Table 10.35).

The big note between Example 1 and Example 2 is the change in food wet mass, since the Example 1 estimate assumes the food is in wet live cyanobacteria cells

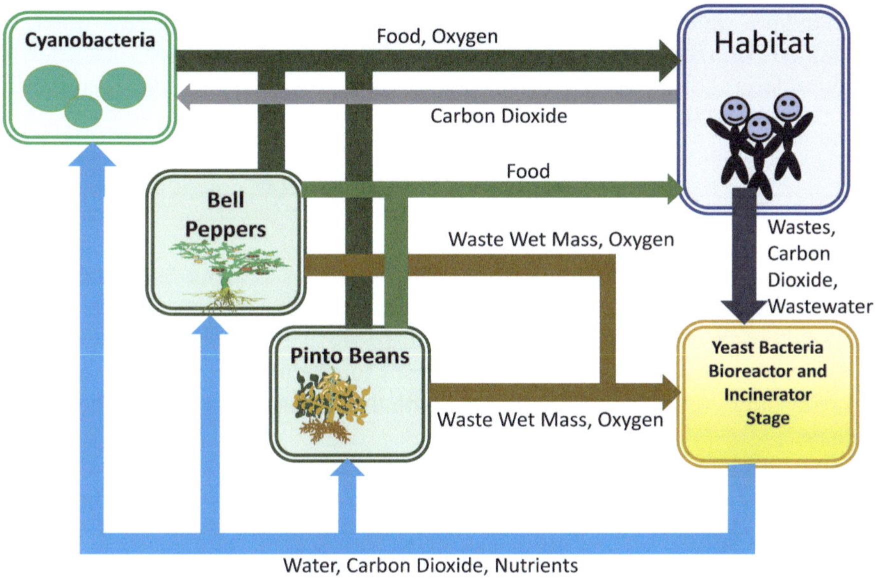

Fig. 10.29 Major mass flows, Example 2

Table 10.34 Example 2 electrical energy estimate

	Watts/day						
Stage	High pressure gas	Fan	Liquid	Photosynthetic light	Work light	Heat balancing and rejection	Total kW
Cyanobacteria	9	1	7615	53	0	384	8
Bell Peppers	7	12,276	595	499,995	347	25,597	539
Pinto Beans	208	16,911	352	2,277,990	1582	114,852	2412
Yeast-Bacteria-Incinerator (Processing)	140	4	3484		0.25	254	4
Total	364	29,192	12,046	2,778,037	1929	141,087	2963

Table 10.35 Pessimistic starting mass estimate for Example 2

	Inputs for a year (kg)				
Stage	CO_2 (kg)	O_2 (kg)	H_2O clean (kg)	Nutrient dry mass (kg)	Food wet mass (kg)
Habitat		30,660	–		65,850
Cyanobacteria	3696		24,584	383	
Bell Peppers	3,322		27,492	112	
Pinto Beans	94,559		50,431	3362	
Total (kg)	101,576	30,660	102,507	3,858	65,850
Total (T)	102	31	103	4	66

that are 90% water, versus the swap out for pinto beans which are 18% water. In Example 1, in practice, the cyanobacteria input would be a powder or thick paste of cyanobacteria at most, and a bunch of drinking water, again if the initial food supply is not something like MREs.

10.3.3.3 Example 3. Adding in the Animals

In this example, we use the same crops and assumptions as above but add in tilapia that eat cyanobacteria. Here, humans no longer eat algae, but do eat fish that eat algae, without wasting any parts, head to tail. To get to a balance here, we will need to work backwards from a diet for the humans to size food crops and then get the algae (cyanobacteria) inputs for the fish. With those masses, we can get the overall balance.

First, let's look at the major mass flow diagram in Fig. 10.30.

The implication here is that the tilapia fish stage dumps fish waste into the overall waste management system as in Fig. 10.31, but if you built this farm, it need not be so. You could pipe the tilapia tank water through the roots of the bell peppers and pinto beans as a first stage cleaner, a.k.a. aquaponics, reducing the need to use the machinery in the yeast-bacteria bioreactor. That said, the worst case means the fish waste ends up processed much like human wastes, which also simplifies balance calculations for our simplified estimates.

Given these mass flow assumptions, we need a human model for the habitat. We start with the assumed human diet in Table 10.36.

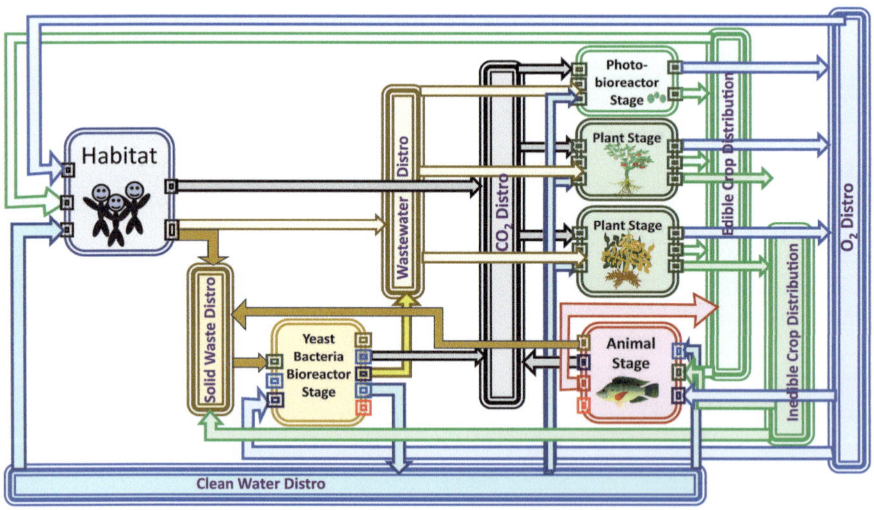

Fig. 10.30 Example 3 farm mass flow schematic with an added animal stage

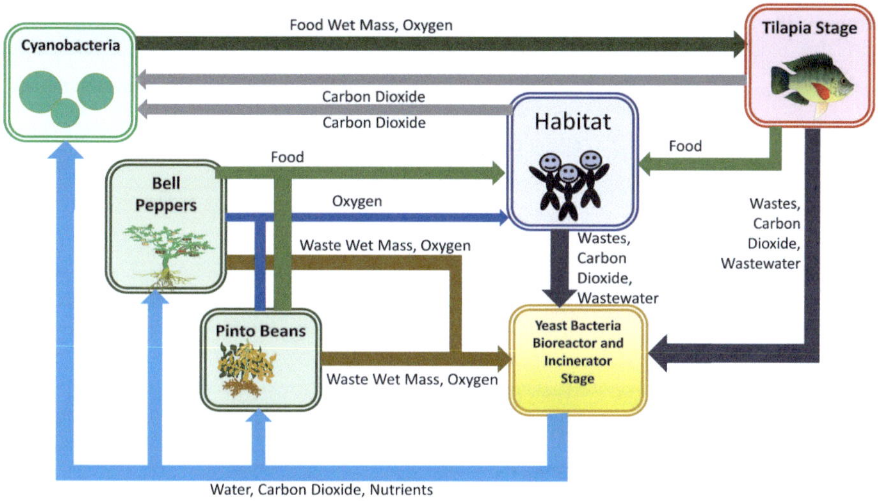

Fig. 10.31 Major mass flows for Example 3

Table 10.36 Assumed human diet for Example 3

Crop food	Food wet mass inputs per person (kg)	kcal/Person
Bell Peppers	0.25	72
Pinto Bean	0.50	1700
Tilapia	0.25	240
Total	*1.00*	*2012*

To eat those pinto beans in this case, we will need some drinking water to boil the beans and do something with the fish. Using Table 10.36 and adding a liter of water per person (1 kg/person per day), we can get the habitat stage mass flow in Table 10.37.

Now we need to know how the tilapia process mass. We estimate starting with data in the tilapia mass flow model in Chap. 3 (Figs. 3.9, 3.10, 3.11 and 3.12). Tilapia inputs and outputs per day per living kilogram of fish. This model assumes algae are the primary feed. With numbers from Table 10.37, we can get Table 10.38.

Table 10.37 Habitat model for Example 3

Stage	Minimum living mass in stage (kg)	CO_2 (kg)	O_2 (kg)	H_2O clean (kg)*	H_2O waste (kg)	Waste dry mass (kg)	Nutrient dry mass (kg)	Food wet mass (kg)
Habitat	100 people	101	−84	−100	150.8	32.2	0	−100

Table 10.38 Tilapia per unit mass flow model for Example 3. This model already assumes continuous harvest

Tilapia	g per day per living kg
Inputs	
Food wet mass (Algae)	10
Clean water	4.7
O_2	1.5
Total inputs	*16.20*
Outputs	
CO_2	2
Living biomass (total)	6.00
% water in biomass	***75%***
Water out in biomass	***4.50***
Dry mass in living biomass	***1.50***
Waste wet mass	***8.20***
Waste dry mass	0.80000
Water in waste wet mass	7.40000
Total outputs	*16.20*
% edible mass produced	38.00%
Living biomass (food)	**6.00**
% water in food	***75.00%***
Dry mass out in food	1.5
Water out in food	4.5

Now we need to back out the tilapia need for the menu, which is 0.25 kg/person/day * 100 people = 25 kg of whole fish per day. Doing some simple multiplication (i.e. multiplying everything by 25 kg/0.006 kg living mass output = multiply all by 4167 which is the rough estimate for stage mass), and knowing that the continuous concept of mass flow is the same as in Table 10.38; we get Table 10.39.

Tilapia to production growth is exponential and takes roughly 180 days from egg to harvest, no counting breeders. The model borrowed from the fish section far above already assumes the curve in Fig. 10.32.

Table 10.39 Tilapia mass flow for Example 3

Stage	Minimum living mass in stage (kg)	CO_2 (kg)	O_2 (kg)	H_2O clean (kg)*	H_2O waste (kg)	Waste dry mass (kg)	Feed wet biomass	Nutrient dry mass (kg)	Food wet mass (kg)
Tilapia	*See below*	*8.33*	*−6.25*	*−19.58*	*30.83*	*3.33*	*−41.67*	*0*	*25.00*

Note that there is a new column: feed wet mass. All aerobic organisms need to feed. We will shortcut the gymnastics of dry v. wet since we know the photobioreactor is the feed source for this example

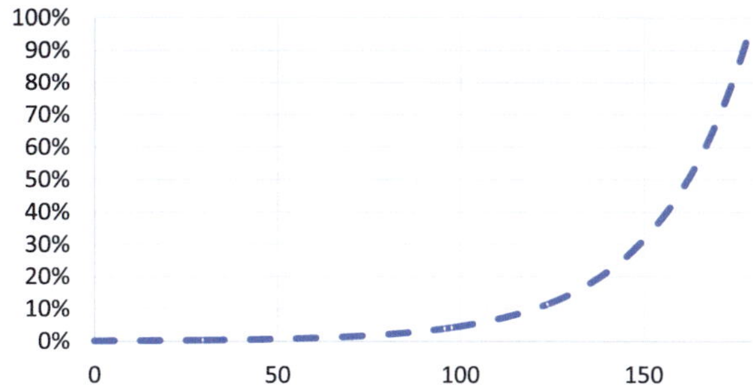

Fig. 10.32 Example 3 tilapia growth curve

There are a few ways to tackle the biomass in the stage; one way is to use the per kg multiplier from the mass flow model, 4167 kg. Using the same big bang analysis, we used for pinto beans, gives a mass of 180 days * 25 kg/day = 4500 kg. Using a perfect exponential mapping of the curve above, 26.5 * 25 kg = 663 kg in stage. A linear model, which is more pessimistic, leads to 2317 kg. The truth is somewhere in between 663 kg and 4500 kg. Spacing for tilapia is somewhere between 30 kg/m^3 for a round tank, to 50kg/m^3 for a perfect raceway. We are talking a small footprint either way, from 220 m^3 to 1500 m^3.

Assuming an exponential model, with 5% mortality, and 0.5% breeders, we could use this stage living mass estimate: 663 kg * (1 + (5% + 0.5%)) = 700 kg. The real-world mortality rate on Earth can be far higher.

Mortality results in waste wet mass, around an added 33.2 kg/day, 75% water. The breeders add more feed, and more waste, but no growth, 3.31 kg added. All told, the mass flow table for the tilapia stage now looks like Table 10.40.

Now add in the algae to feed the fish. We need 43.96 kg/day of algae and scaling Example 2 cyanobacteria data by 43.96/68.09 = 65%; we can get Table 10.41.

Likewise, we scale the data using the same growth models in Example 2 for this example for bell peppers and pinto beans based on human food need as in Table 10.42.

Combining the tables above leads to an overall mass flow as in Table 10.43. As before, the yeast-bacteria bioreactor + incinerator stage balances out the values for the crops and habitat in the rows above. Adding all the rows for each mass type results in 0 kg, showing that the mass flow is balanced.

The gas and liquid systems look like Table 10.44, adding gases and liquids (including waste masses) together.

Sizing all the remaining farm stages as in Table 10.45.

Table 10.40 Tilapia mass flow model total mass, with mortality and breeders Example 3

Stage	Minimum living mass in stage (kg)	CO_2 (kg)	O_2 (kg)	H_2O clean (kg)*	H_2O waste (kg)	Waste dry mass (kg)	Feed wet biomass	Nutrient dry mass (kg)	Food wet mass (kg)
Tilapia	700 kg	8.79	−6.59	−20.66	28.07	9.36	−43.96	0	25.00

Table 10.41 Example 3 cyanobacteria photobioreactor mass flow. Note that the food wet mass column in Examples 1 and 2 becomes feed wet mass (for tilapia) in this table

Stage	Minimum living mass in stage (kg)	CO_2 (kg)	O_2 (kg)	H_2O clean (kg)*	H_2O waste (kg)	Waste dry mass (kg)	Feed wet biomass	Nutrient dry mass (kg)	Food wet mass (kg)
Cyanobacteria	161	−6.54	5.63	−42.82	0.40	0.041	43.96	−0.68	0

Table 10.42 Example 3 bell peppers and pinto beans mass balance

Stage	Minimum living mass in stage (kg)	CO_2 (kg)	O_2 (kg)	H_2O clean (kg)*	H_2O waste (kg)	Waste dry mass (kg)	Feed wet biomass	Nutrient dry mass (kg)	Food wet mass (kg)
Bell Peppers	9942	−6.46	4.67	−53.49	28.12	2.38	0.00	−0.22	25.00
Pinto Beans	5369	−167.98	130.65	−89.59	14.92	67.97	0.00	−5.97	50.00

Table 10.43 Example 3 overall mass flow table

Stage	Minimum living mass in stage (kg)	CO_2 (kg)	O_2 (kg)	H_2O clean (kg)*	H_2O waste (kg)	Waste dry mass (kg)	Feed wet biomass	Nutrient dry mass (kg)	Food wet mass (kg)
Habitat	100 People	101.00	−84.00	−100.00	150.80	32.20		0.00	−100
Tilapia	700	8.79	−6.59	−20.66	28.07	9.36	−43.96	0.00	25
Cyanobacteria	161	−6.54	5.63	−42.82	0.40	0.04	43.96	−0.68	0
Bell Peppers	9,942	−6.46	4.67	−53.49	28.12	2.38		−0.22	25
Pinto Beans	5,369	−167.98	130.65	−89.59	14.92	67.97		−5.97	50
Yeast-Bacteria Bioreactor + Incinerator	769 kg processed per day	71.19	−50.37	306.55	−222.30	−111.95		6.87	0.00
Total		0	0	0	0	0	0	0	0

Table 10.44 Example 3 gas and liquid system mass balance

System	Mass per day (kg)	Multiplier	Volume in m^3	Area in m^2
Gas system	644	0.0010	1	2
Liquid system	1383	0.0769	107	36

Table 10.45 Example 3 sizing multipliers

Stage	Sizing method	Area (m^2)	Volume (m^3)
Cyanobacteria	*Reused from Example 1*	3	9
Bell Peppers	*Multiplier of living kg in stage, 1/1.68 m^2/kg, assume 10% for tanks, walking area,* etc	6509	19,528
Pinto Beans	*4 Cohort Continuous Harvest Model yields 11,505 m2 at 1 m high, add 10% for tanks, walking area,* etc. *No Stacking*	5906	5906
Tilapia	*4 Cohorts: 0–30 days, 31–60 days, 61–120 days, 121–180 days. 10% for input and output tanks and working volume. Assume circular 1 m high tanks (*1.273)*	78	61
Yeast-Bacteria Bioreactor + Incinerator	*769 kg processed, Rough multiplier of (total inputs and outputs)*2*1 kg/liter/1000 L to m^3, rounded up, 2 m high*	1	2

If you are comparing Example 2 and Example 3, you noted a total footprint drop. The farm in Example 3 shows why aquaponics is an efficient system for getting a high protein diet. The cost for Example 3 is in initial supply.

We can use the breakouts for energy as before, but we now add energy types for the tilapia stage as in Fig. 10.33.

The recirculation values here get added to the liquid column in the energy table. Again, we should use coefficients from Chap. 5, with a 7% heat rejection cost to be more pessimistic than in Example 2.

The same multipliers apply in Example 1, except for the fish. From (Meyer, 2023), we can guess that recirculation for the fish is about 1450 kg/day/kg fish in stage. We have 700 kg of fish in the stage, so recirculation mass is 700 kg fish * 1450 kg/day/kg fish = 1,015,000 kg/day. Having fish fed on algae shrinks the area footprint but will cost you a little electricity need, as shown in Table 10.46.

Again, light drives the energy bill for green crops. For initial mass, we compute just like for the previous examples, but need to add extra water to fill the fish tanks as in Table 10.47.

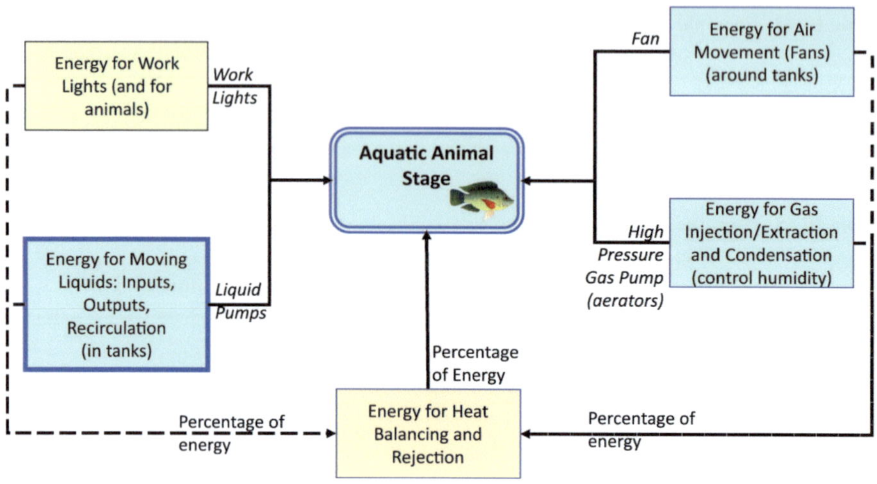

Fig. 10.33 Example 3 aquatic stage energy breakout

Table 10.46 Electrical energy with recirculation

	Watts/day						
Stage	high pressure gas	Fan	Liquid	Photosynthetic light	Work light	Heat balance and rejection	kW Total
Tilapia	15	82	1533	0	9.7	115	1.75
Cyanobacteria	6	1	4916	34	0	347	5
Bell Peppers	5	8717	423	355,057	246	25,511	390
Pinto Beans	135	10,966	228	1,477,112	1026	104,263	1594
Yeast-Bacteria Bioreactor + Incinerator	122	3	71	0	0.1	14	0.21
Total	282	19,768	7170	1,832,202	1282	130,249	1991

10.3.3.4 Comparing Examples 1, 2, and 3

The big metrics for comparison are initial mass, footprint, and energy versus the tradeoff for menu/crop diversity as in Table 10.48. We can compare all three examples by putting the crops produced (which led to menus), the total ground covered (footprint), the volume enclosed for the farm, and the electrical need (load) in wattage. These tradeoffs tell the engineer how to size the power production systems (such as fission or solar), construction effort and material (due to volume and footprint), and nutritionists and cooks how to plan meals.

Table 10.47 Initial mass needs for first year Example 3

Stage	Inputs for a year (kg)				
	CO_2 (kg)	O_2 (kg)	H_2O clean (kg)*	Nutrient dry mass (kg)	Food wet mass (kg)
Habitat		30,660	36,500		36,500
Cyanobacteria	2,386		15,870	247	
Bell Peppers	2359		19,522	80	
Pinto Beans	61,314		32,701	2180	
Tilapia		2406	33,208		
TOTAL (kg)	66,059	30,660	104,594	2507	36,500
TOTAL (mT)	66	31	105	3	37
Overall Total (T)					*240*

Table 10.48 Example comparison table for footprint, volume, and energy

Example	Human food crops	Total footprint (hectares)	Total volume (m^3)	Electricity load (MW)
1	Cyanobacteria, Yeast, Bell Peppers	0.40	11,854	0.8
2	Cyanobacteria, Bell Peppers, Pinto Beans	1.55	21,913	3.0
3	Bell Peppers, Pinto Beans, Tilapia	1.25	25,506	2.0

In Example 3, the cyanobacteria produced are fed entirely to tilapia, so cyanobacteria are excluded from human food

Example 3 needs around 2 MW in power, and in comparison, we had 1 MW for Example 1, and 3 MW for Example 2. Glop diets like in Example 1 save volume and area, but cost in variety (ugh!). Aquaponic systems as in Example 3 cost some space but shrink the footprint and energy over a vegan solution as in Example 2, due to the algae-animal-plant nutrient and water flow, like the natural flow in a pond.

In all these examples, we ignored something critical, the chemical element balance. Balancing elements, especially nitrogen, would be key before going with a farm size for design. The same can be said for minerals. My technical papers go into these balances, but I am going to leave them off here. The need for the trace elements and key minerals may drive in situ mining, food supplements, and import from other settlements or Earth, depending on the element deficiency. Such balancing acts are very much more difficult to compute.

10.4 Going Further

What about a more complex farm, as in Fig. 10.34?

Going past just a few crops makes the work of balancing mass, especially considering element balances, much more complex. Adding more fidelity and results from experiments is good too. We handwaved over the yeast-bacteria bioreactor and incinerator array, but in reality, we would have to include critical chemical pathways and processing rates for each reaction type. Getting balance usually means making a model for each crop to the element level, and the bioreactors and incinerators, then taking a guess at a possible set of farm masses, then running a simulation to get the balance or not, then adjusted guessing until close. There is also a great need for experimental data for possible space farms, and it would be a great idea to make a prototype on Earth before copying the farm in space.

10.4.1 Tracking Elements

The big elements in the farm flow are carbon, oxygen, hydrogen, and nitrogen, in space farm parlance, the 'CHON's. A thorough analysis, however, should add sulfur and potassium (S, K) since they are in amino acids. Even with the 'CHONSK', we also will need to track the stuff in ash too, such as Zn, Fe, Cu, etc. In the habitat, hydrogen and oxygen from water, nitrogen and carbon and hydrogen from food, and oxygen gas, become carbon and oxygen in exhalation, hydrogen and oxygen in wastewater, and nitrogen and everything else in wastes. The waste wet biomass from

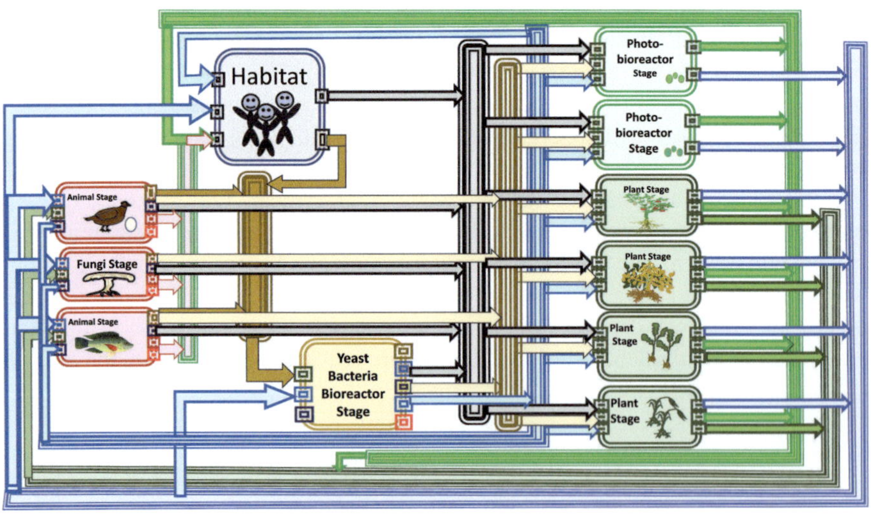

Fig. 10.34 Complex farm mass flow schematic

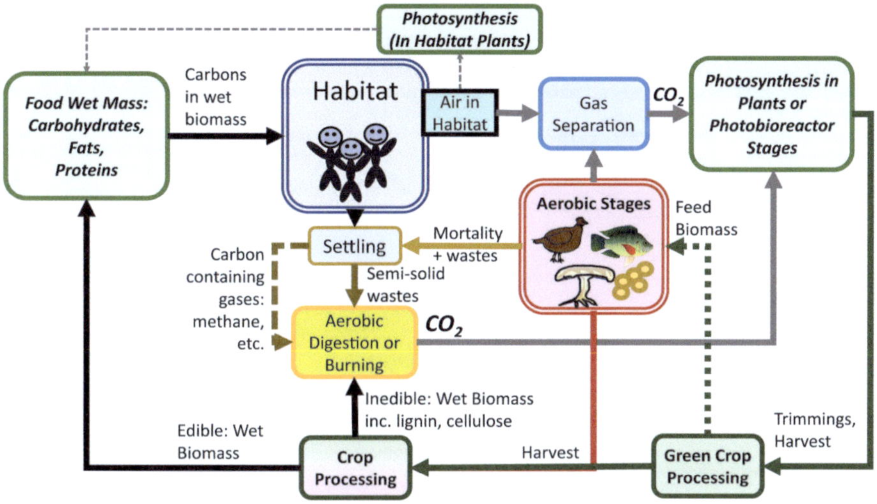

Fig. 10.35 Farm carbon cycle

urine, feces, and other black water sources holds many key elements. This is the same exchange for crop animals. For plants, it is a reverse, where nutrients, assumed processed lightly or filtered from animal stages, or heavily processed in a bioreactor from the habitat, become the nitrogen and other element source for the plants, while the bulk of carbon, oxygen, and hydrogen come from carbon dioxide and water as in Fig. 10.35.

The nitrogen cycle is key, see Fig. 10.36. Nitrogen is therefore another item that must be monitored, especially since it can provide a human-inert gas in partial pressure for breathing. Using helium as an inert gas in plant growing areas may mitigate this issue. Nitrogen broken down by humans and animals via protein digestion becomes available for every plant and algae as ammonia and urea, so urine and sweat must be collected and recycled. Nitrogen compounds and water will be key supply items, and part eventually of an economic web between settlements, because even if it is near ideal cycles, some nitrogen will be lost.

Hydrogen and oxygen move around the farm in CO_2, H_2O, carbohydrates, and O_2, mostly via photosynthesis and the reverse in aerobic respiration. Water cycles (see Fig. 10.13) are internal to each stage via cycles of condensation and evaporation (including transpiration, sweat, breath), and via recirculation. But water is also an input to photosynthesis for green crops and growth in all crops. Water is broken out of biomass using aerobic respiration or burning, for input to all stages as needed.

Oxygen cycles with water and carbon dioxide via respiration and burning, but also in gas and dissolved forms (see Fig. 10.37). Sometimes these forms are locked into soils for use by plant roots or absorbed and confined long term in oxides that can be a loss to the farm as a whole.

Other elements are typically recorded as ash in analysis, or as minor components of some amino acids (cysteine and phenylalanine for sulfur). Calcium, sodium, zinc,

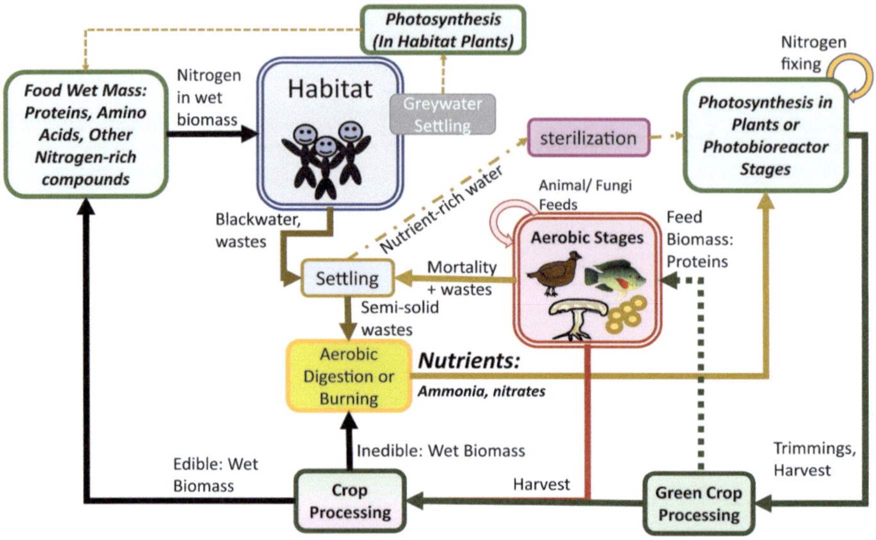

Fig. 10.36 Space farm's nitrogen cycle

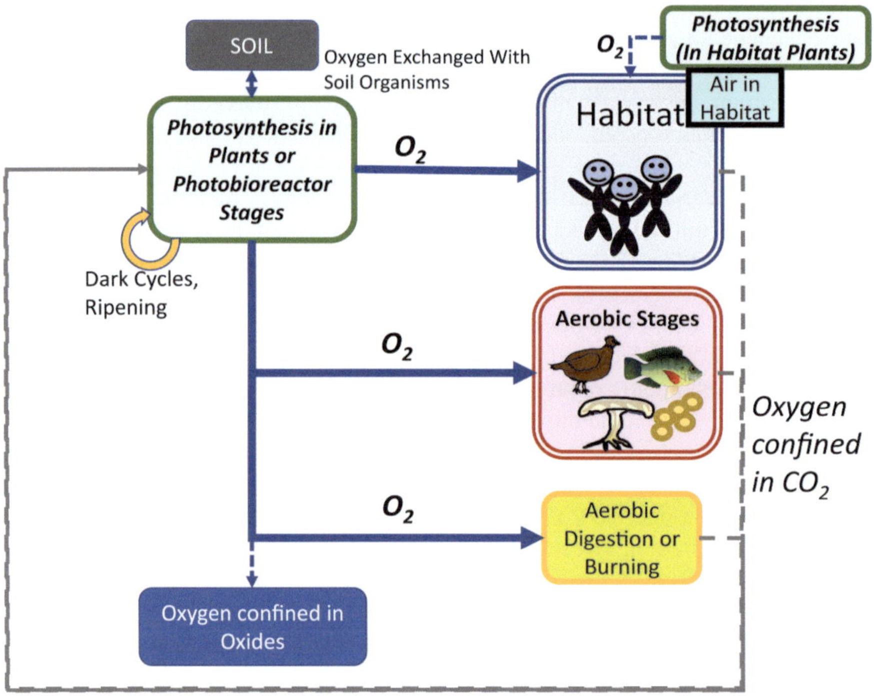

Fig. 10.37 Simplified farm oxygen cycle

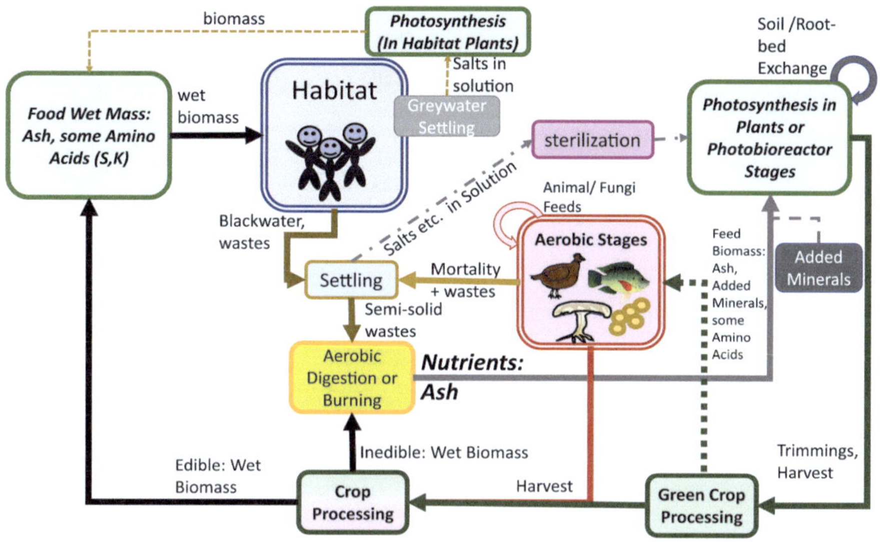

Fig. 10.38 Ash, salts, and mineral farm cycle

copper, and potassium are key for salts and other biological reactions. Overall, the ash components cycle (see Fig. 10.38) similar to nitrogen, usually via wet biomass, is converted to ash or salts by aerobic respiration and burning.

Detailed analysis and experiments are very much needed to track all the element cycles before building a real space farm, which is why a good chemical process engineer and careful sensing and control systems are a must for the space farm planner.

10.4.2 Attempting an Element Balance: Problems with People for Example 3

One of the issues in tracking elements through a food to human to farm process is the fact that we assume either the carbon dioxide exhaled is as in NASA-STD-3001 (2005) or similar, or humans eat a light diet. The problem again shows in carbon. Given all the cool math in Example 3, we can take a stab at how elements track through people, starting with the diet.

First, we need to look at the human diet and the percentage that is dry mass, as in Table 10.49.

Next, we need to look at the dry mass elemental breakout of each food, by dry mass, as in Table 10.50, calculated very loosely from nutritional data for each crop and chemical data in Chap. 2.

Table 10.49 Example 3 human diet. Example 3 diet broken out by dry mass and wet mass per person

Food	Wet mass in (kg)	Dry mass	Water
Pinto Beans	**0.50**	0.410	0.09
Bell Peppers	**0.25**	0.020	0.23
Cyanobacteria	0	0	0
Tilapia	**0.25**	0.063	0.19
Total	1.000	0.492	0.51

Table 10.50 Element percentage estimate per food per person

Crop	C (%)	H (%)	O (%)	N (%)	S (%)	Other (%)
Pinto Beans	42.39	6.76	45.36	3.44	0.14	1.91
Bell Peppers	39.73	6.47	46.36	1.68	0.12	5.64
Cyanobacteria	42.69	7.02	34.23	9.41	0.27	6.38
Tilapia	49.00	7.98	23.10	9.29	0.54	10.10

Multiplying the element table by the dry mass in the diet for each crop, you could get Table 10.51.

Now if you are not careful, you might forget a key fact: people do not digest all the fiber in food. Since I am an engineer, not a dietician or doctor, what follows is a lot of engineering guesses. So, for fiber you have to subtract a bit from C, H, O as per Table 10.52 (this is also in Chap. 2) for each crop.

Each crop has a percentage of fiber, again from nutritional data, in dry mass, as in Table 10.53.

This leads to the next set of calculations which is a guess at the dry mass that as a guess and can be used in the human aerobic metabolism in Table 10.54.

This reveals a problem. If we assume that humans exhale 1.01 kg of carbon dioxide, we would need around 1.01 * 27% = 0.272 kg per person. Looking at Table 10.54, we have 0.167 kg carbon per person. We do not have enough carbon!

Table 10.51 Estimated mass in diet of elements by dry mass per person for each food

Crop	Total dry mass	C	H	O	N	S	Other
Pinto Beans	0.410	0.174	0.028	0.186	0.014	0.001	0.008
Bell Peppers	0.020	0.008	0.001	0.009	0.000	0.000	0.001
Cyanobacteria	–	–	–	–	–	–	–
Tilapia	0.063	0.031	0.005	0.014	0.006	0.000	0.006
Total Dry Mass in diet	***0.492***	***0.213***	***0.034***	***0.209***	***0.020***	***0.001***	***0.015***

Table 10.52 Estimated percentages by element for cellulose

	C	H	O
Fiber %	44	6	49

Table 10.53 Estimated mass of elements for fiber in each food item (dry mass) per person

Crop	Fiber %	Dry mass of fiber	C	H	O
Pinto Beans	24	0.0984	0.0437	0.0061	0.0485
Bell Peppers	11	0.0022	0.0010	0.0001	0.0011
Cyanobacteria	3	–	–	–	–
Tilapia	0	–	–	–	–
Total Dry Mass		***0.101***	***0.045***	***0.006***	***0.050***

Table 10.54 Dry mass elements from diet minus fiber per person

Crop	Total mass	C	H	O	N	S	Other
Pinto Beans	0.31	0.13	0.022	0.138	0.014	0.001	0.008
Bell Peppers	0.02	0.007	0.001	0.008	0.000	0.000	0.001
Cyanobacteria	–	–	–	–	–	–	–
Tilapia	0.06	0.031	0.005	0.014	0.006	0.000	0.006
Sum	***0.391***	***0.168***	***0.028***	***0.160***	***0.020***	***0.001***	***0.015***

Even assuming people magically digested all the fiber, we do not have enough carbon, since carbon is available in Table 10.51 is 0.212 kg.

Working backwards from the diet and elements in Table 10.55 is how the rest works out, ignoring the oxygen input and carbon dioxide output from the Example 3 human model.

These calculations mean either of two calculations in Example 3 need to change, either the human metabolism releases less carbon dioxide on this diet (then implying maybe less oxygen input), or we need more food to increase the carbon in the dry mass of the diet, likely adding a carbohydrate crop like wheat or rice. Increasing the food implies more stages for the additions, while the lower human metabolism means less carbon dioxide to the farm. This becomes a bit of a tail chasing exercise. One solution is to boost the tilapia (i.e. by 2.5 times), which boosts the cyanobacteria by the same ratio, adding about 25% to the flow through the yeast-bacteria-incinerator,

Table 10.55 Summary table estimated from Example 3 elements, showing a revised, estimated, human habitat model, far from ideal

	Total	C	H	O	N	S	Other
Dry mass for metabolism in	(0.391)	(0.167)	(0.028)	(0.160)	(0.020)	(0.001)	(0.015)
Implied CO_2 out	*0.620*	*0.167*		*0.453*	–	–	–
Implied O_2 in	*(0.453)*	–	–	*(0.453)*	–	–	–
Fiber dry mass in	(0.101)	(0.045)	(0.006)	(0.050)	–	–	–
Wastes dry mass out	0.325	0.045	0.034	0.209	0.020	0.001	0.015
Net	–	–	–	–	–	–	–

Table 10.56 Sample balancing scenario. Assumes CO_2 from habitat is 0.8 kg/pp/day and oxygen is 0.6 kg/pp/day

Stage	Minimum living mass in stage (kg)	Volume cubic meters	Footprint hectares
Barley	5000	18,512	0.62
Bell Peppers	2300	781	0.03
Chlorella	500	2695	0.09
Pinto Beans	2000	1918	0.06
Potatoes	1400	407	0.01
Rice	700	1954	0.07
Shrimp	3400	733	0.02
Silver Carp	3000	128	0
Soybeans	5400	8982	0.3
Spirulina	500	2696	0.09
Tilapia	1000	83	0
Tomato	1400	1508	0.05
Yeast-Bacteria-Incinerator Bioreactor		8157	0.27
Total		48,554	1.61

and roughly adding 25% to the footprint and energy needs. This is not guaranteed to element balance across every stage but does mass balance.

10.4.3 So, What Does an Element Balanced Farm Look like?

There are other problems with Example 3, such as nutrition, and just plain awful long-term menus. One nearly totally balanced solution set, with nutritional minimums, and more menu options, looks like Table 10.56, though also assumes a lower metabolism in habitat, via (Meyer, 2018, 2023). Given this larger farm and mass flow, the electrical energy required is roughly 4.6 MW.

To get to this farm, I used the pseudo-genetic algorithm method in the next section, and around 1.4 million iterations.

10.4.4 Pseudo-Genetic Algorithm (A Machine Learning) Balancing Method

Given the wicked number of unknowns if we use any menu diversity at all, especially to minimize farm size, we need a way to seek a solution. One way is what I call the pseudo-genetic algorithm method, as in Fig. 10.39. Solution starts by picking people count and picking a set of possible crops. The algorithm guesses a total living mass in each stage randomly, ranged by a minimum and a maximum for each crop. All the mass sizes are stored, to make a unique 'genetic string' for each possible farm. The software then computes the mass balance for the farm and habitat, and what foods, water, and oxygen are sent to the habitat. If it meets the minimum needs for the habitat, the difference in mass from a zero balance becomes a fitness score. Millions of possible guesses are scored and ranked. The best hundred or so are 'mutated' by randomly picking changes to the masses for each stage, to generate thousands to millions of possible farm 'strings'. The masses are again computed and scored, ranked, and mutated, and the process is iterated until all the masses are close to a balance for at least one possible farm. These remaining possible farms are then graded for footprint size or similar criteria. For 12 crops, this means around 10–20 million guesses across all iterations. Sounds like a lot of crunching, but given access to a cloud, the time is under a week at worst, even assuming human checks before each run.

Note, this method works even if the habitat is not the only mass source. You may be designing a farm that accepts input mass from other sources or in situ, then generating export crops. Or you may have tourists that bring food with them, leaving their wastes in your farm. Economic trade webs create interesting mass flow webs for each space farm on the web. Regardless of your math method, finding a balance gets you started to a design.

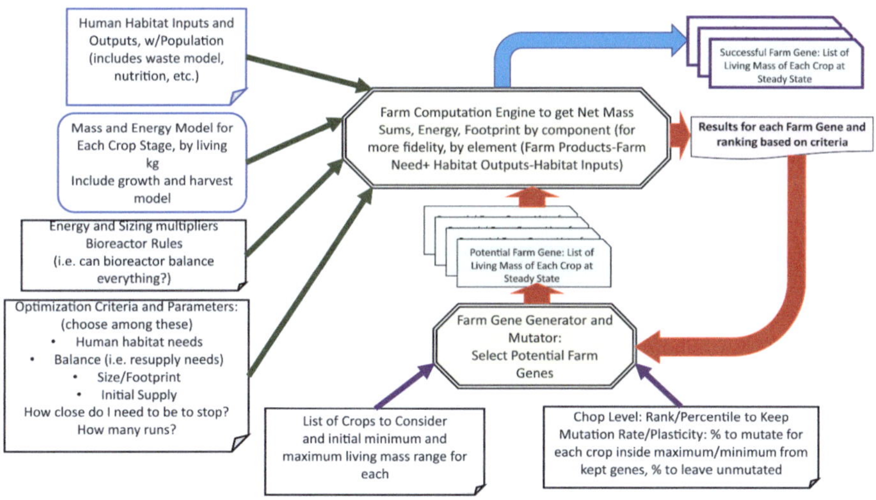

Fig. 10.39 Pseudo-genetic method to optimize space farm masses

10.5 Carbon Dioxide Issue and Getting Chemicals

This is noted before, but worth mentioning again. You do not exhale enough carbon dioxide to grow enough calories to survive, and if the cellulose in vegetable matter is not broken down for water and carbon dioxide, all the carbon will be out of the food cycle, so you starve. See Fig. 10.40. If waste is not processed, or used for other purposes like fuel, construction, soil, or plastics, it will have to be replaced.

Replacement is less of an issue on Mars where the atmosphere can be condensed and used as a carbon dioxide source. In some locales in the solar system, such as some asteroids, comets, and moons, mining dry ice or carbonates can fix the issue. On our closest planetary body however, on the Moon, carbon comes at a premium. Carbon is also a critical resource on free-floating settlements, since everything is either cycled or supplied. Therefore, much care must be taken to monitor and control carbon in the farm-habitat cycle, and inside the farm. Control systems must watch the element balance tightly, among many other factors. Bioreactors and incinerators are again, a must.

Nitrogen is another concern. Nitrogen gas fixing bacteria around roots for some crops like legumes can create a bit of havoc if not controlled. Plants, including legumes, prefer nitrogen sources like ammonia or urea (which becomes ammonia), but under some conditions they use bacteria to grab nitrogen from the air anyhow. Nitrogen ends up in nodules, and the partial pressure balance gets tilted. Controlling the biologically available nitrogen compounds will be a detailed and continuous effort.

Salts are a further issue. Salts can accumulate from urea and urine and must be balanced in liquids and soils and follow the cycle in Fig. 10.38. Using reverse osmosis filters can pull salts and other minerals in concentration of liquids, but in soils salts

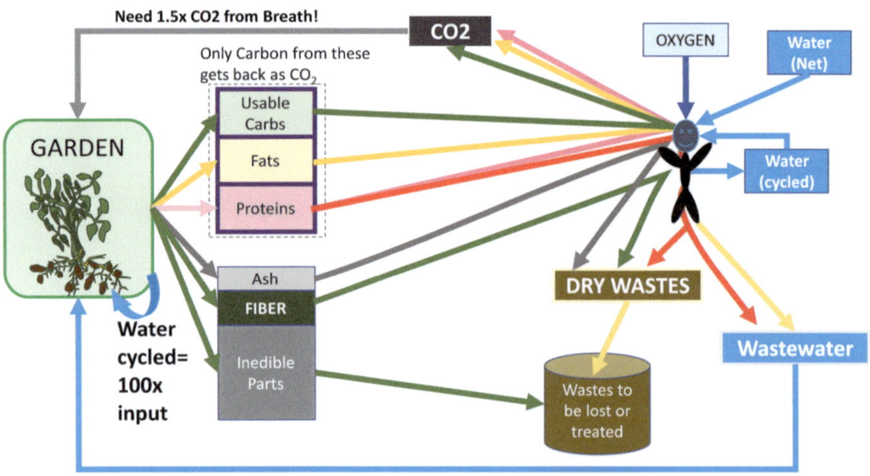

Fig. 10.40 Carbon cycle problem

must be washed out by water or consumed by salt loving plants. The reverse can also occur. Some minerals may need to be provided to the farm to allow proper biological function, and if they accumulate in some stages at the expense of others, the whole mineral cycle becomes perilous. Minerals in food are often recorded in ash, though not always, and minerals are often transported through systems as salts and carbonates.

10.6 Summary

Building the space farm from all the components is like building the elephant, a bite at a time. Looking at troubleshooting gave us the hierarchy of parts. Once decisions are made about how many people will be in the habitat, then what foods you want them to eat, assembling the crops, then seeking balance can be difficult, but not impossible. Simple rules of thumb, like figuring out the large-scale mass flows between stages, then figuring out how big each stage becomes, lead to rough estimates for farm size, energy requirements, and initial supply. Deeper balances are key for not just the major CHON elements, but also all the elements found in the ash components of nutritional analysis. Even with perfect calculations, real space farms will be a constant unsteady control problem, like balancing a pool ball on the tip of the pool cue.

10.7 Questions

1. Troubleshooting shows the controls are needed to keep the farm balanced and working. If the settlement control center received a message that the air around the tanks in the shrimp stage dropped below acceptable levels, what actions or responses may occur to raise the air temperature around the shrimp tanks?
2. Where should a space farm designer put artificial intelligence to maintain the mass balance and health of the space farm?
3. Why might you want a yeast-bacteria-incinerator stage in lieu of just a series of yeast and bacteria bioreactors? What is the role of the yeast-bacteria bioreactor stage or yeast-bacteria-incinerator stage in the overall mass flow of the farm-habitat system?
4. What are the types of mass distribution systems needed by the space farm?
5. How is black water utilized in the overall farm mass flow?
6. How does the decision to make living soil for crops affect the overall mass flow of the farm, especially for black water and other wastes?
7. Why might the farm need an initial settling tank between a cattle stage and the rest of the farm?
8. What is habitat-source balancing, and what rules must be followed using this method to balance the farm mass flow?

9. Based on the three examples in the chapter, what happens when the menu includes more than just glop and glop?
10. Based on the three examples in the chapter, how does adding an animal crop that can eat cyanobacteria or algae affect the energy and mass flow of the farm?
11. What is the big problem with Example 3's mass balance and human model?
12. How could a space farm designer use artificial intelligence to calculate a balanced space farm mass flow?

References

Meyer, B. L. (2018). *Mass and volumes for a spectrum of multistage evolving space farms*. 11/2017, NSS Space Settlement Journal, Issue #3, Issue #3, November, 2017-September, 2018, © 2017 Bryce L. Meyer.

Meyer, B. L. (2023). *Electrical requirements for a spectrum of multi-stage space farms*. NSS Space Settlement Journal, Issue #4, September, 2018-March, 2020, © Bryce L. Meyer.

NASA-STD-3001. (2005). *Volume 2, Revision A NASA space flight human-system standard, Volume 2: Human factors, habitability, and, environmental health*, Approved: 02-10-2015, National Aeronautics and Space Administration, Washington, DC 20546-0001.

SIMOC. (2022). *Scalable, interactive model of an off-world community (SIMOC)*, Arizona State University, School of Earth and Space Exploration, Interplanetary Initiative Pilot Project, and National Geographic Society,© Copyright 2018–2022 | simoc.space, found at: https://simoc.space/

Chapter 11
Living with a Space Farm

Abstract This chapter covers considerations and scenarios that arise from having a fully operational farm that produces crops and recycles mass that must be processed into foods, and the settlement and farm together define the experience of those visiting and living in the settlement. First, crops become edible food menus via processing. Ingredient and food sources are considered as the settlement evolves to include dependents. Finally, there is a user story of how a space settler manages and controls a fully constructed space farm as part of a healthy space community.

Keywords Food · Menu · Food processing · Crop processing · Fermentation

11.1 Introduction

This chapter covers considerations and scenarios that arise from having a fully operational farm that produces crops and recycles mass to support real people. In this chapter, crops become edible food menus, settlement evolution drives interesting tradeoffs for sourcing foods and ingredients, and finally, there is a user story of how a space settler manages and controls a fully constructed space farm as part of a healthy space community.

11.2 Menus and Meals: Beyond Glop

As previously noted, meals, resulting from the possible menus allowed by the space farm, affect human happiness and psychology, to which anyone reading this could attest. The process of converting raw crop outputs into a series of foods that are both edible and enjoyable will require both the science and arts of cooking, along with the science of food processing.

© The Author(s), under exclusive license to Springer Nature Switzerland AG 2026
B. L. Meyer, *Designing Farms in Space*, https://doi.org/10.1007/978-3-032-04395-5_11

11.2.1 Cooking in Space: Equipment, Adaptations, and Ideas

Food affects mood. A good meal can turn a bad day into a fair one, maybe even a good one. One of the core purposes of the farm is to make food, not just bland glop. The more crops that can grow, the more menu options every settler, or tourist, can enjoy. However, even with in-habitat spices and herbs, cooks in early settlements will have a limited set of raw materials to craft menus that are not monotonous. Additionally, not all settlers or tourists will want the same foods or have the same dietary needs. A diet heavy in fats and carbohydrates might be great for the younger set, but not so good for the older residents. Some might have beliefs that limit the consumption of some animals or plants. It will take interesting combinations of ingenuity and machinery to make meals that are something that everyone can enjoy. Menu determination is a core tradeoff for what species end up in the farms. For example, Fig. 11.1 is the classic glop producing farm. You might see this set-up in the spaceship for the long trip to Mars or the outer planets.

To avoid the glop fate, space farm engineers and cooks need to work together. The key goal to getting a good meal is determining pathways from raw crop harvests to tasty meals of various dishes. These pathways may require additional resources like water, energy, and possibly other trace ingredients like salt. A set of these pathways is in Fig. 11.2.

Cooking equipment will also need alterations in space. Boiling is affected by pressure levels in the kitchen. Bubbles from boiling could become exceptionally problematic in lower gravities, since the bubbles will not have the same push to come out of the liquid and could pop out explosively. Bread may not rise the same way in various gravities. Cooking pots for soups may be centrifuges. Bread pans may

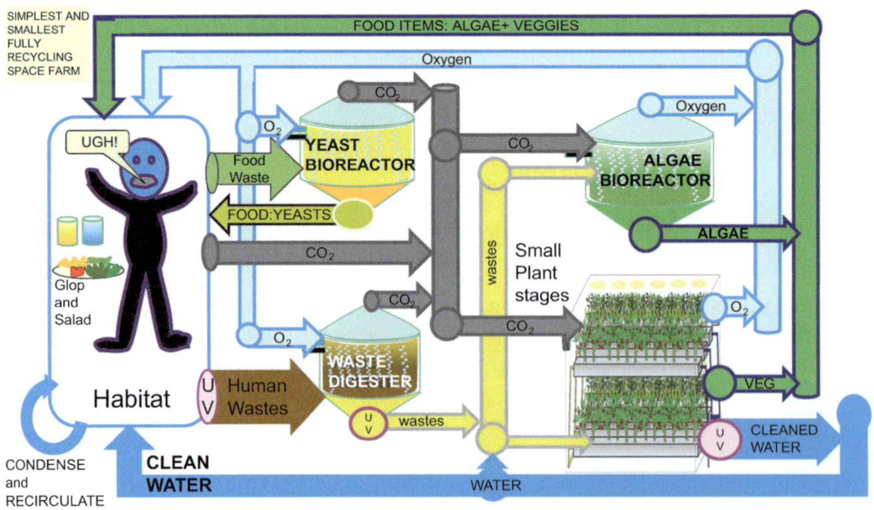

Fig. 11.1 Simple glop and salad space farm schematic

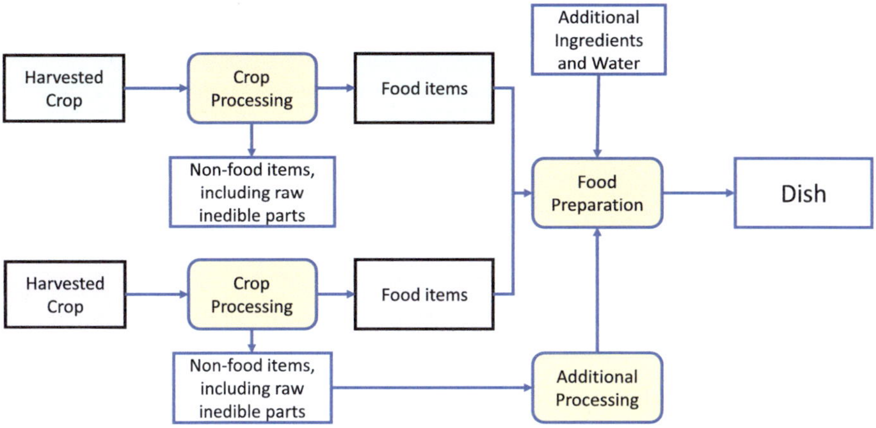

Fig. 11.2 Process from crop to plate

be bread skewers. In artificial gravities, low diameter rotation might make liquids on griddles pool or slide. Cooking with a flame might not be a good idea in high oxygen settlements. These concerns will result in inventions and economic opportunities, no doubt.

11.2.2 Going from Glop to Good: Chow Time in Early Stage Settlements

Early-stage settlements will have the most limited fresh food options. It takes at least six months to get most plant crops other than lettuce and the like, and almost a year for most animals. If the footprint is limited, a large garden or fish tank set might be out. As the preserved food starts running low, most cooks will be limited to algae, yeast, and a few fast-growing herbs, lettuce, or microgreens. Despair not! We have options!

Let's start with the 'algae'. Most of the wet mass made in a farm that just has photobioreactors and aerobic reactors will be either green algae or cyanobacteria. You can eat these straight up, but they have an earthy flavor that will not endear you, nor keep you healthy long term. With a bit of yeast from that bioreactor, more glop, the dietary balance is better. That said there are things that make glop into something a bit more edible.

One option is drying the algae to form thin solid sheets, or nearly dry sheets of flexible greenness. Those familiar with miso soup or seaweed salad are familiar with this route. Dried and cut lengths of algae sheet make good stock for either soup or salad, or a passable pasta substitute once rehydrated. Take the same sheets and powder them, and they can make an additive to other dishes, or be mixed to make a pseudo-flour for baking. The powder also makes a good jelly, like the alginate used

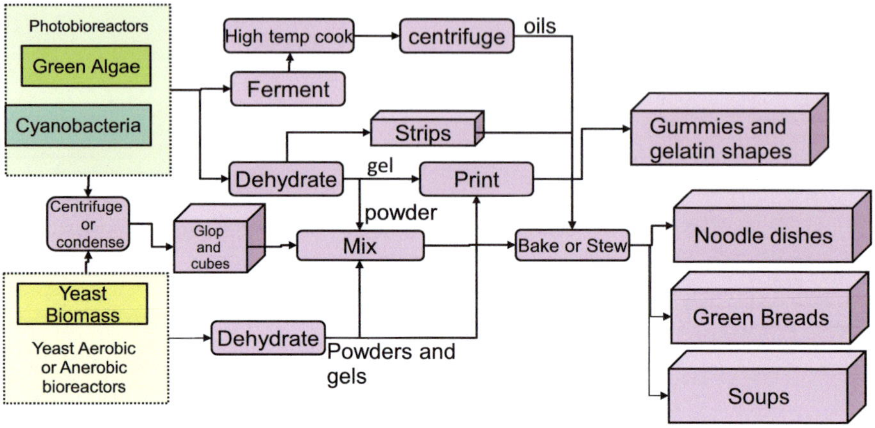

Fig. 11.3 Turning glop into something more edible. The yumminess will rely on spices grown in habitat or supplied, and salt

to thicken ice cream. Add powder to the pseudo-noodles and water, and there is a sort of pasta once some flavors are added, adding in the yeast too, as in Fig. 11.3.

Another option takes advantage of three-dimensional printing. Algae can be concentrated, mixed with colorings, and placed into hoppers that print the algae into shapes. Gummies are a very real option. Printed into shape, aerated, and dried, the algae jelly becomes puffs.

What if we take the algae, and use a small fermenter, process it? Algae can become a familiar kombucha-like drink. Algae can become vinegar with fermentation as well. With a little bit of heat and centrifuge, algae become oil. With a bit more processing, algae can become something that is a bit sweet. Now we are talking!

Now we add in the other glop, yeast. The yeast-bacteria bioreactor array can be used to not only digest wastes and extra algae but make yeast solids. Yeast is a common food additive on Earth, not just to make bread or pizza dough rise, but also as a flavoring to snacks and a wide variety of foods. Dried yeast power and be found in several foods that would surprise the reader. Pizza toppings often contain yeast solids (as confirmed by your formerly pizza-slinging author). Yeast has the benefit of being very high in proteins. In dried form with salts, yeast tastes a bit like cheese. Like algae, yeast can be injected into 3D printers, then made into shapes and sheets. Yeast can be jellied, especially by adding algae, into scoopable dishes or processed using more fermentation into cheese-like, dough-like, or meat-like foods. Given time, an algae wine or beer might be passable. Drink enough green wine, maybe anything tastes good?

Given the discussion above, yeast and algae together, with those early spices, make more foods possible than one might guess. Noodles with sort of meatballs, check. Fried nuggets with sweet and sour sauce, check. Puff algae snacks flavored with yeast sort of cheese, check. Sort of seaweed salads (maybe with that early lettuce) with yeast flavored vinegar and oil dressing, yep again. Greenish breads on the side. Gelatin desserts and pseudo-ice cream round out just a few possibilities.

So, you could live for a while on just the two bioreactor stages, maybe with a few greens too. But add the other farm parts and stages, and some time, and food gets even more interesting. After a year of glop goodies on a ride to Mars, settlers might want something more reminiscent of Earth foods.

11.2.3 Veggies Add Wonderous Diversity

The next step up the farm ladder will be adding a few vegetables to the food stock. Six months or less get you tomatoes, many salad greens like chards and lettuce, and many other veggies like bell peppers, eggplants, and chilis. Real sauces, real salads. About time! If you have ever been without greens for a while, you will begin to crave the crunch of lettuce. The first crops we get will be small, but will be very welcome to settlers eating algae, yeast, and leftover stored supplies. Their bowels will be very thankful. Yeast and algae will still be the main protein sources, but fresh foods will at least add side dishes, such as salads, salsa, sauces, and soups. Figure 11.4 shows a few pathways that incorporate these early veggies to recipe items.

Veggies are mostly water (roughly 90%) and have fiber in addition to simple sugars that can become juices or fermented into other products, as in the tomato example, in Fig. 11.5.

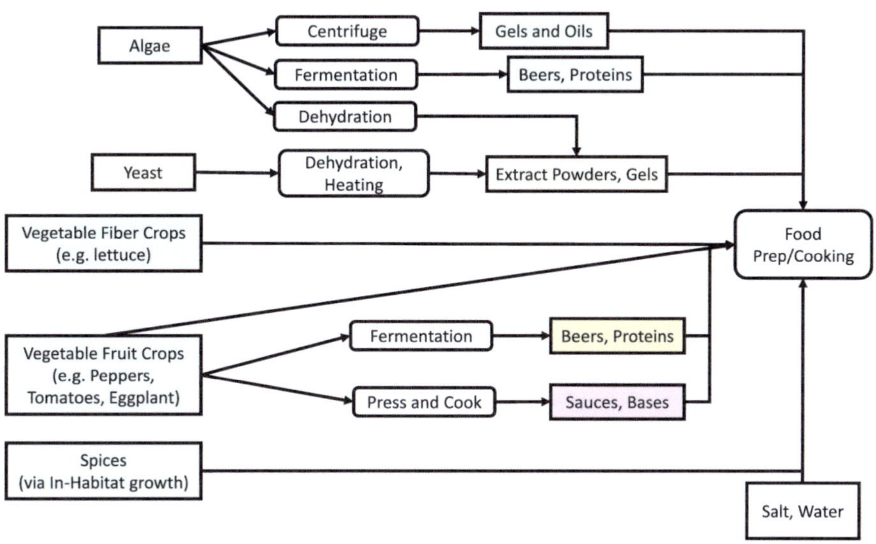

Fig. 11.4 Glop plus veggies food pathways

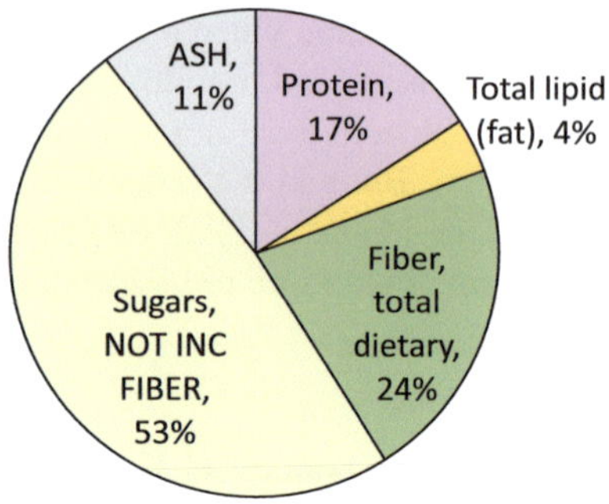

Fig. 11.5 Tomato nutrition by dry mass (Calculated from data in USDAARS (2022))

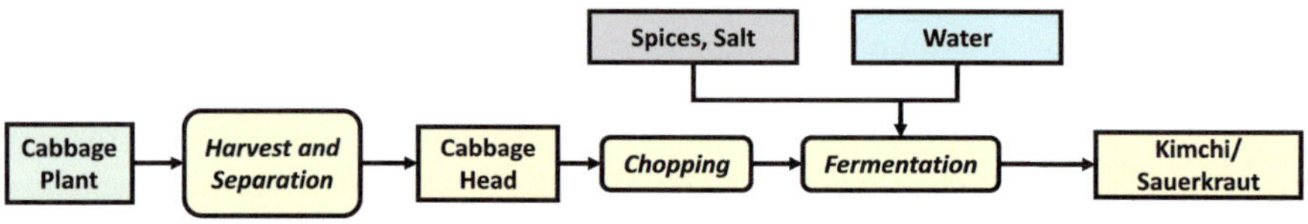

Fig. 11.6 Going from raw cabbage to kimchi and sauerkraut

The sugars become very useful for making menu items with a bit of tangy sweetness. After a long while without refined sugars, the tongue will delight in the subtle sweetness of simple salsa or pasta sauce. We may have green algae noodles, and yeast almost-cheese, but we could have real eggplant parmesan with real tomato sauce.

Given a bit of cabbage, we arrive at important foods, and spicy additives to many recipes as in the pathway in Fig. 11.6.

Cabbage grows just like head lettuce, and many cultures figured out how to stretch it into foods of many types. The sour part of sauerkraut usually turns my nose due to many years of dosing from cafeterias and household dinners, but kimchi on the other hand is a wonder. Yes, one could say the only difference is spices. Processing is about the same for both. After harvest, the cabbage is pulled apart and then chopped. In kimchi, the chopping is course; the remaining large pieces are rubbed with spices and then mixed with water and fermented. In sauerkraut the chopping is more fine, and fermentation occurs in a brine of spices and salt. In both cases, anywhere from a few days to weeks results in the sour, umami-rich result.

11.2.4 Grains, Beans, and Starchy Things

The first high calories crops will take 180 days to mature but take a year to produce in a steady state. Beans, root vegetables, and grains are major caloric sources that humans are wired to want. They also provide staples that wean the colony, or more likely save the colonists, from glop. People will be more than ready for food sans

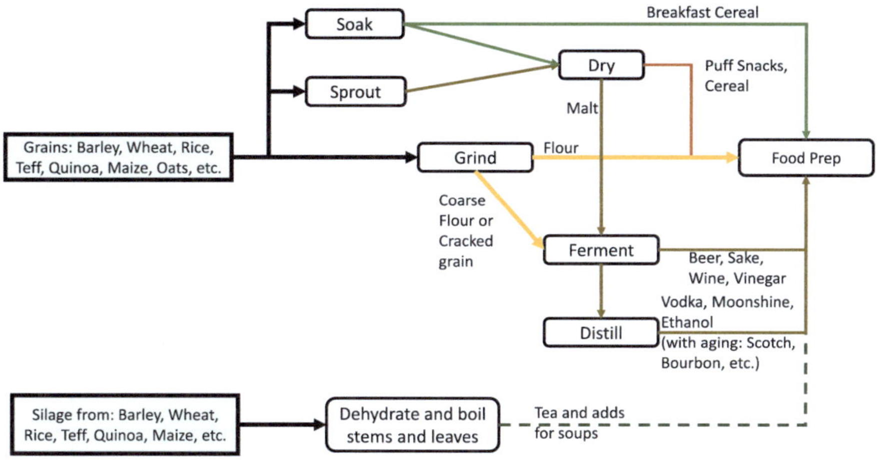

Fig. 11.7 Grain processing to foods

algae after that first year, especially once shipped in foods run low. The first grain and bean crop will be cause for serious celebration; read the section on beer below.

Grains will result in real breads and muffins, sources for better vinegars and beers, pasta, and additives for many dishes. Cookies? You bet. Eggs might take a while yet, but we can fake those (ask vegans). Rice, wheat, barley, and quinoa all work for just about every recipe. Swelled grains make breakfast cereal, something your author had as a child. A little fermentation adds protein and flavor, as tasted in sourdough. Grains can also be ground and processed into milk substitutes, as in Fig. 11.7.

Beans of various types add key amino acids (proteins) and fiber and form the staple of many of the world's foods. Gladiators ate lots of beans. Soybeans are critical as an oil crop as well, adding cooking oil that is easier to make than algae oil. Soybeans are a multitool crop; once heated to reduce sterols that can be made into milk substitutes, flours, cheese-like foods, tofu, curds, and meat substitutes. Common beans have a variety of cultivars and varieties that result in green beans, baked beans, bean burgers, and my favorite Indian and Mexican dishes. Like soybeans, common beans can be dried into flour as well. Dosa is a great breakfast option from bean flour, especially if you had a few peanuts and spices. Beans can be made into Asian sweets and desserts (like mochi) as well, adding sweet tastes to dietary options.

Like grains and beans, potatoes offer many uses beyond the baked and boiled uses. Potatoes can be made into sweeteners, vodka, flour, thickening flakes, and if boiled right and settled made into a candy. Potatoes also have some protein.

These grains, beans, and starchy roots result in a long list of menu possibilities, as shown in Fig. 11.8.

Processing for legume crops like soybeans can be more complex, but results in many commonly eaten foods, especially in Asian cuisines. See Fig. 11.9. Tofu fried in peanut oil, dipped in peanut and soy sauce is a favorite of your author. First, to get the tofu, soybean kernels are separated from the plant and pods. Then the kernels are

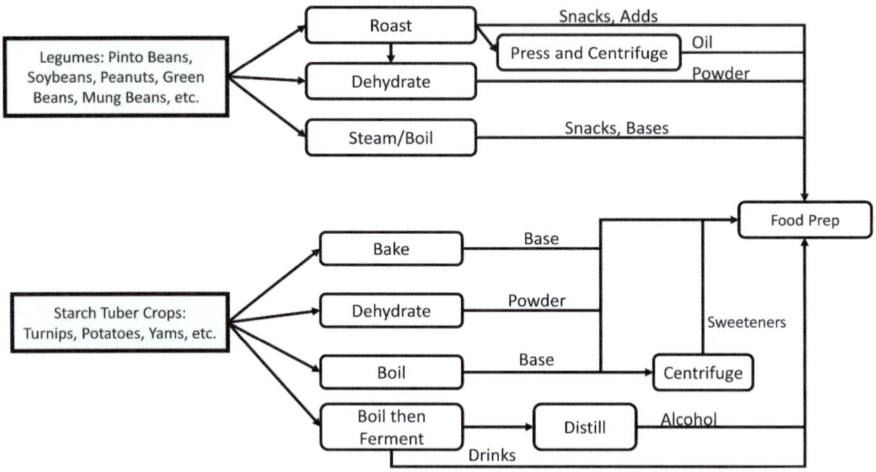

Fig. 11.8 Legumes (and beans) and potato processing to food

soaked and heated. After the kernels swell, the beans are blended, then fermented. The resulting liquid (soy milk) can be used itself or mixed with a bit of enzymes and minerals. The mix is let to ferment again and coagulates into curds (just like when making cheese). The curds are tofu, which can be dehydrated and cooked further to get more firm. Like cheese, tofu can then be fried. The liquids that don't go into tofu are then salted, and further fermented, and aged, resulting in soy sauce, a magic ingredient in many dishes.

What about the peanut oil and sauce parts? What would Thai food be without peanut sauces? Well, now peanuts, another legume, go through processing too. George Washington Carver, known to any USA school kid, invented many processes

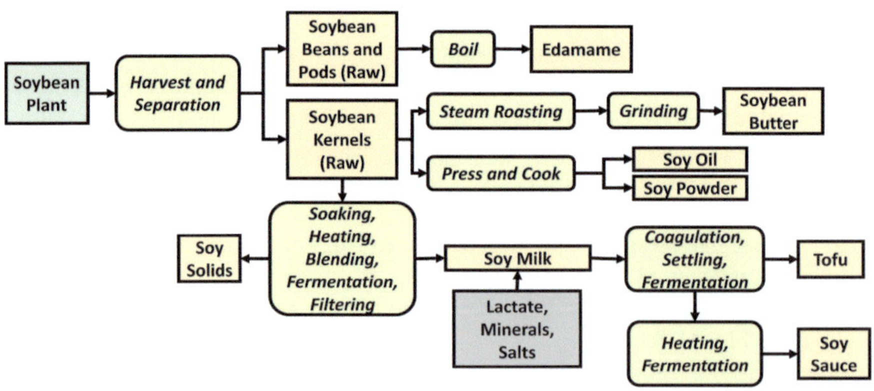

Fig. 11.9 Processing soybeans to food products

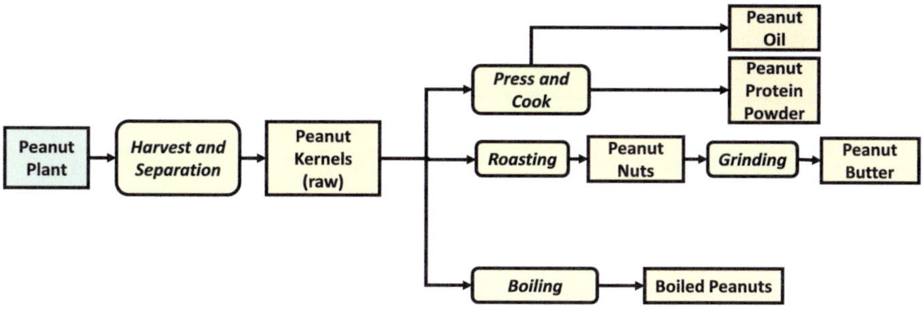

Fig. 11.10 Processing peanuts to foods

for peanuts, though peanut butter was before him. Some of his processes include those to get peanut oil, as in Fig. 11.10.

The Thai peanut sauce usually includes peanut butter, a staple of US children, with curry powder (turmeric and others), and soy sauce. If you were slick enough to grow turmeric in the habitat or in the corners of one of the larger farm stages, the roots become the revered core for curry. Adding a dash of chili powder (from the chili pepper plants) helps too and encourages a bit of forehead sweat.

The protein powders from soybeans and peanuts are a core for many dishes, a supplement, and an additive to add an extra pop of thickness to soups.

11.2.5 Bring on the Meats!

While many meat substitutes will be on the list for smaller settlements, animal meats can be on the menu for larger settlements, opening a variety of possible menu options, and injecting an extra protein source, as in the nutrition data in Fig. 11.11. As a nice benefit, as noted in Example 3 in Chap. 10, the algae to fish (or shrimp) loop is very mass and energy efficient, especially in combination with hydroponics.

Animals add many more options if the whole animal is used, as opposed to just filets or choice cuts, see Fig. 11.12. Many cultures know how to use the whole animal, given drying, fermentation, grinding, and heat. Fish heads, bones, etc., can be ground up to make patties, fish balls, or sausage, fermented to make fish sauce, turned into jerky or powders, rendered for oil, or many other ideas. Insect crops can be ground to flour as well. Mushrooms are processed like animals too, with raw, cooked, and dried options.

Combining animal proteins with veggies, starches, spices, and assuming a source of salts opens a series of menu options fused from a wide mix of human cultures that even a cruise ship passenger would find appetizing, even given a wide variety of dietary restrictions.

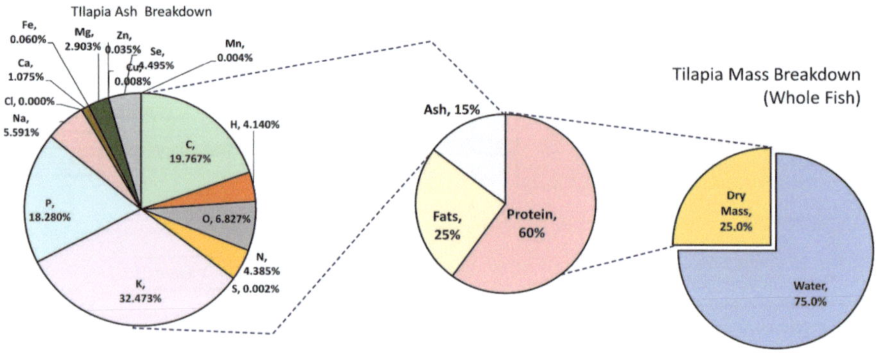

Fig. 11.11 Meats like tilapia and nutrition. Calculated from sources and figures in Chap. 3

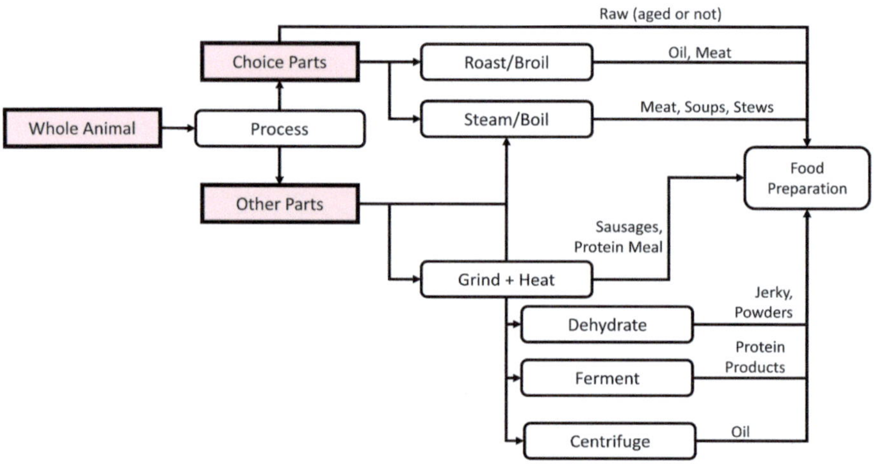

Fig. 11.12 Processing meat crops for food

11.2.6 *Beer, Wine, and Coffee in Space*

An early way to make back some of the cost of a settlement might be specialty items like beer, wine, and coffee. Maybe even space moonshine, vodka, or others. More likely, settlers will make and consume their own beverages like wine and beer, to take the edge off the day, or to celebrate a victory in moon ball. For those unfamiliar with the basics of making beer or wine, the process goes loosely like so: grind and cook the sugar source, cool and add yeast, let sit for a few weeks with a one-way valve on top, then separate out the liquid and enjoy. Algae can be fermented and distilled to make wine or moonshine, which might not taste great, but could ease the mood. Same with tomatoes or other fruit veggies. To concentrate the alcohol from fermented wine or beer, distill by heating, separating ethanol from methanol and water in vapor using a cooling column. The liquor store really opens when the first grains and potatoes hit. It may seem a bit of a waste to spend food on fermentation but fear not my mass concerned farmers. Once fermented, the dregs make great high protein food for people and animals alike. Having eaten the leftovers from sake, beer, and sour mash from whiskey, I can vouch that you could do worse, and dried, the dregs make interesting cereal. Ever have a malted milkshake or malt-based cereal?

Beer and sake take about 30 days, and to note, many popular beers (such as American lagers) are essentially flavored sake (having worked for a major beer company I can name a few). Sake is a combination of overcooked rice and yeast, though it can have some amazingly detailed steps and art. Wheat beers are a big product, and ancient Babylonians and Egyptians made wheat and barley beers and rough wheat drinks as all day fortifiers. While hops take years to grow, other bitter agents work too, many from those in-habitat growths including carbonized grains, lavender, mint, tea, or other spices. Vodka takes about the same rough timeframe and can be made from any starchy source, including of course potatoes. Corn brews are also easy, sour mash a more solid form. Simple wines can be made from pumpkins, carrots, strawberries, algae, and many other sugar sources. Good whiskeys take many years of aging, but a good distiller and filter array can turn out a vodka quick.

Your author has a long history of making wines and beers, using recipes from parents and grandparents. Here is the basic process on Earth. To make a fermented beverage you have to have three things: good water, a sugar source, and a good yeast. Ideally, control of fermentation rate and sugar to acid levels determines flavor. Yeasts such as Saccharomyces cerevisiae are primary, and over time a moon or Mars variety will emerge. The yeasts digest the sugar source anaerobically to make alcohol, carbon dioxide, and water. If bacteria get in with oxygen, you get vinegar (acetic acid) instead, still assuming pH leans to neutral or acidic (maybe something you want, but if looking to beer or wine, not). If pH gets too alkaline you might get things you do not want, like a horrible bad batch of bacteria ooze that smells like sweaty socks. Given the right balance of sugars, yeast, temperatures, and anaerobic conditions, starting in a sterile vessel with uncontaminated water, you usually get what you want. To manage pH, it is common to use citric acid to keep pH below neutral at the start. Fruits and grains lead to wines. To get beers, a malt is added, which adds enzymes that quickly break down the starches in grains. Many would argue that it is not beer without hops. It might not technically be a true beer if it uses any grain but barley. In either event, the process is to mix water and sugar source, and possibly spices, together. Then boil and mix the batch. If pH is above 7, a small amount of acid, such as citric acid, is added. The mixture, aka mash, is put in a sterile container and then cooled to below 40 °C. Yeast is added, a one-way valve included, and a cooling jacket or not is used to control fermentation over a period of a few weeks. For beer, hops and malts are added to the mash when boiling. In the case of beer, the fermentation gets very violent and can shoot out the valves if care is not taken. The yeast consumes as much sugar as possible, generating heat, making bubbles of carbon dioxide that rise to the top, and leaving yeast shells (lees) that accumulate at the bottom, until the alcohol gets too high, the sugar is all used up, or the brewer/vintner stops the reaction. Next the mix might be distilled to make either medicinal alcohols or drinkables. In the case of whiskeys, rums, and bourbons, barrels or barrel wood are used to age the mix, which would have to either be brought from Earth, or chemically faked using another woody crop (maybe bamboo or silage). For vodkas or the like, the distillate is charcoal filtered and mixed with water. Most American Bourbons do filter with charcoal, again hard to get in space unless you cheat with another crop residue burned just right.

In space the process will be similar, but gravity can affect the process. In zero gravity, a membrane mediated yeast might be more effective, but at a minimum active stirring and gas removal are a must to make the yeasts ferment. The bubbles in low gravity can't just rise to the top and might drift according to pressure gradients. Acid may accumulate and kill the yeast and ruin the batch. Likewise, active filtering may be needed to pull out the lees. The closer gravity is to Earth's the less technological the process becomes. Moon beer might have longer fermentation rates and a very different flavor than the same ingredients batched on Earth.

Space tourist locales will want to serve the local brew to tourists seeking the local flavor, and good brews may become exports to generate money. People will pay money for rarer drinks. On Earth people pay lots for water, and even more for brews made from interesting water sources. Who wouldn't pay extra for a moon beer made with crater water? Add to the local truffle crop, and local coffee, and you have a strong gift shop selection.

While drinking is the highest purpose for making distillates in my humble opinion, there are a myriad of other uses, from disinfectants and mouthwash (especially combined with mint) and cleaners to various chemical processes and pathways. Even a settlement of teetotalers will need to distill good old alcohol.

On the other drink spectrum, coffees will also be a good target, especially for export and tourists. Coffee takes at least three years to first crop from a bush that is waist to head high, but coffee bushes can be grown in both habitat and in plant stages. A cheat would be to bring a near mature bush. Like brew, coffee from exotic locales demands a premium, and is a perk for tourists seeking local flavor. On Earth, people pay crazy money to drink coffee from beans digested by jungle cats! Cat poop coffee is a real thing! Given that, coffee again becomes a key export, especially once grown in regolith to acquire the unique flavor of the moon, Mars, or wherever. It would be likely that the locals would drink hydroponic teas and coffees but grow coffees and teas in soils for export.

11.2.7 Storing Foods for Later Meals

Even with continuous harvest of every crop, there are many reasons to store food items. One is to store imports before use or exports before shipping. Another is to store extra production against future deficits, or store the initial supply or a backup food supply against disaster. Some of these methods are very low cost given access to vacuum of space, radiation, and cold temperatures in the shade. Given these concerns, there are many technologies to preserve foods, each with costs in taste and nutritional loss, based on food type, as in Fig. 11.13. If the goal is a backup, food may need to last

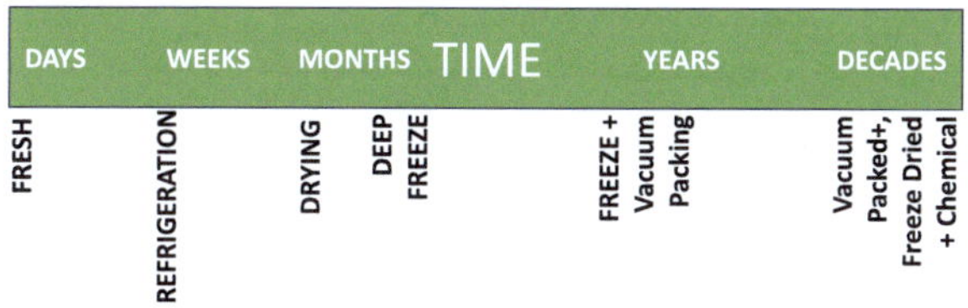

Fig. 11.13 Storage technology with time graph

decades, leading to vacuum freezing with chemical storage. Years can be gained just with freeze drying using vacuum, though recovering the water may be a requirement to maintain mass balance. Simple refrigeration buys a few days to a week and will be common in kitchens and to move high protein crops or crop products. Freezing is needed for many products like desserts, or part of food preparation for some recipes.

11.3 Evolving Menus

It isn't just the farm that grows in complexity over time. As a settlement matures, the workforce and residents go from hardy explorers willing to eat whatever, to people with families, and likely tourists and visitors, who may want a complex menu as well. The addition of non-worker dependents is really what determines if a space farm is at a settlement, or just an outpost. Some outposts, like those in proximity to Earth or larger settlements, will have the option of importing a variety of ingredients and foods. This evolution is seen in Fig. 11.14.

Core trades to get the ingredients for a diversity of menus are cost to grow and source locally (including in situ) versus cost of import. If it is cheap to ship in food, it becomes less critical to recycle mass for a settlement. Likewise, if labor is very expensive or missing at the settlement (maybe the robots refuse to work), but cheap and available on Earth or another settlement, then the trade favors bringing in food. There is no surprise in this analysis, as many modern fast food restaurant chains use food that is mostly prepared in central prep kitchens, then packaged in bulk for final cooking and prep in the fast food restaurant or convenience store kitchen. Did you really think that breakfast burrito, as in Fig. 11.15, from chopped peppers and onions to sausage (or veggie sausage-like stuff) and (real or fake) eggs was entirely prepped locally?

Luxury costs for luxury guests as time progresses will require entirely fresh foods and other farm products made on site, pushing the need for some tourist focused gardening. Local flavor, such as coffees, vegetables or fruits, salads, truffles, or spices, will add uniqueness to resort fare for guests, made by processing and farming wastes from prior guests and settlers. An interesting irony is that the visitors as a result represent a constant mass input, largely used to feed crew fresh foods. Guest poo means crew veggies. Long tour lengths will entail spouses and families, increasing the need for farming, for air and water processing at a minimum. History shows us that as people lay down roots, populations grow fast, and the settlement's average

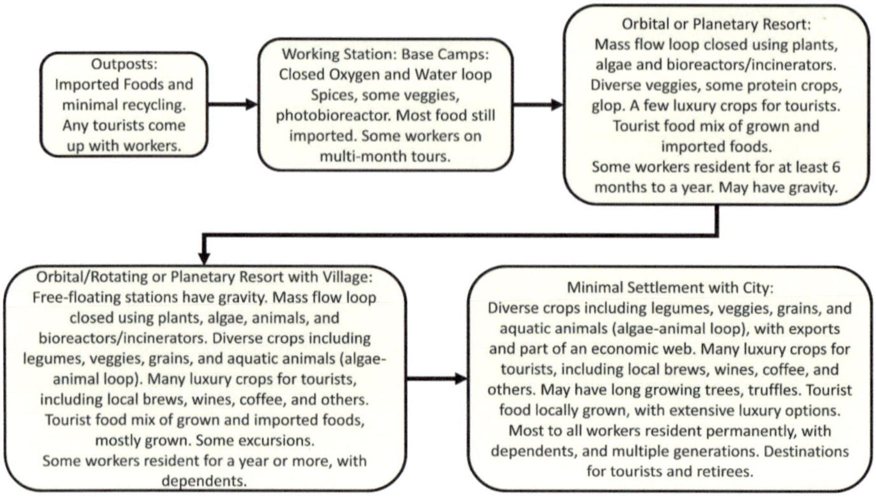

Fig. 11.14 Evolution from outpost to settlement resort

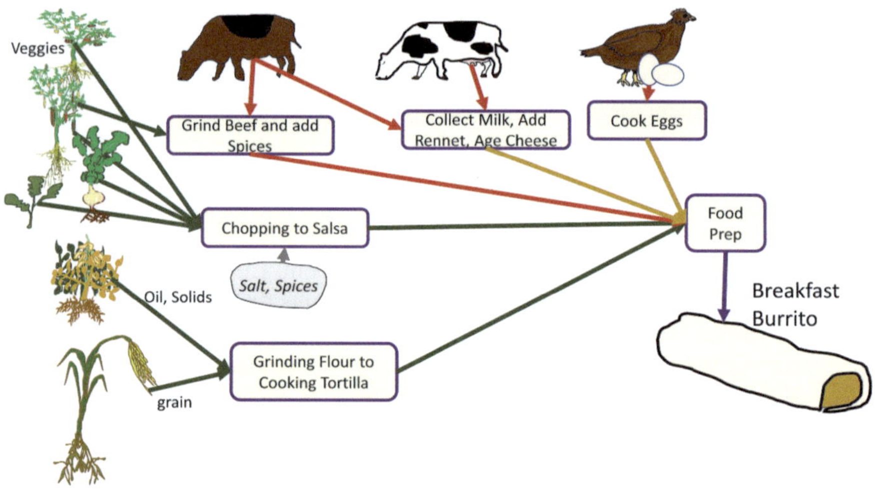

Fig. 11.15 Burrito of many crops. Every food item is the sum of many crops, combined using various processes to get a dish that we want to eat

age trends to the twenty-somethings until the settlement has many generations under its belt.

11.4 User Story: What is It like to Live with a Space Farm?

Here is a vision of what daily life might be like in a space farm. Engineers call these user stories, and they are used to flesh out a design. It is sci-fi but allows us to walk through a future space farm with parks and multiple stages. As you read this story, pay attention to tradeoffs, design choices, processes, and design elements. Some are subtle inclusions and clues.

Janet woke up, grabbed a bit of tea out of the pot and then walked into her main room. She flicked on her handheld, then flicked the screen from the handheld to the wall screen. The schematic web of the farm lay on her screen, mostly green dots connected by green lines, with one dot sort of yellow. "Damn pump must be acting up again" she chimed. A point at the yellow circle and reverse pinch, and it expanded. The yeast-bacteria bioreactor array virtual schematic appeared, and sure enough, the liquid pump from module 25 was overheating again. The machine learning system had already shunted flow using the parallel pump and path, and the alarm to her indicated time to replace the pump. There were a few remaining in stores; hopefully they still worked. If not, she would have to ask the technicians to print her a new one from the stored designs in the settlement's cloud system. "I'll change the pump myself." She left her apartment.

As Janet walked down the well-lit hall tube, she glanced at the narrow trays of zucchinis along the walls, basking in light from overhead and reached over and looked at their leaves. "More salad soon." She continued walking and as she entered the main farm airlock, she glanced at the sage growing in pots in the corner from grey water. The scent invoked mountain scenes and beat the stink of B-O and plastic. The door recognized her and as she waved, the door opened for her, a slight hiss in pressure pushed past.

Labeled doors lead to the gardens, aquatics, photobioreactors, and on the far left a hall to the yeast-bacteria bioreactor. She walked over, waved her hand again, and the door opened, another whoosh. In the antechamber, there were suits, helmets, and air tanks, though the monitors indicated human-breathable air. She waved and the door opened. If a leak made the air bad, or there was low pressure due to a vacuum leak, the door would not have opened until she suited up and her suit told the door she was sealed up and ready to enter.

Inside, tall tanks reached from floor to a dome top well above her reach, and ladders and stairs lead all over the arrays. Pipes and boxes connected it all, and another set of screens showed status, with backup rows of lights. Her smart-pad was already queued to the stage and showed the offending location when waved across her view. Arrows on the screens pointed to the equipment locker, which was awake and blinked. As she walked over to it, she heard it click open, and the shelf holding a spare pump of the correct type lit up, and she knew as soon as she pulled it out, the system would add the pump to the 'need list', to be either manufactured by the accumulative manufacturing area, repaired from the old pump, or ordered for the next resupply ship, whichever was most efficient. A cart next to the locker rolled to her, and she lifted the pump onto the cart, which had arms that cradled the cermet pump like a baby. Like a puppy, the cart rolled behind her, then she pointed to the blinking failing pump, and the cart rolled over to it, and lifted the pump to allow a simple pull off for install. The cart also used an arm to grab the required wrench and test kit and present it to Janet. She reached into the cart and grabbed a simple plastic coverall and gloves, and a face cover, and put them on. Fluids had already rerouted around the pump as its flow dropped, and an overhead fan sucked air, but even so the smell was outstandingly bad. A bucket rolled over caught the fluids that leaked, and she replaced the pump. Nothing wasted. The test kit showed green, and after she took the suit and cover off and dropped them into the cleaning bin, she used her pad to reroute the flow. Currently, the array was acting as in aerobic mode, though Janet also walked over to a separate array that was her favorite. The cart rolled the now bagged pump to the corner, placed it on a shelf and then rolled itself back to the locker. The cask she now touched was also temperature controlled and produced carbon dioxide, consuming rice at the moment. She grabbed a small cup and opened a tap to release a clear milky liquid. She exclaimed "kanpai!" and drained the cup, noting the slightly sweet burn that followed. Almost ready for public consumption on next Apollo Day.

Janet walked out of the airlock, and turned to the garden door, entered its airlock, then walked past the various grow rooms, looking through the windows, waving at her colleagues, then entering one for the rooms that had a less enriched atmosphere. Some rooms had very carbon dioxide-rich air that required a mask and tank or filter, but she went into the soil room. The beginnings of soil filled a few beds, some with green potato leaves and a very few small pine seedlings arching above. The pines were an attempt to grow truffles for export, years hence. Fingers in the dirt exposed the sweet humus smelling black soil around the potatoes, a mix of regolith and digested sterilized wastes. A small earthworm stuck out, and fungi rhizomes webbed the new soil. This was a beginning that in years would lead to a real dirt farm. For now, a few pots and beds with a few plants were all they had. Her soul remembered her upbringing on Earth holding living soil like this, before college, before the long trip to the lunar outposts, and now to the plains and tunnels of Mars. She knew the dirt itself could be an export, to one of the rotating free space settlements that dotted the skies over her head in Mars orbit, or the economic stations further away from Sol that traded with the 'Belt. Never mind, she was the Ag. Chief, and this was her farm.

Rajesh walked behind her and smiled. "Can't keep your hands clean Jan, can you?".

"You know I can't Raj." She replied.

"Neither can I. There is something deep inside that feels joy at the texture and smell of living dirt and growing things." Rajesh was the manager of this stage, and he was happy and dirty all the time.

Janet sighed, wistfully looked at the beds and plants one more time, then strolled out "You are lucky Raj, and I am jealous."

"I know" he said as she left out the door. She could chew up her whole day playing in the dirt, contented and happy, but that was Raj's domain, and she needed to check other parts of her larger responsibility.

Another lock led to the largest stage, the hydroponic soybean area. Once in the outer area for the stage, several locks lead to growth chambers that divided the soy crop, and in each chamber cohorts of plants rotated from seedling trays through exponential growth, linear growth, then dry-out areas. She donned a mask to keep away the carbon dioxide headache inside the chamber. Four months from seed to harvest, though they harvested every day. Stacks and stacks of plants with catwalks and overhead tracks filled the volume of the chamber. She knew the support tanks sat in the outer stage area, filled with many days' worth of inputs. Robot arms and various automata managed the plants, and sensor arrays were everywhere. In a few hours, Tikara would work with robots to harvest today's plants sitting in the dry out chamber. The bean pods would end up in processor, beans extracted and checked, then taken by pipe to the kitchen. The silage would be shunted to the bioreactors. The robots were already seeding their replacements in the seedling room. She exited the soybean chamber and stage.

Another hall led to an area with several chambers under construction. A few shrimp tanks sat in the corner of the cavernous room, pipes leading to a few trays of tomatoes in their embryonic aquaponics grow area. She walked over to the round tanks that came up to her waist, yellow–red light over the tank to keep the shrimp happy. A green pipe lead from the photobioreactor stage and occasionally plopped a thick goop of live green algae for the shrimp to eat. She swung her tablet over the tank and noticed that the tank was minus six shrimp from the few hundred two-inch prawns. "They are cannibals to be sure" she said to the aquaculture technician checking a tank. Ho responded back "nothing as tasty to shrimp as another shrimp". Not even shells were left in most cases. The walls of the round tank had small black snails that kept the tank's algae and bacteria in balance. The bacteria and algae in turn kept the ammonia levels in check. She smelled the water, and sure enough, it had an earthy clean smell.

She finished her morning walk by leaving the farm's main area and went to the habitat's park. Rules dictated that all visitors shed shoes and socks, though some shed all their clothing too. Under a high ceiling with white-yellow light tuned to be very sunlike as on Earth. A walkway followed a flowing shallow stream. The stream had crayfish, clams, snails, and silver carp, and received grey water and other sterilized and oxygenated water from the habitat-farm system. In addition to the animals in the stream, hydrilla, and other water plants grew greenly along the edges, waving in the flow. Small trees started growth to the ceiling, including bananas, and 'multitrees' of grafted fruit branches made to look like trees. Coffee and tea bushes were scattered around, and columns were covered with vines of grapes and berries. Benches and pads were scattered to allow her fellow settlers and a few tourists to lay out and enjoy the 'outdoors'. The park was her idea, though many had inputs to make the park look like a forest. Food for mind and soul. Residents occasionally got a quota of things to pick, but today was just a grow day. Janet grabbed a bench, and for a short while, let her mind wander until her tablet beeped, and another alarm beckoned.

11.5 Summary

This chapter covered the conversion of crops into edible, enjoyable foods, then considerations for sourcing foods and ingredients, and labor, and finally a user story to explore the design elements of a mature space farm. The process to make dinner from raw produce will result in additional wastes, and what people do not eat will create wastes too. The kitchens in the habitat will not be exactly like those on Earth, especially in lower gravity environments. It is likely in a farm of large size, a storage and processing area will have to be added and enclosed to prepare crops for kitchens and to prepare foods for storage.

11.6 Questions

1. What are steps to process a crop to a food item on a menu?
2. What processing can turn algae into algae oil?
3. What are pathways to use yeast solids in various foods?
4. Given a grain, what are a few food and drink options? What processes lead to these foods and drinks?
5. How could I make a sweetener for desserts without sugar cane or sugar beets, given potatoes or rice?
6. Why are oily legumes like soybeans or peanuts a key to unlocking many possible menu items?
7. What are some pathways from crops of fish, soybeans, rice, and bell peppers to a menu that includes fish balls, fried in soybean oil, on rice, with chopped peppers, and fish filets with soy sauce?
8. Other than moonshine, what other reasons may settlers need to distill and ferment grains? What are some issues with fermentation on the moon?
9. How could a space farmer in a new farm shortcut the time to grow coffee, given five years to grow a mature bush from seed?
10. What factors lead to a need to have dependents at a space settlement? What factors also lead to an improved menu set for workers at a settlement?
11. Given the user story in this chapter, what role does AI play in Janet's day? What tradeoffs did the farm designer make in Janet's farm? What changes would Janet see in 5 years in her farm?
12. Given the user story and the rest of the chapter, what level of maturity does Janet's settlement have? What menus would emerge from the farm in the user story?

Reference

USDAARS. (2022). *FoodData Central*. U.S. Department of Agriculture (USDA), Agricultural Research Service (ARS), found at: https://fdc.nal.usda.gov/

Chapter 12
Conclusion: Developing Space Farms and Settlements

Abstract This chapter ties up the themes in the book overall by examining at a top level of the factors that affect space farm designs, then adds top-level concerns for space farms, including choosing locations to build space farms, getting the inputs to start the mass flow cycle, construction of the farm, getting crop organisms to a distant space farm. There is also a discussion of how different crop products and farm input may be part of a larger space economic web and outgrowth of current space efforts. Finally, the chapter covers research topics that will lead to better farm designs.

Keywords Economic web · In situ · 3d printing · Research

12.1 Introduction

This chapter ties up the book overall by examining at a top level of the factors that affect space farm designs. This chapter then adds top-level concerns for space farms, including locations to build space farms, getting the inputs to start the mass flow cycle and for construction of the farm, getting crop organisms to a distant space farm. The role of space farms and their inputs and products will eventually be part of an economic web with Earth and other settlements, and these roles will affect design. As space farms are designed, even before large settlements are fielded, there are research topics that will lead to better farm designs, and many of these can be explored now.

12.2 Recap of Why and What

In the beginning, this book sets out to explain in simple terms how we can build space farms, using current and near-current technologies and methods. In the beginning we discussed the evolution of space settlements. Then we went through the biochemistry of life, leading to the mass flow through various types of living things, from algae

© The Author(s), under exclusive license to Springer Nature Switzerland AG 2026

B. L. Meyer, *Designing Farms in Space*, https://doi.org/10.1007/978-3-032-04395-5_12

and yeasts to plants and animals. We discussed machinery to make the farm work. We covered taking crops and putting them into various structures to assemble the growth chambers into stages. We also covered how mass flow and energy flow in a few examples of space farms. Then we discussed what dinner might look like and how space farms become part of an economic web.

The whole point is so you, the reader, can start designing and planning for space farms as part of space outposts, bases, and settlements from low Earth orbit to the wider solar system.

12.3 Setting It up: How Much, to Where, When

Where you set up a settlement affects many trades and options for a space farm, resulting in various designs that might be suited for a lava tube on Mars, but not for rotating space stations. In addition to the gravity differences between locations, there are concerns for transport time and cost, in situ resources, radiation levels, temperature ranges, and many other variables.

12.3.1 Trades and Concerns

Getting the parts and organisms of the space farm to the settlement, then getting the whole kit and caboodle running is a set of challenges to design, transport, time, and capability. Every farm will be an adaptation to the settlement location and conditions, humans in the habitat, and time from delivery to start-up. Many parts or even whole stages may be assembled elsewhere and delivered ready to go, from computing systems and pumps to valves, robots, and bioreactors. Other parts may be sent as parts and assembled on site. Other items crafted from in situ components or from raw material delivered to the settlement. It is likely that small farms might be mostly pre-constructed, then expanded over time as the settlement grows or needs change (like dietary boredom). The more complex the machinery, and the smaller and lighter the item, the more likely it is delivered in whole. Again, pumps, valves, processors, heaters, sterilizers, and the like are probably coming in from elsewhere, at least to start.

However, anything that is heavy, simple, and can be made of locally derived materials, should be done on site. The big, heavy, simple things also favor pre-assembly by robots, so can be made before people arrive. For planetary type bodies, or settlements with access to asteroids, things like walls, roofs, trays tanks, pipes, and fiber optics can be made on site. It is possible, that lighting might also be printed locally, but for now, LEDs will be sourced from Earth or more developed settlements. Tanks are something that every farm will need, from bioreactors, to settling tanks, to fish and shrimp tanks, and can be made on site on the Moon or Mars from regolith. Ditto for root and filter media, or soil base, which can be as simple as gravel or ceramic shapes. Silica is a common element and is key for most structural uses. Given accumulative manufacturing, local material becomes the low-cost way to make farms.

12.3.2 Elements of Life at Site

The simplest and heaviest part of the farm itself is the water in it. Shipping water is expensive no matter what, but fortunately water is available all over the solar system. A settlement of any size beyond an outpost will need local water sources, if for no other reason than getting oxygen and hydrogen. Purifying the water will cost energy, but distillation can be the method to turn dirty ice into clean water. Likewise, hydrogen and oxygen can be combined to make water, or organics burned to make water. Some of these cycles are in Fig. 12.1.

The next heavy part is carbon. While carbon is great for structures, and maybe rocket fuel, we will need it for life too. Carbon is the building block of organic chemistry, and the living parts of the farm, you included, require lots of carbon. Preferably, carbon dioxide is the best kind of carbon supply for the farm, since photosynthesis turns it into everything else. That said, many forms of carbon can be reacted in chemical processes to be farm useful, including straight up graphite. Graphite can be oxidized (burned) to get carbon dioxide. Comets contain carbon compounds, as does the atmosphere of Mars, some parts of the lunar regolith (though with lots of energy to get it), and many asteroids.

After carbon, nitrogen is next. Amino acids on average are 15% nitrogen, and the nitrogen cycles in the farm are key. Amino acids get assembled into proteins or used

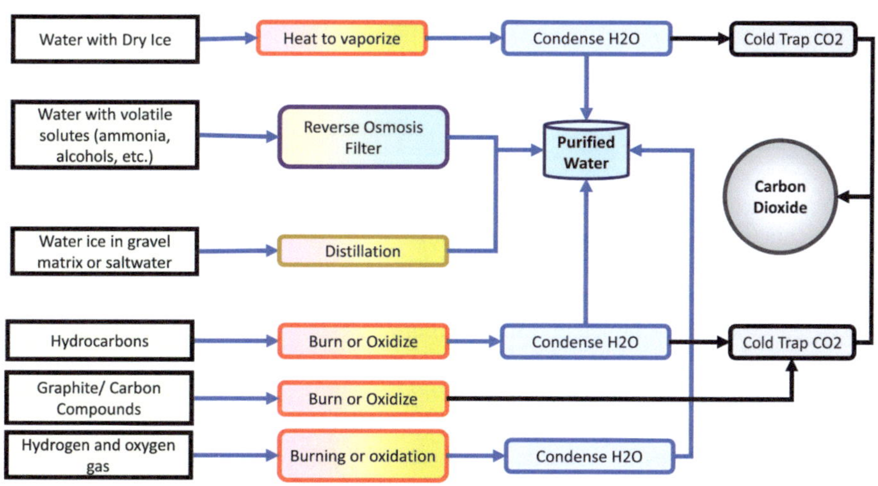

Fig. 12.1 Pathways from raw materials to water and carbon dioxide

in molecules like DNA and RNA, then eaten and digested into ammonia and urea, then reused by plants and algae to make more amino acids. Luckily nitrogen is in many places, from the stable gas form N_2, which some bacteria and plants together can use. Even better, ammonia is good source, which can be diluted to feed algae and plants. Ammonia is found in many places in the solar system. More reactive molecules and simple amino acids can be useful too. Some asteroids, comets, Mars, and the Moon all have nitrogen compounds in some measure, at least in rocks, though at worst the wastes of the first colonists become a source as they eat preserved foods. Pee will become a vital resource!

There are many other materials that may be garnered from local sources that are required by life in the farm, but lower in mass percentage. Sulfur, chlorides, calcium, iron, zinc, and many others, listed as 'ash' in an assay, are needed. Chemists will need to make sure they are either found on site, or brought in. Sulfur can be found in native (pure) form many places, especially on the moons of Jupiter, and on our moon in sulfites (Vaniman et al., 1992). Most metals can be pulled from ores, while salts can hold potassium, sodium, and calcium and separated by electrolysis. In lower gravity, people release lots of calcium, so it may need to be supplied in greater quantities. Perchlorates on Mars for example, have chlorine, but in a toxic form that will need processing to be useful. It is a tradeoff to decide to bring or source compounds. Vitamins included.

12.3.3 In Situ Resources and Accumulative Manufacturing

Honestly, much work already exists for this topic in works and papers by better materials engineers than I, including (NASA, 2022; NASA-SBIR, 2022; Bajpayee et al., 2020; Starr & Muscatello, 2010; Reches, 2019). That said, here are a few thoughts to consider for the space farmer.

12.3.3.1 Getting the Basics

Making settlements of large sizes will require use of material beyond Earth. Planetary settlements must use materials to at least enclose the volume of the habitat and farm. This makes the first materials for settlements structural materials like metals and ceramics. A nice side effect from breaking down metal oxides to get structural metal is oxygen. Making ceramics and metals requires extensive energy, especially heat energy, and will require large solar ovens, or more likely, nuclear power.

Some elements may require supplementation, at least initially, from Earth due to difficulties in getting in situ elements. For example, it may take time to spin up mining capabilities to resource carbon on the moon, or nitrates on Mars. Free-floating settlements need everything of course, depending on where the settlement orbits. Asteroid resources for distant free-floating settlements may be more mass efficient than sending resources from the moon. Dust may need to be scooped from accumulations at LaGrange points. For the farm living mass flow, the priorities are water, then carbon dioxide, then nitrates like ammonia. Most settlement locations will stake out the 'water holes' like Shackleton Crater on the moon, simply due to the need for water for rocket fuel. Given water, any oxygen could be sourced by electrolysis. Lacking carbon dioxide itself, oxygen and carbon compounds can be burned to make carbon dioxide, so methane and other hydrocarbons are important finds, assuming leftover oxygen from smelting. Much work has focused on getting oxygen from Martian air, but the CO_2 is just what I want for a farm! Bring in the Martian atmosphere, then heat it and compress it, and your farm will love it! Nitrogen becomes a bit trickier. Most farm scenarios I have run need around 1% of dry mass as nitrogen. This works out to at least 50 g of nitrogen per day per person scattered in the various food items and crops, as amino acids, urea, and ammonia or similar. For the first year, for 100 adults, 4 metric tons of urea would fit the bill. This is in addition to the carbon, hydrogen, oxygen, and nitrogen from human wastes as they eat stored foods.

12.3.3.2 Making the Machines and Farm Stages

Many of the simple machinery elements in Chap. 5 can be made using ceramics sintered or melted from planetary surfaces, as in Fig. 12.2. Tanks, columns, tubes, and fins are all simple to make from metals or glasses, and even plastics. Accumulative manufacturing can produce most pump bodies and fins and already does so for complex machinery in aerospace. Wire is extruded from metals and can be wrapped to make electric motors or coils. Simple circuit components can again be manufactured on site using ceramics and metals, but more complicated parts unfortunately are likely shipped to site from Earth or orbital factories. Controllers, processors, and many parts inside the bioreactors, including most of the lighting, are probably coming from Earth until truly complex printers are set up at settlement. Smart valves can be made on site, if the electronics are brought in, but they are again, likely sourced from Earth. Most fish tanks and hydroponic trays can be sintered or made from concrete from various rocks already on site on the moon and Mars, and lots of work already exists in this area. The tradeoff for making parts on site versus just shipping depends on time to get the farm producing versus cost to ship the mass of each component.

12.3.4 Embryos and Seeds, Seedlings

Getting the starters for a space farm to the settlement presents challenges. Some settlements may be very close, making whole live plants, live animals, and maybe even mature trees a possibility. The Kalpana (see (Globus et al., 2017; Marotta & Globus, 2021)) Low Earth Orbit (LEO) settlement is within launch of common rockets, and last-minute requests to Earth can be brought up on the order of months, while cargo is only in transit on the order of a few days. Beyond Earth orbital

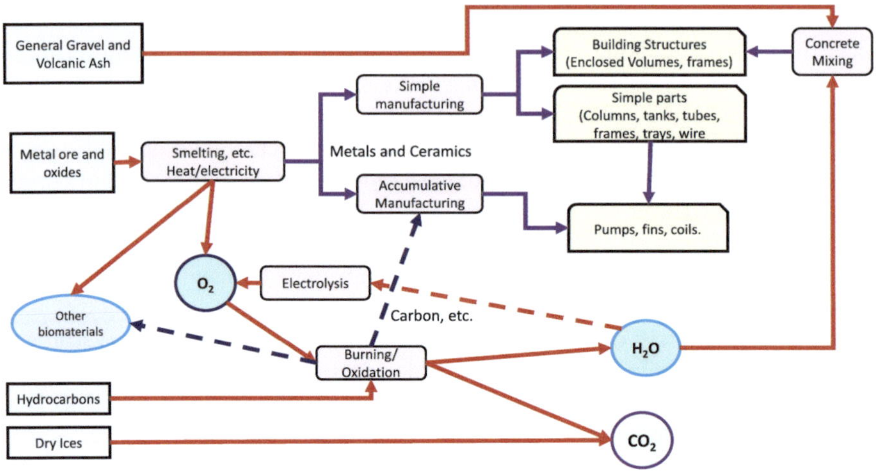

Fig. 12.2 Raw materials to farm needs

settlements, lunar settlements or settlements at LaGrange points will take weeks of transit, with exposure to higher levels of radiation than in LEO. Going further yet, asteroid, Martian, or further settlements require shipment times of many months, with a year of planning to load each shipment. Longer shipment times again expose cargoes to more time in radiation, and in general allow more time for mayhem to occur. Expense to ship mass to settlements is orders of magnitude higher for distance. As a result, more fragile live mature crops might be possible to LEO or the Moon but might be more challenging to Mars. There are options, however. Some crops could act as part of the spaceship life support on longer journeys, then moved to the settlement. An example is where a ship brings a large load of algae, bush, or vine crops grown in transit. These algae and plants consume human carbon dioxide and wastewater on the trip out, slowly trading oxygen, water, and food in storage for green mass ready to move to the settlement. These green items grown over a long transit could jumpstart the settlement's farm.

However, the simplest and easiest way to get crop starters to the settlements is to ship seeds for plants, spores for bacteria, algae, or fungi, and frozen embryos or gametes for animals. These small items can be heavily shielded to prevent radiation damage, and even using early twenty-first-century technology, seeds, spores, and gametes can make the journey even if it takes years. Given the size of seeds, there could be a vast overmatch to account for loss. In many cases, seeds can survive centuries before planting and germination. Algae and bacteria naturally form spores that once dehydrated, can again last for years before use.

Plant cultivars such as stems, to be grafted limbs or other plant parts that can be grown into whole plants could be the crop transport method. These snippets could be maintained by human passengers, adding green sights and pleasant smells to the air of the ship. These ready to transplant crops use transit time for productivity and to reduce start up time for the farm.

Animals are more difficult, but very possible. Difficulty for animals depends on needs for incubation and a womb. Insects have eggs that can wait for conditions, though how long one can ship frozen cricket or mealworm eggs would need more work. Aquatic animals are all very adapted to ship and mix. The easiest animal fertilized eggs to ship are euphausiids like brine shrimp. Brine shrimp eggs (a.k.a. sea monkey seeds) are a commonly mailed item and can last for at least a few years even in nature, where dry lake beds can stay dry for years and decades waiting for rain. Cryopreservation is a science that is decades old but still evolving. Many fish and crustaceans have freezable fertilized eggs and in particular gametes. Since their gametes in nature meet in water already, it would be easy to defrost eggs and milt, then mix to fertilize them in a small tank. Once fertilized, the organisms begin to grow out. Hatcheries are expert at the gamete to grow out process for most commercial crops, from fish and prawns to lobsters. More detailed work is needed to test how long and at what temperatures fit each crop, but extensive work to make gamete banks for fish and shellfish has already been done on Earth (see Hoseinifar et al., 2020).

Birds are next in line for shipability. Chicken eggs, sperm, and simple cells are already a topic of cryopreservation, and are explored on Earth in poultry science to preserve various genetic lines (see Nakamura et al., 2016). It is very possible that a settlement receives poultry popsicle gametes, which are kept for years while the settlement grows enough to support chickens.

Going from birds to mammals is more problematic and more difficult. Unless an artificial womb is developed for rodents, cattle, or other livestock, it may make more sense to ship living female rodents, goats, or the like, then inseminate them at the settlement using preserved sperm, or implant them with frozen embryos or frozen then fertilized embryos. Given the concern for gestation in various gravities, starting with rodents could be a good exploration of what happens in lower gravities.

The final frontier in getting crops to site, may be to store the information in the genomic and epigenomic code and just ship that information. This scenario was explored in sci-fi works by Arthur C. Clarke and others. On site, the genetic information is then decoded using DNA technologies already used on Earth for genetic modification work. The DNA is then planted in a biological scaffold, and grown into the first set of organisms, using shipped and in situ materials. In this way, updates and improvements could be sent, or crafted on-site, though the technological reach to get organisms beyond yeasts and bacteria is still beyond current work. However, current technology could be used to send genetic modifications and modify organisms to get more crop production, or other traits using CISPR or similar modification. It is also possible that sequence data from successful selectively breed cultivars might become a valuable export between settlements.

12.4 The Future Space Farm Economic Web

As settlements, and their farms, develop in many places, from Earth Orbit, to the Moon, Mars, and beyond, trade webs will emerge that will affect how space farms are used and developed. This section dives into how that future may happen.

Getting from where we are now, to where we want to be, requires a series of economic leaps. Once space tourism and more extensive space industry take off, the money will be in space. In the early twenty-first century, space is already a critical economic asset due to mostly communications, navigation, and observation. Looking back down at Earth for weather or other imagery is a money maker. Few outposts in orbit have people, starting with the International Space Station and other outpost stations by a variety of commercial and national players. These outposts are not self-sufficient, nor close the mass flow loop. However, there is starting to be a path forward. Some things can be made best in the space environment, microgravity included. The next wave of space products from space factories has already seen prototypes, including medical, electronic, and chemical products. These initial factories will be largely automated, but eventually also need people to run them due to the time to get repairs and perform troubleshooting from Earth to space. The evolution of the money flow from space is simplified in Fig. 12.3.

Workers need food. Shipping up food is expensive. Workers also want some greenery and nice smelling air, both of which drive at least salad greens and spices. Also, the first tourists are in orbit, and soon cruises in space will be a thing for the most well-heeled. Cruisers want excursions, which will mean at least orbital hotels, maybe moonbases. Science bases on the moon become industrial sites and tourist stops. Free-floating stations from orbit to lunar space start as zero-g but eventually will rotate to generate gravity to allow longer tour length and convenience. These stations

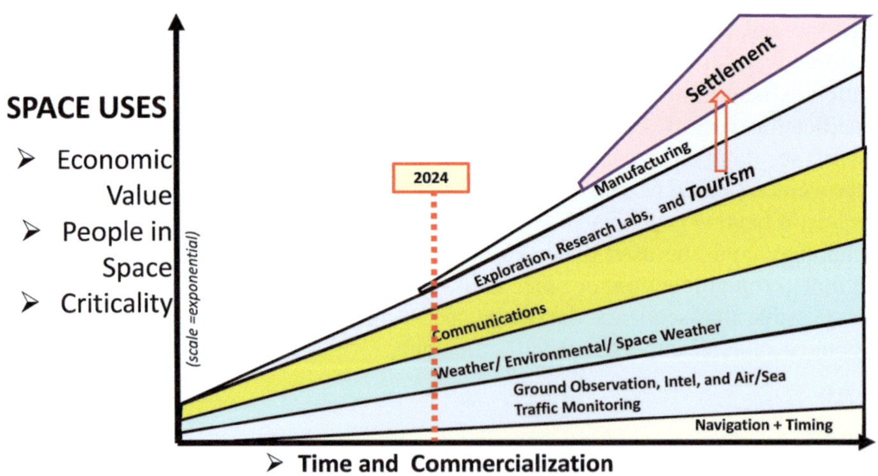

Fig. 12.3 From current space uses to settlement

will drive more tourism and the first long-term settlers, workers, and retirees. Parks must exist for those spending long tours in space, but the parks are also part of the farm. On the moon, bases will expand to become destinations. Expanding bases mean expanding farms. At some point in the free-floating settlements or on the planetary surfaces, dependents follow the workers. Dependents cause the average age to drop, as that population eventually outnumbers retirees and tourists. Multiple stations, bases, and settlements will drive the need for specialized production, including in food. Some bases will become breadbaskets for other stations with less extensive farms. Mars has lots of carbon dioxide and water. Given a cheap way to lift produce up its gravity well, Mars can become the breadbasket for the outer solar system. The moon has water and metals and may have carbon or nitrogen in rocks below the surface, still to be determined. Some locations will specialize in luxury items, or high value crops. At some point, this specialization drives a whole web of economics. Eventually the economics in space exceed those on Earth. At some point as we leap further out, the population off Earth exceeds the population on Earth. See Fig. 12.4.

When the settlements are to a point where a child born off Earth grows up and has children of their own, humanity has become a truly space-faring species. Trillions of happy, smiling, babies require at least 100 Earth equivalents in space, scattered across free-floating settlements, bases on the moon, Mars, moons of outer planets, asteroids, and outer belts. One day, given the number of exoplanets, maybe even beyond.

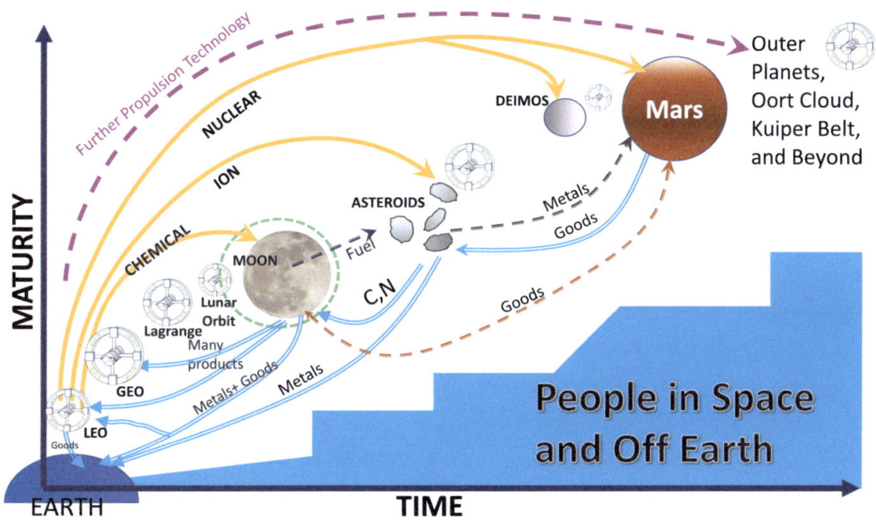

Fig. 12.4 Going from LEO to beyond. Gathered from AIAA Space Settlement Technical Committee meetings and hand drawings. The committee strongly voted to put this diagram in this book

12.5 If I Had Lots of Money Right Now and Wanted Space Farms Soon, What Research Would I Fund?

Several key technologies and datasets are needed! These technologies also benefit Earth right now too. Many are being explored by companies, the Mars Society, National Space Society, national space programs, and universities, but they are just starting to develop key contributing technologies and knowledge banks, and need much more research to achieve a fully developed space farm in free space, or on planetary bodies.

12.5.1 Key Research Area 1: Find and Quantify Ins, Outs, and Volumes, for Each Core Crop Species, Under Various Space Conditions

While there are tons of data from the 1950s to now on how much carbon dioxide and water are absorbed by a crop, we have very little scientific data for key food crops that cover the complete input and output graph to plant crops off Earth. This data is limited for various artificial light levels and spectrums, against growth and food production, especially in various hydroponic and aeroponic conditions for food crops. Cannabis farms have collected much data here, as have some researchers at NASA, but synergizing all this work into repeated experiments at scale to collect good space farm planning data has yet to occur. Pick a few key crops, isolate in a hydroponic chamber with a few hundred plants, and measure all the ins, outs, and recirculates. Duplicate this research under various gravities, especially lunar and Martian gravity. The same level of rigor needs to be added for candidate fungi, insects, and animals, especially aquatic species like shrimp or fish, maybe rodents or chickens as well.

12.5.2 Key Research Area 2: What Optimizations Improve Productivity for Each Crop?

How much carbon dioxide can I use to increase growth for plants, under what humidity factors, using what hydroponic, aeroponic, or soil regimes? What bacteria and fungi can be added to hydroponic or root beds to increase growth? What if I increase carbon dioxide concentration and add oxygen to the roots of vascular plants? Can I alter light spectra, light levels, and timing to make better crops? What photobioreactor architectures for what species can work for small areas to make lots of edible glop? What cohorts can be used to implement continuous harvest ideas? What genotypes, cultivars, or varieties are suitable for what conditions? What happens if I alter mineral content or add radiation?

12.5.3 Key Research Area 3: Implementing and Testing Bioreactor Technologies, Incinerators, and Other Recycling Technologies at Various Scales

Bioreactors and photobioreactors are already extensively used on Earth, so how do we miniaturize them, ship them up the gravity well, and use them to close the loop? What bioreactor or digester designs can take crop wastes, and or human wastes, and make the inputs for algae or plants? What incinerator designs are fit for space use, and how do we use them in combination with bioreactors? How do we connect all the parts and machines to process human and crop wastes at scale, at the rates needed for various farms? What farms or diets lead to what solutions? How do these change with gravity, i.e. freefall, 1/6, 1/3, to full 1 g? How do we tune the arrays of machinery and organisms for variations in human populations, tourists, etc.?

12.5.4 Key Research Area 4: Making Scale Space Farms that Close the Loop

We need to build scale space farms that completely account for all inputs and outputs, for various mixes of crops. Biospheres were a start. We could just assume human wastes and gas mixes, then close off the crops using various designs and see how we can balance the result. Scaling up to fully closed systems are a must before we rely completely on them in space. As a result, there need to be close cycle experiments using facilities from small scale to large scale, using real human inputs and outputs.

12.5.5 Key Research Area 5: Cryogenic Storage and Reconstitution of Various Crop Species

How long can I store seeds under what conditions? What crop species can be stored and still germinate, especially with some (human-tolerable) exposure to radiation? What insect, fish, shrimp, or other animal species can be shipped as embryos, or gametes that are used on site, at what temperatures and conditions? What about bacterial cultures, fungi of various types including yeasts? What crops can make the trip to Mars orbit, Jupiter orbit, or further? Can I archive seeds, spores, or gametes as a backup for years? What about bioreactor cultures? Can I direct assemble seeds or gametes from just DNA/epigenomic data?

12.5.6 Key Research Area 6: Checking How Various Machines and Settling Technologies Work in Various Gravity Levels, Including Artificial Gravity

How do columns settle, fermenters work, aerators aerate, hydroponic trays flow, and toilets flush? What modifications are needed for fluid flow, air rotation, heating and cooling, humidifiers or dehumidifiers work at freefall, 1/6 g, 1/3 g? What about artificial gravity settlements? Will I need centrifuges or extra pumps in various systems at various fluid flow points? This data will be hard to collect on Earth. We will need rotating satellites and research spacecraft at a minimum. We have data for freefall from the ISS, but what about lunar or mars gravity? What about centrifugal effects on machines in rotating settlements of various diameters? If I size the station to proposed Kalpana or Avalon station sizes, how does the rotation affect pumping, settling, aeration, or bubbling as pipes pass from outer edge to inside? What about smaller stations? We will also need experiments on the moon and Mars as well.

12.5.7 Key Research Area 7: Find Out the Effects of Low Gravity and Artificial Gravity on Crop Animal Reproduction

While there have been a few experiments using mice and rats in microgravity as a proxy for humans (see (Walton et al., 2005)), there is very little good data on aquatic crop animals, or crop birds in various gravities. There need to be experiments in larger scales either on the moon or Mars or using spinning free-floating laboratories. Given the efficiency of the aquatic herbivore to algae cycle, some of these experiments should focus on fish and shrimp.

12.5.8 Key Research Area 8: Determine Parameters of Artificial Gravity and Low Gravity as They Affect Plant Crops

Plant crops on Earth grow straight and tall, and vine crops can be directed using string or supports, but how does wheat grow in a rotating settlement a kilometer wide? Do the shafts bend due to Coriolis effects? What heights and twists of plants correspond to rotating habitats of various diameters and spin rates?

12.5.9 Key Research Area 9: Determine What Methods Can Be Used to Optimize Natural Light for Plant Production. What Tradeoffs Apply to Using Natural Light Versus Artificial Lights?

Given the energy for lights in well-developed space farms, using free natural light can reduce both electricity need, and the associated heat rejection from using artificial lights. On the moon or Mars, can I use mirrors or other light pipe technologies to mitigate artificial light use? In rotating habitats in free space, what can I do to mitigate very short light–dark cycles using mirrors, fiber optics, or other methods? What balance of natural light and artificial light can be sent to a growth chamber? Can I have clear growth domes on Mars?

12.5.10 Key Research Area 10: What Methods Need to Be Applied to Allow Pollination of Flowering Crops or Wind Pollinated Crops?

What pollination methods can I use in space farms for wind and insect pollinated crops? Can I shift the fans in a growth area to improve pollination? Do I really need a robot bee or kid with a paintbrush? What factors can be applied to improve pollination in various farm locations?

12.5.11 Key Research Area 11: What Levels of Radiation Are Tolerable to Crop Species?

There is little data to establish what radiation levels of what radiation types are tolerable to crop species, especially when combined with artificial or reduced gravity levels as on the moon or Mars. This radiation tolerance will determine where growing areas can be built, and if they can be built under domes. What breeds and cultivars of plants can tolerate what radiation levels? What effects on the crop are acceptable to farmers?

12.5.12 Key Research Area 12: How Does the Artificial Space Environment Affect Nutrition from Crops?

Given the use of hydroponics, aeroponics, or soil developed from regolith, and the added conditions of artificial light, controlled atmospheres, and gravity, how is the nutritional value of foods from crops affected? Will space rice be lower in proteins? Will space soybeans have less ash? Does soil matter? What can be done for soilless methods to improve nutrition? Will settlers need to import vitamins? These questions are partially being answered by vertical farming on Earth, but there needs to be in space research in reduced gravities, and under space settlement conditions.

12.6 Summary

This chapter covers factors that will influence space farm design, including location, availability and processing of materials, and including farming as part of a space economic web. There are also many open research areas for space farm crop selection and design, and many of these research areas can only be answered in space, though some can be explored right now on Earth.

12.7 Questions

1. How does the location of a space farm affect the availability and cost of initial inputs? What is the heaviest initial input for the farm?
2. Assume a space farm built near the Martian north pole, where there is water and dry ice mixed into a matrix of sand and gravel? What methods could be used to extract water and carbon dioxide for use in the farm?
3. Assume a space farm built near the lunar south pole, near permanently dark craters and peaks of eternal light. What design methods could be used to both gather inputs, and reduce electricity consumption?
4. Why should settlement diameter matter for rotating settlements in free space (like those in science fiction or proposed by Gerard O'Neill), especially for tall crops like corn, wheat, or trees?
5. What factors might be limiting the use of domed space farms on the moon or Mars?
6. Given the Key Research Areas, what experiment would you design to further space farm technologies and designs given an unlimited budget?

References

Bajpayee, A., Farahbakhsh, M., Zakira, U., Pandey, A., Ennab, L. A., Rybkowski, Z., Dixit, M. K., Schwab, P. A., Kalantar, N., Birgisson, B., & Banerjee, S. (2020). In situ resource utilization and reconfiguration of soils into construction materials for the additive manufacturing of buildings. *Frontiers in Materials*, 7, 52. https://doi.org/10.3389/fmats.2020.00052

Globus, A., Arora, N., Bajoria,A., & Straut, J. (2017). *The Kalpana one orbital space settlement revised, 2017, via National Space Society (NSS).* https://space.nss.org/wp-content/uploads/Kalpana-One-2007.pdf

Hoseinifar, S., Hu E., Fatihah, S., Diwan, A., Harke, S., Gopalkrishna. & Panche, A. (2020). Cryobanking of fish and shellfish egg, embryos and larvae: An overview. *Frontiers in Marine Science, 7*. https://doi.org/10.3389/fmars.2020.00251. https://www.researchgate.net/publication/341219024_Cryobanking_of_Fish_and_Shellfish_Egg_Embryos_and_Larvae_An_Overview

Marotta, T., & Globus, A. (2021). *The high frontier: An easier way*, December 21, 2021, ISBN: 978-0464706304.

NASA. (2022). National Aeronautics and Space Administration (NASA), In-Situ Resource Utilization (ISRU) website, Page Last Updated: April 6, 2020, Page Editor: Loura Hall, NASA, NASA Official: Brian Dunba, 2022. https://www.nasa.gov/isru

NASA-SBIR. (2022). Small Business Innovation Research (SBIR) &Small Business Technology Transfer (STTR) program, in-situ resource utilization and space manufacturing, topic number: X9.01, Parent: Lunar and Planetary Surface Operations, EHB Solicitation ID: SBIR_05_P1, 2022. https://sbir.nasa.gov/content/situ-resource-utilization-space-manufacturing

Nakamura, Y., Kawanishi, F., Miyahara, D., Mori, T., Ono, T., Takeda, K., Nirasawa, K., Kagami, H. & Tagami, T. (2016). Efficient system for preservation and regeneration of genetic resources in chicken. https://www.researchgate.net/publication/303843830_Efficient_system_for_preservation_and_regeneration_of_genetic_resources_in_chicken

Reches,Y. (2019). Concrete on mars: options, challenges, and solutions for binder-based construction on the red planet. *Cement and Concrete Composites, 104*, 103349, ISSN 0958-9465. https://doi.org/10.1016/j.cemconcomp.2019.103349. https://www.sciencedirect.com/science/article/pii/S0958946518312885

Starr, S., & Muscatello, A. C. (2010). Mars in situ resource utilization: a review. *Planetary and Space Science, 182*, 104824, ISSN 0032-0633, https://doi.org/10.1016/j.pss.2019.104824. https://www.sciencedirect.com/science/article/pii/S0032063319301618

Vaniman, D., Pettit, D., & Heiken, G. (1992). Uses of luna sulfur. In W. W. Mendell (Ed.), *2nd conference on lunar bases and space activities* (Vol. 1 (Parts 1–4) & Vol. 2 (Parts 5–8), p. 706). NASA Conferences Publication 3166. https://space.nss.org/the-second-conference-on-lunar-bases-and-space-activities-of-the-21st-century/

Walton, K. D., Harding, S., Anschel, D., Harris, Y. T., & Llinás, R. (2005). The effects of microgravity on the development of surface righting in rats. *The Journal of Physiology, 565*(Pt 2), 593–608. https://doi.org/10.1113/jphysiol.2004.074385. Epub 2005 Mar 17. PMID: 15774538; PMCID: PMC1464515

Glossary

Term Definition

accumulative manufacturing (also called 3D printing, three-dimensional printing, additive manufacturing) The process using machines to build up an item using layers and rows of materials usually by spraying or sintering, to make a three-dimensional design. Derived originally from inkjet printer technology.

aerobic metabolism or reactions that use oxygen. The normal way most animals use to gain metabolic energy from food by taking in oxygen.

aeroponics growth of plants using sprays and vapor to the roots, in lieu of soil or liquids

Agriculture planned growth of crops

Anaerobic metabolism or reactions without using oxygen. For example, alcohol is produced using anaerobic respiration.

annual (crop) A crop that puts all its energy into a single season, reproduces, then dies. Wheat is an example annual crop.

aquaponics The combination of aquatic animal crops and hydroponic plant crops, such that wastes from the animal crops are flowed to the roots of plant crops as nutrients in the same water flow.

aquatic animal animals that live in water for most or all of their lifecycle, and typically directly absorb oxygen from water using gills

aquatic stage a division of the space farm where an aquatic animal is grown for harvest, containing tanks, controls, and environment to optimize growth.

ash The part of foods or organisms left after water, carbohydrates including fiber, fats, and amino acids/proteins are removed.

big bang harvest All organisms in the crop are grown at the same time from seed/ egg to harvest, then all organisms are harvested. This is how Earth grains are grown.

biochemistry how living things use elements and compounds

biomass formerly or currently living matter, or wastes, that contain compounds of carbon, hydrogen, oxygen, nitrogen, and other elements. Wet biomass is matter that contains water as in living organisms, or as released from living organisms.

© The Editor(s) (if applicable) and The Author(s), under exclusive license to Springer Nature Switzerland AG 2026

B. L. Meyer, *Designing Farms in Space*, https://doi.org/10.1007/978-3-032-04395-5

Dry biomass is matter where all the water has been extracted (usually using heat, freezing, or vacuum)

bioreactor Machinery that uses organisms to perform a chemical process and/or produce biomass

blackwater (black water) Wastewater that contains a significant load of human or animal wastes and food residues

blower a machine that moves gases at moderate pressure at high rates

capillary action the flow of water in small tubes where the surface tension dictates flow direction

carbohydrates Carbon, oxygen, and hydrogen compounds, usually with the formula CnH2nOn. Includes sugars, starches, cellulose.

Cellular Agriculture cultivation of simple (single cell or few celled) organisms as crops, including growth of yeasts, bacteria, and algae.

cellulose Also known as fiber. A family of compounds of carbon, hydrogen, and oxygen made by plants to form cell walls and structures. Usually broken down in nature by wildfire or by bacteria and fungi.

chemosynthesis direct creation of compounds using chemical reactions without organisms

Choice Pieces parts of an animal or plant traditionally eaten in Western diets

CHON Acronym for Carbon, Hydrogen, Oxygen, and Nitrogen, which are the most prevalent elements in living things from Earth.

circulation the movement of fluids (gases, liquids)

clean water Water that is suitable for human use and consumption

Closed (Loop) System A set of processes where mass is cycled around, but no mass is added or lost.

column a long tube used to separate or combine masses using gravity, density, or flow.

compound chemically boned combinations of elements

condensation (condense) liquids formed from vapor using cooling or change in pressure. Dew is the condensation of water vapor from the air onto cool objects.

continuous harvest continuous cycling of crop grow cycles to have planting and harvest every day.

control system a combination of software and devices designed to regulate a process

crop organisms grown for consumption or use

cryopump, cold trap, cryofilter machinery using a cold plate or fins to separate gases from a mixture using condensation. Carbon dioxide condenses at a warmer temperature than oxygen or nitrogen, so a cryofilter (cryopump, cold filter) can pull carbon dioxide from a mix of oxygen nitrogen, and carbon dioxide.

determinate Bush-shaped plants that produce fruit for a short period

digester a simple tank that contains either bacteria or enzymes to break down wastes from complex compounds to simple compounds like water and carbon dioxide.

distill, distillation using heat or vacuum to convert a mixed liquid into purified liquids and gases. Often used to extract alcohol or water from mixed solutions.

doubling rate the time it takes for an organism make another copy of itself. Used here for the time it takes for a mass of an organism like yeast, algae, or bacteria, to make twice the original amount of living yeast, algae, or bacteria.

dry mass mass of organisms, wastes, or foods with water removed

ECLSS 1. Environmental Control and Life Support System, 2. Ecologically Closed Loop Support System

ecology the study of the series of relationships and processes linking organisms, including the organisms, and their environment

ecosystem a series of relationships and processes linking organisms, including the organisms, and their environment, especially focused on the cycling of mass

edible Can be consumed by people as food

environment The temperature, chemicals, matter, and other conditions that surround a living thing or group of living things

enzyme Chemicals that can break down or convert compounds. Salivary amylase is an enzyme in human saliva that breaks down starches to simpler sugars.

eudicot Plants that have more than 1 seed leaf upon sprouting (germination) from a seed

evaporation conversion of liquids to vapors

exponential (growth) the multiplication of mass of organisms, so that mass at time $t + 2$ is a multiple of mass $t + 1$ which is a multiple of mass at time t.

extractor a mechanism to pull out different masses

fan a machine that moves gases at lower pressures, using having blades

farm A combination of organisms, environment, and machinery to grow food and recycle wastes

fats (lipids, oils) Organic compounds of carbon, hydrogen, and oxygen, that have chains of molecules making them not soluble in water.

fermentation the action of organisms like bacteria or yeasts to consume sugars with or without oxygen to make more organisms and release carbon dioxide, water, and possibly alcohol.

fluids anything that flows, including gases and liquids

footprint the area covered by a stage or crop

free-floating space settlement A place in the universe beyond Earth where humans live that is not attached to a planet, moon, or asteroid. Space ships may fit this definition, but the typical sense is a space station.

fungi Eukaryotic organisms from yeasts to mushrooms, including truffles and morels. Like animals, take in feed and sometimes oxygen to release carbon dioxide.

GEO Geosynchronous Earth Orbit, orbits at 35,786 km above the Earth's surface to match Earth's rotation

germination the emergence of a plant from its seed shell, usually used by farmers as a term for when a plant first pops out of the soil.

grafting the combination of the parts of two different plants

greywater water that has been used for washing food, showering, hand washing, or other uses that only slightly contaminate water

growth budget the percentage (ratio) of inputs allocated by an organism to growth versus metabolism

growth chamber (growing chambers) enclosed volume(s) where a crop (or crops) is (are) grown

Hardness As in water hardness, a measurement of the content of ions, minerals, etc. in solution in water. Harder water has more minerals or ions in it.

harvest the collection of the edible parts of a crop

heat rejection the action to lower the temperature in an area. Using electricity, and due to aerobic metabolism, most stages of the farm produce too much heat that must be pulled away (or rejected) to keep the temperatures within the best range for the crops.

Human Model The mass flow per person, including inputs of oxygen, food, and water, and outputs of wastes in water and carbon dioxide. This model includes nutritional profile of food, such as kilocalories, carbohydrates, fiber, proteins, minerals, vitamins, salts, and fats.

Humidity (relative) the percentage of water vapor in the air in a volume relative to the maximum amount of water vapor the air can hold.

hydroponics A variety of methods to grow plants without soil, using liquids and with or without solid objects (substrates) in the liquid.

in situ resources use of resource on site. Example: using carbon dioxide on Mars gathered from the Martian atmosphere

incinerator A machine that uses heat to convert inputs (wastes) and oxygen to carbon dioxide, water vapor, and ash.

indeterminate sprawling vine plants that grow constantly and push out fruit continuously

injector a mechanism to push together different masses using mixing

input the stuff that goes into a system, machine, stage, or organism

kcal (kilocalories) kcal is an abbreviation for kilocalories, the human usable energy in food. In the USA, sometimes just called calories.

Lactating (lactation) the production of milk by mature female mammals

LEO Low Earth Orbit, any orbit from 100 km to 2,500 km above the Earth's surface

life cycle The process from initiation (fertilized egg/spore/seed or from cells or fragment) through reproduction to death for an organism. For maize, this process is from seed to germination to growth to flowering to seed production to dry-out to harvest.

lignin A complex biochemical compound of carbon, hydrogen, and oxygen, made by some plants to make wood and woody stems. Very hard for bacteria and fungi to break down. In nature, usually wildfire breaks down lignin.

liquid flowing form of matter that is dense enough to be essentially incompressible.

living mass wet mass of living things when alive

mass the property of matter that has weight in gravity

mass balance mass of inputs equals mass of outputs

mass flow the movement of matter in a process, system, or organism

mass flow loop the movement of matter in a process where outputs become inputs

membrane mediated bioreactor aka MMR, a combination of machines with organisms affixed to a sheet or wall (a membrane), in which fluid flow over the membrane and organisms to make biomass or perform a chemical process
Metabolism the process of chemical reactions where organisms gain cellular energy to survive
MMR Abbreviation for Membrane Mediated (bio)Reactor
model an estimated process with numbers for how something works, including inputs, and outputs
mortality (rate) The ratio of organisms that die before maturity or harvest.
nutrients Chemicals, usually mixed, and usually in either dry form or in solution, that is usable by plants and algae for growth
organisms living things composed of cells
output the matter or energy that leaves an organism, process, or system.
PBR Abbreviation for PhotoBioReactor
perennial organisms that live multiple growth seasons or grow continuously for many seasons. Pecan trees are perennials, as are some types of genetically modified tomato plants.
pH The inverse ratio of hydrogen ion ($H+$) activity in a solution (aka potential for hydrogen). Acidic solutions have a $pH<7$, Alkaline solutions have a $pH > 7$. Most plants and animals grow best in soils or water with pH near 7 (neutral).
photobioreactor Machinery that uses organisms to perform a chemical process and/or produce biomass using photosynthesis (i.e. uses light)
photosynthesis Chemical process that uses chlorophyll and light to combine carbon dioxide and water into sugars, releasing oxygen.
photosynthetic Something, usually a plant or algae, that uses photosynthesis
planetary bodies Planets, asteroids, or moons big enough to be mostly round and have perceptible gravity.
pollination (pollinators) plants merge genes with another plant via flowering, pollinators are any animal (like bee, moths, bats, birds) that transfers pollen from one plant to the female parts of the same or another plant
potable water Water that is suitable for human use and consumption
production, productivity The ratio of growth to wastes and inputs. The efficiency of growth of desired products (like food) given inputs.
Proteins (and amino acids) Organic combinations of amino acids in complex shapes. Amino acids are simple compounds of carbon, oxygen, hydrogen, and nitrogen.
pump a machine that uses energy to increase the pressure or move fluids
recirculation the continuous cycling of mass. An example is using a fan to take the same air in a growth chamber and move it inside the chamber.
regolith the dirt of a planet, asteroid, or moon that isn't Earth
Respiration Chemical reactions in a living thing that use food energy
Ruminants animals that use complex digestive structures (like a cow's stomach) to host bacteria to break down cellulose in food. Ruminants often regurgitate a cud of partially digested fiber, chew it, and then swallow it to move it between stomachs.

sensor a device to measure a value or series of values using electronics. Examples include thermometers, pressure gauges, cameras, microphones, and other devices.

soil also known as dirt. A combination of various materials and living things that allow plants to grow.

space The universe beyond the Earth's atmosphere

space farm Combination of organisms, environment, and machinery to grow food and recycle wastes for settlement or exploration of the universe beyond the Earth's atmosphere

space settlement A place in the universe beyond Earth where humans live. Permanent space settlements are places humans go to live indefinitely.

Stage (farm stage) As used here, the combination of crops, containers (including growth chambers), and machines needed to keep the crop growing optimally, and manage the flow of mass into and out of the crop as they grow. Can be centralized into a single location, or virtual to track the mass flow and energy for growing the crop.

steady state The time when the mass flow into and out of a system (or a kilogram of living crop) is consistent over time (i.e. the same between now and a day from now).

tank An enclosed container to hold gases, solids, or liquids

Terrestrial Animal A complex aerobic organism that lives in air on Earth, usually arthropods, mammals, birds, snails, and reptiles

tissue A type of the living or formerly living structure of cells that are part of a complex organism. Green flat pieces of a leaf are a tissue of plants.

transpiration The evaporation of water from a plant's green parts, for example leaves of a maize plant. Plants release hundreds of times the mass of water in transpiration than the water they take in to grow.

urban farm Growing edible crops in confined, artificial, spaces, such as rooftops of buildings, along hallways, between buildings, in warehouses, and in shipping containers.

valves simple machines to control the flow of gases or liquids

vapor gaseous form of matter, that may contain droplets like clouds

vascular plant a type of photosynthetic organism that has root, leaves, and stems and critically has channels for moving liquids from the roots to the leaves via transpiration, and channels to send sugars to the rest of the plant from the leaves.

vat bioreactor A very simple device that has a container (a vat or tank) with a mix of organisms in solution with inputs and outputs to make biomass and/or perform a chemical process

Vegetative reproduction Growing a plant from clippings, or naturally via runners, in lieu of seeds.

vertical farm Stacking of crops to minimize floor area (footprint).

wet mass mass of foods, wastes, or organisms including water as present in nature.

yeast microscopic fungi, commonly used on Earth to make bread, beer, and wines.

Yeast-Bacteria Bioreactor (Stage) The combination of machines, including tanks and membranes that hold bacteria and yeasts, to convert inputs like wastes and oxygen to water, carbon dioxide, alcohol, biomass, or whatever is needed.

yield, crop The percentage (or ratio) of a crop that is edible versus inputs to grow the crop to harvest or per day.

Index

© The Editor(s) (if applicable) and The Author(s), under exclusive license to Springer Nature Switzerland AG 2026

B. L. Meyer, *Designing Farms in Space*, https://doi.org/10.1007/978-3-032-04395-5

T

MIX
Papier aus verantwortungsvollen Quellen
Paper from responsible sources
FSC® C105338

If you have any concerns about our products,
you can contact us on
ProductSafety@springernature.com

In case Publisher is established outside the EU,
the EU authorized representative is:
Springer Nature Customer Service Center GmbH
Europaplatz 3, 69115 Heidelberg, Germany

Printed by Libri Plureos GmbH
in Hamburg, Germany